MATHEMATICS
FOR COMPUTER TECHNOLOGY

PRENTICE-HALL SERIES IN TECHNICAL MATHEMATICS

Frank L. Juszli, editor

MATHEMATICS
FOR COMPUTER TECHNOLOGY

Paul Calter
Professor of Mathematics
Vermont Technical College

Prentice-Hall, Englewood Cliffs, N.J., 07632

Library of Congress Cataloging-in-Publication Data

Calter, Paul.
Mathematics for computer technology.

Includes indexes.
1. Electronic data processing—Mathematics.
I. Title.
QA76.9.M35C35 1985 510 85-24389
ISBN 0-13-562190-9

Editorial/production supervision: Jane Zalenski
Interior design: Dawn Stanley
Cover design: Wanda Lubelska Design
Manufacturing buyer: John Hall

Printed in the United States of America

10 9 8 7 6 5 4 3 2 1

ISBN 0-13-562190-9 025

Prentice-Hall International (UK) Limited, *London*
Prentice-Hall of Australia Pty. Limited, *Sydney*
Prentice-Hall Canada Inc., *Toronto*
Prentice-Hall Hispanoamericana, S.A., *Mexico*
Prentice-Hall of India Private Limited, *New Delhi*
Prentice-Hall of Japan, Inc., *Tokyo*
Prentice-Hall of Southeast Asia Pte. Ltd., *Singapore*
Editora Prentice-Hall do Brasil, Ltda., *Rio de Janeiro*
Whitehall Books Limited, *Wellington, New Zealand*

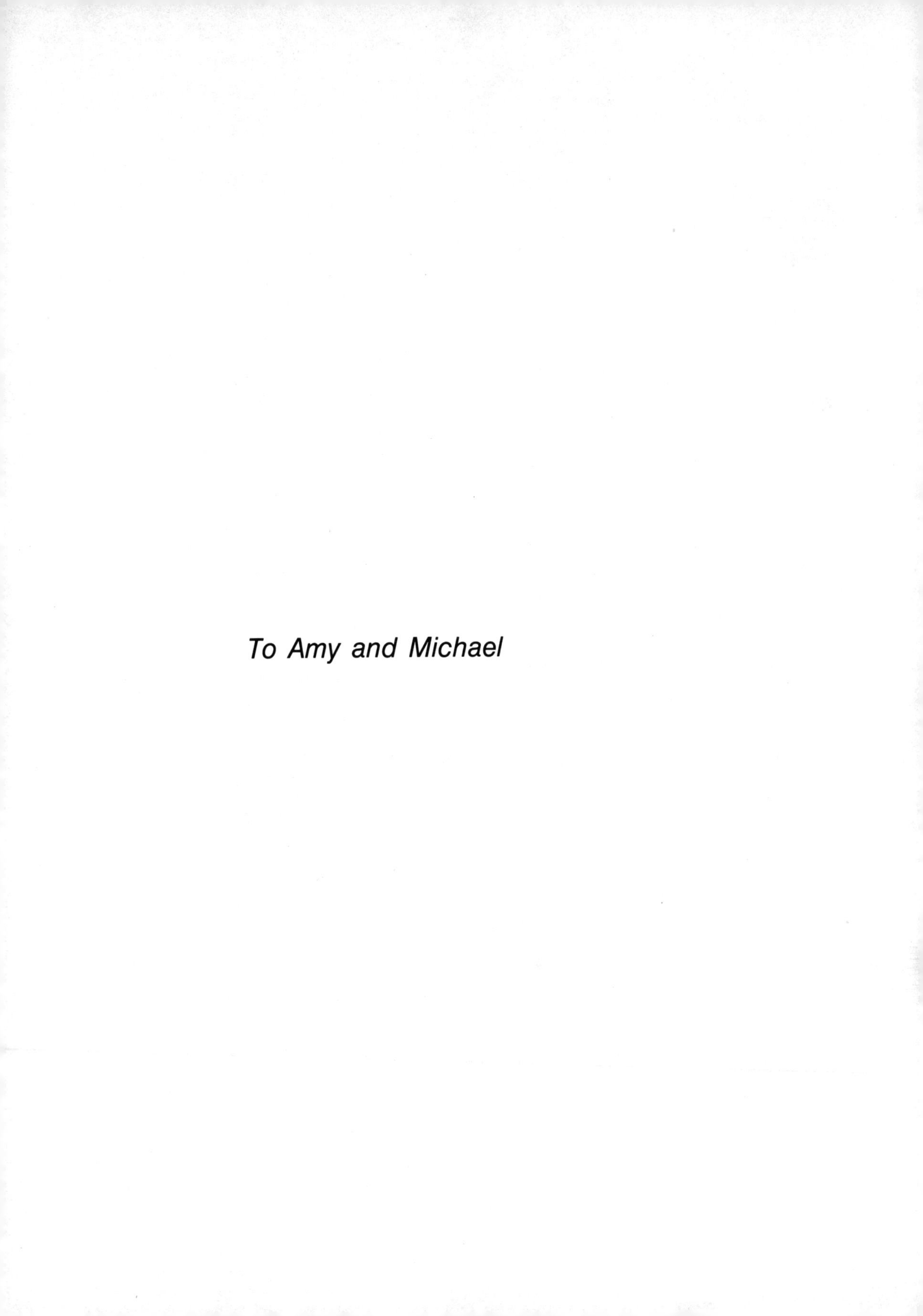

To Amy and Michael

Contents

8

Systems of Linear Equations 233

9

Matrices and Determinants 256

Radicals 307

This material may be covered any time after Chapter 5.

Quadratic Equations 322

This chapter should follow Chapter 10.

12

Right Triangles and Vectors 354

We now start a block of three chapters on trigonometry. They may be covered any time after Chapter 5.

Contents

The right triangle can be covered earlier if needed for other subjects.

Logarithms can be studied any time after Chapter 6, although they are usually left for later.

Inequalities can be studied any time after Chapter 6, but since none of the other sections depend on it, it is often presented late in the course.

Inequalities and Linear Programming 530

18

Boolean Algebra 548

Appendices 581

Index to Applications 665

General Index 669

Preface

Coverage: This book is primarily a technical mathematics textbook. However, it is a tech math book that attempts to cover as many computer-related topics as possible, while retaining the content of a traditional tech math course. Further, those traditional topics are presented, where possible, from an algorithmic point of view. This text may be used for a one- or a two-semester course, and is rigorous enough for a precalculus course.

Audience: We hope to reach students in three areas. First, those students enrolled in a computer technology program at a technical school or two-year technical college, taking a course usually called "Technical Mathematics." Such students need the full range of traditional tech math subjects, as well as a strong introduction to computer-related mathematics. Second, we hope to reach those technology students, primarily from electrical technology, who need a text that is more computer oriented than the traditional technical mathematics text, but who still need the traditional topics to prepare them for calculus. Finally, students enrolled in a "Computer Mathematics" course need, in addition to the "computer" topics, the basic algebra, geometry, and trigonometry that computer math texts often omit or treat sparingly. Even when time does not allow presentation of such topics in class, the student can review them from the material presented here.

Differences: This book differs from a traditional technical mathematics book. While keeping the usual tech math topics, it includes the following topics not normally found in such a book.

1. Binary, hexadecimal, octal, and BCD numbers.
2. Errors in computation.
3. Numerical methods for solving equations, such as simple iteration and the bisection method.
4. Sets.
5. Gauss–Seidel method for solving sets of equations.
6. More thorough treatment of matrices and determinants.
7. Computer methods for evaluating determinants.
8. Computer methods for solving sets of equations by matrix methods.
9. Computer solution of linear programming problems.
10. Boolean algebra.
11. Suggested programming exercises for students, with complete programs given in the answer key. A diskette containing all the programs is available to instructors, who may copy it for their students.
12. A summary of the BASIC language.

A mathematics book is never easy reading, so much care has gone into making the material as clear as possible. We follow an intuitive rather than rigorous approach, and present information in small segments. Marginal notes, numerous illustrations, and careful page layout are designed to make the material easy to follow.

Order of Presentation: No two instructors include the same topics or present them in the same order. Faced with this dilemma, we have chosen a traditional order of presentation, which is as close as possible to the preferences of many instructors. (We made a survey.) You may, of course, change the order of presentation to suit your needs, and the marginal notes in the Contents will help you make such changes.

Exercises: Since practice is as essential in learning mathematics as it is for most other skills, we include practice problems. Each chapter contains several large Exercises that give practice in the material in the preceding sections. Problems are graded by difficulty, and grouped by type to allow concentration on a particular type. Chapter Tests, however, are scrambled as to both type and difficulty. Answers to odd-numbered problems are given in Appendix H, and all others are included in the Solutions Manual. A supplementary *test bank* is also available, which includes chapter objectives, exams for each chapter, and a final examination.

Applications: Many problems are included from the several branches of technology. They are included both for their own sake, and to help answer the (usually unspoken) question: "What's this stuff good for?" The Index to Applications will help the reader find specific types of problems.

Calculator: The pocket calculator is not treated in a separate section, but is integrated throughout the text. Most of the basic operations are explained in Chapter 1, and others are treated as needed. Careful attention is paid to the use (and abuse) of significant digits, a problem that has gotten worse with the use of calculators, which give answers to about 10 (mostly worthless) digits.

Computer: A course in computer mathematics can be enriched by the programs suggested at the end of the Exercises. However, it is not our intention to teach programming here, and the programs may be skipped without harm. Completed programs are given in the answer key (Appendix D), in the Solutions Manual, and on diskette.

Formulas: Each important formula is boxed and numbered in the text. At the risk of having the text classified as a "cookbook," we also list these formulas in Appendix A as the Summary of Facts and Formulas. This list provides a ready reference as well as a common thread between chapters, and may help the student to see interconnections that might otherwise be overlooked.

Common Errors: An instructor quickly learns the pitfalls and traps that "get" students year after year. Some of these are boxed in the text and labeled *Common Error*.

Acknowledgments: For reviewing the manuscript and making suggestions, I thank my colleagues at Vermont Technical College: Byron Angell, Walt Granter, John Knox, Don Nevin, Peter Rasmussen, and Robert Rees. I also thank my son Michael for his careful work in writing and debugging the many computer programs in the text. Much material from my previous book *Technical Mathematics* has been used in this text. For reviews of that manuscript I am indebted to Ellen Kowalczyk, Madison Area Technical College; Donald Reichman, Mercer County Community College; and Ursula Roden, Nashville State Technical Institute. Special thanks go to Frank Juszli, Editor of the Prentice-Hall Series in Technical Mathematics, for his review of the entire text and his valuable suggestions. Finally, I want to thank my editor, Tim McEwen of Prentice-Hall, for his support and good advice during this project.

Paul Calter
Randolph, Vermont

Note to the Student:

Most of the formulas you are likely to need in your work in basic computer technology are presented in Appendix A, Summary of Facts and Formulas. The majority of these are formally introduced in the text, but a certain number of them, particularly those in the "Applications" section of the Appendix, are easy enough to use without discussion. In the exercise sections of the text, you are often asked to apply one of these formulas; to save time, therefore, you should go directly to Appendix A to find the equation referred to, given that not every formula will be found in the text. The page edges in Appendix A are shaded for easy reference.

Numerical Computation

In this chapter we review ordinary arithmetic, but in a different way—by explaining basic operations in terms of the calculator. Calculators make arithmetic easier, but they also introduce a complication: that of knowing how many digits shown in a calculator display should be kept. Sometimes it is incorrect to keep them all. We also learn some rules that will help us in algebra.

We will consider *units* of measurement: how to convert from onc unit to another, and how to substitute into equations and formulas. We will learn *scientific notation*—a simplified way to handle numbers with many zeros. Calculators and computer display such numbers in this way. *Percentage*, one of the basic mathematical ideas in technology, is covered next. Finally, we review some useful facts from geometry, along with practical applications of these facts. Computation of areas and volumes of various geometric figures will be emphasized.

This chapter should enable you to use most of the keys on your calculator to do some fairly difficult computations. The keys for trigonometry and logarithms are discussed in later chapters.

1-1. THE REAL NUMBERS

Before we start our calculator practice, we must get a few definitions out of the way. In mathematics, as in many fields, you will have trouble understanding the material if you do not clearly understand the meanings of the words that are used.

The three dots indicate that the sequence of numbers continues indefinitely.

Integers: The *integers*

$$\ldots, -4, -3, -2, -1, 0, 1, 2, 3, 4, \ldots$$

are the whole numbers, including zero, and including negative values as well.

Rational and Irrational Numbers: The *rational* numbers include the integers and all other numbers that can be expressed as the quotient of two integers: for example,

$$\frac{1}{2}, \quad \frac{3}{5}, \quad \frac{57}{23}, \quad \frac{98}{99}, \quad \text{and} \quad 7$$

Numbers that cannot be expressed as the quotient of two integers are called *irrational*. Some irrational numbers are

$$\sqrt{2}, \quad \sqrt[3]{5}, \quad \text{and} \quad \sqrt{7}, \quad \pi$$

Real Numbers: The rational and irrational numbers together make up the *real numbers*.

Numbers such as $\sqrt{-4}$ do not belong to the real number system. They are called *imaginary numbers* and are discussed in Chapter 16. Except when otherwise noted, all the numbers we work with are real numbers.

Systems having bases other than 10 are used in computer science: binary *(base 2),* octal *(base 8), and hexadecimal (base 16) (see Chapter 2).*

Decimal Numbers: Most of our computations are with numbers written in the familiar *decimal* system. The names of the *places* relative to the *decimal point* are shown in Fig. 1-1. We say that the decimal system uses a *base of 10* because it takes 10 units in any place to equal 1 unit in the next-higher place.

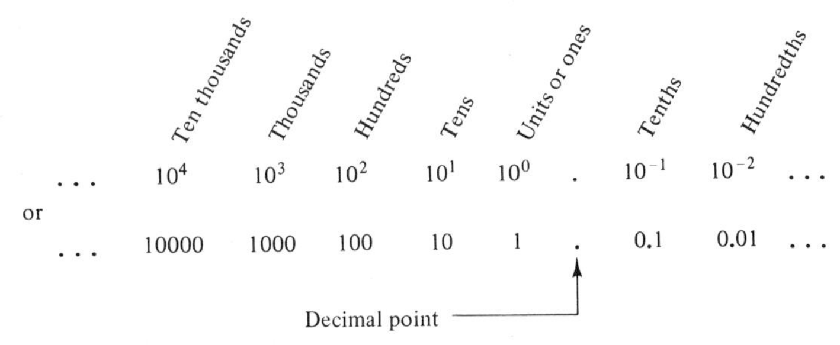

FIGURE 1-1. Values of the positions in a decimal number.

Positional Number Systems: A *positional* number system is one in which the *position* of a digit determines its value. Our decimal system is positional.

Example: In the number 351.3, the digit 3 on the right has the value $\frac{3}{10}$, but the digit 3 on the left has the value 300.

The position to the left of the decimal point is called position 0. The position numbers then increase to the left and decrease to the right of the 0 position.

Place Value: Each position in a number has a *place value* equal to the base of the number system raised to the position number. The place values in the decimal number system, as well as the place names, are shown in Fig. 1-1.

Number Line: We can graphically represent every real number as a point on a line called the *number line* (Fig. 1-2). It is customary to show the positive numbers to the right of zero and the negative numbers to the left.

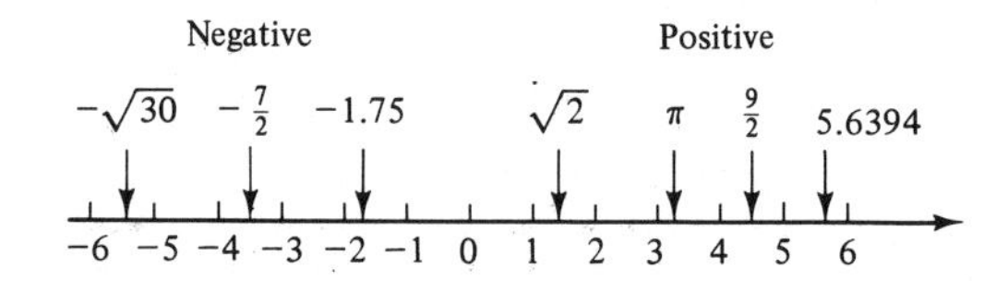

FIGURE 1-2. The number line.

Signs of Equality and Inequality: Several signs are used to show the relative position of two quantities on the number line.

$a = b$ means that a *equals* b and that they occupy the same position on the number line.

$a \neq b$ means that a and b are *not equal* and have different locations on the number line.

$a > b$ means that a is *greater than* b and lies to the right of b on the number line.

$a < b$ means that a is *less than* b and lies to the left of b on the number line.

$a \cong b$ means that a is *approximately equal to* b and that a and b are *near* each other on the number line.

Absolute Value: The *absolute value* of a number n is its *magnitude* regardless of its algebraic sign. It is written $|n|$.

Examples:

(a) $|5| = 5$

(b) $|-9| = 9$

(c) $|3 - 7| = |-4| = 4$

(d) $-|-4| = -4$

Approximate Numbers: Most of the numbers we deal with in technology are approximate.

Examples:

(a) All numbers that represent *measured* quantities are approximate. A certain shaft, for example, is approximately 1.75 in. in diameter.

(b) Many *fractions* can be expressed only approximately in decimal form. Thus, $\frac{2}{3}$ is approximately equal to 0.6667.

(c) *Irrational numbers* can be written only approximately in decimal form. The number $\sqrt{3}$ is approximately equal to 1.732.

Exact Numbers: *Exact* numbers are those that *have no uncertainty*.

Examples:

(a) There are exactly 24 hours in a day, no more, no less.

(b) An automobile has exactly four wheels.

(c) Exact numbers are usually integers, but not always. For example, there are *exactly* 2.54 cm in an inch, by definition.

(d) On the other hand, not all integers are exact. For example, a certain town has a population of *approximately* 3500 people.

Significant Digits: In a decimal number, zeros are often used as placeholders, to locate the decimal point. We say that they are *not significant*. All the remaining digits in the number are called *significant digits,* including any zeros that are not merely placeholders.

Example: The numbers 497.3, 39.05, 8003, and 2.008 each have *four* significant digits.

An overscore is sometimes placed over the last trailing zero that is significant. Thus, the numbers 32.5$\overline{0}$ and 735,$\overline{0}$00 each have four significant digits.

Example: The numbers 1570, 24,900, 0.0583, and 0.000583 each have *three* significant digits. The zeros in these numbers serve only to locate the decimal point.

Example: The numbers 18.50, 1.490, and 2.000 each have *four* significant digits. The zeros here are not needed to locate the decimal point. They are placed there to show that those digits are in fact zeros, and not something else.

Accuracy and Precision: The *accuracy* of a decimal number is given by the number of *significant digits* in the number; the *precision* of a decimal number is given by the number of *decimal places*.

Example: The number 1.255 is accurate to four significant digits, and precise to three decimal places. We also say that it is precise to the nearest thousandth.

Rounding: We will see, in the next few sections, that the numbers we get from a computation often contain *worthless digits* that must be *thrown away*. Whenever we do this, we must *round* our answer.

Round up (increase the last retained digit by one) when the discarded digits are greater than 5. *Round down* (do not change the last retained digit) when the discarded digits are less than 5.

Examples:

Number	*Rounded to three decimal places*
4.3654	4.365
4.3656	4.366
4.365501	4.366
1.764999	1.765
1.927499	1.927

When the discarded portion is 5 *exactly*, it does not usually matter whether you round up or down. The exception is when adding or subtracting a long column of figures. If, when discarding a 5, you always rounded up, you could bias the result in that direction. To avoid that you want to round up about as many times as you round down, and a simple way to do that is to always *round to the nearest even number*.

This is just a convention. We could just as well round to the nearest odd number.

Examples:

Number	*Rounded to two decimal places*
4.365	4.36
4.355	4.36
7.76500	7.76
7.75500	7.76

EXERCISE 1

Equality and Inequality Signs

Insert the proper sign of equality or inequality ($=$, $\cong$, $>$, $<$) between each pair of numbers.

1. 7 and 10 **2.** 9 and -2 **3.** -3 and 4

4. -3 and -5 **5.** $\frac{3}{4}$ and 0.75 **6.** $\frac{2}{3}$ and 0.667

Absolute Value

Evaluate each expression.

7. $-|9 - 23| - |-7 + 3|$

8. $|12 - 5 + 8| - |-6| + |15|$

9. $-|3 - 9| - |5 - 11| + |21 + 4|$

Significant Digits

Determine the number of significant digits in each approximate number.

10. 78.3 **11.** 9274 **12.** 4.008 **13.** 9400

14. 20,000 **15.** 5000.0 **16.** 0.9972 **17.** 1.0000

Round each number to two decimal places.

18. 38.468 **19.** 1.996 **20.** 96.835001 **21.** 55.8650

22. 398.372 **23.** 2.9573 **24.** 2985.339 **25.** 278.382

Round each number to one decimal place.

26. 13.98 **27.** 745.62 **28.** 5.6501 **29.** 0.482

30. 398.36 **31.** 34.927 **32.** 9839.2857 **33.** 0.847

Round each number to the nearest hundred.

34. 28,583 **35.** 7550 **36.** 3,845,240 **37.** 274,837

Round each number to three significant digits.

38. 9.284 **39.** 2857 **40.** 0.04825 **41.** 483,982

42. 0.08375 **43.** 29.555 **44.** 29.45001 **45.** 8372

1-2. ADDITION AND SUBTRACTION

Addition and Subtraction of Integers: Try the following addition problem on your calculator.

Example: Evaluate 7392 + 1147.

Solution: On your calculator:

Press	*Display*
7392 [+]	7392
1147 [=]	8539 (answer)

AOS and RPN: Does your calculator have an = key, so that you were able to do the last problem as written? If, so, your calculator uses algebraic notation or the *algebraic operating system* (AOS). If you have no [=] key but have an [ENTER] key instead, your calculator uses *reverse Polish notation* (RPN). Do not worry if you have one rather than the other; both are good.

Named for the Polish philosopher and mathematician Jan Lukasiewicz (1878–1956).

On an RPN calculator the computation shown above would be:

Press	*Display*
7392 [ENTER]	7392
1147 [+]	8539

Unless otherwise noted, computations in this book will be in AOS. You should have no trouble following the same computations on an RPN calculator.

Addition and Subtraction of Negative Numbers: Let a and b stand for two *positive* quantities. Our first rule of signs is:

All boxed and numbered formulas are tabulated in Appendix A.

Rule of Signs for Addition	$a + (-b) = a - b$	**6**

The operation of adding a negative quantity $(-b)$ *to the quantity* a *is equivalent to the operation of subtracting the positive quantity* b *from* a.

Examples:

(a) $7 + (-2) = 7 - 2 = 5$

(b) $9 + (-15) = 9 - 15 = -6$

(c) $-8 + (-3) = -8 - 3 = -11$

Our second rule is:

Rule of Signs for Subtraction	$a - (-b) = a + b$	**7**

The operation of subtracting negative quantity $(-b)$ *from a quantity* a *is equivalent to the operation of adding the positive quantity* b *to* a.

Note that the minus sign $(-)$ is used for *two different things:*

1. To indicate a *negative quantity*
2. For the operation of *subtraction*

This difference is clear on the calculator, which has *separate keys* for these two functions. The *change-sign* key [+/−] or [CHS] is used to enter a negative number, whereas the [−] key is used for subtraction. More on the change-sign key in Sec. 1-3.

Examples:

(a) $15 - (-3) = 15 + 3 = 18$

(b) $-5 - (-9) = -5 + 9 = 4$

(c) $-25 - (-5) = -25 + 5 = -20$

These laws are surely familiar to you, even if you do not recognize their names. We will run into them again when studying algebra.

Commutative and Associative Laws: The *commutative law*

Commutative Law for Addition	$a + b = b + a$	**1**

simply says that you can add numbers in *any order*.

Example:

$$2 + 3 = 3 + 2 = 5$$

The *associative law*

Associative Law for Addition	$a + (b + c) = (a + b) + c = (a + c) + b$	**3**

says that you can group numbers to be added in several ways.

Example:

$$\begin{aligned} 2 + 3 + 4 &= 2 + (3 + 4) = 2 + 7 = 9 \\ &= (2 + 3) + 4 = 5 + 4 = 9 \\ &= (2 + 4) + 3 = 6 + 3 = 9 \end{aligned}$$

Addition and Subtraction of Approximate Numbers: Addition and subtraction of integers is simple enough. But now let us tackle the problem mentioned earlier: How many digits do we keep in our answer when adding or subtracting *approximate numbers*?

Rule	When adding or subtracting approximate numbers, keep as many decimal places in your answer as contained in the number having the fewest decimal places.

Example:

$$32.4 \text{ cm} + 5.825 \text{ cm} = 38.2 \text{ cm} \quad (\textit{not } 38.225 \text{ cm})$$

We do not use the symbol $\cong$ when dealing with approximate numbers. We would not write, for example, $32.4 \text{ cm} + 5.825 \text{ cm} \cong 38.2 \text{ cm}$.

Students hate to throw away those last digits. Remember that by keeping worthless digits you are telling whoever reads that number that it is more precise than it really is.

Example:

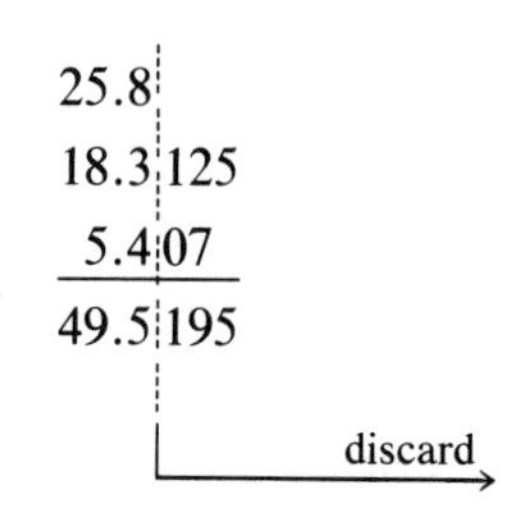

Example: A certain stadium contains about 3500 people. It starts to rain and 372 people leave. How many are left in the stadium?

Solution: Subtracting, we obtain

$$3500 - 372 = 3128$$

which we round to 3100.

Combining Exact and Approximate Numbers: Treat the exact number as if it had *more* decimal places than any of the approximate numbers.

Example: Find the sum, in minutes, of 2 h and 35.8 min.

Solution: We must add an exact number (120) and an approximate number (35.8):

$$\begin{array}{rl} 120 & \text{min} \\ +\ \underline{35.8} & \text{min} \\ 155.8 & \text{min} \end{array}$$

Since 120 is exact, we do *not* round our answer to the nearest 10 minutes, but retain as many decimal places as in the approximate number. Our answer is thus 155.8 min.

Word Problems: Problems in mathematics are often stated in verbal form. We discuss word problems in detail in Chapter 5, but we can start now to sharpen our skills by doing simple verbal problems in arithmetic. When doing the exercise, look for words that indicate *addition* (sum, increase, gain, accumulation, etc.), for words that indicate *subtraction* (difference, loss, decrease, deduction, etc.), and for words that stand for *equality* (equals, is, are, amounts to, etc.).

Example: A bank account having a balance of $5832.95 received two deposits of $359.25 and $38.99, followed by a withdrawal of $836.27. Find the new balance.

Solution: Do not hesitate to use a dictionary if you are unfamiliar with the banking terms *balance, deposit,* and *withdrawal.* It should be clear that we must add the deposits to the initial balance, and subtract the withdrawal.

$$\text{new balance} = \$5832.95 + \$359.25 + \$38.99 - \$836.27$$
$$= \$5394.92$$

EXERCISE 2

Addition and Subtraction of Integers

Combine as indicated.

1.	2.	3.
−955	8275	−748
+212	−2163	−212
−347	− 874	−156

Add each column of figures.

4.	5.	6.
$99.84	96256	98304
24.96	6016	6144
6.24	376	384
1.56	141	24576
12.48	188	3072
.98	1504	144
3.12	752	49152

Combine as indicated.

7. $926 + 863$ **8.** $274 + (-412)$ **9.** $-576 + (-553)$

10. $-207 + (-819)$ **11.** $-575 - 275$ **12.** $-771 - (-976)$

13. $1123 - (-704)$ **14.** $818 - (-207) + 318$

Addition and Subtraction of Approximate Numbers

Combine each set of approximate numbers as indicated. Round your answer.

15. $4857 + 73.8$ **16.** $39.75 + 27.4$ **17.** $296.44 + 296.997$

18. $385.28 - 692.8$ **19.** $0.000583 + 0.0008372 - 0.00173$

20. Mt. Blanc is 15,572 ft high, and Pike's Peak is about 14,000 ft high. What is the difference in their heights?

21. California contains 158,933 square miles, and Texas 237,321 square miles. How much larger is Texas than California?

22. A man willed $125,000 to his wife and two children. To his son he gave $44,675, to his daughter $26,380, and to his wife the remainder. What was his wife's share?

23. A circular pipe has an inside radius of 10.6 cm and a wall thickness of 2.125 cm. It is surrounded by insulation having a thickness of 4.8 cm (Fig. 1-3). What is the outside diameter D of the insulation?

24. A batch of concrete is made by mixing 267 kg of aggregate, 125 kg of sand, 75.5 kg of cement, and 25.25 kg of water. Find the total weight of the mixture.

25. Three resistors, having values of 27.3 ohms (Ω), 4.0155 Ω, and 9.75 Ω, are wired in series. What is the total resistance? (See Eq. A41.)

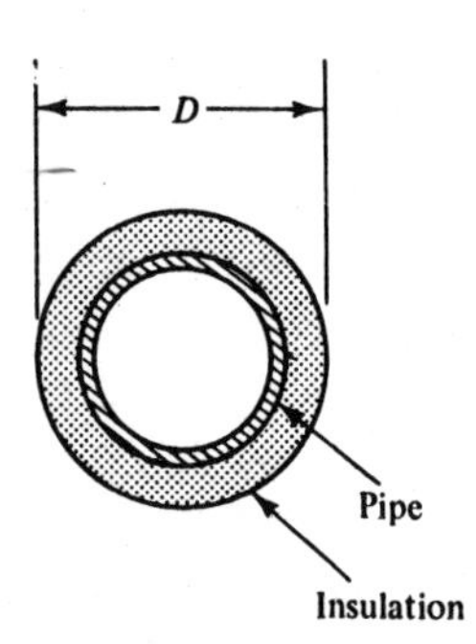

FIGURE 1-3. An insulated pipe.

1-3. MULTIPLICATION

Factors and Product: The numbers we multiply to get a *product* are called *factors*.

$$3 \times 5 = 15$$

factors ↑ ↑ ↑ product

Multiplication by Calculator: The keystrokes used to multiply the factors 23 and 57 are:

Press	*Display*
23 [×]	23
57 [=]	1311

Commutative, Associative, and Distributive Laws: The *commutative law*

Commutative Law for Multiplication	$ab = ba$	**2**

states that the *order* of multiplication is not important.

Example: It is no surprise that

$$2 \times 3 = 3 \times 2$$

The *associative law*

Associative Law for Multiplication	$a(bc) = (ab)c = (ac)b = abc$	**4**

allows us to group the numbers to be multiplied in any way.

Example:

$$2 \times 3 \times 4 = 2(3 \times 4) = 2(12) = 24$$
$$= (2 \times 3)4 = (6)4 = 24$$
$$= (2 \times 4)3 = (8)3 = 24$$

The *distributive law* tells us that

Distributive Law	$a(b + c) = ab + ac$	**5**

Example:

$$2(3 + 4) = 2(7) = 14$$

But the distributive law enables us to do the same computation in a different way:

$$2(3 + 4) = 2(3) + 2(4) = 6 + 8$$
$$= 14 \quad \text{as before}$$

Rule of Signs: The rules of signs for multiplication say that *the product of two quantities of like sign is positive:*

As before, a and b are positive *quantities.*

Rule of Signs for Multiplication	$(+a)(+b) = (-a)(-b) = +ab$	**8**

and that *the product of two quantities of unlike sign is negative:*

Rule of Signs for Multiplication	$(+a)(-b) = (-a)(+b) = -ab$	**9**

Examples:

(a) $$2(-3) = -6$$
(b) $$(-2)3 = -6$$
(c) $$(-2)(-3) = 6$$

When we multiply two negative numbers we get a positive product. So when multiplying a *string* of numbers, if an even number of them are negative, the answer will be positive, and if an odd number of them are negative, the answer will be negative.

Examples:

(a) $$2(-3)(-1)(-2) = -12$$
(b) $$2(-3)(-1)(2) = 12$$

Multiplying Negative Numbers by Calculator: Negative numbers are entered into a calculator by first entering the positive value and then *changing its sign* by pressing a *change-sign key,* marked [+/−] or [CHS]. Check your manual to see how the change-sign key is marked on your calculator.

Example: To multiply −96 and −83, we use the following keystrokes:

Press	*Display*
96 [+/−] [×]	−96
83 [+/−]	−83
[=]	7968

A simpler way to do the last problem would be to multiply +96 and +83 and determine the sign by inspection.

Common Error	Do not try to use the [−] key to enter negative numbers into your calculator. The [−] key is only for subtraction.

Multiplication of Approximate Numbers

Rule	When multiplying two or more approximate numbers, round the result to as many digits as in the factor having the fewest significant digits.

Example:

12.1	×	15.6	=	189
↑ three digits		↑ three digits		↑ three digits

When the factors have different numbers of significant digits, keep the same number of digits in your answer as is contained in the factor that has the *fewest* significant digits.

Example:

123.56	×	2.21	=	273
↑ five digits		↑ three digits		↑ keep three digits

Common Error	Do not confuse *significant digits* with *decimal places.* The number 274.56 has *five* significant digits and *two* decimal places. Decimal places determine how we round after adding or subtracting. Significant figures determine how we round after multiplying, and, as we will soon see, after dividing, raising to a power, or taking roots.

Multiplication with Exact Numbers: When using *exact numbers* in a computation, treat them as if they had *more* significant figures than any of the approximate numbers in that computation.

Example: If a certain car tire weighs 32.2 lb, how much will four such tires weigh?

Solution: Multiplying, we obtain

$$32.2(4) = 128.8 \text{ lb}$$

Since the 4 is an exact number, we retain as many significant figures as contained in 32.2, and round our answer to 129 lb.

EXERCISE 3

Multiplication

Multiply each approximate number and retain the proper number of digits in your answer.

1. 3967×1.84 **2.** 4.900×59.3 **3.** 93.9×0.0055908

4. $4.97 \times 9.27 \times 5.78$ **5.** $69.0 \times (-258)$ **6.** $-385 \times (-2.2978)$

7. $2.86 \times (4.88 \times 2.97) \times 0.553$ **8.** $(5.93 \times 7.28) \times (8.26 \times 1.38)$

Word Problems

9. What is the cost of 52.5 tons of cement at $63.25 a ton?

Financial calculations are usually worked to the nearest penny, regardless of the significant digits in the original numbers.

10. If 108 tons of rail is needed for 1 mile of track, how many tons will be required for 476 miles, and what will be its cost at $925 a ton?

11. Three barges carry 26 tons of gravel each and a fourth carries 35 tons. What is the value of the whole, at $12.75/ton?

12. Two cars start from the same place and travel in opposite directions, one at the rate of 45 km/h, the other at 55 km/h. How far apart will they be at the end of 6 h?

13. What will be the cost of building a telephone line 274 km long, at $5723 per km?

14. The current to a projection lamp is measured at 4.7 A when the line voltage is 115.45 V. Using Eq. A43, find the power dissipated in the lamp.

15. A gear in a certain machine rotates at the speed of 1808 rev/min. How many revolutions will it make in 9.5 min?

16. How much will 1000 washers weigh if each weighs 2.375 g?

17. One inch equals exactly 2.54 cm. Convert 385.84 in. to centimeters.

18. If there are 360 degrees per revolution, how many degrees are there in 4.863 revolutions?

1-4. DIVISION

Definitions: The *dividend,* when divided by the *divisor,* gives us the *quotient:*

$$\text{dividend} \div \text{divisor} = \text{quotient}$$

or

$$\frac{\text{dividend}}{\text{divisor}} = \text{quotient}$$

A quantity a/b is also a fraction, *and can also be referred to as the* ratio *of a to b. Fractions and ratios are treated in Chapter 8.*

Division by Calculator: To divide 861 by 123 on an AOS calculator:

Press	*Display*
861 [÷]	861
123 [=]	7

When we multiplied two integers, we always got an integer for an answer. This is not always the case when dividing.

Example: When we divide 2 by 3, we get 0.666666666 . . . . We must choose how many digits we wish to retain, and round our answer. Rounding to, say, four digits, we obtain

$$2 \div 3 \cong 0.667$$

Here it is appropriate to use the ≅ symbol.

Division of Approximate Numbers: The rule for rounding is almost the same as with multiplication:

Rule	After dividing one approximate number by another, round the quotient to as many digits as there are in the original number having the fewest significant digits.

Example: Divide 846.2 by 4.75.

Solution: By calculator,

$$846.2 \div 4.75 = 178.1473684$$

Since 4.75 has three significant digits, we round our quotient to 178.

Example: Divide 846.2 into three equal parts.

Solution: We divide by the integer 3, and since we consider integers to be exact, we retain in our answer the same number of significant digits as in 846.2.

$$846.2 \div 3 = 282.1$$

Dividing Negative Numbers: *The quotient is positive when dividend and divisor have the same sign;*

Rule of Signs for Division	$\frac{+a}{+b} = \frac{-a}{-b} = \frac{a}{b}$	**10**

and *the quotient is negative when dividend and divisor have opposite signs.*

Rule of Signs for Division	$\frac{+a}{-b} = \frac{-a}{+b} = -\frac{a}{b}$	**11**

Examples:

(a) $8 \div -4 = -2$

(b) $-8 \div 4 = -2$

(c) $-8 \div -4 = 2$

Example: Divide 85.4 by -2.5386 on the calculator.

Solution: The keystrokes are:

Press	*Display*
85.4 [÷]	85.4
2.5386 [+/−] [=]	−33.6405893

which we round to three digits, getting -33.6.

As with multiplication, the sign could also have been found by inspection.

Zero: Zero divided by any quantity (except zero) is zero. But division *by* zero is not defined. It is an illegal operation in mathematics.

Example: Using your calculator, divide 5 by zero.

Solution:

Press	*Display*
5 [÷] 0 [=]	flashing display or other indication of error

Reciprocals: The *reciprocal* of any number n is $1/n$.

Examples:

(a) The reciprocal of 10 is 1/10.

(b) The reciprocal of 1/2 is 2.

Reciprocals by Calculator: Simply enter the number and press the [1/x] key. Keep as many digits in your answer as there are significant digits in the original number.

Examples:

(a) The reciprocal of 6.38 is 0.157.

(b) The reciprocal of −2.754 is −0.3631.

EXERCISE 4

Division

Divide, then round your answer to the proper number of digits.

1. 947 ÷ 5.82 **2.** 0.492 ÷ 0.00478 **3.** −99.4 ÷ 286.5

4. −4.8 ÷ −2.557 **5.** 5836 ÷ 8264 **6.** 5.284 ÷ 3.827

7. 94,840 ÷ 1.33876 **8.** 3.449 ÷ −6.837

9. 2,497,000 ÷ 150,000 **10.** 2.97 ÷ 4.828

Reciprocals

Find the reciprocal of each number, retaining the proper number of digits in your answer.

11. 693 **12.** 0.0063 **13.** −396,000 **14.** 39.74

15. −0.00573 **16.** 938.4 **17.** 4.992 **18.** −6.93

19. 11.1 **20.** −375 **21.** 1.007 **22.** 3.98

Word Problems Involving Reciprocals

23. Using Eq. A42, find the equivalent resistance of a 473-Ω resistor and a 928-Ω resistor, connected in parallel.

24. When an object is placed 126 cm in front of a certain thin lens having a focal length f, the image will be formed 245 cm from the lens. The distances are related by

$$\frac{1}{f} = \frac{1}{126} + \frac{1}{245}$$

Find f.

25. The sine of an angle θ (written $\sin \theta$) is equal to the reciprocal of the cosecant of θ ($\csc \theta$), as in Eq. 152 a. Find $\sin \theta$ if $\csc \theta = 3.58$.

26. If two straight lines are perpendicular, the slope of one line is the negative reciprocal of the slope of the other. If the slope of a line is -2.55, find the slope of a perpendicular to that line.

1-5. POWERS AND ROOTS

Definitions: In the expression

$$2^4$$

the number 2 is called the *base* and the number 4 is called the *exponent*. The expression is read "two to the fourth power." Its value is

$$2^4 = 2 \cdot 2 \cdot 2 \cdot 2 = 16$$

Powers by Calculator: To square a number on the calculator, simply enter the number and press the $\boxed{x^2}$ key. To raise a number to other powers, use the $\boxed{y^x}$ key, as in the following example. Round your result to the number of significant digits contained *in the base,* not the exponent.

Example: Find $(3.85)^3$.

Solution:

Instead of displaying 3.85 as shown here, some calculators may show the natural logarithm of 3.85 (1.348 . . .).

Press	*Display*
3.85 $\boxed{y^x}$	3.85
3 $\boxed{=}$	57.066625

which we round to 57.1.

Negative Base: A negative base raised to an *even* power gives a *positive* number. A negative base raised to an *odd* power gives a *negative* number.

Examples:

(a) $(-2)^2 = (-2)(-2) = 4$

(b) $(-2)^3 = (-2)(-2)(-2) = -8$

(c) $(-1)^{24} = 1$

(d) $(-1)^{25} = -1$

If you try to do these problems on your calculator, you will probably get an error indication. Most calculators will not work with a negative base, even though this is a valid operation.

Then how do you do it? Simply enter the base as *positive,* find the power, and determine the sign by inspection.

We can raise a negative number only to an integral *power.*

Example: Find $(-1.45)^5$.

Solution: From the calculator,

$$(+1.45)^5 = 6.41$$

Since we know that a negative number raised to an odd power is negative, we write

$$(-1.45)^5 = -6.41$$

Negative Exponent: A number can be raised to a negative exponent on the calculator with the [y^x] key. Just remember to change the sign of the exponent before pressing the [=] key.

Example: Evaluate $(3.85)^{-3}$.

Solution: The keystrokes are:

Press			*Display*
3.85	[y^x]		3.85
3	[+/−]	=	0.0175233772

which we round to 0.0175.

Fractional Exponents: We will see later that fractional exponents are another way of writing radicals. For now, we just evaluate fractional exponents on the calculator. For a calculator not having parentheses, first calculate the exponent and store it in the memory. Then enter the base, press [y^x], and recall the exponent from memory.

Example: Evaluate $8^{2/3}$.

Solution:

Press	*Display*
2 [÷] 3 [=] [STO]	0.6666666667
8 [y^x] [RCL] [=]	4

Some calculators use $x \to M$ instead of STO, and RM instead of RCL.

Roots: If $a^n = b$, then

$$\sqrt[n]{b} = a$$

which is read "the *n*th root of *b* equals *a*." The symbol $\sqrt{\ }$ is a *radical sign, b* is the *radicand,* and *n* is the *index* of the radical.

Examples:

(a) $\sqrt{4} = 2$ because $2^2 = 4$

(b) $\sqrt[3]{8} = 2$ because $2^3 = 8$

(c) $\sqrt[4]{81} = 3$ because $3^4 = 81$

When we speak about the root of a number we mean, unless otherwise stated, the principal root. ***The principal root of a number has the same sign as the number itself.***

Principal Root: The *principal root* of a positive number is defined as the *positive* root. Thus, $\sqrt{4} = +2$, not ± 2.

The principal root is *negative* when we take an *odd* root of a *negative* number.

Example:

$$\sqrt[3]{-8} = -2$$

because $(-2)(-2)(-2) = -8$.

Roots by Calculator: To find square roots, enter the number and then press the [$\sqrt{x}$] key. To find other roots, use the [$\sqrt[x]{y}$] key, as in the following example. (See the following section if you have no [$\sqrt[x]{y}$] key.) Retain as many digits in your answer as there are significant digits in the original number.

Example: Find $\sqrt[5]{28.4}$, the fifth root of 28.4.

Solution:

Press	*Display*
28.4 [$\sqrt[x]{y}$]	28.4
5 [=]	1.952826537

which we round to 1.95.

Powers of 10 will be needed for scientific notation in Sec. 1-8. Arrange these powers of 10 in order and try to invent a rule that will enable you to write the value of a power of 10 without doing the computation.

Evaluate each power of 10.

13. 10^2 **14.** 10^1 **15.** 10^0 **16.** 10^{-4}

17. 10^3 **18.** 10^5 **19.** 10^{-2} **20.** 10^{-1}

21. 10^4 **22.** 10^{-3} **23.** 10^{-5}

Evaluate each expression, retaining the correct number of digits in your answer.

24. $(8.55)^3$ **25.** $(1.007)^5$ **26.** $(9.55)^3$ **27.** $(-4.82)^3$

28. $(-77.2)^2$ **29.** $(8.28)^{-2}$ **30.** $(0.0772)^{0.426}$ **31.** $(5.28)^{-2.15}$

32. $(35.2)^{1/2}$ **33.** $(462)^{2/3}$ **34.** $(88.2)^{-2}$ **35.** $(-37.3)^{-3}$

Word Problems Involving Powers

36. The distance traveled by a falling body, starting from rest, is equal to $16t^2$ (Eq. A17), where t is the elapsed time. In 5.448 s, the distance fallen is $16(5.448)^2$ ft. Evaluate this quantity.

37. The power dissipated in a resistance R through which is flowing a current I is equal to I^2R (Eq. A45). Therefore, the power in a 365-Ω resistor carrying a current of 0.5855 A is $(0.5855)^2(365)$ W. Evaluatc this power.

38. From Eq. 122, the volume of a cube of side 35.8 cm is $(35.8)^3$. Evaluate this volume.

39. From Eq. 128, the volume of a 59.4-cm-radius sphere is $\frac{4}{3}\pi(59.4)^3$ cm^3. Find this volume.

40. An investment of \$2000 at a compound interest rate of $6\frac{1}{4}\%$, left for $7\frac{1}{2}$ years, will be worth $2000(1.0625)^{7.5}$ dollars (from Eq. A9). Find this amount.

Roots

Evaluate each radical without using your calculator.

41. $\sqrt{25}$ **42.** $\sqrt[3]{27}$ **43.** $\sqrt{49}$

44. $\sqrt[3]{-27}$ **45.** $\sqrt[3]{-8}$ **46.** $\sqrt[5]{-32}$

Evaluate each radical by calculator, retaining the proper number of digits in your answer.

47. $\sqrt{49.2}$ **48.** $\sqrt{1.863}$ **49.** $\sqrt[3]{88.3}$

50. $\sqrt{772}$ **51.** $\sqrt{3875}$ **52.** $\sqrt[3]{7295}$

53. $\sqrt[3]{-386}$ **54.** $\sqrt[5]{-18.4}$ **55.** $\sqrt[3]{-2.774}$

Roots and Powers Related: If your calculator does not have a $\boxed{\sqrt[x]{y}}$ key, you can find roots by using the $\boxed{y^x}$ key. For this, you need to know the relationship between roots and exponents. It is

We will study this equation in detail later. For now, we just use it to find roots with the $\boxed{y^x}$ key.

$$\sqrt[n]{a} = a^{1/n} \qquad \textbf{36}$$

Example: Find $\sqrt[4]{482}$ using the $\boxed{y^x}$ key.

Solution: By Eq. 36,

$$\sqrt[4]{482} = 482^{1/4}$$

On the calculator:

Press	*Display*
482 $\boxed{y^x}$	482
4 $\boxed{1/x}$	0.25
$\boxed{=}$	4.685562762

which we round to 4.69.

Odd Roots of Negative Numbers by Calculator: An *even* root of a negative number is *imaginary* (such as $\sqrt{-4}$). We will study these in Chapter 16. But an *odd* root of a negative number is *not* imaginary. It is a real, negative, number. As with powers, most calculators will not accept a negative radicand. Fortunately, we can outsmart our calculators and take odd roots of negative numbers anyway.

Example: Find $\sqrt[5]{-875}$.

Solution: We know that an odd root of a negative number is real and negative. So we take the fifth root of +875, by calculator,

$$\sqrt[5]{+875} = 3.88 \quad \text{(rounded)}$$

and we only have to place a minus sign before the number.

$$\sqrt[5]{-875} = -3.88$$

EXERCISE 5

Powers

Evaluate without using a calculator.

1. 2^3 **2.** 5^3 **3.** $(-2)^3$ **4.** 9^2

5. 1^3 **6.** $(-1)^2$ **7.** $(-1)^{40}$ **8.** 3^2

9. 1^8 **10.** $(-1)^3$ **11.** $(-1)^{41}$ **12.** $(-3)^2$

Applications of Roots

56. The period (time for one swing) of a simple pendulum 2.55 ft long is

$$T = 2\pi\sqrt{\frac{2.55}{32}} \quad \text{seconds}$$

Evaluate T.

57. The magnitude of the impedance in a circuit having a resistance of 3540 Ω and a reactance of 2750 Ω is, from Eq. A55,

$$Z = \sqrt{(3540)^2 + (2750)^2} \quad \text{ohms}$$

Find Z.

58. The geometric mean between 3.75 and 9.83 is, by Eq. 59,

$$B = \sqrt{(3.75)(9.83)}$$

Evaluate B.

1-6. UNITS OF MEASUREMENT

Units: A *unit* is a standard of measurement, such as the meter, inch, hour, or pound.

SI stands for Le Système Internationale d'Unités.

The two main systems of units in use are the British system (feet, pounds, gallons, etc.) and the SI (metric) system (meters, kilograms, newtons, etc.). In addition, some special units (such as a *square* of roofing material) and some obsolete units (such as rods and chains) must occasionally be dealt with.

Conversion Factors: Each physical quantity can be expressed in any one of a bewildering variety of units. Length, for example, can be measured in meters, inches, nautical miles, angstroms, feet, and so on.

We can *convert* from any unit of length to any other unit of length by multiplying by a suitable *conversion factor.*

Example: Convert 15.4 in. to centimeters.

Solution: From Appendix B we find the relation between inches and centimeters:

$$2.54 \text{ cm} = 1 \text{ in.}$$

Dividing both sides by 1 in., we get the *conversion factor:*

$$\frac{2.54 \text{ cm}}{1 \text{ in.}} = 1$$

Multiplying yields

$$15.4 \text{ in.} = 15.4 \text{ in.} \times \frac{2.54 \text{ cm}}{\text{in.}} = 39.1 \text{ cm} \quad \text{(rounded)}$$

Suppose, in the preceding example, that we had divided both sides by 2.54 cm instead of by 1 in. We would have gotten another conversion factor:

$$\frac{1 \text{ in.}}{2.54 \text{ cm}} = 1$$

Thus, each relation between two units of measurement gives us *two* conversion factors. Since each of these is equal to 1, we may multiply any quantity by them *without changing the value* of that quantity. We will, however, cause the units to change. But which of the two conversion factors should we use? It is simple. Multiply by the conversion factor that will *cancel the units you wish to eliminate*.

Example: Convert 145 acres to square meters.

Solution: From Appendix B we find the equation

$$1 \text{ acre} = 4047 \text{ square meters}$$

We must write our conversion factor so that the unwanted units (acres) are in the denominator, so that they will cancel. So our conversion factor is

$$\frac{4047 \text{ m}^2}{1 \text{ acre}}$$

Multiplying, we obtain

$$\begin{aligned} 145 \text{ acres} &= 145 \cancel{\text{acres}} \times \frac{4047 \text{ m}^2}{1 \cancel{\text{acre}}} \\ &= 586{,}815 \text{ m}^2 \end{aligned}$$

which we round to three significant digits, getting 587,000 m^2.

Sometimes you may not be able to find a *single* conversion factor linking the units you want to convert. You may have to use *more than one*.

Example: Convert 8825 yd to nautical miles.

Solution: In Appendix B we find no conversion between nautical miles and yards, but we see that

$$1 \text{ nautical mile} = 6076 \text{ ft}$$

and also that

$$3 \text{ ft} = 1 \text{ yd}$$

So

$$\begin{aligned} 8825 \text{ yd} &= 8825 \cancel{\text{yd}} \times \frac{3 \cancel{\text{ft}}}{1 \cancel{\text{yd}}} \times \frac{1 \text{ nau mi}}{6076 \cancel{\text{ft}}} \\ &= 4.357 \text{ nautical miles} \end{aligned}$$

Converting Areas and Volumes: Conversion factors for *areas* or *volumes,* if not in the table, can be obtained by squaring or cubing the conversion factors for *length.* If we take the equation

$$2.54 \text{ cm} = 1 \text{ in.}$$

and square both sides, we get

$$(2.54 \text{ cm})^2 = (1 \text{ in.})^2$$

or

$$6.4616 \text{ cm}^2 = 1 \text{ in.}^2$$

the conversion between square centimeters and square inches.

Example: Convert 864 yd^2 to acres.

Solution: Appendix B has no conversion for square yards. However,

$$1 \text{ yd} = 3 \text{ ft}$$

Squaring yields

$$1 \text{ yd}^2 = (3 \text{ ft})^2 = 9 \text{ ft}^2$$

Also from the table,

$$1 \text{ acre} = 43{,}560 \text{ ft}^2$$

So

$$864 \text{ yd}^2 = 864 \cancel{\text{yd}^2} \times \frac{9 \cancel{\text{ft}^2}}{1 \cancel{\text{yd}^2}} \times \frac{1 \text{ acre}}{43{,}560 \cancel{\text{ft}^2}} = 0.179 \text{ acre}$$

Converting Rates to Other Units: A *rate* is the amount of one quantity expressed *per unit of some other quantity.* Some rates, with typical units, are:

rate of travel (mi/h) or (km/h)	flow rate (gal/min), (m^3/s)
application rate (lb/acre)	unit price (dollars/lb)

Each rate contains *two* units of measure; miles per hour, for example, has *miles* in the numerator and *hours* in the denominator. It may be necessary to convert *either or both* of those units to other units. Sometimes a single conversion factor can be found (such as 1 m/h = 1.466 ft/s), but more often you will have to convert each unit with a *separate* conversion factor.

Be sure to write the original quantity as a built-up fraction, $\frac{a}{b}$, rather than on a single line, a/b. This will greatly reduce your chances of making an error.

Example: A certain chemical is to be added to a pool at the rate of 3.74 oz per gallon of water. Convert this to pounds of chemical per cubic foot of water.

Solution: We write the original quantity as a fraction, and multiply by the appropriate factors, themselves written as fractions.

$$3.74 \text{ oz/gal} = \frac{3.74 \text{ oz}}{\text{gal}} \times \frac{1 \text{ lb}}{16 \text{ oz}} \times \frac{7.481 \text{ gal}}{\text{ft}^3} = 1.75 \text{ lb/ft}^3$$

EXERCISE 6

Conversion of Units

Convert.

Here we make a distinction between mass *(kilograms, slugs) and* weight *(newtons, pounds). Weight has the same units as force.*

1. 185 ft to meters
2. 9.37 mi to kilometers
3. 993 cm to inches
4. 4935 yd to meters
5. 79.2 acres to square meters
6. 148 acres to ares
7. 88.3 cm^2 to square inches
8. 525 yd^3 to cubic meters
9. 994 kg to slugs
10. 1.05 ft^2 to cm^2
11. 77.3 N to ounces
12. 96,500 lb to newtons
13. 5528 N to pounds
14. 2.5 hp to kilowatts

Convert units on the following time rates.

15. 63 ft/s to miles per hour
16. 555 gal/min to cubic meters per hour
17. 55.4 mi/h to kilometers per hour

1-7. SUBSTITUTING INTO EQUATIONS AND FORMULAS

Substituting into Equations: We get an *equation* when two expressions are set equal to each other.

Example: $x = 5a - 2b + 3c$ is an equation that enables us to find x if we know a, b, and c.

We will study equations in detail starting with Chapter 5, but for now we can use our calculators to substitute into equations. To *substitute into an equation* means to replace the letter quantities in an equation by their given numerical values, and to perform the indicated computation.

Example: Substitute the values

$$a = 5$$
$$b = 3$$
$$c = 6$$

into the equation

$$x = \frac{3a + b}{c}$$

Solution: Substituting, we obtain

$$x = \frac{3(5) + 3}{6} = \frac{18}{6} = 3$$

When substituting approximate numbers, be sure to round your answer to the proper number of digits. Treat any integers in the given equation as exact numbers.

Substituting into Formulas: A *formula* is an equation expressing some general mathematical or physical fact, such as the formula for the area of a circle of radius r:

Area of a Circle	$A = \pi r^2$	**114**

We substitute into formulas just as we substituted into equations; except that we now carry *units* along with the numerical values. You will often need conversion factors to make the units cancel properly, so that the answer will be in the desired units.

Example: A tensile load of 4500 lb is applied to a bar that is 5.2 yd long and has a cross-sectional area of 11.6 cm². The elongation is 0.38 mm. Using Eq. A32, find the modulus of elasticity E in pounds per square inch.

Solution: Substituting the values *with units* into Eq. A32, we obtain

$$E = \frac{PL}{ae} = \frac{4500 \text{ lb} \times 5.2 \text{ yd}}{11.6 \text{ cm}^2 \times 0.38 \text{ mm}}$$

If the units to be used in a certain formula are specified, convert all quantities to those specified units before substituting into the formula.

Notice that we have a length (5.2 yd) in the numerator and a length (0.38 mm) in the denominator. To make these units cancel, we use the conversion factors,

$$25.4 \text{ mm} = 1 \text{ in.}$$

and

$$36 \text{ in.} = 1 \text{ yd}$$

Also, our answer is to have square inches in the denominator, not square centimeters. So we use another conversion factor,

$$6.452 \text{ cm}^2 = 1 \text{ in.}^2$$

$$E = \frac{4500 \text{ lb} \times 5.2 \text{ yd}}{11.6 \text{ cm}^2 \times 0.38 \text{ mm}} \times \frac{25.4 \text{ mm}}{\text{in.}} \times \frac{36 \text{ in.}}{\text{yd}} \times \frac{6.452 \text{ cm}^2}{\text{in.}^2}$$

$$= 31{,}000{,}000 \text{ lb/in.}^2 \quad \text{(rounded to two digits)}$$

Common Error	Students often neglect to include *units* when substituting into a formula, with the result that the units often do not cancel properly.

EXERCISE 7

Substituting into Equations

Substitute the given integers into each equation. Do not round your answer.

1. $y = 5x + 2 \quad (x = 3)$

2. $y = 2m^2 - 3m + 5 \quad (m = -2)$

3. $y = 2a - 3x^2 \quad (x = 3, a = -5)$

4. $y = 3x^3 - 2x^2 + 4x - 7 \quad (x = 2)$

5. $y = 2b + 3w^2 - 5z^3 \quad (b = 3, w = -4, z = 2)$

6. $y = \frac{r^2}{x} - \frac{x^3}{r} + \frac{w}{x^2} \quad (x = 5, w = 3, r = -4)$

Substitute the given approximate numbers into each equation. Round your answer to the proper number of digits.

7. $y = 7x - 5 \quad (x = 2.73)$

8. $y = 2w^2 - 3x^2 \quad (x = -11.5, w = 9.83)$

9. $y = 8 - x + 3x^2 \quad (x = -8.49)$

10. $y = \sqrt[3]{8x + 7w} \quad (x = 1.255, w = 2.304)$

11. $y = \sqrt{x^3 - 3x} \quad (x = 4.25)$

12. $y = (w - 2x)^{1.6} \quad (x = 1.8, w = 7.2)$

Substituting into Formulas

13. Use Eq. A8 to find the amount to which \$3000 will accumulate in 5 years at a simple interest rate of 6.5%.

14. Using Eq. A17 find the displacement after 1.3 min of a body thrown downward with a speed of 12 ft/s.

15. Using Eq. A28 convert 128°F to degrees Celsius.

16. The torque T delivered by a motor of horsepower P rotating at N rev/min is

$$T = \frac{33{,}000P}{2\pi N} \quad \text{ft lb}$$

Find the torque delivered by a $\frac{3}{4}$-hp motor rotating at 1800 rev/min.

17. Find the torque delivered by a 3.5-hp motor rotating at 150 rad/s.

18. A bar 15.2 m long having a cross-sectional area of 12.7 cm^2 is subject to a tensile load of 22,500 N. The elongation is 2.75 mm. Use Eq. A32 to find the modulus of elasticity in newtons per square centimeter.

19. Use Eq. A9 to find the amount y obtained when \$9570 is allowed to accumulate for 5 years at a compound interest rate of $6\frac{3}{4}$%.

20. The formula for the pressure loss h in a pipe is

$$h = \frac{6270fLQ^2}{D^5} \quad \text{ft}$$

where f is the friction factor, L the length of pipe in feet, Q the flow rate in cubic feet per second, and D the pipe diameter in inches. Compute the pressure drop in a 3.35-in.-diameter, 125-ft-long pipe, where $f = 0.017$ and the flow rate is 155 gal/min.

21. The resistance of a copper coil is 775 Ω at 20°C. The temperature coefficient of resistance is 0.00393 at 20°C. Use Eq. A48 to find the resistance at 80°C.

1-8. SCIENTIFIC NOTATION

Definitions: Let us multiply two large numbers on the calculator: say, 500,000 and 300,000. On an algebraic calculator:

Press	*Display*
500000 [×]	500000
300000 [=]	1.5 11

We get the strange-looking display, [1.5 11]. What has happened?

Our answer (150,000,000,000) is too large to fit the calculator display, so the machine has automatically switched to *scientific notation*. Our calculator display actually contains *two* numbers: a decimal number (1.5) and an integer (11). Our answer is equal to the decimal number, multiplied by 10 raised to the value of the integer.

Calculator display: [1.5 11]

Scientific notation: $\underline{1.5} \times \underline{10^{11}}$

decimal part ↑ ↑ power of 10

Ten raised to a power (such as 10^{11}) is called a *power of 10.*

A number is said to be in *scientific notation* when it is written as a number between 1 and 10, multiplied by a power of 10.

Examples:

(a) 2.74×10^3

(b) 8.84×10^9

(c) 5.4×10^{-6}

(d) -1.2×10^{-5}

are numbers written in scientific notation.

Evaluating Powers of 10: We did some work with powers in Sec. 1-5. We saw, for example, that 2^3 meant

$$2^3 = 2 \cdot 2 \cdot 2 = 8$$

Powers of 10 are found in the same way.

Examples:

(a) $$10^2 = 10 \times 10 = 100$$

(b) $$10^3 = 10 \times 10 \times 10 = 1000$$

Negative powers, as before, are calculated by means of Eq. 35, $x^{-a} = 1/x^a$:

Examples:

(a) $$10^{-2} = \frac{1}{10^2} = \frac{1}{100} = 0.01$$

(b) $$10^{-5} = \frac{1}{10^5} = \frac{1}{100{,}000} = 0.00001$$

Summarizing some powers of 10 in a table:

Remember these from Exercise 5? Did you find a rule to switch between scientific notation and decimal notation? One such rule is given in the following section.

Positive powers	*Negative powers*
$1{,}000{,}000 = 10^6$	$0.1 = 10^{-1}$
$100{,}000 = 10^5$	$0.01 = 10^{-2}$
$10{,}000 = 10^4$	$0.001 = 10^{-3}$
$1{,}000 = 10^3$	$0.0001 = 10^{-4}$
$100 = 10^2$	$0.00001 = 10^{-5}$
$10 = 10^1$	$0.000001 = 10^{-6}$
$1 = 10^0$	

Converting Numbers to Scientific Notation: Rewrite the given number with just *one digit* to the left of the decimal point, and discard any nonsignificant zeros. Make the power to which 10 is raised equal to the number of places the decimal point was moved from its original location: *positive* if moved to the *left* and *negative* if moved to the *right*.

Example: Convert 2700 to scientific notation.

Solution: We place the decimal point between the 2 and the 7. This is a move of three places to the left from its original position, so the power is +3. Discarding the two nonsignificant zeros, we get

$$2700. = 2.7 \times 10^3$$

moved three places   power = 3

Another way of looking at it is that the movement of the decimal point three places to the left is equivalent to multiplying by 10^{-3}. We must *compensate* for this by multiplying by 10^3. Since $(10^{-3})(10^3)$ equals 1, we will not have changed the value of the original number.

$$2700 = \underbrace{2700 \times 10^{-3}}_{} \times 10^3$$
$$= \quad 2.7 \quad \times 10^3$$

Example: Convert 0.00000950 to scientific notation.

Solution: We move the decimal point six places to the right, so the power is -6. We drop the nonsignificant zeros to the left of the 9 and retain the significant digit following the 5.

$$0.00000950 = 9.50 \times 10^{-6}$$

six places

To convert *from* scientific notation *to* decimal notation, simply reverse the process.

Example: Convert 4.82×10^5 to decimal form.

Solution: Moving the decimal point five places to the right, we obtain

$$4.82 \times 10^5 = 482{,}000$$

five places

Example: Convert 8.25×10^{-3} to decimal form.

Solution: Moving the decimal point three places to the left, we obtain

$$8.25 \times 10^{-3} = 0.00825$$

three places

Addition and Subtraction: If two or more numbers to be added or subtracted have the *same power of 10,* simply combine the numbers and keep the same power of 10.

Examples:

This is another application of the distributive law, Eq. 5: $ab + ac = a(b + c)$

(a) $$2 \times 10^5 + 3 \times 10^5 = 5 \times 10^5$$

(b) $$8 \times 10^3 - 5 \times 10^3 + 3 \times 10^3 = 6 \times 10^3$$

If the powers of 10 are different, *they must be made equal* before the numbers can be combined. A shift of the decimal point of one place to the *left* will *increase* the exponent by 1. Conversely, a shift of the decimal point one place to the *right* will *decrease* the exponent by 1.

Examples:

(a) $$1.5 \times 10^4 + 3 \times 10^3 = 1.5 \times 10^4 + 0.3 \times 10^4 = 1.8 \times 10^4$$

(b) $$1.25 \times 10^5 - 2 \times 10^4 + 4 \times 10^3 = 1.25 \times 10^5 - 0.2 \times 10^5 + 0.04 \times 10^5 = 1.09 \times 10^5$$

Multiplication: We multiply powers of 10 by *adding their exponents.*

We are really using Eq. 29, $x^a \cdot x^b = x^{a+b}$. This equation is one of the laws of exponents that we will study in Chapter 4.

Examples:

(a) $$10^3 \cdot 10^4 = 10^{3+4} = 10^7$$

(b) $$10^{-2} \cdot 10^5 = 10^{-2+5} = 10^3$$

To multiply two numbers in scientific notation, multiply the decimal parts and the powers of 10 *separately.*

Example:

$$(2 \times 10^5)(3 \times 10^2) = (2 \times 3)(10^5 \times 10^2) = 6 \times 10^{5+2} = 6 \times 10^7$$

Division: We divide powers of 10 by subtracting their exponents:

Examples:

(a) $$\frac{10^5}{10^3} = 10^{5-3} = 10^2$$

(b) $$\frac{10^{-4}}{10^{-2}} = 10^{-4-(-2)} = 10^{-2}$$

As with multiplication, we divide the decimal parts and the powers of 10 separately.

Examples:

(a) $$\frac{8 \times 10^5}{4 \times 10^2} = \frac{8}{4} \times \frac{10^5}{10^2} = 2 \times 10^{5-2} = 2 \times 10^3$$

(b) $$\frac{12 \times 10^3}{4 \times 10^5} = 3 \times 10^{3-5} = 3 \times 10^{-2}$$

Scientific Notation on Computer or Calculator: A computer or calculator will switch to scientific notation by itself when the printout or display gets too long. For example, a computer might print the number

2.75E-6

which we interpret as

$$2.75 \times 10^{-6} \quad \text{or} \quad 0.00000275$$

We may also *enter* numbers in scientific notation. On a computer, simply type the number followed by E, followed by the exponent to which 10 is to be raised.

Example: To enter the number 5,824,000,000, which in scientific notation is 5.824×10^9, we type

```
5.824E9
```

On a *calculator* we use the *enter exponent* key, usually marked EE or EEX.

Example: Enter the number 3.5×10^4.

Solution: The keystrokes are:

Press	*Display*
3.5 [EE]	3.5 00
4	3.5 04

Example: Enter the number 9.5×10^{-7}.

Solution:

Press	*Display*
9.5 [EE]	9.5 00
7 [+/−]	9.5 −07

Computing in Scientific Notation by Calculator: Simply follow the same calculator procedures as you would for decimal numbers and let the calculator worry about the powers of 10.

Example: Multiply 5.38×10^3 and 3.973×10^{-2}.

Solution:

Press		*Display*
5.38	[EE]	5.38 00
3	[X]	5.38 03
3.973	[EE]	3.973 00
2	[+/−]	3.973 − 02
	[=]	2.137474 02

We now round 2.137474 to three digits, getting 2.14×10^2, or 214.

A prefix is a group of letters placed at the beginning of a word to modify the meaning of that word.

Metric Prefixes: Multiples and submultiples of metric units are sometimes indicated by a *prefix* rather than by scientific notation. For example, the prefix *kilo* means 1000, or 10^3. Thus, a *kilo*meter is 1000 meters.

Example: Express 15,485 m in kilometers.

Solution:

$$\begin{aligned} 15{,}485 \text{ m} &= 15.485 \times 10^3 \text{ m} \\ &= 15.485 \text{ km} \end{aligned}$$

Other frequently used metric prefixes are

mega	10^6	milli	10^{-3}
centi	10^{-2}	micro	10^{-6}

Example: Convert 1875 microseconds (μs) to milliseconds (ms).

Solution:

$$\begin{aligned} 1875\ \mu\text{s} &= 1875 \times 10^{-6} \text{ s} \\ &= 1.875 \times 10^{-3} \text{ s} \\ &= 1.875 \text{ ms} \end{aligned}$$

EXERCISE 8

Powers of 10

Write each number as a power of 10.

1. 100 **2.** 1,000,000 **3.** 0.0001

4. 0.001 **5.** 100,000,000

Write each power of 10 as a decimal number.

6. 10^5 **7.** 10^{-2} **8.** 10^{-5}

9. 10^{-1} **10.** 10^4

Scientific Notation

Write each number in scientific notation.

11. 186,000 **12.** 0.0035 **13.** 25,742

14. $80\bar{0}0$ **15.** 98.3×10^3 **16.** 0.0775×10^{-2}

17. $115{,}000 \times 10^{-3}$ **18.** 0.4593×10^5 **19.** 145.2

20. 0.075

Convert each number from scientific notation to decimal notation.

21. 2.85×10^3
22. 1.75×10^{-5}
23. 9×10^4
24. 9.00×10^4
25. 3.667×10^{-3}

Multiplication and Division

Multiply the following powers of 10.

26. $10^5 \cdot 10^2$
27. $10^4 \cdot 10^{-3}$
28. $10^{-5} \cdot 10^{-4}$
29. $10^{-2} \cdot 10^5$
30. $10^{-1} \cdot 10^{-4}$

Divide the following powers of 10.

31. $10^8 \div 10^5$
32. $10^4 \div 10^6$
33. $10^5 \div 10^{-2}$
34. $10^{-3} \div 10^5$
35. $10^{-2} \div 10^{-4}$

Multiply without using a calculator.

36. $(3 \times 10^3)(5 \times 10^2)$
37. $(5 \times 10^4)(8 \times 10^{-3})$
38. $(2 \times 10^{-2})(4 \times 10^{-5})$
39. $(7 \times 10^4)(30{,}000)$
40. $(5 \times 10^{-3})(9000)$
41. $(0.003)(4 \times 10^{-5})$
42. $(90{,}000)(200{,}000)$
43. $(150{,}000)(0.002)$

Divide without using a calculator.

44. $(8 \times 10^4) \div (2 \times 10^2)$
45. $(6 \times 10^4) \div 0.03$
46. $(3 \times 10^3) \div (6 \times 10^5)$
47. $(8 \times 10^{-4}) \div 400{,}000$
48. $(9 \times 10^4) \div (3 \times 10^{-2})$
49. $49{,}000 \div (7.0 \times 10^{-2})$
50. $(8 \times 10^{-3}) \div (4 \times 10^5)$
51. $0.0009 \div (3 \times 10^{-2})$

Addition and Subtraction

Combine without using a calculator.

52. $(3.5 \times 10^4) + (2.1 \times 10^5)$
53. $(75 \times 10^2) + 3200$
54. $(1.557 \times 10^2) + (9 \times 10^{-1})$
55. $0.037 - (6 \times 10^{-3})$
56. $(7.2 \times 10^4) + (1.1 \times 10^4)$
57. $(2.15 \times 10^4) + (3.4 \times 10^3) - (1.1 \times 10^2)$
58. $(3.9 \times 10^{-2}) - (2.4 \times 10^{-2})$
59. $(9.55 \times 10^{-5}) - (6.2 \times 10^{-6}) + (3 \times 10^{-7})$

Scientific Notation on the Calculator

Perform the following computations in scientific notation. Combine the powers of 10 by hand or with your calculator.

60. $(1.58 \times 10^2)(9.82 \times 10^3)$
61. $(9.83 \times 10^5) \div (2.77 \times 10^3)$

62. $(3.87 \times 10^{-2})(5.44 \times 10^{5})$

63. $(2.74 \times 10^{3}) \div (9.13 \times 10^{5})$

64. $(5.6 \times 10^{2})(3.1 \times 10^{-1})$

65. $(7.72 \times 10^{8}) \div (3.75 \times 10^{-9})$

66. $(4.79 \times 10^{3})^2$

67. $\dfrac{7.29 \times 10^{-2}}{3.54 \times 10^{-5}} - 1.14 \times 10^{3}$

68. $(-5.60 \times 10^{-2})^2$

69. $\dfrac{(6.35 \times 10^{5}) - (9.87 \times 10^{4})}{(1.27 \times 10^{2}) + (2.25 \times 10^{3})}$

70. $(2.75 \times 10^{2})^3$

71. $(-3.12 \times 10^{-2})^3$

72. $(7.28 \times 10^{4}) + (3.9 \times 10^{3})$

73. $\dfrac{(7.14 \times 10^{2})(2.73 \times 10^{3})}{(7.75 \times 10^{5}) + (2.58 \times 10^{4})}$

Applications

74. Three resistors, having resistances of $4.98 \times 10^5\ \Omega$, $2.47 \times 10^4\ \Omega$, and $9.27 \times 10^6\ \Omega$, are wired in series. Find the total resistance, using Eq. A41.

75. Find the equivalent resistance if the three resistors of Problem 74 are wired in parallel. Use Eq. A42.

76. Find the power dissipated in a resistor if a current of 3.75×10^{-3} A produces a voltage drop of 7.24×10^{-4} V across the resistor. Use Eq. A43.

77. The voltage across a 8.35×10^5-Ω resistor is 2.95×10^{-3} V. Find the power dissipated in the resistor, using Eq. A44.

78. Three capacitors, 8.26×10^{-6} farad (F), 1.38×10^{-7} F, and 5.93×10^{-5} F, are wired in parallel. Find the equivalent capacitance by Eq. A51.

79. A tensile load of 2.85×10^4 lb is supported by a member having a cross-sectional area of 1.86×10^2 in.^{2}. Find the stress in the member, using Eq. A30.

80. A wire 4.75×10^3 cm long when loaded is seen to stretch 9.55×10^{-2} cm. Find the strain in the wire, using Eq. A31.

81. A bar having a cross-sectional area of 1.45 in.^{2} and a length of 2.55×10^2 in. is subjected to a tensile load of 1.24×10^4 lb. The elongation is 9.55×10^{-2} in. Find the modulus of elasticity E of the material, using Eq. A32.

82. The wind pressure against the side of a certain building is 8.5×10^{-2} lb/in.^{2}. If the area of the side of the building is 5.8×10^6 in.^{2}, find the total force exerted by the wind, using Eq. A26.

83. How long will it take a rocket traveling at a rate of 3.2×10^6 m/h to travel the 3.8×10^8 m from the earth to the moon? Use Eq. A16.

84. A new memory unit containing 4.55×10^4 bits of memory is added to a computer that already has a memory of 7.25×10^6 bits. Find the total number of bits after the new unit is installed.

85. The oil shale reserves of the United States are estimated at 2×10^9 tons. How long would this last at a rate of consumption of 9×10^7 tons/yr?

Metric Prefixes

86. Convert.

In these problems write your answer in scientific notation if the numerical value is greater than 1000 or less than 0.1. Be careful to retain the proper number of significant digits.

(a) 174,000 m to kilometers
(b) 0.00537 V to millivolts
(c) 39,400 g to kilograms
(d) 4.83×10^{-5} kW to watts
(e) 1.45×10^9 Ω to megohms
(f) 396×10^4 N to kilonewtons
(g) 2584 picofarads to microfarads
(h) 35,960 nanoseconds to milliseconds

1-9. PERCENTAGE

Definition of Percent: The word *percent* means *by the hundred,* or *per hundred.* A percent thus gives the number of parts in every hundred.

Example: If we say that a certain concrete mix is 12% cement by weight, we mean that 12 out of every 100 lb of mix will be cement.

Rates: The word *rate* is often used to indicate a percent, or percentage rate, as in "rejection rate," "rate of inflation," or "growth rate."

Example: A failure rate of 2% means that 2 parts out of every 100 would be expected to fail.

Percent as a Fraction: Percent is another way of expressing a *fraction* having 100 as the denominator.

Example: If we say that a builder has finished 75% of a house, we mean that he has finished 75/100 (or 3/4) of the house.

Converting Decimals to Percent: Before working some percentage problems, let us first get some practice in converting decimals and fractions to percents, and vice versa. To convert decimals to percent, simply move the decimal point two places to the right and affix the percent symbol (%).

Examples:

(a)
$$0.75 = \frac{75}{100} = 75\%$$

(b) $$3.65 = \frac{365}{100} = 365\%$$

(c) $$0.003 = \frac{0.3}{100} = 0.3\%$$

Converting Fractions or Mixed Numbers to Percent: First write the fraction or mixed number as a decimal, and then proceed as above.

Examples:

(a) $$\frac{1}{4} = 0.25 = \frac{25}{100} = 25\%$$

(b) $$\frac{5}{2} = 2.5 = \frac{250}{100} = 250\%$$

(c) $$1\frac{1}{4} = 1.25 = \frac{125}{100} = 125\%$$

Converting Percent to Decimals: Move the decimal point two places to the left and remove the percent sign.

Examples:

(a) $$13\% = \frac{13}{100} = 0.13$$

(b) $$4.5\% = \frac{4.5}{100} = 0.045$$

(c) $$155\% = \frac{155}{100} = 1.55$$

(d) $$27\tfrac{3}{4}\% = \frac{27.75}{100} = 0.2775$$

Converting Percent to a Fraction: Write a fraction with 100 in the denominator and the percent in the numerator. Remove the percent sign and reduce the fraction to lowest terms.

Examples:

(a) $$75\% = \frac{75}{100} = \frac{3}{4}$$

(b) $$87.5\% = \frac{87.5}{100} = \frac{875}{1000} = \frac{7}{8}$$

(c) $$125\% = \frac{125}{100} = \frac{5}{4} = 1\frac{1}{4}$$

Amount, Base, and Rate: Percentage problems always involve three quantities:

1. The *percent rate, P*
2. The *base, B*: the quantity we are taking the percent of
3. The *amount A* we get when we take the percent of the base

In a percentage problem, you will know two of these three quantities (amount, base, or rate) and be required to find the third. This is easily done, for the rate, base, and amount are related by the equation

Percentage	amount = base × rate or $A = BP$ where P is expressed as a decimal.	**12**

Finding the Amount When the Base and Rate are Known: We substitute the given base and rate into Eq. 12 and solve for the amount:

Example: What is 35 percent of 80?

Solution: In this problem the rate is 35%, so

$$P = 0.35$$

But is 80 the amount or the base?

Tip	In a particular problem, if you have trouble telling which number is the base and which is the amount, look for the key phrase "*percent of.*" The quantity following this phrase is *always the base.*

We look for the key phrase "percent of."

base

What is 35 percent of [80]?

Since 80 immediately follows *percent of,*

$$B = 80$$

From Eq. 12,

$$A = BP = 80(0.35) = 28$$

Common Error	Do not forget to convert the percent rate to a *decimal* when using Eq. 12.

Example: Find 0.1% of 5600.

Solution: We substitute into Eq. 12, with

$$B = 5600$$

and

$$P = 0.001 \quad (\textit{not } 0.1!)$$

So

$$A = BP = 5600(0.001) = 5.6$$

Finding Amounts by Calculator: We make use of the [%] key, as in the following examples.

Example: Find 18% of 248 by calculator.

Solution: The keystrokes are:

Press	*Display*
248 [×]	248
18 [%] [=]	44.64

Finding the Base When a Percent of It Is Known: We see from Eq. 12 that the base equals the amount divided by the rate (expressed as a decimal), or $B = A/P$.

Example: 12% of what number is 72?

Solution: Finding the key phrase,

12 percent of [what number] is 72?
base

it is clear that we are looking for the base. So

$$A = 72 \qquad \text{and} \qquad P = 0.12$$

By Eq. 12,

$$B = \frac{A}{P} = \frac{72}{0.12} = 600$$

Example: 125 is 25% of what number?

Solution: From Eq. 12,

$$B = \frac{A}{P} = \frac{125}{0.25} = 500$$

Finding the Percent One Number Is of Another Number: From Eq. 12, the rate equals the amount divided by the base, or $P = A/B$.

Example: 42 is what percent of 400?

Solution: By Eq. 12, with $A = 42$ and $B = 400$,

$$P = \frac{A}{B} = \frac{42}{400} = 0.105 = 10.5\%$$

Example: What percent of 1.4 is 0.35?

Solution: From Eq. 12,

$$P = \frac{A}{B} = \frac{0.35}{1.4} = 0.25$$

$$= 25\%$$

EXERCISE 9

Conversions

Convert each decimal to a percent.

1. 3.72 **2.** 0.877 **3.** 0.0055

4. 0.563 **5.** 0.234 **6.** 0.0573

Convert each fraction to a percent. Round to three significant digits.

7. $\frac{2}{5}$ **8.** $\frac{3}{4}$ **9.** $\frac{7}{10}$

10. $\frac{3}{7}$ **11.** $\frac{7}{8}$ **12.** $\frac{5}{12}$

Convert each percent to a decimal.

13. 23% **14.** 2.97% **15.** $287\frac{1}{2}\%$

16. $6\frac{1}{4}\%$ **17.** $\frac{3}{4}\%$ **18.** $\frac{3}{5}\%$

Convert each percent to a fraction.

19. 37.5% **20.** $12\frac{1}{2}\%$ **21.** 150%

22. 3% **23.** $16\frac{2}{3}\%$ **24.** $33\frac{1}{3}\%$

Finding the Amount

Find:

25. 40% of 250 tons **26.** 15% of 300 mi **27.** $33\frac{1}{3}\%$ of 660 kg

28. $12\frac{1}{2}\%$ of 72 gal **29.** 30% of 400 liters **30.** 50% of $240

31. A resistance, now 7250 Ω, is to be increased by 15%. How much resistance should be added?

32. It is estimated that $\frac{1}{2}\%$ of the earth's surface receives more energy than the total projected needs for the year 2000. Assuming the earth's surface area to be 1.97×10^8 mi^2, find the required area in acres.

33. As an incentive to install solar equipment, a tax credit of 40% of the first \$1000 and 25% of the next \$6400 spent on solar equipment is proposed. How much credit would a homeowner get when installing \$5000 worth of solar equipment?

34. It is estimated that the United States has tar sands containing 3.3×10^{10} barrels of oil and that 90% of the tar sands are in Utah. If it is possible to extract 9% of the oil from the sands, how much oil could be extracted from the Utah tar sands?

35. How much metal will be obtained from 375 tons of ore if the metal is $10\frac{1}{2}$% of the ore?

Finding the Base

Find the number of which:

36. 86.5 is $16\frac{2}{3}$%
37. 45 is 1.5%
38. $\frac{3}{8}$ is $1\frac{1}{3}$%
39. 50 is 25%
40. 60 is 60%
41. 60 is 75%

42. A Department of Energy report on an experimental electric car gives the range of the car as 160 km and states that this is "50% better than on earlier electric vehicles." What was the range of earlier electric vehicles?

43. A man withdrew 25% of his bank deposits and spent $33\frac{1}{3}$% of the money drawn in the purchase of a radio worth \$250. How much money had he had in the bank?

44. Solar panels provide 55% of the heat for a certain building. If \$225/yr is now spent for heating oil, what would have been spent if the solar panels were not used?

45. If the United States imports 9.14 billion barrels of oil per day and if this is 48.2% of its needs, how much oil is needed per day?

Finding the Rate

What percent of:

46. 24 is 12?
47. 36 is 12?
48. 40 is 8?
49. 840 men are 420 men?
50. 450 h are 150 h?
51. 450 tons are 300 tons?

52. A 50,500-liter-capacity tank contains 5840 liters of water. Express the amount of water in the tank as a percentage of the total capacity.

53. In a journey of 1560 km, a person traveled 195 km by car and the rest of the distance by rail. What percent of the distance was traveled by rail?

54. A power supply has a dc output of 50.0 V with a ripple of 0.75 V peak to peak. Express the ripple as a percentage of the dc output voltage.

55. The construction of a factory cost $136,000 for materials and $157,000 for labor. The labor cost was what percentage of the total?

56. If the United States needs 17.8 billion barrels of oil per day and imports 9.12 billion barrels per day, what percent of its oil needs are imported?

57. By insulating, a homeowner's yearly fuel consumption dropped from 628 gal to 405 gal. His present oil consumption is what percent of the former?

1-10. PERCENT CHANGE AND PERCENT EFFICIENCY

We cover percent error in Sec. 3-4, and percent concentration in Sec. 5-5.

Comparisons between Two Amounts: Percentages are often used to compare two quantities. You often hear statements like the following:

The price of steel rose 3% over last year's price.
The weights of two cars differed by 20%.
Production dropped 5% from last year.

In each of these, *two quantities* are being compared:

steel price this year	vs.	steel price last year
weight of first car	vs.	weight of second car
production this year	vs.	production last year

Percent Change: When the two numbers being compared involve a *change* from one to the other, the *original value* is usually taken as the base.

$$\text{percent change} = \frac{\text{new value} - \text{original value}}{\text{original value}} \times 100\% \qquad \textbf{13}$$

Example: A certain price rose from $1.55 to $1.75. Find the percentage change in price.

Be sure to show the direction of change with a plus or minus sign, or with words like "increase" or "decrease."

Solution: We use the original value, $1.55, as the base. From Eq. 13,

$$\text{percent change} = \frac{1.75 - 1.55}{1.55} \times 100\% = 12.9\% \text{ increase}$$

A common type of problem is to *find the new value* when the original value is changed by a given percent. We see from Eq. 13 that

$$\text{new value} = \text{original value} + (\text{original value}) \times (\text{percent change})$$

Example: Find the cost of a $156 suit after the price increases by $2\frac{1}{2}\%$.

Solution: The original value is 156, and the percent change, expressed as a decimal, is 0.025. So

$$\text{new value} = 156 + 156(0.025) = \$159.90$$

The calculator is convenient for this type of problem.

Example: What will be the cost of a $5500 machine after an increase of 5% in price?

Solution:

Press	*Display*
5500 [+] 5 [%] [=]	5775

Percent Efficiency: The power output of any machine or device is always *less* than the power input, because of inevitable power losses within the device. The *efficiency* of the device is a measure of those losses.

$$\text{percent efficiency} = \frac{\text{output}}{\text{input}} \times 100\% \qquad \textbf{16}$$

Example: A certain electric motor consumes 865 W and has an output of 1.12 hp. Find the efficiency of the motor. (1 hp = 746 W.)

Solution: Since output and input must be in the same units, we must convert either to horsepower or to watts. Converting the output to watts, we obtain

$$\text{output} = 1.12\ \text{hp}\left(\frac{746\ \text{W}}{\text{hp}}\right) = 836\ \text{W}$$

By Eq. 16,

$$\text{percent efficiency} = \frac{836}{865} \times 100\% = 96.6\%$$

EXERCISE 10

Percent Change

What number increased by:

1. 12% of itself = 560?

2. 17% of itself = 702?

3. $16\frac{2}{3}$% of itself = 700?

4. $2\frac{1}{2}$% of itself = 820?

What number decreased by:

5. 10% of itself = 90?

6. 20% of itself = 48?

7. 15% of itself = 544?

8. 90% of itself = 12.6?

Find the percent change when a quantity changes:

9. from 29.3 to 57.6

10. from 107 to 23.75

11. from 227 to 298

12. from 0.774 to 0.638

13. The temperature in a building rose from 19°C to 21°C during the day. Find the percent change in temperature.

14. A casting initially weighing 115 lb has 22% of its material machined off. What is its final weight?

15. A certain common stock rose from a value of \$35$\frac{1}{2}$ per share to \$37$\frac{5}{8}$ per share. Find the percent change in value.

16. A house that costs \$635 per year to heat has insulation installed in the attic, causing the fuel bill to drop to \$518 per year. Find the percent change in fuel cost.

17. A mechanic who received \$82 a day had his salary increased 8%. What were his daily wages then?

18. The 1979 U.S. energy consumption of 37 million barrels oil equivalent per day is expected to climb to 48 million by 1985. Find the percent increase in consumption.

19. The population of a certain town is 8118, which is 12$\frac{1}{2}$% more than it was 3 years ago. What was the population then?

20. The temperature of a room rose from 68°F to 73°F. Find the percent increase.

21. A train running at 25 mi/h increases its speed 12$\frac{1}{2}$%. How fast does it then go?

Percent Efficiency

22. A certain device consumes 18 hp and delivers 12 hp. Find its efficiency.

23. An electric motor consumes 1250 W. Find the horsepower it can deliver if it is 85% efficient. (1 hp = 746 W.)

24. A water pump requires an input of $\frac{1}{2}$ hp and delivers 10,000 lb of water per hour to a house 72 ft above the pump. Find its efficiency.

25. A certain speed reducer delivers 1.7 hp with a power input of 2.2 hp. Find the percent efficiency of the speed reducer.

26. A certain motor consumes 1100 W and has an efficiency of 85%. Find the output, in horsepower.

1-11. CALCULATING AREAS AND VOLUMES

Area of a Triangle: The *altitude* of a triangle is the perpendicular distance from a vertex to the opposite side called the *base*, or an extension of that side (Fig. 1-4).

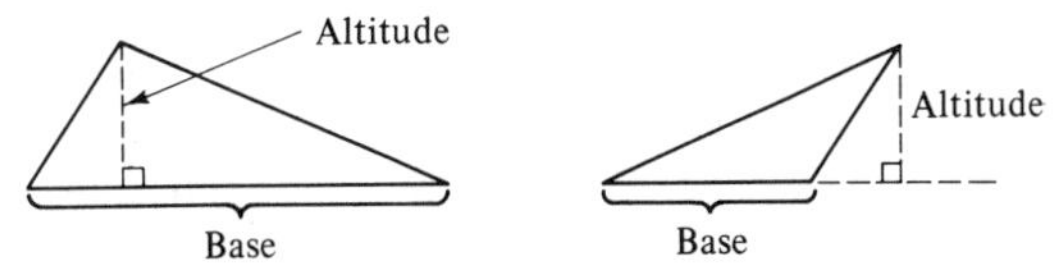

FIGURE 1-4. Altitude and base of a triangle.

For a triangle of height h and base b,

The measuring of areas, volumes, and lengths, is sometimes referred to as mensuration.

Area of a Triangle	$A = \dfrac{bh}{2}$	**137**

The area of a triangle equals one-half the product of the base and the altitude to that base.

Example: Find the area of the triangle in Fig. 1-4 if the base is 52.0 and the altitude is 48.0.

Solution: By Eq. 137,

$$\text{area} = \frac{52.0(48.0)}{2} = 1248$$

If the altitude is not known but we have instead the lengths of the three sides, we may use *Hero's formula*. If a, b, and c are the lengths of the sides,

Hero's Formula	area of triangle $= \sqrt{s(s-a)(s-b)(s-c)}$ where s is half the perimeter, $s = \dfrac{a+b+c}{2}$	**138**

Example: Find the area of a triangle having sides of lengths 3.25, 2.16, and 5.09.

Solution: From Eq. 138,

$$s = \frac{3.25 + 2.16 + 5.09}{2} = 5.25$$

$$\text{area} = \sqrt{5.25(5.25 - 3.25)(5.25 - 2.16)(5.25 - 5.09)} = 2.28$$

Sum of the Angles: An extremely useful relationship exists between the interior angles of any triangle.

Sum of the Angles	The sum of the three interior angles of any triangle is 180 degrees.	**139**

Example: Find angle A in a triangle if the other two interior angles are 38° and 121°.

Solution: By Eq. 139

$$A = 180 - 121 - 38 = 21°$$

Right Triangles: In a right triangle, the side opposite the right angle is called the *hypotenuse,* and the other two sides are called *legs.* The legs and the hypotenuse are related by the well-known *Pythagorean theorem:*

Pythagorean Theorem	The square of the hypotenuse of a right triangle is equal to the sum of the squares of the two legs: $a^2 + b^2 = c^2$	**145**

Example: A right triangle has legs of length 6 units and 11 units. Find the length of the hypotenuse.

Solution: By Eq. 145, letting c = the length of the hypotenuse,

$$c^2 = 6^2 + 11^2 = 36 + 121 = 157$$

$$c = \sqrt{157} = 12.5 \quad \text{(rounded)}$$

Some Special Triangles: In a *30–60–90* right triangle (Fig. 1-5a), the side opposite the 30° angle is half the length of the hypotenuse.

A *45° right triangle* (Fig. 1-5b) is also *isosceles,* and the hypotenuse is $\sqrt{2}$ times the length of either side.

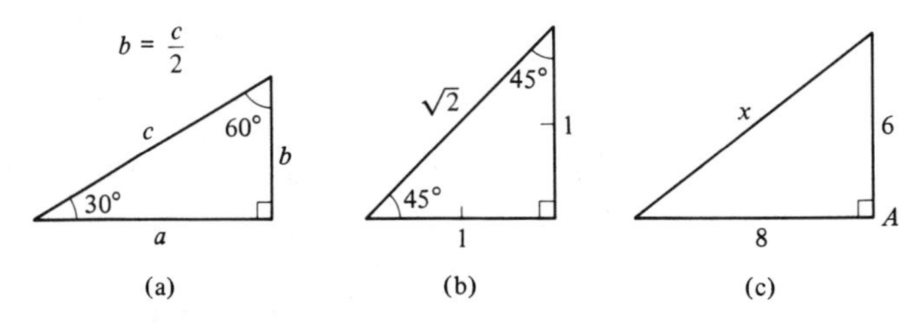

FIGURE 1-5. Some special right triangles.

A *3–4–5 triangle* is a right triangle in which the sides are in the ratio of 3 to 4 to 5.

Example: In Fig. 1-5c, side x is 10 cm.

Quadrilaterals: A *quadrilateral* is a polygon having four sides. They are the familiar figures shown in Fig. 1-6. The formula for the area of each region is given right on the figure.

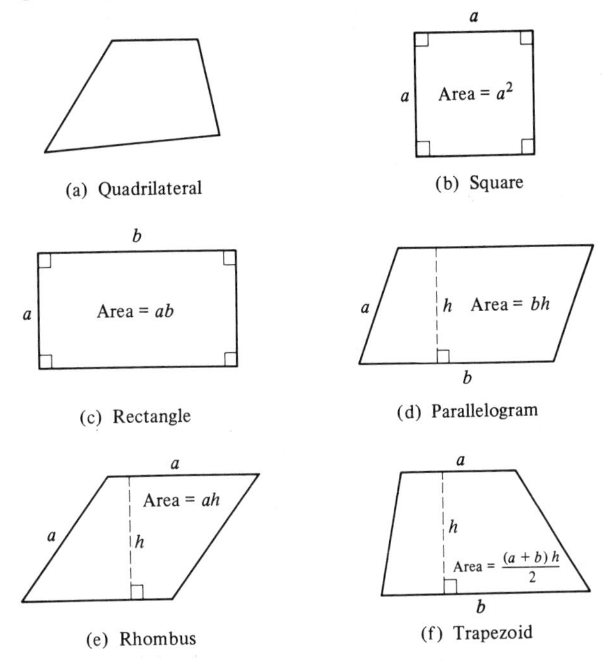

FIGURE 1-6. Quadrilaterals.

Example: Find the area of a trapezoidal parcel of land whose two parallel sides are 162 m apart and have lengths of 225 m and 312 m.

Solution:

$$\text{area} = \frac{(a + b)h}{2} = \frac{(225 + 312)(162)}{2} = 43{,}497 \text{ m}^2$$

or 43,500 m^2, rounded.

The Circle: For a circle of diameter d and radius r:

Circle of radius r and diameter d $\pi \cong 3.1416$	circumference $= 2\pi r = \pi d$	**113**
	area $= \pi r^2 = \dfrac{\pi d^2}{4}$	**114**

Example: Find the circumference and area of a 25.7-cm-radius circle.

Solution: By Eq. 113,

$$C = 2\pi(25.7) = 161 \text{ cm}$$

and by Eq. 114,

$$A = \pi(25.7)^2 = 2075 \text{ cm}^2$$

Volumes and Areas of Solids: The volume of a solid is a measure of the space it occupies or encloses. It is measured in cubic units (cm^3, ft^3, m^3, etc.) or, usually for liquids, in liters or gallons. One of our main tasks in this chapter is to compute the volumes of various solids.

We speak about three different kinds of areas. *Surface area* will be the *total* area of the surface of a solid, including the ends, or bases. The *lateral area,* which will be defined later for each solid, does *not* include the areas of the bases. The *cross-sectional area* is the area of the plane figure obtained when we slice the solid in a specified way.

The formulas for the areas and volumes of some common solids are given in Fig. 1-7.

Example: Find the volume of a cone having a base area of 125 cm^2 and an altitude of 11.2 cm.

122	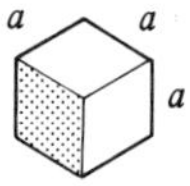	Cube	Volume = a^3
123			Surface area = $6a^2$
124	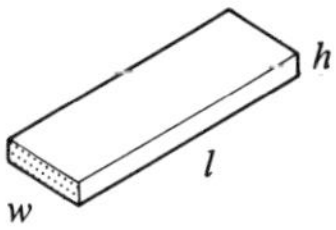	Rectangular parallel-epiped	Volume = lwh
125			Surface area = $2(lw + hw + lh)$
126	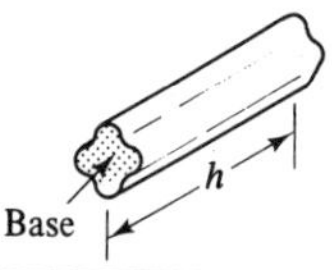	Any cylinder or prism	Volume = (area of base)(altitude)
127		Right cylinder or prism	Lateral area = (perimeter of base)(altitude) (not incl. bases)
128	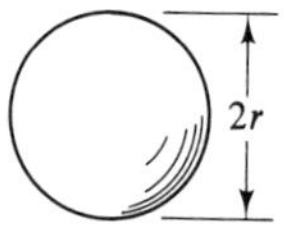	Sphere	Volume = $\frac{4}{3}\pi r^3$
129			Surface area = $4\pi r^2$
130	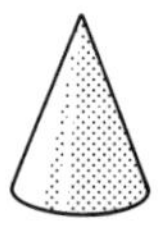	Any cone or pyramid	Volume = $\frac{1}{3}$(area of base)(altitude)
131		Right circular cone or regular pyramid	Lateral area = $\frac{1}{2}$(perimeter of base) × (slant height)
132	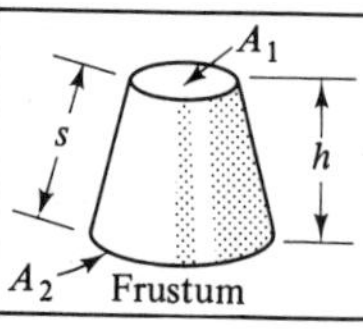	Any cone or pyramid	Volume = $\frac{h}{3}(A_1 + A_2 + \sqrt{A_1 A_2})$
133		Right circular cone or regular pyramid	Lateral area = $\frac{s}{2}$(sum of base perimeters) = $\frac{s}{2}(P_1 + P_2)$

FIGURE 1-7. Some solids.

Solution: By Eq. 130,

$$\text{volume} = \frac{125(11.2)}{3} = 467 \text{ cm}^3$$

EXERCISE 11

Triangles

For practice, do these problems using only geometry, even if you know some trigonometry.

1. What is the cost of a triangular piece of land whose base is 828 ft and altitude 412 ft, at $1125 an acre?
2. At $3.50 a square yard, find the cost of paving a triangular court, its base being 105 ft and its altitude 21 yd.
3. A new highway divides a parcel of land into two pieces (Fig. 1-8). Find the number of acres in the triangular lot ABC. (1 acre = 43,560 ft^2.)

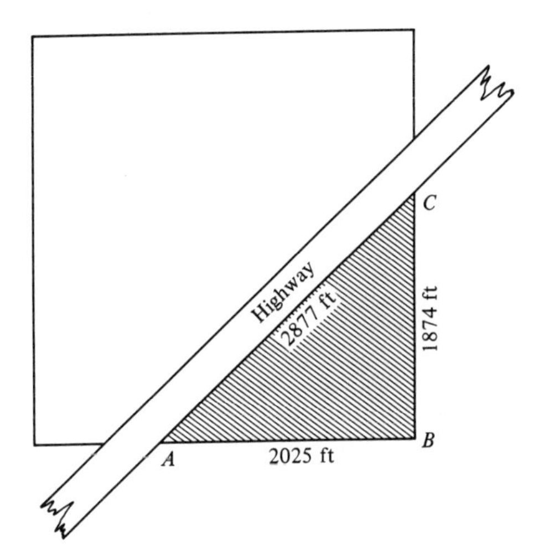

FIGURE 1-8

Quadrilaterals

4. Find the area of the parallelogram in Fig. 1-9.

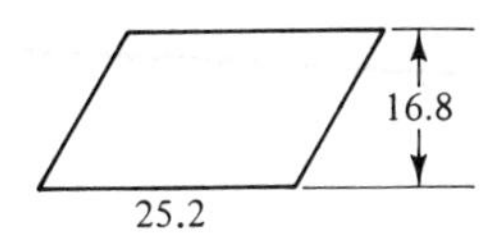

FIGURE 1-9

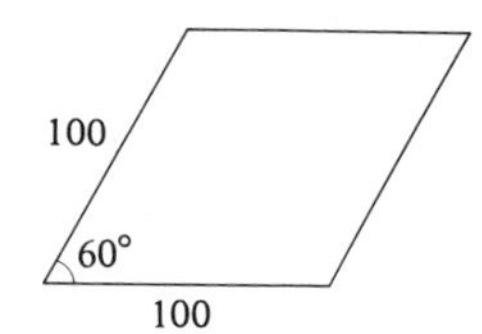

FIGURE 1-10

5. Find the area of the rhombus in Fig. 1-10.

6. Find the area of the trapezoid in Fig. 1-11.

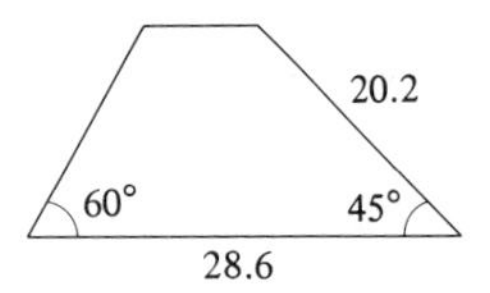

FIGURE 1-11

7. Find the cost of shingling a roof, 64 ft 9 in. long and 45 ft wide, at 95 cents/ft^2.

8. What will be the cost of flagging a sidewalk 312 ft long and 6.5 ft wide, at \$13.50/yd^2?

9. How many tiles 9 in. square will cover a floor 48 ft by 12 ft?

10. What will it cost to carpet a floor, 6.25 m by 7.18 m, at \$7.75/m^2?

11. How many rolls of paper, each 8 yd long and 18 in. wide, will paper the sides of a room 16 ft by 14 ft and 10 ft high, deducting 124 ft^2 for doors and windows?

12. What is the cost of plastering the walls and ceiling of a room 40 ft long, 36 ft wide, and 22 ft high, at \$8.50/yd^2, allowing 1375 ft^2 for doors, windows, and baseboard?

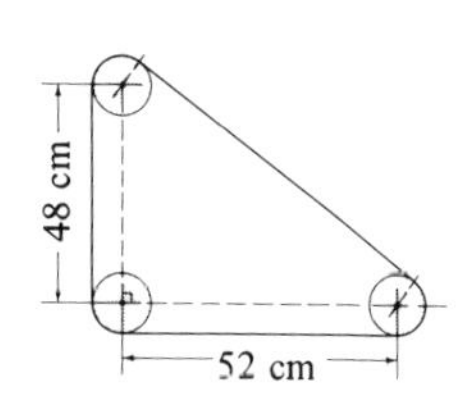

FIGURE 1-12

13. What will it cost to cement a cellar 25.3 ft long and 18.4 ft wide, at \$0.35/ft^2?

14. Find the cost of lining a topless rectangular tank 68 in. long, 54 in. wide, and 48 in. deep, with zinc, weighing 5.2 lb per square foot, at \$1.55/lb installed.

Circle

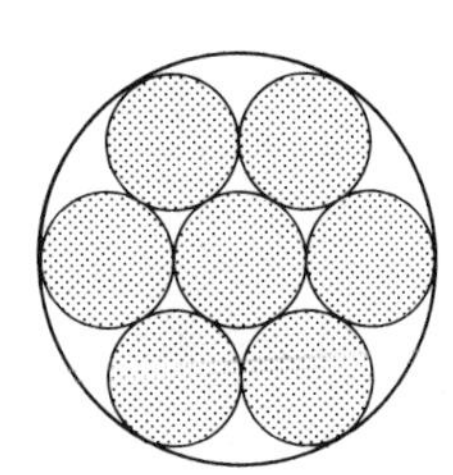

FIGURE 1-13

15. The area of the bottom of a circular pan is 196 in.^{2}. What is its diameter?

16. Find the diameter of a circular solar pond containing 125 m^2.

17. What is the circumference of a circular lake 33 m in diameter?

18. The radius of a circle is 5 m. Find the diameter of another circle containing four times the area of the first.

19. The distance around a circular park is 2.5 mi. How many acres does it contain?

By cutting from both sides, it is possible to fell a tree whose diameter is twice the length of the chain-saw bar.

20. A woodcutter uses a tape measure and finds the circumference of a tree to be 95 in. Assuming the tree to be circular, what length of chain-saw bar is needed to fell the tree?

21. What must be the diameter d of a cylindrical piston so that a pressure of 125 lb/in.^{2} on its circular end will result in a total force of 3600 lb?

Hint: *The total* curved *portion of the belt is equal to the circumference of one pulley.*

22. Find the length of belt needed to connect the three 15-cm-diameter pulleys in Fig. 1-12.

23. Seven cables of equal diameter are contained within a circular conduit, as in Fig. 1-13. If the inside diameter of the conduit is 25.6 cm, find the cross-sectional area *not* occupied by the cables.

24. A circular tile walk is 1.0 m wide and surrounds a circular garden that is 21.0 m in diameter. Find its area.

25. A certain car tire is 78.5 cm in diameter. How far will the car move forward with one revolution of the wheel?

Prism and Rectangular Parallelepiped

26. Find the volume of the triangular prism in Fig. 1-14.

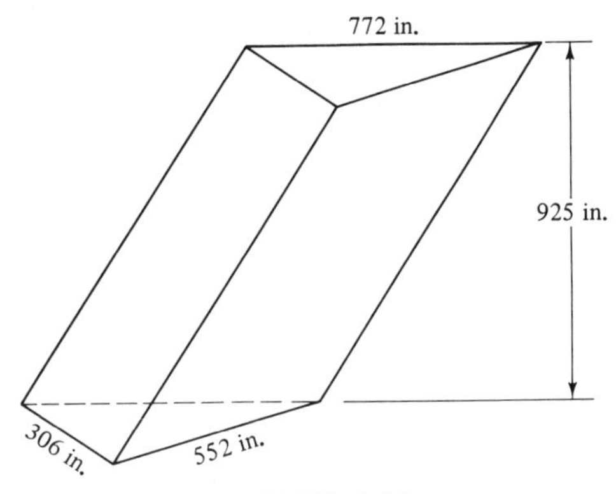

FIGURE 1-14

Take the density of steel as 450 lb/ft³.

27. A 5.00-in. cube of steel is placed in a surface grinding machine, and the vertical feed set so that 0.0050 in. of metal is removed at each cut. How many cuts will be needed to reduce the weight of the cube by 1.53 lb?

28. A rectangular tank is being filled with liquid, with each cubic meter of liquid increasing the depth by 2 cm. The length of the tank is 12 m.
    (a) What is the width of the tank?
    (b) How many cubic meters will be required to fill the tank to a depth of 3 m?

29. How many loads of gravel will be needed to cover 2.0 mi of roadbed, 35 ft wide, to a depth of 3.0 in. if one truckload contains 8 yd^3 of gravel?

Cylinder

30. Find the volume and lateral area of a right circular cylinder having a base radius of 128 and a height of 285.

31. A 7.0-ft-long piece of iron pipe has an outside diameter of 3.5 in. and weighs 126 lb. Find the wall thickness.

32. A certain bushing is in the shape of a hollow cylinder 18 mm in diameter and 25 mm long, with an axial hole 12 mm in diameter. If the density of the material from which they are made is 2.7 g/cm^3, find the mass of 1000 bushings.

33. A steel gear is to be lightened by drilling holes through the gear. The gear is 3.5 in. thick. Find the diameter d of the holes if each is to remove 12 oz.

The engine displacement *is the total volume swept out by all the pistons.*

34. A certain gasoline engine has four cylinders, each with a bore of 82 mm and a piston stroke of 95 mm. Find the engine displacement in liters.

Cone and Pyramid

35. The circumference of the base of a right circular cone is 40 in. and the slant height is 38 in. What is the area of the lateral surface?

36. Find the volume of a circular cone whose altitude is 24 cm and whose base diameter is 30 cm.

37. Find the weight of the mast of a ship, its height being 50 ft, the circumference at one end 5 ft and at the other 3 ft, if the density of the wood is 58.5 lb/ft^3.

38. The slant height of a right pyramid is 11 in., and the base is 4 in. square. Find the area of the entire surface.

39. How many cubic feet are in a piece of timber 30 ft long, one end being a 15 by 15 in. square, and the other a 12×12 in. square?

40. Find the total surface area and volume of a tapered steel roller 12 ft long and having end diameters of 12 in. and 15 in.

Sphere

41. Find the volume and surface area of a sphere having a radius of 744.

42. Find the volume and radius of a sphere having a surface area of 462.

A great circle *of a sphere is one that has the same diameter as the sphere.*

43. Find the surface area and radius of a sphere that has a volume of 5.88.

44. How many great circles of a sphere would have the same area as that of the surface of the sphere?

45. Find the weight in pounds of 100 steel balls (density $= 450$ lb/ft^3) each 2.5 in. in diameter.

46. A spherical radome encloses a volume of 9000 m^3.

(a) Find the radome radius, r.

(b) If constructed of a material weighing 2.0 kg/m^2, find the weight of the radome.

Computer

47. Write a program that will accept as input the three sides of any triangle and compute and display the area by Hero's formula.

48. A certain building supplies store has stocks of window glass having widths of 10 in., 12 in., 14 in., . . . , 28 in. Each of these widths is available in lengths of 20 in., 24 in., 28 in., . . . , 40 in. The price of glass is $\$2.18/ft^2$. Write a program to compute and display the area in square feet and the price of each size of glass available.

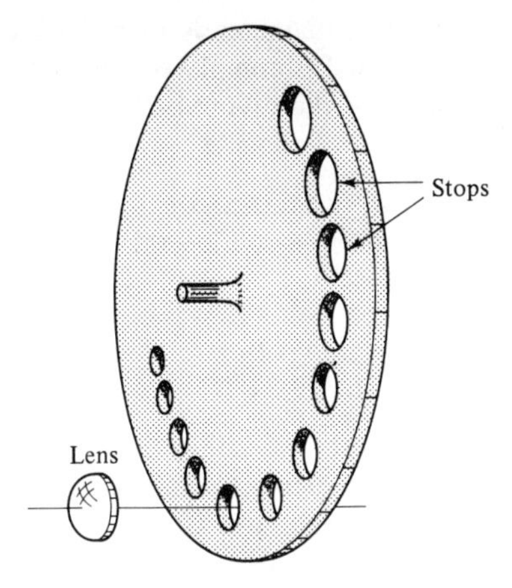

FIGURE 1-15

49. A lens in a certain optical instrument has a diameter of 75 mm. This diameter is reduced by a series of circular "stops" (Fig. 1-15), with each stop reducing the area of the opening by 15% of its previous area. Write a program that will compute the required diameter of each stop, and will display both the stop diameter and the area in mm^2. Continue until the stop diameter falls below 20 mm.

CHAPTER TEST

1. Combine: $1.435 - 7.21 + 93.24 - 4.1116$

2. Give the number of significant digits in:
(a) 9.886 **(b)** 1.002 **(c)** 0.3500 **(d)** $15,\bar{0}00$

3. Multiply: $21.8(3.775 \times 1.07)$

4. A regular triangular pyramid has an altitude of 12 m, and the base is 4 m on a side. Find the area of a section made by a plane parallel to the base and 4 m from the vertex.

5. A fence parallel to one side of a triangular field cuts a second side into segments of 15 m and 21 m long. The length of the third side is 42 m. Find the length of the shorter segment of the third side.

6. The average solar radiation in the continental United States is about 7.4×10^5 J/m^2·h. How many kilowatts would be collected by 15 acres of solar panels?

7. An item rose in price from \$29.35 to \$31.59. Find the percent increase.

8. Find the percent concentration of alcohol if 2 liters of alcohol is added to 15 gallons of gasoline.

9. A rocket ascends at an angle of 60° with the horizontal. After 1 min it is directly over a point that is a horizontal distance of 12 mi from the launch point. Find the speed of the rocket.

10. A rectangular beam 16 in. thick is cut from a log. Find the greatest depth that can be obtained from a log 20 in. in diameter.

11. Divide: $88.25 \div 9.15$

12. Find the reciprocal of 2.89.

13. Evaluate: $-|-4 + 2| - |-9 - 7| + 5$

14. Evaluate: $(9.73)^2$

15. A cylindrical tank 4.0 m in diameter is placed with its axis vertical and is partially filled with water. A spherical diving bell is completely immersed in the tank, causing the water level to rise 1.0 m. Find the diameter of the diving bell.

16. Evaluate: $\sqrt[3]{842}$

17. Two vertical piers are 240 ft apart and support a circular bridge arch. The highest point of the arch is 30 ft higher than the piers. Find the radius of the arch.

18. Evaluate: $(7.75)^{-2}$

19. Evaluate: $\sqrt{29.8}$

20. Evaluate: $(123)(2.75) - (81.2)(3.24)$

21. Evaluate: $(91.2 - 88.6)^2$

22. The Department of Energy estimates that there are 700 billion barrels of oil in the oil shale deposits of Colorado, Wyoming, and Utah.
   (a) Express this amount in scientific notation.
   (b) Express this amount in gallons (1 barrel = 42 gal.)
   (c) Express this amount in liters.
   (d) The Department of Energy estimates that each ton of shale will yield 25 gal of oil. How many tons of oil shale are there in these deposits?
   (e) If the density of oil is 57.2 lb/ft^3, compute the percent concentration (by weight) of the oil in the shale.
   (f) A pilot plant is to be built that will process 10^4 barrels of oil per day. How many tons of shale must be processed each week?
   (g) How many tons of oil shale are needed to yield a barrel of oil?

23. Two antenna masts arc 10 m and 15 m high and are 12 m apart. How long a wire is needed to connect the tops of the two masts?

24. Find the area and side of a rhombus whose diagonals are 100 and 140.

25. Evaluate: $\left(\dfrac{77.2 - 51.4}{21.6 - 11.3}\right)^2$

26. Evaluate: $y = 3x^2 - 2x$ when $x = -2.88$

27. Evaluate: $y = 2ab - 3bc + 4ac$ when $a = 5$, $b = 2$, and $c = -6$

28. Evaluate: $y = 2x - 3w + 5z$ when $x - 7.72$, $w = 3.14$, and $z = 2.27$

29. Round to two decimal places.
   (a) 7.977 (b) 4.655 (c) 11.845 (d) 1.004

30. Two concentric circles have radii of 5 and 12. Find the length of a chord of the larger circle, which is tangent to the smaller circle.

31. A belt that does not cross goes around two pulleys, each with a radius of 4.0 in. and whose centers are 9.0 in. apart. Find the length of the belt.

32. Round to three significant digits.
   (a) 179.2 (b) 1.076 (c) 4.8550 (d) 45,725

33. A news report states that a new hydroelectric generating station in Holyoke, Massachusetts, will produce 47 million kWh/yr, and that this power, for 20 years of operation, is equivalent to 2 million barrels of oil. Using these figures, how many kilowatthours is each barrel of oil equivalent to?

34. A certain generator has a power input of 2.5 hp and delivers 1310 W. Find its percent efficiency.

35. Using Eq. A30, find the stress in pounds per square inch for a force of 1.17×10^3 N distributed over an area of $3.14 \times 10^3\ \text{mm}^2$.

36. Combine: $(8.34 \times 10^5) + (2.85 \times 10^6) - (5.29 \times 10^4)$
37. Divide: -39.2 by -0.003826
38. Convert 6930 Btu/h to foot-pounds per minute.
39. Divide: 8.24×10^{-3} by 1.98×10^7
40. What percent of 40.8 is 11.3?
41. Evaluate: $\sqrt[5]{82.8}$
42. Multiply: $(4.92 \times 10^6) \times (9.13 \times 10^{-3})$
43. Insert the proper sign of equality or inequality between $-\frac{2}{3}$ and -0.660.
44. Convert 0.000426 milliampere to microamperes.
45. Find 49.2% of 4827.
46. Combine: $-385 - (227 - 499) - (-102) + (-284)$
47. Find the reciprocal of -0.582.
48. Find the percent change in a voltage that increased from 110 V to 118 V.
49. A right triangle has legs of length 28.4 cm and 37.6 cm. Find the length of the hypotenuse, and the area.
50. Write in decimal notation: 5.28×10^4
51. Convert 49.3 pounds to newtons.
52. Evaluate: $(45.2)^{-0.45}$
53. Using Eq. A17, find the distance in feet traveled by a falling object in 5.25 s, thrown downward with an initial velocity of 284 m/min.
54. Write in scientific notation: 0.000374
55. 8463 is what percent of 38,473?

Binary Numbers

The decimal number system that we reviewed in the preceding chapter is fine for calculations done by human beings, but it is not the easiest system for a computer to use. A *digital* computer contains elements that can be in either of two states: on or off, magnetized or not magnetized, and so on, and for such devices we introduce *binary numbers*.

In this chapter we learn how to convert between binary and decimal, and how to do simple arithmetic with binary numbers. In Chapter 18 we will see how this is actually accomplished using logic gates.

We then examine the way in which binary numbers are stored in a computer so that we can later understand how this can affect the accuracy of a computation.

2-1. THE BINARY NUMBER SYSTEM

Binary Numbers: A *binary number* is a sequence of the digits 0 and 1, such as

$$1101001$$

The number shown has no fractional part, and so is called a *binary integer*. A binary number having a fractional part contains a *binary point* (also called a *radix point*), as in the number

1001.01

Base or Radix: The *base* of a number system (also called the *radix*) is equal to the number of digits used in the system.

Examples:

(a) The decimal system uses the ten digits

0 1 2 3 4 5 6 7 8 9

and has a base of 10.

(b) The binary system uses two digits

0 1

and has a base of 2.

Bits, Bytes, and Words: Each of the digits is called a *bit,* from

Binary digIT

A *byte* is a group of 8 bits, and a *word* is the largest string of bits that a computer can handle in one operation. The number of bits in a word is called the *word length.* Different computers have different word lengths, with 8 or 16 bits being common for desktop or personal computers. The longer words are often broken down into bytes for easier handling. Half a byte (4 bits) is called a *nibble.*

Note that this is different than the usual meaning of these prefixes, where kilo means 1000 and mega means 1,000,000.

A *kilobyte* (Kbyte or KB) is 1024 (2^{10}) bytes, and a *megabyte* (Mbyte or MB) is 1,048,575 (2^{20}) bytes.

Writing Binary Numbers: A binary number is sometimes written with a subscript 2 when there is a chance that the binary number would otherwise be mistaken for a decimal number.

Example: The binary number 110 could easily be mistaken for the decimal number 110, unless we write it

110_2

Similarly, a decimal number that may be mistaken for binary is often written with a subscript 10, as in

101_{10}

Long binary numbers are sometimes written with its bits in groups of four for easier reading.

Example: The number 100100010100.001001 is easier to read when written

1001 1110 1011.0010 01

The leftmost bit in a binary number is called the *high-order* or *most significant bit (MSB)*. The bit at the extreme right of the number is the *low-order* or *least significant bit (LSB)*.

Place Value: We saw in Chapter 1 that a positional number system is one in which the position of a digit determines its value, and that each position in a number has a *place value* equal to the base of the number system raised to the position number.

The place values in the binary number system are

Compare these place values to those in the decimal system shown in Fig. 1-1.

$$\ldots\ 2^4 \quad 2^3 \quad 2^2 \quad 2^1 \quad 2^0 \ . \ 2^{-1} \quad 2^{-2} \ \ldots$$

or

$$\ldots\ 16 \quad 8 \quad 4 \quad 2 \quad 1 \ . \ \tfrac{1}{2} \quad \tfrac{1}{4} \ \ldots$$

binary point ↑

A more complete list of place values for a binary number is given in Table 2-1.

TABLE 2-1

Position	*Place value*
10	$2^{10} = 1024$
9	$2^9 = 512$
8	$2^8 = 256$
7	$2^7 = 128$
6	$2^6 = 64$
5	$2^5 = 32$
4	$2^4 = 16$
3	$2^3 = 8$
2	$2^2 = 4$
1	$2^1 = 2$
0	$2^0 = 1$
−1	$2^{-1} = 0.5$
−2	$2^{-2} = 0.25$
−3	$2^{-3} = 0.125$
−4	$2^{-4} = 0.0625$
−5	$2^{-5} = 0.03125$
−6	$2^{-6} = 0.015625$

Expanded Notation: Thus the value of any digit in a number is the product of that digit and the place value. The value of the entire number is then the sum of these products.

Examples:

(a) The decimal number 526 can be expressed as

$$5 \times 10^2 + 2 \times 10^1 + 6 \times 10^0$$

or

$$5 \times 100 + 2 \times 10 + 6 \times 1$$

or

$$500 \quad + \quad 20 \quad + \quad 6$$

(b) The binary number 1 0 1 1 can be expressed as

$$1 \times 2^3 + 0 \times 2^2 + 1 \times 2^1 + 1 \times 2^0$$

or

$$1 \times 8 + 0 \times 4 + 1 \times 2 + 1 \times 1$$

or

$$8 \quad + \quad 0 \quad + \quad 2 \quad + \quad 1$$

Numbers written in this way are said to be in *expanded notation.*

Converting Binary Numbers to Decimal: Simply write the binary number in expanded notation (omitting those where the bit is 0), and add the resulting values.

Example: Convert the binary number

1001.011

to decimal.

Solution: In expanded notation,

$$\begin{aligned} 1001.011 &= 1 \times 8 + 1 \times 1 + 1 \times \frac{1}{4} + 1 \times \frac{1}{8} \\ &= 8 + 1 + \frac{1}{4} + \frac{1}{8} \\ &= 9\frac{3}{8} \\ &= 9.375 \end{aligned}$$

Largest Decimal Number Obtainable with *n* Bits: The larget possible 3-bit binary number is

$$1\ 1\ 1 = 7$$

or

$$2^3 - 1$$

The largest possible 4-bit number is

$$1\ 1\ 1\ 1 = 15$$

or

$$2^4 - 1$$

Similarly, the largest n-bit binary number is

$2^n - 1$	**17**

Example: If a computer stores numbers with 15 bits, what is the largest decimal number that can be represented?

Solution: The largest decimal number is

$$2^{15} - 1 = 32{,}767$$

Significant Digits: In the preceding example, we saw that a 15-bit binary number could represent a decimal number no greater than 32,767. Thus, it took 15 binary digits to represent five decimal digits. As a rule of thumb, *it takes about 3 bits for each decimal digit.*

Stated another way, if we have a computer that stores numbers with 15 bits, we should assume that the numbers it prints do not contain more than five significant digits.

We will see in Sec. 2-4 that many computers use 15 bits to store an integer, and 23 bits to store a number that has a binary point (plus 9 more bits to give the sign and the position of the binary point).

Example: If we want a computer to print numbers containing seven significant digits, how many bits must it use to store those numbers?

Solution: Using our rule of thumb, we need

$$3 \times 7 = 21 \text{ bits}$$

Converting Decimal Numbers to Binary: To convert a decimal *integer* to binary, we first divide it by 2, obtaining a quotient and a remainder. We write down the remainder, and divide the quotient by 2, getting a new quotient and remainder. We then repeat this process until the quotient is zero.

This is sometimes called the ***dibble-dobble*** *method.*

Example: Convert the decimal integer 59 to binary.

Solution: We divide 59 by 2, getting a quotient of 29 and a remainder of 1. Then dividing 29 by 2 gives a quotient of 14 and a remainder of 1. These calculations, and those that follow can be arranged in a table, as follows.

	Remainder		
2 ⟌59			↑
2 ⟌29	1	LSB	
2 ⟌14	1		
2 ⟌7	0		read up
2 ⟌3	1		
2 ⟌1	1	MSB	
2 ⟌0	1		

The conversion can, of course, be checked by converting back to decimal.

Our binary number then consists of the digits in the remainders, with those at the *top* of the column appearing to the *right* in the binary number. Thus,

$$59_{10} = 111011_2$$

To convert a decimal *fraction* to binary, we first multiply it by 2, remove the integral part of the product, and multiply by 2 again. The procedure is then repeated.

Example: Convert the decimal fraction 0.546875 to binary.

Solution: We multiply the given number by 2, getting 1.09375. We remove the integral part, 1, leaving 0.09375, which we again multiply by 2, getting 0.1875. We repeat the computation until we get a product that has a fractional part of zero, as in the following table.

Notice that we remove the integral part, shown in boldface, before multiplying by 2.

		Integral Part	
	0.546875		
×	2		
	1.09375	1	MSB ↓ read down
×	2		
	0.1875	0	
×	2		
	0.375	0	
×	2		
	0.75	0	
×	2		
	1.50	1	
×	2		
	1.00	1	LSB ↓

Note that this is the reverse of what we did when converting a decimal integer to binary.

We stop now that the fractional part of the product (1.00) is zero. The column containing the integral parts is now our binary number, with the digits at the top appearing at the *left* of the binary number. So

$$0.546875_{10} = .100011_2$$

In the preceding example we were able to find an exact binary equivalent of a decimal fraction. This is not always possible, as shown in the following example.

Example: Convert the decimal fraction 0.743 to binary.

Solution: We follow the same procedure as before and get the following values.

		Integral Part	
	0.743		
×	2		
	1.486	1	MSB
×	2		
	0.972	0	
×	2		
	1.944	1	
×	2		
	1.888	1	
×	2		
	1.776	1	
×	2		
	1.552	1	
×	2		
	1.104	1	
×	2		
	0.208	0	LSB

It is becoming clear that this computation can continue indefinitely. This shows that *not all decimal fractions can be exactly converted to binary*. This inability to make an exact conversion is an unavoidable source of inaccuracy in some computations.

The result of our conversion is, then,

$$0.743_{10} \cong .10111110_2$$

To convert a decimal number having both an integer part and a fractional part to binary, convert each part separately as shown above, and combine.

Example: Convert the number 59.546875 to binary.

Solution: From the preceding examples,

$$59 = 111011$$

and

$$0.546875 = .100011$$

So

$$59.546875 = 111011.100011$$

Converting Binary Fractions to Decimal: We use Table 2-1 to find the decimal equivalent of each binary bit located to the right of the binary point, and add.

Example: Convert the binary fraction 0.101 to decimal.

Solution: From Table 2-1,

$$\begin{aligned} 0.1 &= 2^{-1} = 0.5 \\ 0.001 &= 2^{-3} = \underline{0.125} \end{aligned}$$

Adding, $0.101 = 0.625$

The procedure is no different when converting a binary number that has both a whole and a fractional part.

Example: Convert the binary number 10.01 to decimal.

Solution: From Table 2-1,

$$\begin{aligned} 10 &= 2 \\ \underline{0.01} &= \underline{0.25} \end{aligned}$$

Adding, $10.01 = 2.25$

EXERCISE 1

Conversion between Binary and Decimal

Convert each binary number to decimal.

1. 10 **2.** 01 **3.** 11

4. 1001 **5.** 0110 **6.** 1111

7. 1101 **8.** 0111 **9.** 1100

10. 1011 **11.** 0101 **12.** 11010010

13. 01100111 **14.** 10010011 **15.** 01110111

16. 1001 1010 1000 0010

Convert each decimal number to binary.

17. 5 **18.** 9 **19.** 2

20. 7 **21.** 72 **22.** 28

23. 93 **24.** 17 **25.** 274

26. 937 **27.** 118 **28.** 267

29. 8375 **30.** 2885 **31.** 82740

32. 72649

Convert each decimal fraction to binary. Retain 8 bits to the right of the binary point.

33. 0.5 **34.** 0.25 **35.** 0.75

36. 0.375 **37.** 0.3 **38.** 0.8

39. 0.55 **40.** 0.35 **41.** 0.875

42. 0.4375 **43.** 0.3872 **44.** 0.8462

Convert each binary fraction to decimal.

45. 0.1 **46.** 0.11 **47.** 0.01

48. 0.011 **49.** 0.1001 **50.** 0.0111

Convert each decimal number to binary. Keep 8 bits to the right of the binary point.

51. 5.5 **52.** 2.75 **53.** 4.375

54. 29.381 **55.** 948.472 **56.** 2847.22853

Convert each binary number to decimal.

57. 1.1 **58.** 10.11 **59.** 10.01

60. 101.011 **61.** 11001.01101 **62.** 101001.1001101

2-2. ARITHMETIC OPERATIONS WITH BINARY NUMBERS

Algorithms: Before a computer can do a computation, it must be given a precise step-by-step set of instructions telling it what to do. Such a set of instructions is called a *program* and is written in a language that the particular computer can understand.

A similar set of instructions written in ordinary English is called an *algorithm*. In the following section we give an algorithm for the familiar operation of adding two decimal numbers, and then use the algorithm to add binary numbers.

Algorithm for Addition: We will illustrate the algorithm by adding 128 to 785.

Step 1: Add the digits in the right-hand columns of each number.

```
1 2 8
7 8 5
-----
   13
```

Step 2: Leave the right digit of this sum in place, and *carry* the left digit (if greater than zero) to the next column.

```
  1
1 2 8
7 8 5
-----
    3
```

Step 3: If there is another column, add it. Otherwise, stop.

```
  1
1 2 8
7 8 5
-----
 11 3
```

Step 4: Go to step 2.

$$\begin{array}{ccc} \overset{1}{1} & \overset{1}{2} & 8 \\ 7 & 8 & 5 \\ \hline & 1 & 3 \end{array}$$

Step 3 (repeated):

$$\begin{array}{ccc} \overset{1}{1} & \overset{1}{2} & 8 \\ 7 & 8 & 5 \\ \hline 9 & 1 & 3 \end{array}$$

Step 4 (repeated): We go to step 2, but there is nothing to carry.

Step 3 (repeated): There are no more columns, so we stop.

Adding Binary Numbers: We now use the addition algorithm for binary numbers. But what is the sum of two binary digits, say 1 and 0? The sum of any two binary digits is given in the following table.

Table of Sums		18
	$0 + 0 = 0$ $0 + 1 = 1$ $1 + 0 = 1$ $1 + 1 = 0$ with a carry to the next column	

Example: Add

$$\begin{array}{rcccc} & 1 & 1 & 0 & 1 \\ + & & & 1 & 1 \\ \hline \end{array}$$

Solution:

Step 1: We add 1 and 1 in the right column and get 0 with a carry of 1.

$$\begin{array}{rcccc} & 1 & 1 & \overset{1}{0} & 1 \\ + & & & 1 & 1 \\ \hline & & & & 0 \end{array}$$

Step 2: We add the next column, getting 0 with a carry of 1.

$$\begin{array}{rcccc} & 1 & \overset{1}{1} & \overset{1}{0} & 1 \\ + & & & 1 & 1 \\ \hline & & & 0 & 0 \end{array}$$

We repeat the procedure. As with decimals, we treat any missing digits to the left of our number as zeros.

$$\begin{array}{rcccc} & \overset{1}{1} & \overset{1}{1} & \overset{1}{0} & 1 \\ + & & & 1 & 1 \\ \hline & & 0 & 0 & 0 \end{array}$$

and again,

$$\begin{array}{rrrrrr} & {\scriptstyle 1} & {\scriptstyle 1} & {\scriptstyle 1} & {\scriptstyle 1} & \\ & & 1 & 1 & 0 & 1 \\ + & & & & 1 & 1 \\ \hline & & 0 & 0 & 0 & 0 \end{array}$$

and finally,

$$\begin{array}{rrrrrr} & {\scriptstyle 1} & {\scriptstyle 1} & {\scriptstyle 1} & {\scriptstyle 1} & \\ & & 1 & 1 & 0 & 1 \\ + & & & & 1 & 1 \\ \hline & 1 & 0 & 0 & 0 & 0 \end{array}$$

To add more than two binary numbers, add the first two and then the others one at a time.

Example: Calculate 110 + 100 + 101 + 1000.

Solution: Adding the first two numbers,

$$\begin{array}{r} 110 \\ +\ \underline{100} \\ 1010 \end{array}$$

and the third,

$$\begin{array}{r} +\ \underline{101} \\ 1111 \end{array}$$

and the fourth,

$$\begin{array}{r} +\ \underline{1000} \end{array}$$

giving

$$10111$$

as the total sum.

When the numbers contain fractional parts, align the binary points vertically and add as shown above.

Example: Evaluate 101.01 + 100.1

Solution: We get

$$\begin{array}{r} 101.01 \\ +\ \underline{100.1} \\ 1001.11 \end{array}$$

The commutative, associative, and distributive laws, Eqs. 1 to 5, hold for binary numbers as well as for decimals.

Multiplying Binary Numbers: The product of any two binary digits is given in the following table.

As with decimal numbers, the product of 0 and any other number is 0.

Table of Products		21
	$0 \times 0 = 0$	
	$0 \times 1 = 0$	
	$1 \times 0 = 0$	
	$1 \times 1 = 1$	

Binary multiplication is very similar to decimal multiplication. We show the method with an example.

Example: Multiply the binary numbers 101 and 110.

Solution: We write one above the other.

```
    101
  × 110
```

Then we multiply 101 by 0.

```
    101
  × 110
    ---
    000
```

Then by 1,

```
    101
  × 110
    ---
    000
   101
```

and by 1 again,

```
    101
  × 110
    ---
    000
   101
  101
  -----
```

and then add,

```
    101
  × 110
    ---
    000
   101
  101
  -----
  11110
```

which gives the final product.

Common Error	You may choose to omit multiplication by zero, since this gives zero anyway. If so, *be sure to line up the digits in the correct columns.*

Numbers containing a binary point are multiplied in the same way. As in decimal multiplication, *the number of places to the right of the binary point in the product will be the sum of the numbers of binary points in the factors.*

Example: Multiply the binary numbers 10.1 and 1.01.

Solution: Multiplying,

```
      1 0.1
      1.0 1
      -----
      1 0 1
      0 0 0
  1 0 1
  ---------
  1 1 0 0 1
```

and adding, 1 1 0 0 1

Since one factor has one binary place and the other has two, we place the binary point in the product to the left of the third digit, getting 11.001.

Subtracting Binary Numbers: Let us first review subtraction of decimal numbers. To do the subtraction

```
   7 0 3 9
 - 2 8 6 4
 ---------
```

we first subtract the 4 from the 9,

```
   7 0 3 9
 - 2 8 6 4
 ---------
         5
```

Binary subtraction is more often done using two's complements, *which is described in Sec. 2-3.*

Then, to subtract the 6 from 3, we try to *borrow* from the 0. Since we cannot, we borrow from the 7. The 3 then becomes 13, and the 7 is reduced to 6, and the 0 becomes 9.

```
   6 9
   7̸ 0̸¹ 3 9
 - 2 8  6 4
 ----------
        7 5
```

Finally, we subtract the 8 from the 9 and the 2 from the 6.

```
   6 9
   7̸ 0̸¹ 3 9
 - 2 8  6 4
 ----------
   4 1  7 5
```

We will deal with negative *binary numbers in Sec. 2-3.*

We subtract binary digits as follows.

Table of Differences	0 − 0 = 0 0 − 1 = 1, with borrow from next column 1 − 0 = 1 1 − 1 = 0 10 − 1 = 1	**19**

Example: Subtract the binary number 1001 from 1101.

Solution: We write one number above the other and subtract column by column.

$$\begin{array}{cccc} 1 & 1 & 0 & 1 \\ 1 & 0 & 0 & 1 \\ \hline 0 & 1 & 0 & 0 \end{array}$$

or 100.

When we must subtract a 1 from a 0, we must *borrow* as in decimal subtraction. When we borrow from a 1 it becomes a 0.

Example: Subtract the binary number 101 from 110.

Solution: We write

$$\begin{array}{rccc} & 1 & 1 & 0 \\ - & 1 & 0 & 1 \\ \hline \end{array}$$

Borrowing, the 0 in the first column becomes 10, and the 1 in the middle column becomes 0.

$$\begin{array}{rccc} & & 0 & \\ & 1 & \not{1} & {}^{1}0 \\ - & 1 & 0 & 1 \\ \hline \end{array}$$

Subtracting,

$$\begin{array}{rccc} & & 0 & \\ & 1 & \not{1} & {}^{1}0 \\ - & 1 & 0 & 1 \\ \hline & 0 & 0 & 1 \end{array}$$

We can borrow only from a 1. If a zero is encountered we continue left until we find a 1 from which to borrow. Any zeros passed along the way get changed to ones.

Example: Evaluate

$$\begin{array}{rccccc} & 1 & 1 & 0 & 0 & 0 \\ - & & 1 & 0 & 0 & 1 \\ \hline \end{array}$$

Solution: To perform the subtraction in the first column, we must borrow from the 1 in the fourth column.

$$\begin{array}{rccccc} & & & 1 & 1 & \\ & 1 & \not{1} & \not{0} & \not{0} & {}^{1}0 \\ - & & 1 & 0 & 0 & 1 \\ \hline \end{array}$$

Subtracting,

$$\begin{array}{rccccc} & & 0 & 1 & 1 & \\ & 1 & \not{1} & \not{0} & \not{0} & {}^{1}0 \\ - & & 1 & 0 & 0 & 1 \\ \hline & & & 1 & 1 & 1 \end{array}$$

We borrow again,

```
    0 ¹0 1  1
    1  1 0  0 ¹0
  -    1 0  0  1
```

and subtract,

```
       1 1  1  1
```

When the numbers to be subtracted contain *fractional parts,* align the binary points vertically before subtracting.

Example: Evaluate 111.011 − 100.11.

Solution: We write one above the other, with binary points aligned, and supply the missing 0 to the right of the second number.

```
   01 1 1 . 0 1 1
  - 1 0 0 . 1 1 0
```

Subtracting gives

```
          0
    1 1  1 . ¹0 1 1
  - 1 0  0 .  1 1 0
      1  0 .  1 0 1
```

Dividing Binary Numbers: The quotient of two binary digits is shown in the following table.

Table of Quotients	$0 \div 0$ is not defined $1 \div 0$ is not defined $0 \div 1 = 0$ $1 \div 1 = 1$	**22**

Example: Divide the binary numbers, 100001110 ÷ 101.

Solution: We use the same format as for long division,

```
101 ⟌100001110
```

Then we divide 101 into 1000 and get 1. We multiply the divisor by 1 and write the product underneath the dividend.

```
       1
101 ⟌100001110
     101
```

Now subtract and bring down the next digit.

```
       1
101 ⟌100001110
     101
      110
```

We now repeat. 101 divides into 110 one time, so

```
       11
101 | 100001110
      101
       110
       101
```

Then

```
       11
101 | 100001110
      101
       110
       101
         11
```

It is not surprising that many people convert binary numbers to decimal in order to do division or multiplication, and then convert back.

Now 101 will not divide into 11, so we place a 0 in our quotient and bring down the next digit. Continuing gives

```
       110110
101 | 100001110
      101
       110
       101
         111
         101
          101
          101
            0
```

So 100001110 ÷ 101 = 110110.

As with decimal division, we sometimes get a remainder.

Example: Evaluate 1010 ÷ 11.

Solution: Using the same procedure as before gives

```
     11
11 | 1010
     11
      100
       11
        1    ← remainder
```

When either the divisor or dividend contain a binary point, move the point the same number of places in both so that each becomes an integer. Then divide as before.

Example: Divide 100001.11 by .101.

Solution: We move the binary point three places to the right in both divisor and dividend and get

$$100001110 \div 101$$

From a preceding example, the quotient is

$$110110$$

EXERCISE 2

Binary Addition

Add each pair of binary numbers.

1. 1 + 0

2. 1 + 1

3. 10 + 11

4. 01 + 10

5. 11 + 11

6. 11 + 01

7. 1010 + 1001

8. 0101 + 1011

9. 1111 + 1111

10. 1110 + 0110

11. 1100 0110 + 0110 1110

12. 0110 0110 + 1100 1001

13. 1001 0110 1010 1000 + 1101 0111 0101 0001

14. 0110 0111 1010 1100 + 1110 0001 0101 1001

15. 0.10 + 0.11

16. 0.01 + 0.10

17. 0.1101 + 0.1001

18. 0.0110 + 0.1010

19. 1.11 + 0.01

20. 1.01 + 1.10

21. 1010.0110 + 0110.1101

22. 1100.1100 + 1010.1100

Add each group of binary numbers.

23. 1101 + 0101 + 1000

24. 0110 + 0111 + 1110

25. 11.01 + 01.01 + 10.10

26. 11.00 + 10.01 + 10.11

Binary Multiplication

Multiply each pair of binary numbers.

27. 1 × 0

28. 1 × 1

29. 10 × 10

30. 10 × 01

31. 11 × 10

32. 11 × 11

33. 1001 × 1100

34. 0110 × 1101

35. 0.1 × 0.11

36. 1.1 × 0.1

37. 10.1 × 11.0

38. 1.11 × 1.01

39. 1100.01 × 1.11

40. 1.1 × 1101.01

Binary Subtraction

Subtract.

41. 10 − 01

42. 11 − 10

43. 1101 − 1011

44. 1110 − 1001

45. 1101 1001 − 1011 1000

46. 1110 0101 − 1101 0110

47. 1.10 − 1.01

48. 0.110 − 0.011

49. 1011.01 − 1001.11

50. 1100.110 − 1010.101

Binary Division

Divide.

51. 101000 ÷ 10

52. 10100 ÷ 11

53. 11110 ÷ 110

54. 111000 ÷ 111

55. 101101110 ÷ 110

56. 111110001 ÷ 111

57. 1011011100 ÷ 1100

58. 10001111 ÷ 1101

59. 11010 ÷ 110

60. 1100.01 ÷ 1.1

61. 11101.11 ÷ 1.01

62. 101110.011 ÷ 11.1

2-3. COMPLEMENTS

One's Complement: The *one's complement* of a 4-bit binary number is obtained by subtracting the number from the largest possible 4-bit number, 1111 (which is equal to $2^4 - 1$).

Example: To find the one's complement of 1011, we subtract it from 1111,

$2^4 - 1 =$	1111
Subtracting,	− 1011
One's complement	0100

Similarly, the one's complement of an 8-bit number is obtained by subtracting the number from the largest possible 8-bit number, 1111 1111 (which is equal to $2^8 - 1$). In general, the one's complement of an n-bit binary number is found by subtracting the number from a binary number consisting of n 1's, (which is equal to $2^n - 1$).

n-Bit One's Complement of x	$(2^n - 1) - x$	**23**

Example: The 8-bit one's complement of 01011001 is

$2^8 - 1 =$	11111111
Subtracting,	− 01011001
One's complement	10100110

Notice that in both of the preceding examples the one's complement could have been obtained from the original number simply by changing the 1's to 0's and the 0's to 1's. In fact, this always works, and gives us a fast way to find one's complements.

One's Complement	The one's complement of a number can be written by changing the 1's to 0's and the 0's to 1's.	**24**

Two's Complement: The n-bit *two's complement* of a number is found by subtracting the number from 2^n.

n-Bit Two's Complement of x	$2^n - x$	**25**

Example: The two's complement of 1011 is

$2^4 =$	10000
Subtracting,	− 1011
Two's complement	0101

Now the one's complement of that same number 1011 is 0100. We see that in this case, the two's complement 0101 is 1 greater than the one's complement. This will, in fact, be true for any binary number, and leads to the following rule.

Two's Complement	To find the two's complement of a number, first write the one's complement and then add 1.	**26**

Example: Find the two's complement of 1001 0010.

Solution:

Original number	1001 0010
One's complement	0110 1101
Add 1	0110 1110 = two's complement

Before finding the complement, place 0's to the left of the number to obtain the required number of bits.

Example: Find the 8-bit two's complement of 101.

Solution:

Original number	101
Add bits to make an 8-bit number	0000 0101
One's complement	1111 1010
Add 1 to get two's complement	1111 1011

A comparison of the one's complement and the two's complement for the numbers from 1 to 20 is given in the following table.

Two's complements are the most often used method for storing negative numbers in computers.

Number	*Binary*	*One's complement*	*Two's complement*
0	0 0 0 0 0 0 0 0	1 1 1 1 1 1 1 1	0 0 0 0 0 0 0 0
1	0 0 0 0 0 0 0 1	1 1 1 1 1 1 1 0	1 1 1 1 1 1 1 1
2	0 0 0 0 0 0 1 0	1 1 1 1 1 1 0 1	1 1 1 1 1 1 1 0
3	0 0 0 0 0 0 1 1	1 1 1 1 1 1 0 0	1 1 1 1 1 1 0 1
4	0 0 0 0 0 1 0 0	1 1 1 1 1 0 1 1	1 1 1 1 1 1 0 0
5	0 0 0 0 0 1 0 1	1 1 1 1 1 0 1 0	1 1 1 1 1 0 1 1
6	0 0 0 0 0 1 1 0	1 1 1 1 1 0 0 1	1 1 1 1 1 0 1 0
7	0 0 0 0 0 1 1 1	1 1 1 1 1 0 0 0	1 1 1 1 1 0 0 1
8	0 0 0 0 1 0 0 0	1 1 1 1 0 1 1 1	1 1 1 1 1 0 0 0
9	0 0 0 0 1 0 0 1	1 1 1 1 0 1 1 0	1 1 1 1 0 1 1 1
10	0 0 0 0 1 0 1 0	1 1 1 1 0 1 0 1	1 1 1 1 0 1 1 0
11	0 0 0 0 1 0 1 1	1 1 1 1 0 1 0 0	1 1 1 1 0 1 0 1
12	0 0 0 0 1 1 0 0	1 1 1 1 0 0 1 1	1 1 1 1 0 1 0 0
13	0 0 0 0 1 1 0 1	1 1 1 1 0 0 1 0	1 1 1 1 0 0 1 1
14	0 0 0 0 1 1 1 0	1 1 1 1 0 0 0 1	1 1 1 1 0 0 1 0
15	0 0 0 0 1 1 1 1	1 1 1 1 0 0 0 0	1 1 1 1 0 0 0 1
16	0 0 0 1 0 0 0 0	1 1 1 0 1 1 1 1	1 1 1 1 0 0 0 0
17	0 0 0 1 0 0 0 1	1 1 1 0 1 1 1 0	1 1 1 0 1 1 1 1
18	0 0 0 1 0 0 1 0	1 1 1 0 1 1 0 1	1 1 1 0 1 1 1 0
19	0 0 0 1 0 0 1 1	1 1 1 0 1 1 0 0	1 1 1 0 1 1 0 1
20	0 0 0 1 0 1 0 0	1 1 1 0 1 0 1 1	1 1 1 0 1 1 0 0

We say that the lost bits have been tossed into the bit bucket.

Negative Binary Numbers: First, we assume that the binary numbers we are working with are stored with a fixed number n of bits, and that any overflow is *lost*. Therefore if some binary number x has n bits, then adding 2^n to x *will make no change*. For example, if $n = 4$ and

$$x = 1011$$

Then $2^n = 2^4 = 10000$

Adding, 1 1011

overflow is lost (the 1) — number is unchanged (1011)

In other words,

$$x = x + 2^n$$

Now consider a negative number $-M$. Its value is unchanged by adding 2^n,

$$-M = -M + 2^n$$
$$= 2^n - M$$

But $2^n - M$ is the two's complement of M. So

Negative Binary Numbers	If M is a positive binary number, then $-M$ is the two's complement of M.	**27**

Example: Convert the decimal number -5 to a 4-bit binary number.

Solution:

Binary equivalent of $+5$	0101
One's complement	1010
Add 1	1011

Thus, $-5_{10} = 1011_2$.

The Sign Bit: Let us list the decimal numbers from -4 to $+4$, together with the 8-bit binary code for each, with the negative numbers given by the two's complement.

Decimal	*Binary*	*Decimal*	*Binary*
0	00000000	0	00000000
+1	00000001	−1	11111111
+2	00000010	−2	11111110
+3	00000011	−3	11111101
+4	00000100	−4	11111100

We must be careful to reserve *the leftmost bit for this purpose. Thus, in 8-bit binary, positive numbers larger than 01111111 are not allowed, as they would put a 1 into the leftmost bit position.*

Now look at the leftmost bit for each number. We see that *this bit is 0 when the number is positive, and 1 when the number is negative.* Thus, the leftmost bit automatically gives us the sign of the number, and hence is called the *sign bit.*

Example: Determine whether each 8-bit binary number is positive or negative.

(a) 1000 0100
(b) 0100 1010
(c) 0011 0100
(d) 1001 0100

Solution: We simply look at the leftmost bit. Thus

(a) 1000 0100 is negative
(b) 0100 1010 is positive
(c) 0011 0100 is positive
(d) 1001 0100 is negative

Binary Subtraction Using Complements: To subtract a number B from A is no different from adding the negative of B to A. But since the negative of B is given by its two's complement, we see that

Subtracting a binary number B from the binary number A is equivalent to adding the two's complement of B to A.	**20**

Example: Subtract 0010 0100 (decimal 36) from 0011 0101 (decimal 53).

Solution: The number to be subtracted has an 8-bit one's complement of 1101 1011, so its two's complement (and hence its negative) is 1101 1100. Adding gives

```
                                 Decimal check
            0011 0101                 53
          + 1101 1100             + (-36)
       1    0001 0001                 17
overflow ↑
```

Discarding the overflow, we get

0011 0101 − 0010 0100 = 0001 0001

When we subtract a larger number from a smaller, we get a negative result.

Example: Assuming the leftmost bit to be a sign bit, subtract 0010 0100 (decimal 36) from 0000 1110 (decimal 14).

Solution: Instead of subtracting 0010 0100, we add its two's complement (1101 1100).

```
                        Decimal check
     0000 1110               14
   + 1101 1100           + (-36)
     1110 1010              -22
```

We see that our result, 1110 1010, is negative because its sign bit is 1. To find its decimal value, we first find its two's complement 0001 0110. This is equal to +22, so our answer is equal to −22.

EXERCISE 3

Complements

Write the 8-bit one's complement of each binary number.

1. 1101 **2.** 011011

3. 100110 **4.** 01110101

Write the 8-bit two's complement of each binary number.

5. 1110 **6.** 100010

7. 0111011 **8.** 1100111

Write each negative decimal number as an 8-bit binary number.

9. −7 **10.** −9 **11.** −48

12. −92 **13.** −120 **14.** −72

15. −57 **16.** −114

Subtract using two's complements.

17. 1101 − 0110 **18.** 1110 − 1001

19. 10101010 − 10001001 **20.** 11010111 − 10111010

21. 1001 − 1101 **22.** 0110 − 1010

23. 1101 − 1111 **24.** 0111 − 1001

2-4. COMPUTER STORAGE OF BINARY NUMBERS

Computer Representation of Integers: The exact way in which an integer is stored depends on the computer, and is limited by its word length. Here we assume a computer that uses two bytes (16 bits) to store each integer, as follows:

0 0 0 0 0 0 0 0 0 0 0 0 0 0 0 0

sign bit ——↑ (points to the leftmost bit)

We further assume that the number is in two's complement form, with the leftmost bit being the sign bit. With 15 bits available, the computer can store numbers as large as

$$2^{15} - 1 = 32{,}767$$

Example: The decimal integer 11 is represented by

0 0 0 0 0 0 0 0 0 0 0 0 1 0 1 1

Floating-Point Representation: Recall from Sec. 1-8 that we used *scientific notation* to write very large or very small numbers. Thus,

$$5,372,000 = 5.372 \times 10^6$$

We may write binary numbers in a similar way, as a binary number (called the *mantissa*) multiplied by 2 raised to a power (called the *exponent* or *characteristic*).

Example: The binary number 1111 0000 can be written as

$$1111\ 0000 = .1111\ 0000 \times 2^{8}$$

(.1111 0000 is the mantissa; 8 is the exponent)

A binary number written with its binary point to the left of the number is said to be normalized.

Notice that the number of places moved by the binary point is equal to the number in the exponent, just as in scientific notation with decimal numbers.

For each floating-point number, a computer must store the mantissa, the sign of the mantissa, the exponent, and the sign of the exponent. We assume a computer that uses four bytes (32 bits) to store a floating-point number, and uses one byte for the exponent and three bytes for the mantissa.

Mantissa	Exponent
(three bytes)	(one byte)

Computers differ in the ways that they represent the mantissa and exponent. A common way to represent the mantissa is in two's-complement form, when negative, with the binary point located to the right of the sign bit. The exponent, however, is often a 7-bit binary integer, plus an eighth bit to give the sign of the exponent; positive if the sign bit is 0 and negative if the sign bit is 1.

Example: What decimal number is represented by the following?

Note that the exponent is not *stored in two's-complement form.*

Mantissa	Exponent
0 010 1000 0000 0000 0000 0000	0 000 0011

(the first bit of the mantissa and the first bit of the exponent are the sign bits)

Solution: The mantissa is positive, and its value is

$$+.010\ 1000\ 0000\ 0000\ 0000\ 0000$$

The exponent is also positive, and has a value

$$000\ 0011 = +3$$

Our binary number is then

$$+.0101 \times 2^3 = 10.1$$
$$= 2.5_{10}$$

Example: How would our assumed computer store the decimal number 0.00725?

Solution: We first convert to binary,

$$0.00725 = .0000\ 0001\ 1101\ 1011\ 0010\ 0000\ 0000\ 0000$$

or, in floating-point form,

$$.1110\ 1101\ 1001\ 0000\ 0000\ 0000 \times 2^{-7}$$

In binary, the decimal expoonent 7 is equal to 000 0111. The entire number is then

Mantissa	Exponent
0 111 0110 1100 1000 0000 0000	1 000 0111

Double Precision: Many computers allow us to use twice as many bits as usual to store a number, when greater accuracy is needed. This is called *double precision*. Thus a computer that normally uses four bytes to store a number will print about twice as many significant digits when in double precision (16 instead of 7).

Double precision is usually specified when writing the variable name for some quantity. Some computers use the symbol # to specify a double-precision number, so that if X is a number in single precision, X# is that number in double precision.

EXERCISE 4

Assuming a computer that stores numbers as shown in Sec. 2-4, show how the following numbers would be stored.

1. 482 **2.** −37 **3.** 7.5

4. −23.6 **5.** 4,825,285 **6.** 0.00000475

Again assuming the same computer as in Sec. 2-4, find the decimal number represented by each of the following.

7.

Mantissa	Exponent
0 001 0011 0000 0000 0000 0000	0 000 0010

8.

Mantissa	Exponent
1 001 0011 0100 0000 0000 0000	0 000 0011

9.

Mantissa	Exponent
0 010 1100 0000 0000 0000 0000	1 000 0011

10.

Mantissa	Exponent
1 001 0101 1000 0000 0000 0000	1 010 0100

CHAPTER TEST

For this test, assume a computer that stores numbers as in Sec. 2-4.

1. Add the binary numbers 10101.10 and 11100.110.

2. What decimal number is represented by

Mantissa	Exponent
0 011 1101 0000 0000 0000 0000	1 000 0101

3. Write the 8-bit one's complement of 1100101.

4. How would the number 592.7 be represented in a computer?

5. Subtract the binary number 1001.011 from 11001.1.

6. What decimal number is represented by

Mantissa	Exponent
0 101 0111 0000 0000 0000 0000	0 000 0110

7. Write the 8-bit two's complement of 01001011.

8. Convert the binary number 11001010 to decimal.

9. Convert the decimal number 26.875 to binary.

10. Subtract the binary number 0110.100 from 1001.011 using two's complements.

11. What decimal number is represented by

Mantissa	Exponent
1 011 0010 1100 0000 0000 0000	0 000 0111

12. Write the decimal number −247 as a 12-bit binary number.

13. Divide the binary number 110010.1 by 10.01.

14. How would the decimal number 0.000875 be stored?

15. Convert the binary number 110.001 to decimal.

Hexadecimal, Octal, Errors in Computation and BCD Numbers.

Binary numbers are useful in a computer because each bit can be represented by one state of a binary switch which is either on or off. However, they are hard to read, partly because of their great length. To represent a nine-digit social security number, for example, would require a binary number *29 bits long*.

In this chapter we study other ways in which numbers can be represented. Hexadecimal, octal, and binary-coded-decimal systems allow us to express binary numbers more compactly, and make the transfer of data between computers and people much easier.

Also in this chapter we give a brief discussion of how errors can creep into a computation, and some general rules for minimizing their harmful effect.

3-1. THE HEXADECIMAL NUMBER SYSTEM

Hexadecimal Numbers: Hexadecimal numbers (or *hex* for short) are obtained by grouping the bits in a binary number into sets of four, and representing each such set by a single number or letter. A hex number one-fourth the length of the binary number is thus obtained.

Base 16: Since a 4-bit group of binary digits can have a value between 0 and 15, we need 16 symbols to represent all these values. The base of hexadecimal numbers is thus 16. We use the digits from 0 to 9, and the capital letters A to F, as in Table 3-1.

Table 3-1 also shows octal numbers, which we will use later.

TABLE 3-1
Hexadecimal and Octal Number Systems

Decimal	*Binary*	*Hexadecimal*	*Octal*
0	0000	0	0
1	0001	1	1
2	0010	2	2
3	0011	3	3
4	0100	4	4
5	0101	5	5
6	0110	6	6
7	0111	7	7
8	1000	8	10
9	1001	9	11
10	1010	A	12
11	1011	B	13
12	1100	C	14
13	1101	D	15
14	1110	E	16
15	1111	F	17

Converting Binary to Hexadecimal: Group the bits into sets of four starting at the binary point, adding zeros as needed to fill out the groups. Then assign to each group the appropriate letter or number from Table 3-1.

Example: Convert 10110100111001 to hexadecimal.

Solution: Grouping, we get

10 1101 0011 1001

or 0010 1101 0011 1001

From Table 3-1, 2 D 3 9

So the hexadecimal equivalent is 2D39. Hexadecimal numbers are sometimes written with the subscript 16, as in

$$2D39_{16}$$

The procedure is no different for binary fractions.

Example: Convert 101111.0011111 to hexadecimal.

Solution: Grouping,

10 1111 . 0011 111

or 0010 1111 . 0011 1110

From the table, 2 F . 3 E or 2F.3E

Converting Hexadecimal to Binary: Here we simply reverse the procedure.

Example: Convert 3B25.E to binary.

Solution: We write the group of 4 bits corresponding to each hexadecimal symbol.

3	B	2	5	.	E
0011	1011	0010	0101	.	1110

or 11 1011 0010 0101.111.

Instead of converting directly between decimal and hex, many find it easier to first convert to binary.

Converting Hexadecimal to Decimal: As with decimal and binary numbers, each hex digit has a *place value,* equal to the base, 16, raised to the position number. Thus, the place value of the first position to the left of the decimal point is

$$16^0 = 1$$

and the next is

$$16^1 = 16$$

and so on.

To convert from hex to decimal, first replace each letter in the hex number by its decimal equivalent. Write the number in expanded notation, multiplying each hex digit by its place value. Add the resulting numbers.

Example: Convert the hex number 3B.F to decimal.

Solution: We replace the hex B with 11, and the hex F with 15, and write the number in expanded form.

$$\begin{array}{ccccccc} & 3 & & B & . & F & \\ & 3 & & 11 & . & 15 & \\ & (3 \times 16^1) & + & (11 \times 16^0) & + & (15 \times 16^{-1}) & \\ = & 48 & + & 11 & + & 0.9375 & \\ = & 59.9375 & & & & & \end{array}$$

Converting Decimal to Hexadecimal: We repeatedly divide the given decimal number by 16, and convert each remainder to hex. The remainders form our hex number, the last remainder obtained being the most significant digit.

Example: Convert the decimal number 83759 to hex.

Solution:

$83759 \div 16 = 5234$	with remainder of	15 ↑
$5234 \div 16 = 327$	with remainder of	2
$327 \div 16 = 20$	with remainder of	7 read up
$20 \div 16 = 1$	with remainder of	4
$1 \div 16 = 0$	with remainder of	1

Changing the number 15 to hex F and reading the remainders from bottom up, our hex equivalent is 1472F.

Hexadecimal Addition: We make use of Table 3-2, which shows the sum of any two hex digits.

TABLE 3-2 Sums of Hexadecimal Numbers[a]

	0	*1*	*2*	*3*	*4*	*5*	*6*	*7*	*8*	*9*	*A*	*B*	*C*	*D*	*E*	*F*
0	0	1	2	3	4	5	6	7	8	9	A	B	C	D	E	F
1	1	2	3	4	5	6	7	8	9	A	B	C	D	E	F	10
2	2	3	4	5	6	7	8	9	A	B	C	D	E	F	10	11
3	3	4	5	6	7	8	9	A	B	C	D	E	F	10	11	12
4	4	5	6	7	8	9	A	B	C	D	E	F	10	11	12	13
5	5	6	7	8	9	A	B	C	D	E	F	10	11	12	13	14
6	6	7	8	9	A	B	C	D	E	F	10	11	12	13	14	15
7	7	8	9	A	B	C	D	E	F	10	11	12	13	14	15	16
8	8	9	A	B	C	D	E	F	10	11	12	13	14	15	16	17
9	9	A	B	C	D	E	F	10	11	12	13	14	15	16	17	18
A	A	B	C	D	E	F	10	11	12	13	14	15	16	17	18	19
B	B	C	D	E	F	10	11	12	13	14	15	16	17	18	19	1A
C	C	D	E	F	10	11	12	13	14	15	16	17	18	19	1A	1B
D	D	E	F	10	11	12	13	14	15	16	17	18	19	1A	1B	1C
E	E	F	10	11	12	13	14	15	16	17	18	19	1A	1B	1C	1D
F	F	10	11	12	13	14	15	16	17	18	19	1A	1B	1C	1D	1E

[a] Locate the sum of two hexadecimal digits at the intersection of the row and column containing the numbers. Thus, the sum of C and E is 1A.

Examples:

(a) $4 + 5 = 9$

(b) $4 + 7 = B$

(c) $A + 5 = F$

(d) $B + F = 1A$

To add hex numbers having more that one digit, add the right column first. If the sum has more than one digit, carry the 1 to the next column.

Example: Add A5B to 2C9.

Solution: In the right column, B + 9 = 14. We write down the 4 and carry the 1.

$$\begin{array}{r} {}^{1} \\ A\ 5\ B \\ +\ 2\ C\ 9 \\ \hline 4 \end{array}$$

In the next column, 1 + 5 + C = 12. We write the 2 and carry the 1.

$$\begin{array}{r} {}^{1\ 1} \\ A\ 5\ B \\ +\ 2\ C\ 9 \\ \hline 2\ 4 \end{array}$$

In the last column, 1 + A + 2 = D,

$$\begin{array}{r} A\ 5\ B \\ +\ 2\ C\ 9 \\ \hline D\ 2\ 4 \end{array}$$

So A5B + 2C9 = D24.

Hexadecimal Subtraction: To subtract a single hex digit Y from another single hex digit X, locate Y along the left edge of the hex addition table. Then search the row to the right until X is found. The difference $X - Y$ is then located at the top of the table in the same column as X.

Example: Subtract D − 5.

Solution: We find 5 at the left of Table 3-2, and search that same row for D. Directly above that D we find 8, so

$$D - 5 = 8$$

It is often easier to convert the numbers to decimal, do the subtraction, and convert back to hex.

Example: Subtract 4AB from 7F2.

Solution: Converting to decimal,

$$4AB = 1195$$

and

$$7F2 = 2034$$

Subtracting, $2034 - 1195 = 839$

Converting to hex, $839_{10} = 347_{16}$

Hexadecimal Multiplication: To multiply hex numbers we use Table 3-3, which gives the product of any two hex digits.

TABLE 3-3 Products of Hexadecimal Numbers[a]

	0	*1*	*2*	*3*	*4*	*5*	*6*	*7*	*8*	*9*	*A*	*B*	*C*	*D*	*E*	*F*
0	0	0	0	0	0	0	0	0	0	0	0	0	0	0	0	0
1	0	1	2	3	4	5	6	7	8	9	A	B	C	D	E	F
2	0	2	4	6	8	A	C	E	10	12	14	16	18	1A	1C	1E
3	0	3	6	9	C	F	12	15	18	1B	1E	21	24	27	2A	2D
4	0	4	8	C	10	14	18	1C	20	24	28	2C	30	34	38	3C
5	0	5	A	F	14	19	1E	23	28	2D	32	37	3C	41	46	4B
6	0	6	C	12	18	1E	24	2A	30	36	3C	42	48	4E	54	5A
7	0	7	E	15	1C	23	2A	31	38	3F	46	4D	54	5B	62	69
8	0	8	10	18	20	28	30	38	40	48	50	58	60	68	70	78
9	0	9	12	1B	24	2D	36	3F	48	51	5A	63	6C	75	7E	87
A	0	A	14	1E	28	32	3C	46	50	5A	64	6E	78	82	8C	96
B	0	B	16	21	2C	37	42	4D	58	63	6E	79	84	8F	9A	A5
C	0	C	18	24	30	3C	48	54	60	6C	78	84	90	9C	A8	B4
D	0	D	1A	27	34	41	4E	5B	68	75	82	8F	9C	A9	B6	C3
E	0	E	1C	2A	38	46	54	62	70	7E	8C	9A	A8	B6	C4	D2
F	0	F	1E	2D	3C	4B	5A	69	78	87	96	A5	B4	C3	D2	E1

[a] The product of two hexadecimal digits is located at the intersection of the row and column containing those digits. Thus the product of C and E is A8.

Example: Multiply the hex numbers 2B and 5.

Solution: From Table 3-3, the product of 5 and B is 37. We write down the 7 and carry the 3.

```
 3
 2 B
   5
 ---
   7
```

The product of 5 and 2 is A. To A we add the 3 that was carried, getting A + 3 = D.

```
 3
 2 B
   5
 ---
 D 7
```

Example: Multiply the hex numbers 1E3 and A4.

Solution: Try to follow the steps in this solution.

```
    1 E 3
      A 4
    -----
    7 8 C
  1 2 D E
  -------
  1 3 5 6 C
```

Hexadecimal Division: We again use Table 3-3 and proceed as for long division of decimal numbers, as shown in the following example.

Example: Divide 5E3 by A.

Solution: We first see that we cannot divide 5 by A. We then divide 5E by A. We search down the A column in the table for an entry of 5E. Not finding one, we then locate the largest entry that is still smaller than 5E. We find 5A, which is the product of 9 and A. So

```
     9
A ) 5 E 3
```

We multiply 9 by A and place the product, 5A, underneath, and subtract.

```
     9
A ) 5 E 3
    5 A
    ---
      4
```

We bring down the 3, and then divide A into 43. We get a quotient of 6.

```
      9 6
A ) 5 E 3
    5 A
    ---
      4 3
      3 C
      ---
        7    rcmainder
```

Multiplying 6 by A gives 3C, which, when subtracted from 43, leaves a remainder of 7.

EXERCISE 1

Binary–Hex Conversions

Convert each binary number to hexadecimal.

1. 1101

2. 1010

3. 1001

4. 1111

5. 1001 0011

6. 0110 0111

7. 1101 1000

8. 0101 1100

9. 1001 0010 1010 0110

10. 0101 1101 0111 0001

11. 1001.0011

12. 10.0011

13. 1.00111

14. 101.101

Convert each hex number to binary.

15. 6F

16. B2

17. 4A

18. CC

19. 2F35

20. D213

21. 47A2 **22.** ABCD **23.** 5.F

24. A4.E **25.** 9.AA **26.** 6D.7C

Decimal–Hex Conversions

Convert each hex number to decimal.

27. F2 **28.** 5C **29.** 33

30. DF **31.** 37A4 **32.** A3F6

33. F274 **34.** C721 **35.** 3.F

36. 22.D **37.** ABC.DE **38.** C.284

Convert each decimal number to hex.

39. 39 **40.** 13 **41.** 921

42. 554 **43.** 2741 **44.** 9945

45. 1736 **46.** 2267

Hexadecimal Addition and Subtraction

Add each pair of hex numbers.

47. 3 + E **48.** A + 2 **49.** C + D

50. B + 9 **51.** 39 + A2 **52.** DD + C6

53. 1F + D8 **54.** 9D + 35 **55.** 8D15 + F7A2

56. 28B3 + DD7A

Subtract.

57. E − 3 **58.** B − 6 **59.** F − A

60. E − B **61.** EA − 26 **62.** 7C − 3D

63. DD − AA **64.** A3 − F **65.** A82 − 3FF

66. 9D3 − 4AB **67.** 72D4 − EE5 **68.** D8A2 − 7ACF

Hex Multiplication and Division

Multiply.

69. 3 × E **70.** A × B **71.** 9 × D

72. E × F **73.** D × 27 **74.** 2D × 8

75. A5 × 9C **76.** 23 × AA **77.** A4 × 826

Divide.

78. C ÷ 2 **79.** F ÷ 3 **80.** E ÷ 7

81. A ÷ 5 **82.** A5 ÷ 5 **83.** FF ÷ 5

84. 8C ÷ A

3-2. THE OCTAL NUMBER SYSTEM

The Octal System: The *octal* number system uses eight digits, 0 to 7, and hence has a base of eight. A comparison of the decimal, binary, hex, and octal digits is given in Table 3-1.

Binary–Octal Conversions: To convert from binary to octal, write the bits of the binary number in groups of three, starting at the binary point. Then write the octal equivalent for each group.

Example: Convert the binary number 1101000110110 to octal.

Solution: Grouping the bits,

1 101 000 110 110

and the octal equivalents, 1 5 0 6 6

so the octal equivalent of 1101000110110 is 15066.

To convert from octal to binary, simply reverse the procedure shown in the preceding example.

Example: Convert the octal number 7364 to binary.

Solution: We write the group of 3 bits corresponding to each octal digit.

7 3 6 4

111 011 110 100

or 1110 1111 0100.

EXERCISE 2

Binary–Octal Conversions

Convert each binary number to octal.

1. 110 **2.** 010 **3.** 111

4. 101 **5.** 110011 **6.** 011010

7. 01101101 **8.** 110110100 **9.** 1001001

10. 11001001

Convert each octal number to binary.

11. 26 **12.** 35 **13.** 623

14. 621 **15.** 5243 **16.** 1153

17. 63150 **18.** 2346

3-3. BCD CODES

In Sec. 2-1 we converted decimal numbers into binary. Recall that each bit had a place value equal to 2 raised to the position number. Thus, the binary number 10001 is equal to

$$2^4 + 2^0 = 16 + 1 = 17$$

We shall now refer to numbers such as 10001 as *straight binary*.

With a BCD or *binary-coded-decimal* code, a decimal number is not converted *as a whole* to binary, but rather *digit by digit*. For example, the 1 in the decimal number 17 is 0001 in 4-bit binary, and the 7 is equal to 0111. Thus,

17 = 10001 in straight binary

and

17 = 0001 0111 in BCD

When we converted the digits in the decimal number 17 in the preceding example, we used the same 4-bit binary equivalents as in Table 3-1. The bits have the place values 8, 4, 2, and 1, and BCD numbers written in this manner are said to be in *8421 code*.

Converting Decimal Numbers to BCD: Simply convert each decimal digit to its equivalent 4-bit code.

Example: Convert the decimal number 25.3 to 8421 BCD code.

Solution: We find the BCD equivalent of each decimal digit from Table 3-4.

2	5	.	3
0010	0101	.	0011

or 0010 0101.0011.

TABLE 3-4
Binary-Coded-Decimal Numbers

Decimal	*8421*	*2421*	*5211*
0	0000	0000	0000
1	0001	0001	0001
2	0010	0010	0011
3	0011	0011	0101
4	0100	0100	0111
5	0101	1011	1000
6	0110	1100	1010
7	0111	1101	1100
8	1000	1110	1110
9	1001	1111	1111

Converting from BCD to Decimal: Separate the BCD number into 4-bit groups, and write the decimal equivalent of each group.

Example: Convert the 8421 BCD number 1 0011.0101 to decimal.

Solution: We have

	0001	0011	.	0101
From Table 3-4,	1	3	.	5

or 13.5.

Other BCD Codes: There are other BCD codes in use which represent each decimal digit by a 6-bit binary number or an 8-bit binary number, and other 4-bit codes in which the place values are other than 8421. Some of these are shown in Table 3.4. Each is used in the same way as shown for the 8421 code.

Example: Convert the decimal number 825 to 2421 BCD code.

Solution: We convert each decimal digit separately,

8	2	5
1110	0010	1011

or 111000101011.

Other Computer Codes: In addition to the binary, hexadecimal, octal, and BCD codes we have discussed, there are many other codes used in the computer industry. Numbers can be represented by the *excess-3 code* or the *Gray code*. There are error-detecting and parity-checking codes. Other codes are used to represent letters of the alphabet and special symbols as well as numbers. Some of these are the *Morse code*, the American Standard Code for Information Interchange or *ASCII* (pronounced *as-key*) *code*, the Extended Binary-Coded-Decimal Interchange Code or *EBCDIC* (pronounced *eb-si-dik*) *code*, and the *Hollerith code* used for punched cards.

Space does not permit discussion of each code. However, if you have understood the manipulation of the codes we have described, you should have no trouble when faced with a new one.

EXERCISE 3

BCD–Decimal Conversions

Convert each decimal number to 8421 BCD.

1. 62 **2.** 25 **3.** 274

4. 284 **5.** 42.91 **6.** 5.014

Convert each 8421 BCD number to decimal.

7. 1001 **8.** 101 **9.** 1100001

10. 1000111 **11.** 110110.1000 **12.** 111000.10001

BCD Addition

Add each pair of 8421 BCD numbers. First convert to decimal, add, then convert back.

13. 1100110111 + 1001110101 **14.** 011101011001 + 100001010011

15. 100111001 + 1001110101 **16.** 100101110101 + 100001110100

3-4. ERRORS

The types of errors discussed in this section are extremely small and can safely be ignored for most computations.

Any computation is subject to errors of various types. Some of these may be outright mistakes made by the person doing the computation, such as using incorrect data or pushing the wrong key on a calculator or computer. In this section we first show how the size of an error is usually expressed, and then briefly discuss other types of errors that can occur even when you do not make a "mistake."

Absolute, Relative, and Percent Error: There are several ways of expressing the size of a particular error. The magnitude of the error is called the *absolute error,* and is found by subtracting the "true" value from the "actual" value. The absolute error divided by the known or "true" value is called the *relative error.* The relative error, expressed as a percent, is called the *percent error.* Thus

$$\text{percent error} = \frac{\text{absolute error}}{\text{true value}} \times 100\% \qquad \textbf{14}$$

Example: A laboratory weight that is certified to be 500 g is placed on a scale. The reading is 507 g. Assuming the weight to be marked correctly, what are the absolute, relative, and percent errors in the scale reading?

Solution:

$$\text{absolute error} = 507 - 500 = 7 \text{ g}$$

$$\text{relative error} = \frac{7}{500} = 0.014$$

$$\text{percent error} = 0.014 \times 100\% = 1.4\% \text{ high}$$

Round-off Errors: Errors are caused in many calculations simply because we are not able or willing to carry all the significant digits in a number. When using a computer, we are limited by the word length of the particular machine, and when calculating by hand we often have to round numbers to a limited number of digits.

Examples:

(a) The number π has an unlimited number of digits and must be rounded for use in a computation.

(b) When the fraction $\frac{2}{3}$ is converted to decimal, it must be rounded to a fixed number of digits, causing round-off error.

(c) As we saw in Sec. 2-4, every computer has a fixed word length, and any mantissa longer than the available space (which was 23 bits in our examples) is *truncated* to fit.

Even though the number is truncated, it is still referred to as round-off error. The term truncation *error is used for another type of error.*

Ordinary division involving exact numbers can sometimes produce round-off errors.

Example: When we divide a circle, which contains exactly 360°, into exactly seven equal parts, each part will contain

$$\frac{360°}{7} = 51.42857143 \ldots$$

Since the original numbers are exact, we are entitled to keep *all* the digits in the answer. However, if our computer retains say, seven digits, we get 51.42857, with some loss of accuracy.

Adding or Subtracting a Large and a Small Number: When adding a large number to a small number using a computer, it is easy to lose the smaller number.

Example: If we add

$$120{,}523.80 + 0.04$$

we get 120,523.84. Assuming that our computer can handle seven decimal digits, this number will be rounded to

$$120{,}523.8$$

where the last digit, 4, has been lost. But our rule for addition of approximate numbers in Sec. 1-2 says that we are entitled to *keep* that digit. Thus, a small error has been introduced.

Errors Caused by Subtracting Nearly Equal Numbers: When subtracting nearly equal numbers, the number of significant digits can sometimes be drastically reduced.

Example: In subtracting 83754 from 83756, both of which have five significant digits,

$$\begin{array}{r} 83756 \\ -\ \underline{83754} \\ 2 \end{array}$$

we get a result with only one significant digit. This type of error arises when computing manually as well as by computer.

Most computers will give an error message whenever overflow or underflow occurs.

Overflow and Underflow: Sometimes an operation may result in a number that is too large or too small for the computer to handle, called *overflow* and *underflow,* respectively.

Example: If we assume that we have a computer that can only handle numbers up to 10^{38}, and try to multiply

$$2.35 \times 10^{30} \times 1.83 \times 10^{20}$$

we get 4.30×10^{50}, which is beyond the range of the computer.

Errors Caused by Converting to and from Binary: As we mentioned when studying binary fractions, not every decimal fraction can be converted exactly to binary. Usually, the error is too small to worry about, but it can be a problem when asking the computer to compare two numbers.

This is sometimes called a quantizing error.

Example: Consider the BASIC statements

```
10 LET A = 0
20 LET A = A + .1
30 IF A = 10 THEN STOP
40 GO TO 20
```

Now A should eventually become equal to 10, at which time the instruction in line 30 would stop the computation. But because of the error caused when 0.1 is converted to binary, A *will never be exactly equal to 10,* and the computation *will not stop.*

General Rules for Error Reduction: When any of the errors described above occur, they will affect the accuracy of any subsequent steps in the computation. The manner in which errors are propagated throughout a computation, growing or shrinking along the way, is a complicated subject that we cannot cover here. Fortunately, for practical computations in engineering technology where three or four significant digits are usually enough, the small errors discussed in this section can be safely ignored. For more accurate work, the following suggestions will help to reduce errors.

1. When adding a string of numbers that differ greatly in size, *add*

the smaller numbers first, and then combine their sum with the larger numbers.

2. Try to arrange the computation so as to *avoid subtracting two nearly equal numbers*. If this is not possible, try to do the subtraction *last*.
3. Since each step in a calculation may introduce an error, try to *simplify the calculation* so that the number of steps is reduced. The chapters that follow will show how to simplify complicated algebraic and trigonometric expressions.
4. Often we are free to choose the numbers we use in a computation, as when setting up an experiment. If so, *avoid those decimal fractions that will not convert exactly to binary*.
5. Do not round your more accurate data down to the number of digits contained in your least accurate data, but *retain as many digits as are reliably known*. When entering constants such as π or e or accurate data from published tables which are known to many significant digits, retain at least *one more* digit than in your most accurate data. Let the computer carry the full number of digits and round only when you *print* your answers. Do not hesitate to round while printing, to avoid reporting an exaggerated number of significant digits.
6. Use *double precision* (Sec. 2-4) when great accuracy is needed.

EXERCISE 4

1. A certain quantity is measured at 125 units but is known actually to be 128 units. Find the percent error in the measurement.

2. A shaft is known to have a diameter of 35.000 mm. You measure it and get a reading of 34.725 mm. What is the percent error of your reading?

3. A certain capacitor has a working voltage of 125 V dc -10%, $+150\%$. Between what two voltages would the actual working voltage lie?

4. A resistor is labeled as 5500 Ω with a tolerance of $\pm5\%$. Between what two values is the actual resistance expected to lie?

5. A voltmeter is checked by measuring the EMF of a standard cell, known to be 1.0186 V, and the reading obtained is 1.0197 V. What is the percent error in the reading?

We will see from this problem that when quantities are multiplied, their percent errors add.

6. If the measurement of the length of a rectangle is too high by 1%, and the width measurement is too high by 2%, by what percent will the calculated area be if these inaccurate measurements are used?

CHAPTER TEST

1. Add the hex numbers 5DA2 and A4B.
2. Convert the binary number 1100.0111 to hex.
3. Convert the hex number 2D to BCD.
4. Subtract the hex number 3A5 from F24.
5. Convert the hex number 2B4 to octal.
6. Convert the binary number 1100111.011 to octal.
7. A bar, known to be 2.0000 in. in diameter, is measured at 2.0064 in. Find the percent error in the measurement.
8. Convert the hexadecimal number 5E.A3 to binary.
9. Multiply the hex numbers B37 and 2D6.
10. Convert the BCD number 1001 to hex.
11. Convert the octal number 276 to hex.
12. Add the 8421 BCD numbers 011101010111 and 100001110110.
13. Convert the decimal number 8362 to hex.
14. Divide the hex number F8 by hex 29.
15. Convert the BCD number 1001 0010 to octal.
16. Convert the decimal number 482 to octal.
17. Convert the octal number 47135 to binary.
18. Convert the hexadecimal number 3D2.A4F to decimal.
19. Convert the BCD number 0111 0100 to straight binary.
20. Convert the decimal number 382.4 to 8421 BCD.
21. Convert the straight binary number 1010 0010 to BCD.
22. Convert the 8421 BCD number 1001 0100 0011 to decimal.
23. Convert the octal number 2453 to BCD.
24. Convert the octal number 534 to decimal.

Introduction to Algebra

Algebra is a generalization of arithmetic. For example, the statement

$$2 \cdot 2 \cdot 2 \cdot 2 = 2^4$$

can be generalized to

$$a \cdot a \cdot a \cdot a = a^4$$

where a can be *any* number, not just 2. We can go further and say that

$$\underbrace{a \cdot a \cdot a \cdot a \ldots a}_{n \text{ factors}} = a^n$$

where a can be any number, as before, and n can be *any* positive integer, not just 4.

We learn a lot of new words in this chapter. Every field has its special terms, and algebra is no exception. But since algebra is generalized arithmetic, some of what was said in Chapter 1 (such as rules of signs) will be repeated here.

We will redo the basic operations of addition, subtraction, and so on, but now with symbols rather than numbers. Learn this material well, for it is the foundation on which later chapters rest.

4-1. ALGEBRAIC EXPRESSIONS

Mathematical Expressions: A mathematical *expression* is a grouping of mathematical symbols, such as signs of operation, numbers, and letters.

Examples:

(a) $x^2 - 2x + 3$

(b) $4 \sin 3x$

(c) $5 \log x + e^{2x}$

are mathematical expressions.

Algebraic Expressions: An *algebraic expression* is one containing only algebraic symbols and operations (addition, subtraction, multiplication, division, roots, and powers), such as in example (a) above. All other expressions are called *transcendental,* such as examples (b) and (c).

We mentioned equations briefly in Chapter 1 when we substituted into equations. We treat them in detail in Chapter 4.

Equations: None of the expressions above contains an equal sign (=). When two expressions are set equal to each other, we get an *equation.*

Examples:

(a) $2x^2 + 3x - 5 = 0$

(b) $6x - 4 = x + 1$

(c) $y = 3x - 5$

(d) $3 \sin x = 2 \cos x$

are equations.

Constants: *Constants* are quantities that do not change in value in a particular problem. They are usually numbers, such as 8, 4.67, or π, but are often represented by letters. Such letters are usually chosen from the beginning of the alphabet (a, b, c, etc.).

Example: The constants in the expression

$$3x^2 + 4x + 5$$

are 3, 4, and 5.

Example: The constants in the expression

$$ax^2 + bx + c$$

are a, b, and c.

An expression in which the constants are represented by letters, as in the preceding example, is called a *literal* expression.

Variables: The quantities that may change during a problem are called *variables*. They are usually represented by letters from the end of the alphabet (x, y, z, etc.).

Example: In the expression

$$ax + by + cz$$

the letters a, b, and c would usually represent constants, and x, y, and z would represent variables.

We will show how to remove symbols of grouping in Sec. 4-2. We will also see in Sec. 4-4 that symbols of grouping are used to indicate multiplication.

Symbols of Grouping: Mathematical expressions will often contain parentheses ( ), brackets [ ], and braces { }. These are used to group parts of the expression together, and affect the meaning of the expression.

Example:

$$\{[(3 + x^2) - 2] - (2x + 3)\}$$

shows the use of symbols of grouping.

Example: The value of the expression

$$2 + 3(4 + 5)$$

is *different* from

$$(2 + 3)4 + 5$$

The placement of symbols of grouping *is* important.

Terms: The plus and minus signs divide an expression into *terms*.

Example: The expression

$$4x^2 + 3x + 5$$

$4x^2$	$3x$	5
first term	second term	third term

has *three* terms.

An exception is when the plus or minus sign is *within* a symbol of grouping.

Example: The expression

$$\underbrace{[(x + 2x^2) + 3]}_{\text{first term}} - \underbrace{(2x + 2)}_{\text{second term}}$$

has *two* terms.

Factors: Any divisor of a term is called a *factor* of that term.

Example: Some factors of $3axy^2$ are 3, a, x, y, and y.

The number 1, as well as the entire expression $3axy^2$, are also factors. They are not usually stated. See Chapter 7 for a more thorough discussion of factors.

Example: The expression

$$2x + 3yz$$

has two *terms*, $2x$ and $3yz$. The first term has the *factors* 2 and x, and the second term has the factors 3, y, and z.

Coefficient: The *coefficient* of a term is (usually) the constant part of the term.

Example: The coefficient in the term $3axy^2$ is $3a$. We say that $3a$ is the coefficient of xy^2. *But* it is also possible to say that

$3ax$ is the coefficient of y^2

$3ay^2$ is the coefficient of x

$3xy^2$ is the coefficient of a

or even

axy^2 is the coefficient of 3

But when we use the word "coefficient" in this book, it will almost always refer to the constant part of the term ($3a$ in the preceding example), also called the *numerical* coefficient. If there is no numerical coefficient written before a term, it is understood to be 1.

Example: In the expression $\frac{w}{2} + x - y - 3z$

the coefficient of w is $\frac{1}{2}$

the coefficient of x is 1

the coefficient of y is -1

the coefficient of z is -3

Do not forget to include the minus sign with the coefficient.

Degree: The *degree* of a term refers to the power to which the variable is raised.

Examples:

(a) $2x$ is a first-degree term.

(b) $3x^2$ is a second-degree term.

(c) $5y^9$ is a ninth-degree term.

The term degree *is used only when the exponents are* positive integers. *You would not say that $x^{1/2}$ is a "half-degree" term.*

If there is more than one variable, their powers are added to obtain the degree of the term.

Examples:

(a) $2x^2y^3$ is of fifth degree.

(b) $3xyz^2$ is of fourth degree.

The degree of an *expression* is the same as that of the term having the highest degree.

Example: $3x^2 - 2x + 4$ is a second-degree expression.

4-2. ADDITION AND SUBTRACTION OF ALGEBRAIC EXPRESSIONS

Like Terms: *Like terms* are those that differ only in their coefficients.

Example: $2wx$ and $-3wx$ are like terms.

This process is also referred to as collecting terms.

Algebraic expressions are added and subtracted by *combining like terms*. Like terms are added by adding their coefficients.

Examples:

(a) $7x + 5x = 12x$

(b) $8w + 2w - 4w - w = 5w$

(c) $9y - 3y = 6y$

(d) $2x - 3y - 5x + 2y = -3x - y$

Commutative Law of Addition: We mentioned the commutative law (Eq. 1) when adding numbers in Chapter 1. All it means is that the order of addition does not affect the sum $(2 + 3 = 3 + 2)$. The same law applies, of course, when we have letters instead of numbers.

Examples:

(a) $$a + b = b + a$$

(b) $$\begin{aligned} 2x + 3x &= 3x + 2x \\ &= 5x \end{aligned}$$

This law enables us to *rearrange the terms* of an expression for our own convenience.

Example: Simplify the expression

$$4y - 2x + 3z - 7z + 4x - 6y$$

Solution: Using Eq. 1, we rearrange the expression so as to get like terms together.

$$-2x + 4x + 4y - 6y + 3z - 7z$$

Collecting terms, we obtain

$$2x - 2y - 4z$$

With practice, you will soon be able to omit the first step, and collect terms by inspecting the original expression.

Removal of Parentheses: If the parentheses (or other symbol of grouping) is preceded only by a + sign (or no sign), the parentheses may be removed.

Examples:

(a) $$(a + b - c) = a + b - c$$

(b) $$(x + y) + (w - z) = x + y + w - z$$

When the parentheses are preceded only by a (−) sign, change the sign of every term within the parentheses before removing them.

Examples:

(a) $$-(a + b - c) = -a - b + c$$

(b) $$-(x + y) - (x - y) = -x - y - x + y = -2x$$

(c) $$\begin{aligned} &(2w - 3x - 2y) - (w - 4x + 5y) - (3w - 2x - y) \\ &= 2w - 3x - 2y - w + 4x - 5y - 3w + 2x + y \\ &= 2w - w - 3w - 3x + 4x + 2x - 2y - 5y + y \\ &= -2w + 3x - 6y \end{aligned}$$

When there are groups within groups, start simplifying the innermost groups and work outward.

Example:

$$\begin{aligned} &\{3x - [2y + 5z - (y + 2)] + 3\} - 7z \\ &= \{3x - [2y + 5z - y - 2] + 3\} - 7z \\ &= \{3x - [y + 5z - 2] + 3\} - 7z \\ &= \{3x - y - 5z + 2 + 3\} - 7z \\ &= 3x - y - 5z + 5 - 7z \\ &= 3x - y - 12z + 5 \end{aligned}$$

Good Working Habits: Start now to develop careful working habits. Why make things even harder for yourself by scribbling? Form the symbols with care; work in sequence, from top of page to bottom; do not crowd; use a sharp pencil, not a pen; erase mistakes instead of crossing them out. What chance do you have if you cannot read your own work?

Common Error	Do not switch between capitals and lowercase letters without reason. Although b and B are the same alphabetic letter, in a math problem they could stand for entirely different quantities.

EXERCISE 1

Definitions

How many terms are there in each expression?

1. $x^2 - 3x$

2. $7y - (y^2 + 5)$

3. $(x + 2)(x - 1)$

4. $3(x - 5) + 2(x + 1)$

Write the coefficients of each term. Assume that letters from the beginning of the alphabet ($a, b, c, \ldots$) are constants.

5. $5x^3$

6. $2ay^2$

7. $\dfrac{bx^3}{4}$

8. $\frac{1}{4}(x - 5)(b)$

9. $4y^3(a + 4)$

10. $b\left(\dfrac{x - 3}{a}\right)$

Addition and Subtraction

Combine as indicated, and simplify.

11. $7x + 5x$

12. $2x + 5x - 4x + x$

13. $6ab - 7ab - 9ab$

14. $9.4x - 3.7x + 1.4x$

15. Add $7a - 3b + m$ and $3b - 7a - c + m$.

16. What is the sum of $6ab + 12bc - 8cd$, $3cd - 7cd - 9bc$, and $12cd - 2ab - 5bc$?

17. What is the sum of $3a + b - 10$, $c - d - a$, and $-4c + 2a - 3b - 7$?

18. Add $7m + 3n - 11p$, $3a - 9n - 11m$, $8n - 4m + 5p$, and $6n - m + 3p$.

19. Add $8ax + 2(x + a) + 3b$, $9ax + 6(x + a) - 9b$, and $11x + 6b - 7ax - 8(x + a)$.

20. Add $a - 9 - 8x^2 + 16a^3$, $5 + 15a^3 - 12a - 2a^2$, and $6a^2 - 10a^3 + 11a - 13$.

21. Subtract $41x^3 - 2x^2 + 13$ from $15x^3 + x - 18$.

22. Subtract $3b - 6d - 10c + 7a$ from $4d + 12a - 13c - 9b$.

23. Add $x - y - z$ and $y - x + z$.

24. What is the sum of $a + 2b - 3c - 10$, $3b - 4a + 5c + 10$, and $5b - c$?

25. Add $72ax^4 - 8ay^3$, $-38ax^4 - 3ay^4 + 7ay^3$, $8 + 12ay^4$, $-6ay^3 + 12$, and $-34ax^4 + 5ay^3 - 9ay^4$.

26. Add $2a(x - y^2) - 3mz^2$, $4a(x - y^2) - 5mz^2$, and $5a(x - y^2) + 7mz^2$.

27. What is the sum of $9b^2 - 3ac + d$, $4b^2 + 7d - 4ac$, $3d - 4b^2 + 6ac$, $5b^2 - 2ac - 12d$, and $4b^2 - d$?

28. From the sum of $2a + 3b - 4c$ and $3b + 4c - 5d$, subtract the sum of $5c - 6d - 7a$ and $-7d + 8a + 9b$.

29. What is the sum of $7ab - m^2 + q$, $-4ab - 5m^3 - 3q$, $12ab + 14m^2 - z$, and $-6m^2 - 2q$?

30. Add $14(x + y) - 17(y + z)$, $4(y + z) - 19(z + x)$, and $-7(z + x) - 3(x + y)$.

31. Add $3.52(a + b)$, $4.15(a + b)$, and $-1.84(a + b)$.

Symbols of Grouping

Remove symbols of grouping and collect terms.

32. $(3x + 5) + (2x - 3)$

33. $3a^2 - (3a - x + b)$

34. $40xy - (30xy - 2b^2 + 3c - 4d)$

35. $(6.4 - 1.8x) - (7.5 + 2.6x)$

36. $7m^2 + 2bc - (3m^2 - bc - x)$

37. $a^2 - a - (4a - y - 3a^2 - 1)$

38. $(2x^4 + 3x^3 - 4) + (3x^4 - 2x^3 - 8)$

39. $a + b - m - (m - a - b)$

40. $(3xy - 2y + 3z) + (-3y + 8z) - (-3xy - z)$

41. $3m - z - y - (2z - y - 3m)$

42. $(2x - 3y) - (5x - 3y - 2z)$

43. $(w + 2z) - (-3w - 5x) + (x + z) - (z - w)$

44. $2a + \{-6b - [3c + (-4b - 6c + a)]\}$

45. $4a - \{a - [-7a - [8a - (5a + 3)] - (-6a - 2a - 9)]\}$

46. $9m - \{3n + [4m - (n - 6m)] - [m + 7n]\}$

4-3. INTEGRAL EXPONENTS

Definitions: In this section we deal only with expressions that have integers (positive or negative, and zero) as exponents.

Example: We study such expressions as

$$x^3 \qquad a^{-2} \qquad y^0 \qquad x^2y^3w^{-1} \qquad (m+n)^5$$

A positive *exponent* shows how many times the *base* is to be multiplied by itself.

Example: In the expression 3^4, 3 is the base and 4 is the exponent.

$$\text{base} \longrightarrow 3^4 \longleftarrow \text{exponent}$$

Its meaning is

$$3^4 = 3(3)(3)(3) = 81$$

In general,

Positive Integral Exponent	a^n means $\underbrace{a \cdot a \cdot a \cdot a \cdot \ldots \cdot a}_{n \text{ factors}}$	**28**

Common Error	An exponent applies *only* to the symbol directly before it. Thus, $2x^2 \neq 2^2x^2$ but $(2x)^2 = 2^2x^2 = 4x^2$

Multiplying Powers: Let us multiply the quantity x^2 by x^3. From Eq. 28, we know that

$$x^2 = x \cdot x$$

and that

$$x^3 = x \cdot x \cdot x$$

Multiplying, we obtain

$$\begin{aligned} x^2 \cdot x^3 &= (x \cdot x)(x \cdot x \cdot x) \\ &= x \cdot x \cdot x \cdot x \cdot x \\ &= x^5 \end{aligned}$$

Notice that the exponent in the result is the *sum of the two original exponents*.

$$x^2 \cdot x^3 = x^{2+3} = x^5$$

This will always be so. We summarize this rule as our first *law of exponents*.

Products	$x^a \cdot x^b = x^{a+b}$	**29**

When multiplying powers, ***add the exponents.***

Example:

$$x^4(x^3) = x^{4+3} = x^7$$

The "invisible 1" appears again. We saw it before as the unwritten coefficient of every term, and now as the unwritten exponent. It is also in the denominator.

$$x = \frac{1x^1}{1}$$

Do not forget about those invisible 1's. We use them all the time.

Every quantity has an exponent of 1, even though it is not usually written.

Examples:

(a) $x(x^3) = x^{1+3} = x^4$

(b) $a^2(a^4)(a) = a^{2+4+1} = a^7$

(c) $x^a(x)(x^b) = x^{a+1+b}$

Quotients: Let us divide x^5 by x^3. By Eq. 28,

$$\frac{x^5}{x^3} = \frac{x \cdot x \cdot x \cdot x \cdot x}{x \cdot x \cdot x} = \frac{x \cdot x \cdot x}{x \cdot x \cdot x} \cdot x \cdot x$$
$$= x \cdot x = x^2$$

The same result could have been obtained by *subtracting exponents,*

$$\frac{x^5}{x^3} = x^{5-3} = x^2$$

This always works, and we state it as another law of exponents:

Quotients	$\dfrac{x^a}{x^b} = x^{a-b} \qquad (x \neq 0)$	**30**

When dividing powers, ***subtract the exponents.***

Examples:

(a) $$\frac{a^4b^3}{ab^2} = a^{4-1}b^{3-2} = a^3b$$

(b) $$\frac{y^{n-m}}{y^{n+m}} = y^{n-m-(n+m)} = y^{-2m}$$

Power Raised to a Power: Let us take a quantity raised to a power, say, x^2, and raise the entire expression to *another* power, say, 3.

$$(x^2)^3$$

By Eq. 28,

$$(x^2)^3 = (x^2)(x^2)(x^2)$$
$$= (x \cdot x)(x \cdot x)(x \cdot x)$$
$$= x \cdot x \cdot x \cdot x \cdot x \cdot x = x^6$$

a result that could have been obtained by *multiplying the exponents.*

$$(x^2)^3 = x^{2(3)} = x^6$$

In general,

Powers	$(x^a)^b = x^{ab}$	**31**

When raising a power to a power, ***multiply the exponents.***

Examples:

(a) $$(w^5)^2 = w^{5(2)} = w^{10}$$

(b) $$(a^{-3})^2 = a^{(-3)(2)} = a^{-6}$$

(c) $$(10^4)^3 = 10^{4(3)} = 10^{12}$$

(d) $$(b^{x+2})^3 = b^{3x+6}$$

Product Raised to a Power: We now raise a product, such as xy, to some power, say 3.

$$(xy)^3$$

By Eq. 28,

$$\begin{aligned}(xy)^3 &= (xy)(xy)(xy)\\ &= x \cdot y \cdot x \cdot y \cdot x \cdot y\\ &= x \cdot x \cdot x \cdot y \cdot y \cdot y\\ &= x^3y^3\end{aligned}$$

In general,

Product Raised to a Power	$(xy)^n = x^ny^n$	**32**

When a product is raised to a power, each factor may be ***separately*** *raised to the power.*

Examples:

(a) $$(xyz)^5 = x^5y^5z^5$$

(b) $$(2x)^3 = 2^3x^3 = 8x^3$$

(c) $$\begin{aligned}(3.75 \times 10^3)^2 &= (3.75)^2 \times (10^3)^2\\ &= 14.1 \times 10^6\end{aligned}$$

(d) $$\begin{aligned}(3x^2y^n)^3 &= 3^3(x^2)^3(y^n)^3\\ &= 27x^6y^{3n}\end{aligned}$$

(e) $$(-x^2y)^3 = (-1)^3(x^2)^3y^3 = -x^6y^3$$

A good way to test a "rule" that you are not sure of is to try it with numbers. *In this case, does* $(2 + 3)^2$ *equal* $2^2 + 3^2$*? Evaluating each expression, we obtain*

$$(5)^2 \stackrel{?}{=} 4 + 9$$

$$25 \neq 13$$

Common Error	There is *no* similar rule for the *sum* of two quantities raised to a power. $$(x + y)^n \neq x^n + y^n$$

Quotient Raised to a Power: Using the same steps as in the preceding section, see if you can show that

$$\left(\frac{x}{y}\right)^3 = \frac{x^3}{y^3}$$

Or, in general,

Quotient Raised to a Power	$\left(\frac{x}{y}\right)^n = \frac{x^n}{y^n} \quad (y \neq 0)$	**33**

When a quotient is raised to a power, the numerator and denominator may be ***separately*** *raised to the power.*

Examples:

(a) $$\left(\frac{x}{5}\right)^2 = \frac{x^2}{5^2} = \frac{x^2}{25}$$

(b) $$\left(\frac{3a}{2b}\right)^3 = \frac{3^3a^3}{2^3b^3} = \frac{27a^3}{8b^3}$$

(c) $$\left(\frac{2x^2}{5y^3}\right)^3 = \frac{2^3(x^2)^3}{5^3(y^3)^3} = \frac{8x^6}{125y^9}$$

(d) $$\left(-\frac{a}{b^2}\right)^3 = (-1)^3\frac{a^3}{(b^2)^3} = -\frac{a^3}{b^6}$$

Zero Exponent: If we divide x^n by itself we get, by Eq. 30,

$$\frac{x^n}{x^n} = x^{n-n} = x^0$$

But any expression divided by itself equals 1, so

Zero Exponent	$x^0 = 1 \quad (x \neq 0)$	**34**

Any quantity (except 0) raised to the zero power equals 1.

Examples:

(a) $(xyz)^0 = 1$

(b) $3862^0 = 1$

(c) $(x^2 - 2x + 3)^0 = 1$

(d) $5x^0 = 5(1) = 5$

Negative Exponent: We now divide x^0 by x^a. By Eq. 30,

$$\frac{x^0}{x^a} = x^{0-a} = x^{-a}$$

Since $x^0 = 1$, we get

Negative Exponent	$x^{-a} = \dfrac{1}{x^a} \quad (x \neq 0)$	**35**

When taking the reciprocal of a power, ***change the sign of the exponent.***

Examples:

(a) $5^{-1} = \dfrac{1}{5}$

(b) $x^{-4} = \dfrac{1}{x^4}$

(c) $\dfrac{1}{x^{-a}} = x^a$

(d) $\dfrac{1}{xy^{-2}} = \dfrac{y^2}{x}$

(e) $\dfrac{w^{-3}}{z^{-2}} = \dfrac{z^2}{w^3}$

EXERCISE 2

Laws of Exponents

Evaluate each expression.

1. 2^4 **2.** $(-3)^2$ **3.** $(-3)^3$

4. $(|-3|)^3$ **5.** $(0.1)^4$ **6.** $(-2)^5$

Multiply.

7. $a^3 \cdot a^5$ **8.** $x^a \cdot x^2$ **9.** $y^{a+1} \cdot y^{a-3}$

10. $10^4 \cdot 10^3$ **11.** $10^a \cdot 10^b$ **12.** $10^{n+2} \cdot 10^{2n-1}$

Divide. Write your answers without negative exponents.

13. $\dfrac{y^5}{y^2}$ **14.** $\dfrac{5^5}{5^3}$ **15.** $\dfrac{x^{n+2}}{x^{n+1}}$

16. $\dfrac{10^5}{10}$ **17.** $\dfrac{10^{x+5}}{10^{x+3}}$ **18.** $\dfrac{10^2}{10^{-3}}$

19. $\dfrac{x^{-2}}{x^{-3}}$ **20.** $\dfrac{a^{-5}}{a}$

Simplify.

21. $(x^3)^4$ **22.** $(9^2)^3$ **23.** $(a^x)^y$

24. $(x^{-2})^{-2}$ **25.** $(x^{a+1})^2$

Raise to the power indicated.

26. $(xy)^2$ **27.** $(2x)^3$ **28.** $(3x^2y^3)^2$

29. $(3abc)^3$ **30.** $\left(\frac{3}{5}\right)^2$ **31.** $\left(-\frac{1}{3}\right)^3$

32. $\left(\frac{x}{y}\right)^5$ **33.** $\left(\frac{2a}{3b^2}\right)^3$ **34.** $\left(\frac{3x^2y}{4wz^3}\right)^2$

Write each expression with positive exponents only.

35. a^{-2} **36.** $(-x)^{-3}$ **37.** $\left(\frac{3}{y}\right)^{-3}$

38. $a^{-2}bc^{-3}$ **39.** $\left(\frac{2a}{3b^3}\right)^{-2}$ **40.** xy^{-4}

41. $2x^{-2} + 3y^{-3}$ **42.** $\left(\frac{x}{y}\right)^{-1}$

Express without fractions, using negative exponents where needed.

43. $\frac{1}{x}$ **44.** $\frac{3}{y^2}$ **45.** $\frac{x^2}{y^2}$

46. $\frac{x^2y^{-3}}{z^{-2}}$ **47.** $\frac{a^{-3}}{b^2}$ **48.** $\frac{x^{-2}y^{-3}}{w^{-1}z^{-4}}$

Evaluate.

49. $(a + b + c)^0$ **50.** $8x^0y^2$ **51.** $\frac{a^0}{9}$

52. $\frac{y}{x^0}$ **53.** $\frac{x^{2n} \cdot x^3}{x^{3+2n}}$ **54.** $5\left(\frac{x}{y}\right)^0$

4-4. MULTIPLICATION OF ALGEBRAIC EXPRESSIONS

Symbols and Definitions: Multiplication is indicated in several ways: by the usual $\times$ symbol, by a dot, or by parentheses, brackets, or braces. Thus, the product of b and d could be written

$$b \cdot d \qquad b \times d \qquad b(d) \qquad (b)d \qquad (b)(d)$$

Avoid using the $\times$ symbol because it could get confused with the letter x.

Most common of all is to use no symbol at all. The product of b and d would usually be written bd.

We get a *product* when we multiply two or more *factors*.

$$(\text{factor})(\text{factor})(\text{factor}) = \text{product}$$

Rules of Signs: The product of two factors having the *same* sign is *positive;* the product of two factors having *different* signs is *negative.* If a and b are positive quantities, we have

Rules of Signs		
	$(+a)(+b) = (-a)(-b) = +ab$	**8**
	$(+a)(-b) = (-a)(+b) = -(+a)(+b) = -ab$	**9**

When multiplying more than two factors, the product of every *pair* of negative terms will be positive. Thus, if there is an even number of negative factors, the final result will be positive: if there is an odd number of negative factors, the final result will be negative.

Examples:

(a) $$x(-y)(-z) = xyz$$

(b) $$(-a)(-b)(-c) = -abc$$

(c) $$(-p)(-q)(-r)(-s) = pqrs$$

Commutative Law: As we saw when multiplying numbers, the *order* of multiplication does not make any difference. In symbols,

Commutative Law	$ab = ba$	**2**

Example: Multiply $3y$ by $-2x$.

Solution:

$$(3y)(-2x) = 3(y)(-2)(x)$$

By Eq. 2,

$$= 3(-2)(y)(x)$$

$$= -6yx$$

or

$$= -6xy$$

since it is common practice to write the letters in alphabetical order.

Multiplying Monomials: A *monomial* is an algebraic expression having *one term.*

Examples:

(a) $2x^3$ (b) $\dfrac{3wxy}{4}$ (c) $(2b)^3$

To multiply monomials, we make use of Eqs. 2, 4, and 29, and the rule of signs, Eqs. 8 and 9.

Example:

$$\begin{aligned}(4a^3b)(-3ab^2) &= (4)(-3)(a^3)(a)(b)(b^2)\\ &= -12a^{3+1}b^{1+2}\\ &= -12a^4b^3\end{aligned}$$

With some practice you should be able to omit the middle step in many of these solutions. But do not rush it. Use as many steps as you need to get it right.

Example: Multiply $5x^2$, $-3xy$, and $-xy^3z$.

Solution: By Eq. 2,

$$5x^2(-3xy)(-xy^3z) = 5(-3)(-1)(x^2)(x)(x)(y)(y^3)(z)$$

By Eq. 29,

$$= 15x^{2+1+1}y^{1+3}z$$
$$= 15x^4y^4z$$

Example: Multiply $3x^2$ by $2ax^n$.

Solution: By Eq. 2,

$$3x^2(2ax^n) = 3(2a)(x^2)(x^n)$$

By Eq. 29,

$$= 6ax^{2+n}$$

Multiplying a Multinomial by a Monomial: A *multinomial* is an algebraic expression having *more than one term.*

Examples: Some multinomials are

(a) $3x + 5$ (b) $2x^3 - 3x^2 + 7$ (c) $\frac{1}{x} + \sqrt{2x}$

A *polynomial* is a monomial or multinomial in which the powers to which the unknown is raised are all *positive integers*. The first two expressions in the preceding example are polynomials, but the third is not.

A *binomial* is a polynomial with *two* terms, and a *trinomial* is a polynomial having *three* terms. In the preceding example, the first expression is a binomial and the second is a trinomial.

To multiply a monomial and a multinomial, we use the *distributive law* (Eq. 5):

In Sec. 7-1 we use this law to remove common factors.

Distributive Law	$a(b + c) = ab + ac$	**5**

The product of a monomial and a polynomial is equal to the sum of the products of the monomial and each term of the polynomial.

Example:

$$2x(x - 3x^2) = 2x(x) + 2x(-3x^2)$$
$$= 2x^2 - 6x^3$$

The distributive law, although written for a monomial times a binomial, can be extended for a multinomial having any number of terms. Simply *multiply every term in the multinomial by the monomial.*

Examples:

(a) $-3x^3(3x^2 - 2x + 4) = -3x^3(3x^2) + (-3x^3)(-2x) + (-3x^3)(4)$
$= -9x^5 + 6x^4 - 12x^3$

(b) $4xy(x^2 - 3xy + 2y^2) = 4x^3y - 12x^2y^2 + 8xy^3$

Removing Symbols of Grouping: Symbols of grouping, in addition to indicating that the enclosed expression is to be taken as a whole, also indicate multiplication.

Examples:

(a) $x(y)$ is the product of x and y.
(b) $x(y + 2)$ is the product of x and $(y + 2)$.
(c) $(x - 1)(y + 2)$ is the product of $(x - 1)$ and $(y + 2)$.
(d) $-(y + 2)$ is the product of -1 and $(y + 2)$.

The parentheses may be removed after the multiplication has been performed.

Examples:

(a) $x(y + 2) = xy + 2x$

(b) $5 + 3(x - 2) = 5 + 3x - 6 = 3x - 1$

Common Error	Do not forget to multiply *every* term within the grouping by the preceding factor. $-(2x + 5) \neq -2x + 5$
Common Error	Multiply the terms within the parentheses *only* by the factor directly preceding it. $x + 2(a + b) \neq (x + 2)(a + b)$

If there are groupings within groupings, start simplifying with the *innermost* groupings.

Examples:

(a)
$$x - 3[2 - 4(y + 1)] = x - 3[2 - 4y - 4]$$
$$= x - 3[-2 - 4y]$$
$$= x + 6 + 12y$$

(b) $3\{2[4(w - 4) - (x + 3)] - 2\} - 6x$

$$= 3\{2[4w - 16 - x - 3] - 2\} - 6x$$
$$= 3\{2[4w - x - 19] - 2\} - 6x$$
$$= 3\{8w - 2x - 38 - 2\} - 6x$$
$$= 3\{8w - 2x - 40\} - 6x$$
$$= 24w - 6x - 120 - 6x$$
$$= 24w - 120 - 12x$$

Multiplying a Multinomial by a Multinomial: Multiply every term in one multinomial by every term in the other, and combine like terms.

Example:

$$(x + 4)(x - 3) = x(x) + x(-3) + 4(x) + 4(-3)$$
$$= x^2 - 3x + 4x - 12 = x^2 + x - 12$$

Some prefer a vertical arrangement for problems like the last one.

Example:

$$\begin{array}{r} x + 4 \\ \underline{x - 3} \\ -3x - 12 \\ \underline{x^2 + 4x } \\ x^2 + x - 12 \end{array} \qquad \text{as before}$$

Examples:

(a) $(x - 3)(x^2 - 2x + 1)$

$$= x(x^2) + x(-2x) + x(1) - 3(x^2) - 3(-2x) - 3(1)$$
$$= x^3 - 2x^2 + x - 3x^2 + 6x - 3$$
$$= x^3 - 5x^2 + 7x - 3$$

(b) $(2x + 3y - 4z)(x - 2y - 3z)$

$$= 2x^2 - 4xy - 6xz + 3xy - 6y^2 - 9yz - 4xz + 8yz + 12z^2$$
$$= 2x^2 - 6y^2 + 12z^2 - xy - 10xz - yz$$

Product of the Sum and Difference of Two Terms: Let us multiply the binomials $(a - b)$ and $(a + b)$. We get

$$(a - b)(a + b) = a^2 + ab - ab - b^2$$

When we multiply two binomials, we get expressions that occur frequently enough in our later work for us to take special note of here. These are called special products. *We will need them especially when we study factoring.*

The two middle terms cancel, leaving

Difference of Two Squares	$(a - b)(a + b) = a^2 - b^2$	**41**

Thus, the product is the *difference of two squares:* the squares of the original numbers.

Example:

$$(x + 2)(x - 2) = x^2 - 2^2 = x^2 - 4$$

Product of Two Binomials Having the Same First Term: When we multiply $(x + a)$ and $(x + b)$, we get

$$(x + a)(x + b) = x^2 + ax + bx + ab$$

Combining like terms, we obtain

Trinomial, Leading Coefficient = 1	$(x + a)(x + b) = x^2 + (a + b)x + ab$	**45**

Example:

$$\begin{aligned}(w + 3)(w - 4) &= w^2 + (3 - 4)w + 3(-4) \\ &= w^2 - w - 12\end{aligned}$$

Some people like the **FOIL** *rule. Multiply the* **F***irst terms, then* **O***uter,* **I***nner, and* **L***ast terms.*

General Quadratic Trinomial: When we multiply two binomials of the form $ax + b$ and $cx + d$, we get

$$(ax + b)(cx + d) = acx^2 + adx + bcx + bd$$

Combining like terms, we obtain

General Quadratic Trinomial	$(ax + b)(cx + d) = acx^2 + (ad + bc)x + bd$	**46**

Example:

$$\begin{aligned}(3x - 2)(2x + 5) &= 3(2)x^2 + [(3)(5) + (-2)(2)]x + (-2)(5) \\ &= 6x^2 + 11x - 10\end{aligned}$$

Powers of Multinomials: We see from Eq. 28 that raising an expression to a power is the same as multiplying the expression by itself the proper number of times, provided that the power is a positive integer.

Example:

$$(x - 3)^2 = (x - 3)(x - 3)$$
$$= x^2 - 3x - 3x + 9$$
$$= x^2 - 6x + 9$$

Example:

$$(x - 3)^3 = (x - 3)(x - 3)(x - 3)$$
$$= (x - 3)(x^2 - 6x + 9) \quad \text{(from the last example)}$$
$$= x^3 - 6x^2 + 9x - 3x^2 + 18x - 27$$

Combining like terms,

$$= x^3 - 9x^2 + 27x - 27$$

Perfect Square Trinomial: When we *square a binomial* such as $(a + b)$, we get

$$(a + b)^2 = (a + b)(a + b) = a^2 + ab + ab + b^2$$

Collecting terms, we obtain

Perfect Square Trinomial	$(a + b)^2 = a^2 + 2ab + b^2$	**47**

This trinomial is called a *perfect square trinomial,* because it is the square of a binomial. Its first and last terms are the squares of the two terms of the binomial; its middle term is twice the product of the terms of the binomial. Similarly, when we square $(a - b)$, we get

Perfect Square Trinomial	$(a - b)^2 = a^2 - 2ab + b^2$	**48**

Example:

$$(3x - 2)^2 = (3x)^2 - 2(3x)(2) + 2^2$$
$$= 9x^2 - 12x + 4$$

EXERCISE 3

Multiplication of Algebraic Expressions

Multiply the following monomials, and simplify.

1. $x^3(x^2)$

2. $x^2(3axy)$

3. $(-5ab)(-2a^2b^3)$

4. $(3.52xy)(-4.14xyz)$

5. $(6ab)(2a^2b)(3a^3b^3)$

6. $(-2a)(-3abc)(ac^2)$

7. $2p(-4p^2q)(pq^3)$

8. $(-3xy)(-2w^2x)(-xy^3)(-wy)$

Multiply the polynomial by the monomial, and simplify.

9. $2a(a - 5)$

10. $(x^2 + 2x)x^3$

11. $xy(x - y + xy)$

12. $3a^2b^3(ab - 2ab^2 + b)$

13. $-4pq(3p^2 + 2pq - 3q^2)$

14. $2.8x(1.5x^3 - 3.2x^2 + 5.7x - 4.6)$

15. $3ab(2a^2b^2 - 4a^2b + 3ab^2 + ab)$

16. $-3xy(2xy^3 - 3x^2y + xy^2 - 4xy)$

Multiply the following binomials, and simplify.

17. $(a + b)(a + c)$

18. $(2x + 2)(5x + 3)$

19. $(x + 3y)(x^2 - y)$

20. $(a - 5)(a + 5)$

21. $(m + n)(9m - 9n)$

22. $(m^2 + 2c)(m^2 - 5c)$

23. $(7cd^2 + 4y^3z)(7cd^2 - 4y^3z)$

24. $(3a + 4c)(2a - 5c)$

25. $(x^2 + y^2)(x^2 - y^2)$

26. $(3.5a^2 + 1.2b)(3.8a^2 - 2.4b)$

Multiply the following binomials and trinomials, and simplify.

27. $(a - c - 1)(a + 1)$

28. $(x - y)(2x + 3y - 5)$

29. $(m^2 - 3m - 7)(m - 2)$

30. $(-3x^2 + 3x - 9)(x + 3)$

31. $(x^6 + x^4 + x^2)(x^2 - 1)$

32. $(m - 1)(2m - m^2 + 1)$

33. $(1 - z)(z^2 - z + 2)$

34. $(3a^3 - 2ab - b^2)(2a - 4b)$

35. $(a^2 - ay + y^2)(a + y)$

36. $(y^2 - 7.5y + 2.8)(y + 1.6)$

37. $(a^2 + ay + y^2)(a - y)$

38. $(a^2 + ay - y^2)(a - y)$

Multiply the following binomials and polynomials, and simplify.

39. $[x^2 - (m - n)x - mn](x - p)$

40. $(x^2 + ax - bx - ab)(x + b)$

41. $(m^4 + m^3 + m^2 + m + 1)(m - 1)$

Multiply the following monomials and binomials, and simplify.

42. $(2a + 3x)(2a + 3x)9$

43. $(3a - b)(3a - b)x$

44. $c(m - n)(m + n)$

45. $(x + 1)(x + 1)(x - 2)$

46. $(m - 2.5)(m - 1.3)(m + 3.2)$

47. $(x + a)(x + b)(x - c)$

48. $(x - 4)(x - 5)(x + 4)(x + 5)$

49. $(1 + c)(1 + c)(1 - c)(1 + c^2)$

Multiply the following trinomials, and simplify.

50. $(a + b - c)(a - b + c)$

51. $(2a^3 + 5a)(2a^3 - 5ac^2 + 2a)$

52. $[3(a + b)^2 - (a + b) + 2][4(a + b)^2 - (a + b) - 5]$

53. $(a + b + c)(a - b + c)(a + b - c)$

Multiply the following polynomials, and simplify.

54. $(x^3 - 6x^2 + 12x - 8)(x^2 + 4x + 4)$

55. $(n - 5n^2 + 2 + n^3)(5n + n^2 - 10)$

56. $(mx + my - nx - ny)(mx - my + nx - ny)$

Powers of Algebraic Expressions

Square each binomial.

57. $(a + c)$
58. $(p + q)$
59. $(m - n)$
60. $(x - y)$
61. $(A + B)$
62. $(A - C)$
63. $(3.8a - 2.2x)$
64. $(m + c)$
65. $(2c - 3d)$
66. $(x^2 - x)$
67. $(a - 1)$
68. $(a^2x - ax^3)$
69. $(y^2 - 20)$
70. $(x^n - y^2)$

Square each trinomial.

71. $(a + b + c)$
72. $(a - b - c)$
73. $(1 + x - y)$
74. $(x^2 + 2x - 3)$
75. $(x^2 + xy + y^2)$
76. $(x^3 - 2x - 1)$

Cube each binomial.

77. $(a + b)$
78. $(2x + 1)$
79. $(p - 3q)$
80. $(x^2 - 1)$
81. $(a - b)$
82. $(3m - 2n)$

Symbols of Grouping

Remove symbols of grouping, and simplify.

83. $b - (d - b)$
84. $3 + (x - 2)$
85. $-(M - N) + N$
86. $-(x - 2) - (x + 3)$
87. $-5[-2(3y - z) - z] + y$
88. $[(m + 2n) - (2m - n)][(2m + n) - (m - 2n)]$
89. $(x - 1) - \{[x - (x - 3)] - x\}$
90. $(a - b)(a^3 + b^3)[a(a + b) + b^2]$
91. $y - \{4x - [y - (2y - 9) - x] + 2\}$
92. $[2x^2 + (3x - 1)(4x + 5)][5x^2 - (4x + 3)(x - 2)]$

4-5. DIVISION OF ALGEBRAIC EXPRESSIONS

Symbols for Division: Division may be indicated by any of the symbols

$$x \div y \qquad \frac{x}{y} \qquad x/y$$

The names of the parts are

$$\text{quotient} = \frac{\text{dividend}}{\text{divisor}} = \frac{\text{numerator}}{\text{denominator}}$$

The quantity x/y is also called a *fraction,* and is also spoken of as the *ratio* of x to y.

Reciprocals: As we saw in Sec. 1-4, the *reciprocal* of a number is 1 divided by that number. The reciprocal of n is $1/n$. We can use the idea of a reciprocal to show how division is related to multiplication. We may write the quotient of $x \div y$ as

$$\frac{x}{y} = \frac{x}{1} \cdot \frac{1}{y} = x \cdot \frac{1}{y}$$

We see that to *divide by a number* is the same thing as to *multiply by its reciprocal.* This fact will be especially useful in Sec. 7-8 for dividing by a fraction.

If division by zero were allowed, we could, for example, divide 2 by zero and get a quotient x.

$$\frac{2}{0} = x$$

or $\quad 2 = 0 \cdot x$

But there is no number x which, when multiplied by zero, gives 2, so we cannot allow this operation.

Division by Zero: Division by zero is not a permissible operation.

Example: In the fraction

$$\frac{x + 5}{x - 2}$$

x cannot equal 2, or the illegal operation of division by zero will result.

Rules of Signs: The quotient of two terms of *like* sign is *positive.*

$$\frac{+a}{+b} = \frac{-a}{-b} = \frac{a}{b}$$

The quotient of two terms of *unlike* sign is *negative.*

$$\frac{+a}{-b} = \frac{-a}{+b} = -\frac{a}{b}$$

The fraction itself carries a third sign, which, when negative, reverses the sign of the quotient. These three ideas are summarized in the following rules:

Rule of Signs for Division	$\frac{+a}{+b} = \frac{-a}{-b} = -\frac{-a}{+b} = -\frac{+a}{-b} = \frac{a}{b}$	**10**
	$\frac{+a}{-b} = \frac{-a}{+b} = -\frac{-a}{-b} = -\frac{a}{b}$	**11**

These rules show that any *pair* of negative signs may be removed without changing the value of the fraction.

Example: Simplify $-\frac{ax^2}{-y}$.

Solution: Removing the pair of minus signs, we obtain

$$-\frac{ax^2}{-y} = \frac{ax^2}{y}$$

Dividing a Monomial by a Monomial: Any quantity (except zero) divided by itself equals *one*. So if the same factor appears in both the divisor and the dividend, it may be eliminated. This proccss is called *canceling*.

Example: Divide $6ax$ by $3a$.

Solution:

$$\frac{6ax}{3a} = \frac{6}{3} \cdot \frac{a}{a} \cdot x = 2x$$

To divide quantities having exponents, we make use of Eq. 30, the law of exponents for division:

Quotients	$\frac{x^a}{x^b} = x^{a-b}$	**30**

Example: Divide y^5 by y^3.

Solution: By Eq. 30,

$$\frac{y^5}{y^3} = y^{5-3} = y^2$$

When there are numerical coefficients, divide them separately.

Example: Divide $15x^6$ by $3x^4$.

Solution:

$$\frac{15x^6}{3x^4} = \frac{15}{3} \cdot \frac{x^6}{x^4} = 5x^{6-4} = 5x^2$$

If there is more than one unknown, treat each separately.

Example: Divide $18x^5y^2z^4$ by $3x^2yz^3$.

Solution:

$$\frac{18x^5y^2z^4}{3x^2yz^3} = \frac{18}{3} \cdot \frac{x^5}{x^2} \cdot \frac{y^2}{y} \cdot \frac{z^4}{z^3}$$

$$= 6x^3yz$$

Sometimes negative exponents will be obtained.

Example:

$$\frac{6x^2}{x^5} = 6x^{2-5} = 6x^{-3}$$

The answer may be left in this form, or, as is usually done, use Eq. 35 to eliminate the negative exponent. Thus,

$$6x^{-3} = \frac{6}{x^3}$$

The process of dividing a monomial by a monomial is also referred to as simplifying a fraction, *or* reducing a fraction to lowest terms. *We discuss this in Sec. 7-7.*

Example: Simplify the fraction

$$\frac{3x^2yz^5}{9xy^4z^2}$$

Solution: The procedure is no different than if we had been asked to divide $3x^2yz^5$ by $9xy^4z^2$.

$$\frac{3x^2yz^5}{9xy^4z^2} = \frac{3}{9}x^{2-1}y^{1-4}z^{5-2}$$

$$= \frac{1}{3}xy^{-3}z^3$$

or

$$= \frac{xz^3}{3y^3}$$

Do not be dismayed if the expressions to be divided have negative exponents. Apply Eq. 30, as before.

Example: Divide $21x^2y^{-3}z^{-1}$ by $7x^{-4}y^2z^{-3}$.

Solution: Proceeding as before, we obtain

$$\frac{21x^2y^{-3}z^{-1}}{7x^{-4}y^2z^{-3}} = \frac{21}{7}x^{2-(-4)}y^{-3-2}z^{-1-(-3)}$$
$$= 3x^6y^{-5}z^2$$
$$= \frac{3x^6z^2}{y^5}$$

Dividing a Multinomial by a Monomial: Divide each term of the multinomial by the monomial.

Example: Divide $9x^2 + 3x - 2$ by $3x$.

Solution:

$$\frac{9x^2 + 3x - 2}{3x} = \frac{9x^2}{3x} + \frac{3x}{3x} - \frac{2}{3x}$$

This is really a consequence of the distributive law (Eq. 5).

$$\frac{9x^2 + 3x - 2}{3x} = \frac{1}{3x}(9x^2 + 3x - 2) = \left(\frac{1}{3x}\right)(9x^2) + \left(\frac{1}{3x}\right)(3x) - \left(\frac{1}{3x}\right)(2)$$

Each of these terms is now simplified as in the preceding section.

$$\frac{9x^2}{3x} + \frac{3x}{3x} - \frac{2}{3x} = 3x + 1 - \frac{2}{3x}$$

Common Error	Do not forget to divide *every* term of the multinomial by the monomial. $$\frac{x^2 + 2x + 3}{x} \neq x + 2 + 3$$

Examples:

(a) $$\frac{3ab + 2a^2b - 5ab^2 + 3a^2b^2}{15ab} = \frac{3ab}{15ab} + \frac{2a^2b}{15ab} - \frac{5ab^2}{15ab} + \frac{3a^2b^2}{15ab}$$
$$= \frac{1}{5} + \frac{2a}{15} - \frac{b}{3} + \frac{ab}{5}$$

(b) $$\frac{6x^3y - 8xy^2 - 2x^4y^4}{-2xy} = -3x^2 + 4y + x^3y^3$$

Common Error	There is no similar rule for dividing a monomial by a multinomial. $$\frac{a}{b + c} \neq \frac{a}{b} + \frac{a}{c}$$

Dividing a Polynomial by a Polynomial: Write the divisor and the dividend in the order of descending powers of the variable. Supply any missing terms, using a coefficient of zero. Set up as a long-division problem, as in the following example.

This method is used only for polynomials, *expressions in which the powers are all* positive integers.

Example: Divide $x^2 + 4x^4 - 2$ by $x + 1$.

Solution:

1. Write the dividend in descending order of the powers.

$$4x^4 + x^2 - 2$$

2. Supply the missing terms with coefficients of zero.

$$4x^4 + 0x^3 + x^2 + 0x - 2$$

3. Set up in long-division format.

$$(x + 1)\overline{)\,4x^4 + 0x^3 + x^2 + 0x - 2}$$

4. Divide the first term in the dividend ($4x^4$) by the first term in the divisor (x). The result ($4x^3$) is written above the dividend, in line with the term having the same power. It is the first term of the quotient.

$$\begin{array}{r} 4x^3 \qquad\qquad\qquad \\ (x + 1)\overline{)\,4x^4 + 0x^3 + x^2 + 0x - 2} \end{array}$$

5. Multiply the divisor by the first term of the quotient. Write the result below the dividend. Subtract it from the dividend.

$$\begin{array}{r} 4x^3 \qquad\qquad\qquad \\ (x + 1)\overline{)\,4x^4 + 0x^3 + x^2 + 0x - 2} \\ \underline{4x^4 + 4x^3 \qquad\qquad\qquad} \\ -4x^3 + x^2 + 0x - 2 \end{array}$$

6. Repeat steps 4 and 5, each time using the new dividend obtained, until the degree of the remainder is less than the degree of the divisor.

$$\begin{array}{r} 4x^3 - 4x^2 + 5x - 5 \\ (x + 1)\overline{)\,4x^4 + 0x^3 + x^2 + 0x - 2} \\ \underline{4x^4 + 4x^3 \qquad\qquad\qquad} \\ -4x^3 + x^2 + 0x - 2 \\ \underline{-4x^3 - 4x^2 \qquad\qquad} \\ 5x^2 + 0x - 2 \\ \underline{5x^2 + 5x \qquad} \\ -5x - 2 \\ \underline{-5x - 5} \\ 3 \end{array}$$

The result is written

$$\frac{4x^4 + x^2 - 2}{x + 1} = 4x^3 - 4x^2 + 5x - 5 + \frac{3}{x + 1}$$

or

$$4x^3 - 4x^2 + 5x - 5 \quad \text{(R3)}$$

Common Error	Errors are often made during the subtraction step (step 5 in the preceding example). $$\begin{array}{r} 4x^3 \qquad\qquad\qquad\qquad \\ (x+1)\overline{\big)\, 4x^4 + 0x^3 + x^2 + 0x - 2} \\ \underline{4x^4 + 4x^3 \qquad\qquad\qquad} \\ 4x^3 + x^2 + 0x - 2 \end{array}$$ ↗ no! Should be $-4x^3$

Synthetic Division: The division in the preceding section can be done more quickly and written more compactly by a method known as *synthetic division*. Using the same example as before, we first remove all the x's:

In Sec. 11-8 we use synthetic division when solving third- and fourth-degree equations.

$$\begin{array}{r} 4 - 4 + 5 - 5 \\ 1\overline{\big)\, 4 + 0 + 1 + 0 - 2} \\ \underline{\mathbf{4} + 4 \qquad\qquad\quad} \\ -4 + \mathbf{1} + \mathbf{0} - \mathbf{2} \\ \underline{-\mathbf{4} - 4 \qquad\qquad} \\ 5 + \mathbf{0} - \mathbf{2} \\ \underline{\mathbf{5} + 5 \qquad} \\ -5 - \mathbf{2} \\ \underline{-\mathbf{5} - 5} \\ 3 \end{array}$$

The same information can be conveyed if we now eliminate those numbers (shown in boldface type) which are merely repeats of the numbers directly above them:

$$\begin{array}{r} 4 - 4 + 5 - 5 \\ 1\overline{\big)\, 4 + 0 + 1 + 0 - 2} \\ \underline{+4} \qquad\qquad\quad\;\; \\ -4 \qquad\qquad\quad\;\; \\ \underline{-4} \qquad\qquad \\ 5 \qquad\qquad \\ \underline{+5} \qquad \\ -5 \qquad \\ \underline{-5} \\ 3 \end{array}$$

Synthetic division, like the long division we did earlier, works only for polynomials.

which can now be written compactly as

$$\begin{array}{r|rrrrr} 1 & 4 & 0 & 1 & 0 & -2 \\ & & 4 & -4 & 5 & -5 \\ \hline & 4 & -4 & 5 & -5 & 3 \end{array}$$

Finally, since errors are less likely to occur when adding than when subtracting, we *change the sign of the divisor,* and add rather than subtract the numbers in the second line.

$$\text{divisor} \quad \begin{array}{r|rrrrr} -1 & 4 & 0 & 1 & 0 & -2 \\ & & -4 & 4 & -5 & 5 \\ \hline & \underbrace{4 \quad -4 \quad 5 \quad -5}_{\text{quotient}} & & & & 3 \leftarrow \text{remainder} \end{array}$$

Example: Using synthetic division, divide $2x^4 - 5x^2 - 7x + 3$ by $x - 2$.

Solution: We set up the division, including a zero coefficient for the missing x^3 term:

$$\begin{array}{r|rrrrr} 2 & 2 & 0 & -5 & -7 & 3 \\ & \downarrow & & & & \\ \hline & 2 & & & & \end{array}$$

and bring down the first term. We then multiply that number by the divisor and place the product under the second number in the dividend. Then add.

$$\begin{array}{r|rrrrr} 2 & 2 & 0 & -5 & -7 & 3 \\ & & 4 & & & \\ \hline & 2 & 4 & & & \end{array}$$

Repeat this procedure until the last column is reached.

$$\begin{array}{r|rrrrr} 2 & 2 & 0 & -5 & -7 & 3 \\ & & 4 & 8 & 6 & -2 \\ \hline & 2 & 4 & 3 & -1 & 1 \end{array}$$

The quotient is thus $2x^3 + 4x^2 + 3x - 1$ with a remainder of 1.

EXERCISE 4

Division of Algebraic Expressions

Divide the following monomials.

1. z^5 by z^3

2. $45y^3$ by $15y^2$

3. $-25x^2y^2z^3 \div 5xyz^2$

4. $20a^5b^5c \div 10abc$

5. $30cd^2f \div 15cd^2$

6. $36ax^2y \div 18ay$

7. $-18x^2yz \div 9xy$

8. $(x - y)^5$ by $(x - y)^3$

9. $15axy^3$ by $-3ay$

10. $-18a^3x$ by $-3ay$

11. $6acdxy^2$ by $2adxy^2$

12. $12a^2x^2$ by $-3a^2x$

13. $15ay^2$ by $-3ay$

14. $45(a - x)^3$ by $15(a - x)^2$

15. $-21vwz^2 \div 7vz^2$

16. $-33.6rs^2 \div 10.8r^2s$

17. $35m^2nx \div 5m^2x$

18. $20x^3y^3z^3 \div 10x^3yz^3$

19. $16ab$ by $4a$

20. $21acd$ by $7c$

Divide the polynomial by the monomial.

21. $2a^3 - a^2$ by a

22. $42a^5 - 6a^2$ by $6a$

23. $21x^4 - 3x^2$ by $3x^2$

24. $35m^4 - 7p^2$ by 7

25. $27x^6 - 45x^4$ by $9x^2$

26. $24x^6 - 8x^3$ by $-8x^3$

27. $34x^3 - 51x^2$ by $17x$

28. $5x^5 - 10x^3$ by $-5x^3$

29. $-3a^2 - 6ac$ by $-3a$

30. $-5.8x^3 + 2.7x^2y$ by $-3.5x^2$

31. $ax^2y - 2xy^2$ by xy

32. $3xy^2 - 3x^2y$ by xy

33. $4x^3y^2 + 2x^2y^3$ by $2x^2y^2$

34. $3a^2b^2 - 6ab^3$ by $3ab$

35. $abc^2 - a^2b^2c$ by $-abc$

36. $9x^2y^2z + 3xyz^2$ by $3xyz$

37. $x^2y^2 - x^3y - xy^3$ by xy

38. $a^3 - a^2b - ab^2$ by $-a$

39. $a^2b - ab + ab^2$ by $-ab$

40. $xy - x^2y^2 + x^3y^3$ by $-xy$

Divide the polynomial by the binomial.

Use synthetic division for some of these, as directed by your instructor.

41. $x^2 + 15x + 56$ by $x + 7$

42. $x^2 - 15x + 56$ by $x - 7$

43. $x^2 + x - 56$ by $x - 7$

44. $x^2 - x - 56$ by $x + 7$

45. $2a^2 + 11a + 5$ by $2a + 1$

46. $6a^2 - 7a - 3$ by $2a - 3$

47. $3a^2 - 4a - 4$ by $2 - a$

48. $x^4 + x^2 + 1$ by $x^2 + 1$

CHAPTER TEST

1. Multiply: $(b^4 + b^2x^3 + x^4)(b^2 - x^2)$

2. Square: $(x + y - 2)$

3. Evaluate: $(7.28 \times 10^4)^3$

4. Square: $(xy + 5)$

5. Multiply: $(3x - m)(x^2 + m^2)(3x - m)$

6. Divide: $-x^6 - 2x^5 - x^4$ by $-x^4$

7. Cube: $(2x + 1)$

8. Divide: $16.9x^3$ by $3.82x$

9. Divide: $(a - c)^m$ by $(a - c)^2$

10. Simplify: $\left(\dfrac{3a^2}{2b^3}\right)^3$

11. Square: $(4a - 3b)$

12. Multiply: $(x^3 - xy + y^2)(x + y)$

13. Multiply: $(2.7xy - 4.9)(6.2xy - 1.8)$

14. Divide: $x^{m+1} + x^{m+2} + x^{m+3} + x^{m+4}$ by x^4

15. Square: $(2a - 3b)$

16. Multiply: $(a^2 - 3a + 8)(a + 3)$

17. Divide: $a^3b^2 - a^2b^5 - a^4b^2$ by a^2b

18. Multiply: $(2x - 5)(x + 2)$

19. Multiply: $(2m - c)(2m + c)(4m^2 + c^2)$

20. Simplify: $\left(\dfrac{8x^3y^{-2}}{4x^3y^{-3}}\right)^3$

21. Divide: $2a^6$ by a^4

22. Multiply: $(a^2 + a^2y + ay^2 + y^3)(a - y)$

23. Simplify: $y - 3[y - 2(4 - y)]$

24. Divide: $-a^7$ by a^5

25. Multiply: $(2x^2 + xy - 2y^2)(3x + 3y)$

26. Divide: $x^4 - \frac{1}{2}x^3 - \frac{1}{3}x^2 - 2x - 1$ by $2x$

27. Simplify: $-2[w - 3(2w - 1)] + 3w$

28. Divide: $7 - 8c^2 + 5c^3 + 8c$ by $5c - 3$

Simple Equations and Word Problems

When two mathematical expressions are set equal to each other, we get an *equation*. Much of our work in technical mathematics is devoted to solving equations. We start in this chapter with the simplest types, and in later chapters cover more difficult ones (quadratic equations, exponential equations, trigonometric equations, and so forth).

And why is it so important to solve equations? Because equations are used to describe the way certain things happen in the world. For example, the note produced by a guitar string is not a whim of nature, but can be predicted (solved for) when you know the length, the mass, and the tension in the string (when you know the *equation* relating the pitch, length, mass, and tension).

Thousands of equations exist that link together the various quantities in the physical world, in chemistry, finance, and so on, and their number is still increasing. To be able to solve and to manipulate such equations is essential for anyone who has to deal with these quantities on the job.

5-1. EQUATIONS

Equations: An equation has two *sides* or members, and an *equal sign*.

$$\underbrace{3x^2 - 4x}_{\text{left side}} \underset{\text{equal sign}}{=} \underbrace{2x + 5}_{\text{right side}}$$

Conditional Equation: A *conditional equation* is one whose sides are equal only for certain values of the unknown.

Example: The equation

$$x - 5 = 0$$

is a conditional equation because the sides are equal only when $x = 5$.

The symbol $\equiv$ is used for identities. We would write $x(x + 2) \equiv x^2 + 2x$.

When we say "equation," we will mean "conditional equation." Other equations that are true for any value of the unknown, such as $x(x + 2) = x^2 + 2x$, are called *identities*.

The Solution: The *solution* to an equation is that value (or values) of the unknown that make the sides equal.

Example: The solution to the equation

$$x - 5 = 0$$

is

$$x = 5$$

The value $x = 5$ is also called the *root* of the equation. We also say that it *satisfies* the equation.

Checking: Check an apparent solution by substituting it back into the original equation.

Example: Is the value $x = 17$ a solution to the equation

$$\frac{2x + 1}{5} = 7?$$

Solution: Substituting 17 for x in the equation, we get

$$\frac{2(17) + 1}{5} \stackrel{?}{=} 7$$

Get into the habit of checking your work, for errors creep in everywhere.

$$\frac{34 + 1}{5} \stackrel{?}{=} 7$$

$$\frac{35}{5} = 7 \qquad \text{checks}$$

Common Error	Check your solution only in the *original* equation. Later versions may already contain errors.

5-2. SOLVING FIRST-DEGREE EQUATIONS

First-Degree Equations: A *first-degree* equation (also called a *linear* equation) is one in which the terms containing the unknown are all of first degree.

Examples: The equations $2x + 3 = 9 - 4x$, $3x/2 = 6x + 3$, and $3x + 5y - 6z = 0$ are all of first degree.

Solving an Equation: To solve an equation, we *perform the same mathematical operation on each side of the equation.* The object is to get the unknown standing alone on one side of the equation.

Example: Solve $3x = 8 + 2x$.

Solution: Subtracting $2x$ from both sides, we obtain

$$3x - 2x = 8 + 2x - 2x$$

Combining like terms yields

$$x = 8$$

When we subtracted $2x$ from both sides, it vanished from the right side of the equation and appeared on the left side as $(-2x)$. In general, any term can be moved to the other side of the equal sign, provided that you change its sign. This is called *transposing.*

Some teachers object to transposing because it is a "trick," not a true mathematical operation like addition or subtraction. Others find it useful and fast.

Example: Solve the equation $3x - 5 = x + 1$.

Solution: We first subtract x from both sides, and add 5 to both sides.

$$\begin{array}{rcr} 3x - 5 & = & x + 1 \\ \underline{-x + 5} & & \underline{-x + 5} \\ 2x & = & 6 \end{array}$$

Dividing both sides by 2, we obtain

$$x = 3$$

Check: Substituting into the original equation yields

$$3(3) - 5 \stackrel{?}{=} 3 + 1$$

$$9 - 5 = 4 \quad \text{checks}$$

Symbols of Grouping: When the equation contains symbols of grouping, remove them early in the solution.

Example: Solve the equation $3(3x + 1) - 6 = 5(x - 2) + 15$.

Solution: Removing the parentheses, we obtain

$$9x + 3 - 6 = 5x - 10 + 15$$

Combining like terms, $\quad 9x - 3 = 5x + 5$

Adding $-5x + 3$ to both sides,

$$\begin{array}{rcr} 9x - 3 & = & 5x + 5 \\ \underline{-5x + 3} & & \underline{-5x + 3} \\ 4x & = & 8 \end{array}$$

Dividing by 4, $x = 2$

Check:

$$3(6 + 1) - 6 \stackrel{?}{=} 5(2 - 2) + 15$$
$$21 - 6 \stackrel{?}{=} 0 + 15$$
$$15 = 15 \quad \text{checks}$$

Common Error	The mathematical operations you perform must be done *to both sides* of the equation in order to preserve the equality.
Common Error	The mathematical operations you perform on both sides of an equation must be done to each side *as a whole*—not term by term.

Example: Solve for x:

$$\frac{1}{x} = \frac{1}{4} + \frac{1}{5}$$

Solution: Taking reciprocals of both sides, being sure to take the reciprocal of the right side as a whole, we obtain

$$x = \frac{1}{\frac{1}{4} + \frac{1}{5}} = \frac{1}{\frac{9}{20}} = \frac{20}{9}$$

Incorrect Solution: Taking reciprocals of both sides, but taking the reciprocal of the right side term by term, we obtain

$$x \neq 4 + 5 = 9$$

Fractional Equations: A *fractional equation* is one that contains one or more fractions. When an equation contains a single fraction, the fraction can be eliminated by *multiplying both sides by the denominator of the fraction.*

Example: Solve:

$$\frac{x}{3} - 2 = 5$$

Solution: Multiplying both sides by 3,

$$3\left(\frac{x}{3} - 2\right) = 3(5)$$

$$3\left(\frac{x}{3}\right) - 3(2) = 3(5)$$

$$x - 6 = 15$$

Adding 6 to both sides,

$$x = 15 + 6 = 21$$

Check:

$$\frac{21}{3} - 2 \stackrel{?}{=} 5$$

$$7 - 2 = 5 \qquad \text{checks}$$

In Chapter 7 we treat fractional equations in detail. There we will see how to find the least *common denominator.*

When there are two or more fractions, multiplying both sides by the product of the denominators (called a *common* denominator) will clear the fractions.

Example: Solve:

$$\frac{x}{2} + 3 = \frac{x}{3}$$

Solution: Multiplying by 2(3) or 6,

$$6\left(\frac{x}{2} + 3\right) = 6\left(\frac{x}{3}\right)$$

$$6\left(\frac{x}{2}\right) + 6(3) = 6\left(\frac{x}{3}\right)$$

$$3x + 18 = 2x$$

Transposing,

$$3x - 2x = -18$$

$$x = -18$$

Check:

$$\frac{-18}{2} + 3 \stackrel{?}{=} \frac{-18}{3}$$

$$-9 + 3 \stackrel{?}{=} -6$$

$$-6 = -6 \qquad \text{checks}$$

Strategy: While solving an equation, you should keep in mind the objective of *getting x by itself on one side of the equal sign, with no x on the other side*. Use any valid operation toward this end.

Equations can be of so many different forms that we cannot give a step-by-step procedure for their solution, but the following tips should help.

1. Eliminate fractions by multiplying both sides by the common denominator of both sides.
2. Remove any parentheses by performing the indicated multiplication.

3. All terms containing x should be moved to one side of the equation, and all other terms moved to the other side.
4. Like terms on the same side of the equation should be combined at any stage of the solution.
5. Any coefficient of x may be removed by dividing both sides by that coefficient.
6. Check the answer.

Example: Solve:

$$\frac{3x - 5}{2} = \frac{2(x - 1)}{3} + 4$$

Solution:

1. We eliminate the fractions by multiplying by 6,

$$6\left(\frac{3x - 5}{2}\right) = 6\left[\frac{2(x - 1)}{3}\right] + 6(4)$$

$$3(3x - 5) = 4(x - 1) + 6(4)$$

2. Removing parentheses,

$$9x - 15 = 4x - 4 + 24$$

$$9x - 15 = 4x + 20$$

3. We get all x terms on one side by adding 15 and subtracting $4x$ from both sides,

$$9x - 4x = 20 + 15$$

4. Combining like terms,

$$5x = 35$$

5. Dividing by the coefficient of x,

$$x = 7$$

6. Check:

$$\frac{3(7) - 5}{2} \stackrel{?}{=} \frac{2(7 - 1)}{3} + 4$$

$$\frac{21 - 5}{2} \stackrel{?}{=} \frac{12}{3} + 4$$

$$\frac{16}{2} \stackrel{?}{=} 4 + 4$$

$$8 = 8 \qquad \text{checks}$$

Example: Solve:

$$3(x + 2)(2 - x) = 3(x - 2)(x - 3) + 2x(4 - 3x)$$

Solution: We remove parentheses in two steps. First,

$$3(2x - x^2 + 4 - 2x) = 3(x^2 - 5x + 6) + 8x - 6x^2$$

Then

$$-3x^2 + 12 = 3x^2 - 15x + 16 + 8x - 6x^2$$

Combining like terms,

$$\cancel{-3x^2} + 12 = \cancel{-3x^2} - 7x + 16$$

Subtracting 16 from both sides,

$$-7x = 12 - 16 = -4$$

Dividing by -7,

$$x = \frac{4}{7}$$

EXERCISE 1

Simple Equations

Solve and check each equation.

Treat the integers in these equations as exact numbers. Leave your answers in fractional, rather than decimal, form.

1. $x - 5 = 28$

2. $4x + 2 = 18$

3. $8 - 3x = 7x$

4. $9x + 7 = 25$

5. $9x - 2x = 3x + 2$

6. $x + 8 = 12$

7. $x + 4 = 0$

8. $3y - 2 = 2y$

9. $3x = x + 8$

10. $3x = 2x + 5$

11. $3x + 4 = x + 10$

12. $y - 4 = 7$

13. $2.8 - 1.3y = 4.6$

14. $3.75x + 7.24 = 5.82x$

15. $4x + 6 = x + 9$

16. $7x - 19 = 5x + 7$

17. $p - 7 = -3$

18. $-3x = -15$

19. $4x - 11 = 2x - 7$

20. $3x - 8 = 12x - 7$

21. $8x + 7 = 4x + 27$

22. $3x + 10 = x + 20$

23. $\frac{1}{2}x = 14$

24. $\frac{x}{5} = -25$

25. $5x + 22 - 2x = 31$

26. $4x + 20 - 6 = 34$

27. $\frac{x}{3} + 5 = 2x$

28. $5x - \frac{2x}{5} = 3$

29. $2x + 5(x - 4) = 6 + 3(2x + 3)$

30. $6x - 5(x - 2) = 12 - 3(x - 4)$

31. $4(3x + 2) - 4x = 5(2 - 3x) - 7$

32. $6 - 3(2x + 4) - 2x = 7x + 4(5 - 2x) - 8$

33. $3(6 - x) + 2(x - 3) = 5 + 2(3x + 1) - x$

34. $3x - 2 + x(3 - x) = (x - 3)(x + 2) - x(2x + 1)$

35. $(2x - 5)(x + 3) - 3x = 8 - x + (3 - x)(5 - 2x)$

36. $x + (1 + x)(3x + 4) = (2x + 3)(2x - 1) - x(x - 2)$

5-3. SOLVING WORD PROBLEMS

Math skills are important, but so are reading skills. Most students find word problems difficult, but if you have an unusual *amount of trouble with them you should seek out a reading teacher and get help.*

Problems in Verbal Form: We have already done some simple word problems in earlier chapters, and we will have more of them with each new topic, but their main treatment will be here. After looking at these word problems you may ask: "When will I ever have to solve problems like that on the job?"

Probably never. But you will most likely have to deal with technical material in written form: instruction manuals, specifications, contracts, building codes, handbooks, and so on. This chapter will aid you in coping with precise technical documents, where a mistaken meaning of a word could cost dollars, lives, or your job.

Some tips on how to approach any word problem are given in the following paragraphs.

Picture the Problem: Try to *visualize* the situation described in the problem. Form a picture in your mind. *Draw a diagram* showing as much of the given information as possible, including the unknown.

Understand the Words: Look up the meanings of unfamiliar words in a dictionary, handbook, or textbook.

Example: A certain problem contains the following statement:

> A simply supported beam carries a concentrated load midway between its center of gravity and one end.

To understand this statement, you must know what a "simply supported beam" is, and the meanings of "concentrated load" and "center of gravity."

Locate the words and expressions standing for mathematical operations, such as

. . . the difference of . . .

. . . is equal to . . .

. . . the quotient of . . .

Many problems will contain a *rate,* either as the unknown or as one of the given quantities. The word *per* is your tipoff. Look for "miles *per* hour," "dollars *per* pound," "pounds *per* gallon," "dollars *per* mile," and the like. The word *per* also suggests *division.* Miles per gallon, for example, means miles traveled *divided* by gallons consumed.

Identify the Unknown: Be sure that you know exactly *what is to be found* in a particular problem. Find the sentence that ends with a question mark, or the sentence that starts with "Find . . ." or "Calculate . . ." or "What is . . ." or "How much . . ." or similar phrases.

Label the unknown with a statement such as: "Let x = rate of the train, mi/h." Be sure to include *units of measure* when defining the unknown.

Example: Fifty-five liters of a mixture that is 24% alcohol by volume is mixed with 74 liters of a mixture that is 76% alcohol. Find the amount of alcohol in the final mixture.

You may be tempted to label the unknown in the following ways:

This step of labeling the unknown annoys many students; they see it as a nitpicking thing that teachers make them do just for show, when they would rather start solving the problem. But realize that this step is *the start of the problem. It is the crucial first step and* should never be omitted.

Let x = alcohol	This is too vague. Are we speaking of volume of alcohol, or weight, or some other property?
Let x = volume of alcohol	The units are missing.
Let x = liters of alcohol	Do we mean liters of alcohol in one of the *initial* mixtures or in the *final* mixture?
Let x = number of liters of alcohol in final mixture	Good.

We will do complete solutions of this type of problem in Sec. 5-5.

Define Other Unknowns: If there is more than one unknown, you will often be able to define the additional unknowns *in terms of the original unknown,* rather than introducing another symbol.

Example: A word problem contains the following statement: "Find two numbers whose sum is 80 and, . . ."

Here we have two unknowns. We can label them with separate symbols:

$$\text{let } x = \text{first number}$$
$$\text{let } y = \text{second number}$$

but another way would be to label the second unknown in terms of the first:

$$\text{let } x = \text{first number}$$
$$\text{let } 80 - x = \text{second number}$$

Write and Solve the Equation: The unknown quantity must now be related to the given quantities by means of an *equation*. Where does one get an equation? Either the equation is given verbally, right in the problem statement, or else the relationship between the quantities in the problem is one you are expected to know or to be able to find. These could be mathematical relationships or they could be from any branch of technology. For example:

How is the area of a circle related to its radius?

How is the current in a circuit related to the circuit resistance?

How is the deflection of a beam related to its depth?

How is the distance traveled by a car related to its speed?

and so on. The relationships you will need for the problems in this book can be found in the Summary of Facts and Formulas (Appendix A). Of course, each branch of technology has many more formulas than can be shown there. Problems that need some of these formulas are treated later in this chapter.

In the following number problems, the equation is given verbally in the statement of the problem.

Number Problems: Number problems, although of little practical value, will give you practice at picking out the relationship between the quantities in a problem.

Example: If twice a certain number is increased by 9, the result is equal to 2 less than triple the number. Find the number.

Solution: Let x = the number. From the problem statement,

$$2x + 9 = 3x - 2$$

Solving for x,

$$9 + 2 = 3x - 2x$$
$$x = 11$$

Check Your Answer: You should check your answer in the original problem statement. For example, let us check the answer (11) to the last problem.

Twice 11 increased by 9 is 22 plus 9, or 31.

Triple 11 is 33, less 2 is 31. checks

Common Error	Checking an answer by substituting in your *equation* is not good enough. The equation itself may be wrong. Check your answer with the problem statement.

EXERCISE 2

Verbal Statements

Rewrite each expression as an algebraic expression.

1. Eight more than 5 times a certain number.
2. Four consecutive integers.
3. Two numbers whose sum is 83.
4. Two numbers whose difference is 17.
5. The amounts of antifreeze and of water in 6 gal of an antifreeze-water solution.
6. A fraction whose denominator is 7 more than 5 times its numerator.
7. The angles in a triangle, if one angle is half another.
8. The number of gallons of antifreeze in a radiator containing x gallons of a mixture that is 9% antifreeze.
9. The distance traveled in x hours by a car going 82 km/h.
10. If x equals the length of a rectangle, write an expression for the width of the rectangle if (a) the perimeter is 58; (b) the area is 200.

Number Problems

11. Nine more than 4 times a certain number is 29. Find the number.
12. Find a number such that the sum of 15 and that number is 4 times that number.
13. Find three consecutive integers whose sum is 174.
14. The sum of 56 and some number equals 9 times that number. Find it.
15. Twelve less than 3 times a certain number is 30. Find the number.
16. When 9 is added to 7 times a certain number, the result is equal to subtracting 3 from 10 times that number. Find the number.
17. The denominator of a fraction is 2 greater than 3 times its numerator, and the reduced value of the fraction is $\frac{3}{10}$. Find the numerator.
18. Find three consecutive odd integers whose sum is 63.
19. Seven less than 4 times a certain number is 9. Find the number.
20. Find the number whose double exceeds 70 by as much as the number itself is less than 80.
21. If $2x - 3$ stands for 29, for what number does $4 + x$ stand?

5-4. UNIFORM-MOTION PROBLEMS

We turn now from number problems to types that have more application to technology: uniform motion, mixtures, and financial.

Uniform Motion: Motion is called *uniform* when the speed does not change. The distance traveled at constant speed is related to the rate of travel and the elapsed time by

Be careful not to use this formula for anything but uniform motion.

Uniform Motion	distance = rate × time $D = Rt$	**A16**

Example: A plane flies for 3.5 h at 320 km/h. How far does it travel?

Solution: From Eq. A16,

$$D = 3.5(320) = 1120 \text{ km}$$

A Typical Motion Problem: A train departs at noon traveling at a speed of 64 km/h, and a car leaves the same station $\frac{1}{2}$ h later to overtake the train. If the car's speed is 96 km/h, at what time and at what distance from the station will it overtake the train?

Solution: Let x = number of hours traveled by car. Then $x + \frac{1}{2}$ = time traveled by train (hours). The distance traveled by the car is, by Eq. A16,

$$\text{distance} = \text{rate} \times \text{time} = 96x$$

and for the train,

$$\text{distance} = 64(x + \tfrac{1}{2})$$

But the distance is the same for car and train, so

$$96x = 64(x + \tfrac{1}{2})$$

Removing parentheses,

$$96x = 64x + 32$$

$$96x - 64x = 32$$

$$32x = 32$$

$$x = 1 \text{ h} = \text{time for car}$$

$$x + \tfrac{1}{2} = 1\tfrac{1}{2} \text{ h} = \text{time for train}$$

So the car overtakes the train at 1:30 P.M. at a distance of 96 km from the station.

EXERCISE 3

Motion Problems

1. A bus travels 95 km to another town at a speed of 82 km/h. What must be its return rate if the total time for the round trip is to be 2.0 h?

2. A certain power hacksaw has a forward cutting speed of 120 ft/min and a straight stroke of 11 in. It is observed to make 420 cuts (and returns) in 5.0 min. What is the return speed?

3. Two ships start from the same spot and travel in opposite directions, one at 100 km/day and the other at 180 km/day. How long will it take for the ships to be 1500 km apart?

4. An oil slick from a runaway offshore oil well is advancing toward a beach 540 mi away at the rate of 12 mi/day. Two days after the spill, cleanup ships leave the beach and steam toward the slick at a rate of 50 mi/day. At what distance from the beach will they reach the slick?

5. Spacecraft A is over Houston at noon on a certain day and traveling at a rate of 300 km/h. Spacecraft B, attempting to overtake and dock with A, is over Houston at 1:15 P.M. and is traveling in the same direction as A, at 410 km/h. At what time will B overtake A? At what distance from Houston?

6. A ship leaves port at a rate of 24 km/h. After 8 h a launch leaves the same port to overtake the ship, and travels at a rate of 35 km/h. Find the time it will take for the launch to overtake the ship and the distance from port at which they will meet.

7. A certain submarine can travel at a rate of 30 km/h submerged and 40 km/h on the surface. How far can it travel submerged, and return to the starting point on the surface, if the total round trip is to take 8 h?

8. A certain space probe traveled at a speed of 1280 km/h, and then, by firing retro-rockets, slowed to 950 km/h. The total distance traveled was 75,000 km in a time of 63 h. Find the distance from the starting point at which the retro-rockets were fired.

9. An airplane left an airport at 1 P.M. traveling at a rate of 240 mi/h. After traveling for some distance, one engine failed, and the airplane returned to the field at a reduced rate of 110 mi/h, arriving at 3:15 P.M. At what distance from the field did it turn around?

5-5. MIXTURE PROBLEMS

Basic Relationships: The total amount of mixture is, obviously, equal to the sum of the amounts of the ingredients:

These two ideas are so obvious that it may seem unnecessary to write them down. But because they are obvious, they are often overlooked. They state that the whole is equal to the sum of its parts.

amount of mixture = amount of A + amount of B + $\cdots$	**A1**

and for each ingredient,

final amount of each ingredient = initial amount + amount added − amount removed	**A2**

Example: From 100 lb of solder, half lead and half zinc, 20 lb is removed. Then 30 lb of lead is added. How much lead is contained in the final mixture?

Solution:

$$\text{initial weight of lead} = 0.5(100) = 50 \text{ lb}$$
$$\text{amount removed} = 0.5(20) = 10 \text{ lb}$$
$$\text{amount added} = 30 \text{ lb}$$

By Eq. A1,

$$\text{final amount of lead} = 50 - 10 + 30 = 70 \text{ lb}$$

Percent Concentration: The percent concentration of each ingredient is

$$\text{percent concentration of ingredient} = \frac{\text{amount of ingredient in mixture}}{\text{total amount of mixture}} \times 100\% \quad \textbf{15}$$

Example: The percent concentration of lead in the preceding example is, by Eq. 15,

$$\text{percent lead} = \frac{70}{110} \times 100\% = 63.6\%$$

Two Mixtures: When *two mixtures* are combined into a *third* mixture, the amount of any ingredient A in the final mixture is

$$\text{final amount of } A = \text{amount of } A \text{ in first mixture} + \text{amount of } A \text{ in second mixture} \quad \textbf{A3}$$

Example: One hundred liters of gasohol containing 12% alcohol is mixed with 200 liters of gasohol containing 8% alcohol. How many liters of alcohol are in the final mixture?

Solution: By Eq. A3,

$$\begin{aligned}\text{final amount of alcohol} &= 0.12(100) + 0.08(200)\\ &= 12 + 16\\ &= 28 \text{ liters}\end{aligned}$$

A Typical Mixture Problem

Example: How much steel containing 5% nickel and 95% iron must be combined with another steel containing 2% nickel and 98% iron to make 3 tons of steel containing 4% nickel and 96% iron?

Solution: Let x = tons of 5% steel needed. Figure 5-1 shows the three alloys, and the amount of nickel in each. By Eq. A1 the weight of the 2% steel is $3 - x$. The weight of nickel it contains is, by Eq. 15,

$$0.02(3 - x)$$

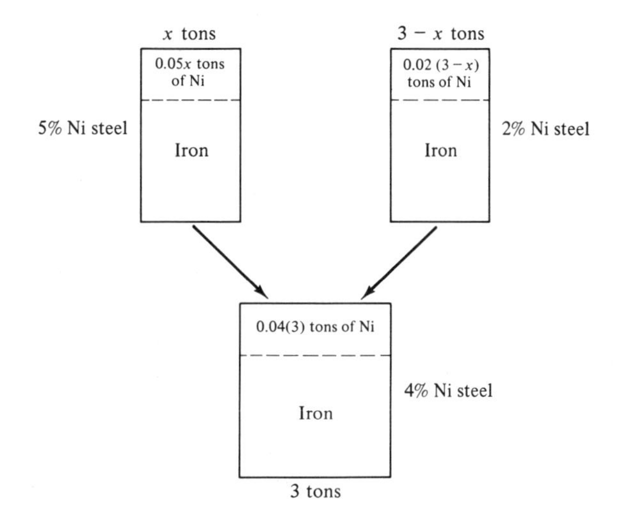

FIGURE 5-1

Notice that we are ignoring the iron *in the alloy and writing our equations only for the nickel. Of course, we could have written our equations for the amount of iron and not for the nickel. The point is that you need to write Eq. A2 for* only one of the ingredients.

The weight of nickel in the 5% steel is

$$0.05x$$

By Eq. A2, the sum of these must give the weight of nickel in the final mixture:

$$0.05x + 0.02(3 - x) = 0.04(3)$$

Multiplying by 100,

$$5x + 2(3 - x) = 4(3)$$

Clearing parentheses,

$$5x + 6 - 2x = 12$$

$$3x = 6$$

$$x = 2 \text{ tons of 5\% steel}$$

Check:

$$3 - x = 1 \text{ ton of 2\% steel}$$

Total weight:

$$2 + 1 = 3 \text{ tons}$$

$$\text{final tons of nickel} = 0.05(2) + 0.02(1) = 0.1 + 0.02 = 0.12 \text{ ton}$$

$$\text{percent nickel} = \frac{0.12}{3} \times 100 = 4\% \qquad \text{checks}$$

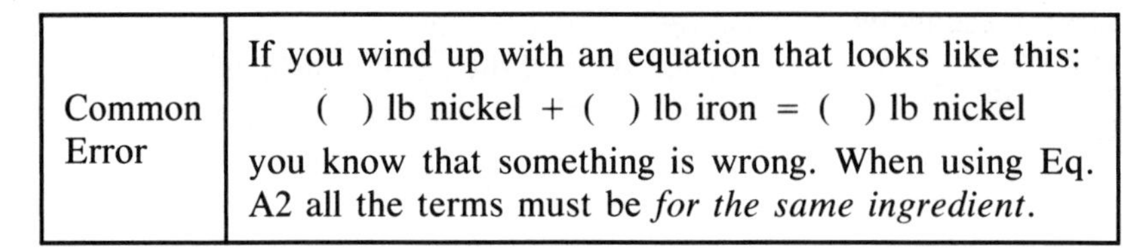

Common Error	If you wind up with an equation that looks like this: ( ) lb nickel + ( ) lb iron = ( ) lb nickel you know that something is wrong. When using Eq. A2 all the terms must be *for the same ingredient*.

EXERCISE 4

Mixture Problems

1. Two different mixtures of gasohol are available, one with 5% alcohol and the other containing 12% alcohol. How many gallons of the 12% mixture must be added to 250 gal of the 5% mixture to produce a mixture containing 9% alcohol?

2. How many metric tons of chromium must be added to 2.5 metric tons of stainless steel to raise the percent of chromium from 11% to 18%?

3. How many kilograms of nickel silver containing 18% zinc and how many kilograms of nickel silver containing 31% zinc must be melted together to produce 700 kg of a new nickel silver containing 22% zinc?

4. A certain bronze alloy containing 4% tin is to be added to 350 lb of bronze containing 18% tin to produce a new bronze containing 15% tin. How many pounds of the 4% bronze are required?

5. How many kilograms of brass containing 63% copper must be melted with 1100 kg of brass containing 72% copper to produce a new brass containing 67% copper?

6. A certain chain saw requires a fuel mixture of 5.5% oil and the remainder gasoline. How many liters of 2.5% mixture and how many of 9% mixture must be combined to produce 40 liters of 5.5% mixture?

7. A certain automobile cooling system contains 11 liters of coolant that is 15% antifreeze. How many liters of mixture must be removed so that, when replaced with pure antifreeze, a mixture of 25% antifreeze will result?

8. A van contains 4000 liters of wine with an alcohol content of 10%. How much of this wine must be removed so that, when replaced with wine with a 17% alcohol content, the alcohol content in the final mixture will be 12%?

9. Fifteen liters of fuel containing 3.2% oil is available for a certain two-cycle engine. This fuel is to be used for another engine requiring a 5.5% oil mixture. How many liters of oil must be added?

10. A certain automatic soldering machine requires a solder containing half tin and half lead. How much pure tin must be added to 55 kg of a solder containing 61% lead and 39% tin to raise the tin content to 50%?

5-6. FINANCIAL PROBLEMS

A Typical Financial Problem: A consultant had to pay income taxes of \$1720 plus 22% of the amount by which her taxable income exceeded \$18,000. Her tax bill was \$1842. What was her taxable income?

Solution: Let x = taxable income (\$). The amount by which her income exceeded \$18,000 is then

$$x - 18{,}000$$

Her tax is 22% of that amount, plus \$1720. So

$$\text{tax} = 1720 + 0.22(x - 18{,}000) = 1842$$

Solving for x,

$$0.22(x - 18{,}000) = 1842 - 1720 = 122$$

$$x - 18{,}000 = \frac{122}{0.22} = 554.55$$

$$x = 554.55 + 18{,}000 = \$18{,}554.55$$

EXERCISE 5

Financial Problems

1. The labor costs for a certain project were \$1501 per day for 17 technicians and helpers. If the technicians earned \$92/day and the helpers \$85/day, how many technicians were employed on the project?
2. A company has \$86,500 invested in bonds, and earns \$6751 in interest annually. Part of the money is invested at 7.4% and the remainder at 8.1%, simple interest. How much is invested at each rate?
3. How much must a company earn in order to have \$95,000 left after paying 27% in taxes?
4. Three equal batches of fiberglass insulation were bought for \$408: the first for \$17/ton, the second for \$16/ton, and the third for \$18/ton. How many tons of each were bought?
5. A water company changed its rates from \$1.95/1000 gal to \$1.16/1000 gal plus a service charge of \$45/month. How much water can you purchase before your monthly bill would equal the bill under the former rate structure?
6. What salary should a person receive in order to take home \$20,000 after deducting 23% for taxes?
7. A person was \$450 in debt, owing a certain sum to a brother, twice that amount to an uncle, and twice as much to a bank as to the uncle. How much was owed to each?
8. A student sold used skis and boots for \$210, getting four times as much for the boots as for the skis. What was the price of each?

9. A person used part of a \$100,000 inheritance to build a house and invested the remainder for 1 year, one-third of it at 6% and two-thirds at 5%, simple interest. The income from both investments was \$320. Find the cost of the house.

10. A used Jeep and a snowplow attachment are worth \$1200, the Jeep being worth seven times as much as the plow. Find the value of each.

11. A person spends one-fourth of her annual income for board, one-twelfth for clothes, one-half for other expenses, and saves \$2000. What is her income?

12. An estate is to be divided among four children in the following manner: to the first, \$2000 more than one-fourth of the whole; to the second, \$3400 more than one-fifth of the whole; to the third, \$3000 more than one-sixth of the whole; and to the fourth, \$4000 more than one-eighth of the whole. What is the value of the estate?

CHAPTER TEST

Solve for x.

1. $2x - (3 + 4x - 3x + 5) = 4$
2. $5(2 - x) + 7x - 21 = x + 3$
3. $3(x - 2) + 2(x - 3) + (x - 4) = 3x - 1$
4. $x + 1 + x + 2 + x + 4 = 2x + 12$
5. $(2x - 5) - (x - 4) + (x - 3) = x - 4$
6. $4 - 5x - (1 - 8x) = 63 - x$
7. $3x - (x + 10) - (x - 3) = 14 - x$
8. $(2x - 9) - (x - 3) = 0$
9. $3x + 4(3x - 5) = 12 - x$
10. $6(x - 5) = 15 + 5(7 - 2x)$
11. $x^2 - 2x - 3 = x^2 - 3x + 1$
12. $(x^2 - 9) - (x^2 - 16) + x = 10$
13. $x^2 + 8x - (x^2 - x - 2) = 5(x + 3) + 3$
14. $x^2 + x - 2 + x^2 + 2x - 3 = 2x^2 - 7x - 1$
15. $10x - (x - 5) = 2x + 47$
16. $7x - 5 - (6 - 8x) + 2 = 3x - 7 + 106$
17. $3x + 2 = \dfrac{x}{5}$
18. $\dfrac{4}{x} = 3$
19. $\dfrac{2x}{3} - 5 = \dfrac{3x}{2}$
20. $5.9x - 2.8 = 2.4x + 3.4$
21. $4.5(x - 1.2) = 2.8(x + 3.7)$
22. $\dfrac{x - 4.8}{1.5} = 6.2x$
23. Subdivide a meter of tape into two parts so that one part shall be 6 cm longer than the other part.
24. A certain mine yields low-grade oil shale containing 18 gal of oil per ton of rock, and another mine has shale yielding 30 gal/ton. How many tons of each must be sent each day to a processing plant which processes 25,000 tons of rock per day, so that the overall yield will be 23 gal/ton?
25. Glazing a square solar collector at \$4/m^2 will cost as much as framing it at \$10/m. What is the length of a side of the collector?

26. A train travels from P to Q at a rate of 24 km/h. After it has been gone $3\frac{1}{2}$ h, an express train leaves P for Q traveling at 90 km/h, and reaches Q $1\frac{1}{4}$ h ahead of the first train. Find the distance from P to Q, and the time taken by the express train.

27. A person owed to A a certain sum, to B four times as much, to C eight times as much, and to D six times as much. A total of \$570 would pay all the debts. What was the debt to A?

Sets, Functions, and Graphs

This chapter contains several different but related ideas. We begin with a study of *sets,* first defining different types of sets and then going on to see how sets can combine to form new sets.

We then look at *functions* and see how they are related to sets. The idea of a *function* provides us with a different way of speaking about mathematical relationships. We could say, for example, that the formula for the area of a circle *as a function of* its radius is $A = \pi r^2$.

It may become apparent as you study this chapter that these same problems could be solved without ever introducing the idea of a function. Does the function concept, then, merely give us new jargon for the same old ideas? Not really.

A new way of *speaking* about something can lead to a new way of *thinking* about that thing, and so it is with functions. It will also lead to the powerful and convenient *functional notation,* which will be especially useful when you study calculus and computer programming.

Finally, we learn about *graphing,* a skill that is indispensable in all branches of technology and in many nontechnical fields, such as business. Also, when studying graphing, we learn how to find an *approximate solution* to any type of equation, and then implement this method on the computer.

6-1. SETS

Sets and Elements: A set is a collection of particular things, such as the set of all automobiles. The objects or members of the set are called *elements*. We usually label a set with a capital letter, and an element of that set with a lowercase letter.

One way to specify a set is to list its members between braces.

Example: The notation

$$Q = \{a, b, c, d, e\}$$

specifies a set named Q whose elements are a, b, c, d, and e. Note that we separate the elements by commas.

Such a rule must, of course, define each member of the set in an unambiguous way.

If a set has many members, it is not practical to list them all. Instead, we can give a *rule* for finding members of the set.

Example: We could specify the set P of all positive integers x as

$$P = \{x : x \text{ is an integer}, x > 0\}$$

Here we read the colon as "such that" and the comma as "and." The statement then reads "P is a set of elements called x, *such that* x is an integer *and* x is greater than zero."

This is sometimes called set-building notation. *A vertical bar is sometimes used instead of the colon.*

The same set can also be written

$$P = \{x : x = 1, 2, 3, \ldots\}$$

Solution Set: Sometimes the rule that defines the members of a set is given in the form of an equation.

Example: The members of the set

$$Q = \{x : 3x - 15 = 0\}$$

are those numbers, called the *solution set* that satisfy the given equation, in this case, $x = 5$. So we may also say that

$$Q = \{5\}$$

Membership Notation: One way to indicate that some item is an element of a certain set is by means of the *special symbol* $\in$.

Example: The expression

$$3 \in P$$

means that 3 is an element of set P, or that 3 "belongs to" P.

Example: The notation

$$-3 \notin P$$

means that -3 is *not* a member of set P.

Equal Sets: Two sets are equal if and only if *they have the same elements.*

Example: If

$$A = \{1, 2, 3, 3, 4\}$$

Note that a set remains the same even if its elements are rearranged or repeated.

and

$$B = \left\{3, 1, \frac{8}{4}, 2^2\right\}$$

then

$$A = B$$

It is also called the universe of discourse.

The Universal Set: If we were discussing several sets, then that set containing all the members of each of these sets is called the *universal set*. It is given the symbol

$$U$$

Example: If we were discussing sets (teams) of baseball players, the universal set would have as its elements all the baseball players in the world that are members of teams.

The Empty Set: A set with no elements is called the *empty* set, or *null* set, and is given the symbol

$$\emptyset$$

Such a set results when its elements are defined by a rule that is always false.

Example: The set

$$A = \{x : x > 10, x < 5\}$$

is empty, since there are no numbers that are both greater than 10 and less than 5. Thus $A = \emptyset$.

Subsets: When every element of a set P is also an element of set Q, we say that P is a *subset* of Q. This is written

$$Q \supset P$$

or

$$P \subset Q$$

We say that Q *contains* P, or that P *is contained in* Q.

Example: If

$$A = \{2, 4, 6, 8, 10\}$$

and if $B = \{4, 6, 8\}$

and if $C = \{6, 8\}$

then $C \subset B$ and $B \subset A$ and $C \subset A$

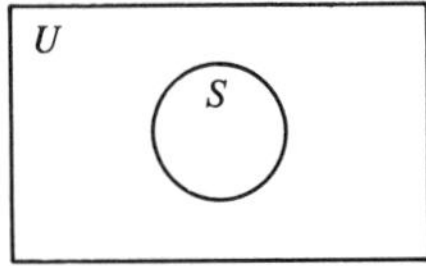
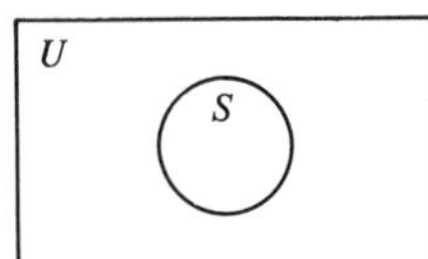

FIGURE 6-1. A Venn diagram.

Example: If

$$P = \{m, n, r\}$$

and $Q = \{m, k\}$

then we write $Q \not\subset P$

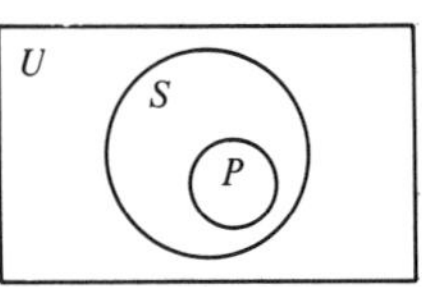

FIGURE 6-2. $P \subset S$.

Every set A is a subset of the universal set U. Further, although less obvious, the empty set $\emptyset$ is a subset of any set A. Thus,

$$U \supset A \supset \emptyset$$

Venn Diagrams: A *Venn diagram* is a picture of a set or sets. If we represent a set S by a circle, the elements of that set are the points within the circle.

Named for John Venn, an English logician (1834–1923).

The universal set U is usually pictured as a rectangle (Fig. 6-1) inside which the other sets are located. If set P is a subset of S, it will be shown lying entirely within the circle representing S (Fig. 6-2).

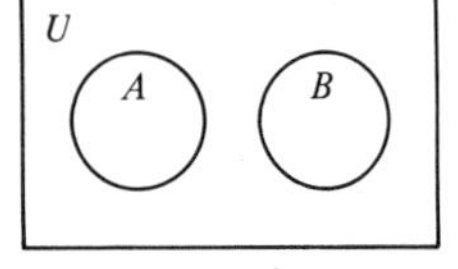

FIGURE 6-3. Disjoint sets.

Disjoint Sets: Two sets A and B are called *disjoint* if they have no elements in common (Fig. 6-3).

Example: The set of all cats is disjoint with the set of all dogs.

Intersection: The *intersection* of two sets A and B consists of those elements common to both A and B. It is written

$$A \cap B$$

and is pictured in Fig. 6-4.

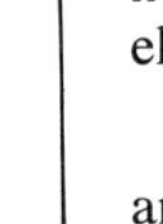

FIGURE 6-4. Intersection. The shaded area is $A \cap B$.

Example: If

$$A = \{1, 3, 5, 7\}$$

and if $B = \{2, 3, 4, 8\}$

then $A \cap B = \{3\}$

In general,

Intersection of Two Sets	$A \cap B = \{x : x \in A, x \in B\}$	**224**

Intersection corresponds to the AND operation in logic. Thus, $A \cap B$ means A AND B. When we study Boolean algebra in Chapter 18, we will see that it is written $A \cdot B$.

Union: The *union* of two sets A and B consists of all the elements contained either in A or in B. It is written

$$A \cup B$$

and is pictured in Fig. 6-5.

Union corresponds to the logical OR *operation. Thus,* $A \cup B$ *means* A OR B, *which we write* $A + B$.

Example: If

$$A = \{1, 3, 5, 7\}$$

and $$B = \{2, 4, 6, 8\}$$

then $$A \cup B = \{1, 2, 3, 4, 5, 6, 7, 8\}$$

In general,

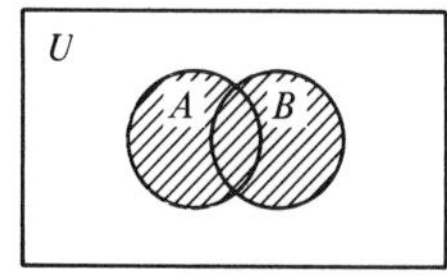

FIGURE 6-5. Union. The shaded area is $A \cup B$.

Union of Two Sets	$A \cup B = \{x : x \in A \text{ or } x \in B\}$	**225**

Complements: The *complement* of a set A are those elements that *do not* belong to A, but that do belong to the universal set U. The complement (sometimes called the absolute complement) of set A is written

$$A^c$$

and is pictured in the Venn diagram (Fig. 6-6).

This corresponds to the NOT *operation in logic. Thus,* A^c *means* NOT B. *It is written* A' *or* $\overline{A}$.

FIGURE 6-6. Complement. The shaded area is A^c.

Example: If

$$U = \text{the set of positive integers}$$

and if $$A = \text{the set of even integers}$$

then $$A^c = \text{the set of odd integers}$$

In general

Absolute Complement	$A^c = \{x : x \in U, x \notin A\}$	**226**

The *relative complement* of a set P with respect to another set Q consists of the elements that belong to Q but not to P. It is written

$$Q - P$$

(sometimes $Q \setminus P$ or $Q \sim P$) and is read "Q minus P."

Example: If

$$M = \{a, b, c, d, e\}$$

and $$N = \{d, e, f, g, h\}$$

then $$M - N = \{a, b, c\}$$

In general, the relative complement is defined as

$$A - B = \{x : x \in A, x \notin B\}$$

A Venn diagram for the relative complement $A - B$ is shown in Fig. 6-7.

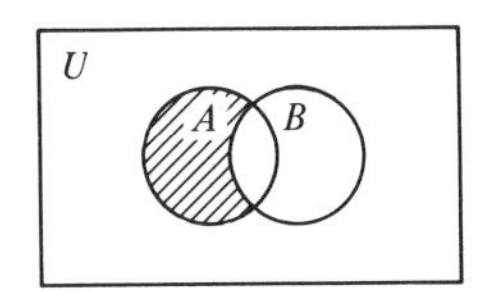

FIGURE 6-7. Relative complement. The shaded area is $A - B$.

Finite Sets: A set that contains a limited number of elements is said to be *finite;* otherwise, it is called infinite. The number of elements in a finite set A will be given the symbol

$$n(A)$$

The number of elements in the *union* of two disjoint sets is obviously the sum of the elements in the original sets. In symbols,

$$n(A \cup B) = n(A) + n(B)$$

Example: If a C class contains 14 men and 15 women, the total number of men and women in the class is

$$n(C) = 29$$

When finding the number of elements in two sets that are not disjoint, we add the elements of the two sets, but then subtract the elements held in common by both sets to avoid counting them twice.

$$n(A \cup B) = n(A) + n(B) - n(A \cap B)$$

Example: On a certain evening, 25 students in a particular math class did homework and 15 watched TV, and of these, 5 did both. How many students were in the class?

Solution: The number of students in the class is

$$25 + 15 - 5 - 35$$

This can be seen in the Venn diagram (Fig. 6-8), where it is clear that the number of students who did homework but watched no TV is 20, the number who watched TV but did no homework is 10, and the number who did both is 5. Adding these three groups gives 35, as before.

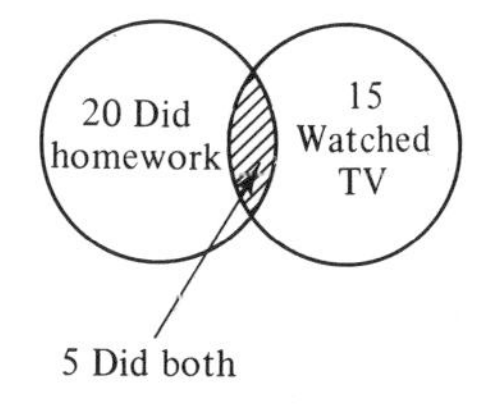

FIGURE 6-8

EXERCISE 1

Sets and Elements

1. Which, if any, of the following sets are equal?

$$A = \{n, n, o, p\}$$
$$B = \{o, p, q\}$$
$$C = \{p, p, o, n, n\}$$
$$D = \{o, q, q, p\}$$

2. Are sets P and Q equal if

$$P = \{x : 5x - 15 = 0\}$$

and

$$Q = \{3\} \quad ?$$

3. If

$$A = \{2, 4, 6, 8, \ldots\}$$
$$B = \{x : x \in A, x < 7\}$$

list the elements of set B.

4. If

$$P = \{1, 2, 3, \ldots\}$$
$$Q = \{x : x \in P, x \text{ is odd}, x < 5\}$$

list the elements of set Q.

5. If

$$M = \{3, 5, 7, 9\}$$
$$N = \{x : x \in M, x - 2 = 4\}$$

list the elements of set N.

Subsets

6. If

$$S = \{a, e, i, o, u\}$$
$$T = \{u, e, i\}$$
$$R = \{a, e, i, n\}$$
$$U = \{a, b, c, d, \ldots, z\}$$

which of the following statements are true, and which false?

(a) $S \subset U$ (b) $R \supset S$ (c) $U \not\supset S$
(d) $R \subset S$ (e) $T \subset S$ (f) $\emptyset \subset T$

Union, Intersection, and Complements

7. If

$$A = \{2, 4, 6, 8, 10\}$$
$$B = \{1, 2, 3, 4, 5, 6\}$$

find (a) $A \cap B$, (b) $A \cup B$, (c) $n(A \cup B)$, and (d) $n(A)$.

8. If

$$U = \{0, 1, 2, 3, 4, 5, 6, 7, 8, 9, 10\}$$

find A^c and B^c for sets A and B of Problem 7.

9. If

$$P = \{a, b, c, d, e\}$$
$$Q = \{d, e, f, g, h\}$$
$$R = \{e, f, g\}$$

find (a) $P - Q$, (b) $Q - P$, (c) $R - Q$, and (d) $n(R - Q)$.

10. If

$$K = \{2, 4, 6, 8, 10\}$$
$$L = \{1, 2, 5, 7, 9\}$$
$$M = \{1, 3, 6, 9\}$$

find (a) $K \cap (L \cup M)$, (b) $(K \cap L) \cup (L \cap M)$, (c) $(K \cup L) - M$, and (d) $n(K \cup L \cup M)$.

6-2. RELATIONS AND FUNCTIONS

Relations between Two Variables: The equations we have been solving in the last few chapters have contained only *one variable*. For example,

$$x(x - 3) = x^2 + 7$$

contains the single variable x.

But many situations involve *two* (or more) variables that are somehow related to each other.

Example: Look at Fig. 6-9 and suppose that you leave your campsite and walk a path up the hill. As you walk, both the horizontal distance x from your camp, and the vertical distance y above your camp, will change. You cannot change x without changing y, and vice versa (unless you jump into the air or dig a hole). The variables x and y are *related*.

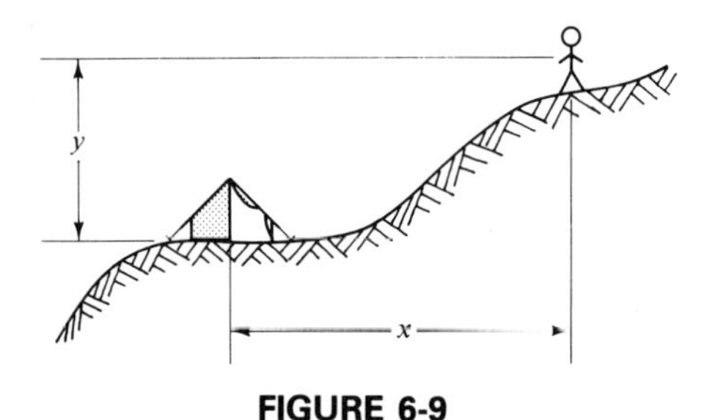

FIGURE 6-9

If we have an *equation* or *rule* linking x and y that enables us to find y for any given value of x, that equation or rule is called a *relation* between x and y.

Example: The equation $x^2 + y^2 = 25$ is a relation between x and y.

Example: The rule "square x to obtain y" is a relation between x and y.

Functions: If the relation between x and y yields *only one* value of y for any x, we say that *y is a function of x*.

Example: In the equation $y = 3x - 5$, we say that y is a function of x because any x gives us only one y. For example, when $x = 2$, $y = 3(2) - 5 = 1$.

Some older textbooks do not distinguish between relations and functions, but between single-valued *functions and* two-valued *or* multiple-valued *functions (what we now call relations).*

Example: In the equation $y = \pm\sqrt{x}$, y is *not* a function of x, because any x gives *two* values of y. This is a relation, but *not* a function.

The fact that a relation is not a function does not imply second-class status. Relations are no less useful when they are not functions. In fact, for practical work, we do not usually care whether our relation is a function or not.

The word *function* is commonly used to describe certain types of equations.

Examples:

$y = 5x + 3$	is a linear function
$y = 3x^3$	is a power function
$y = 7x^2 + 2x - 5$	is a quadratic function
$y = 3 \sin 2x$	is a trigonometric function
$y = 5^{2x}$	is an exponential function
$y = \log(x + 2)$	is a logarithmic function

Verbal Form: As with relations, functions can be given in *verbal form*.

Example: "The shipping charges are 55 cents/lb for the first 50 lb and 45 cents/lb thereafter." This verbal statement is a function relating the shipping costs to the weight of the item.

We sometimes want to switch from verbal form to an equation, or vice versa.

Example: Write y as a function of x if y equals twice the cube of x diminished by half of x.

Solution: We replace the verbal statement by the equation

$$y = 2x^3 - \frac{x}{2}$$

There are, of course, many different ways to express this same relationship.

Example: The equation $y = 5x^2 + 9$ can be stated verbally as: "y equals the sum of 9 and 5 times the square of x."

Example: Express the volume V of a cone having a base area of 75 units *as a function of* its altitude H.

Solution: This is another way of saying: "Write an equation for *the volume of a cone in terms of its base area and altitude.*"

The formula for the volume of a cone is given in the Summary of Facts and Formulas (Appendix A).

$$V = \tfrac{1}{3}(\text{base area})(\text{altitude}) \qquad (130)$$
$$= \tfrac{1}{3}(75)H$$

So

$$V = 25H$$

is the required expression.

We cover graphs in detail later in this chapter.

Function Given Graphically: Relations and functions can also be given in the form of a *graph.*

Example: The graph in Fig. 6-10 is a function. For any value of x, a single value of y can be found.

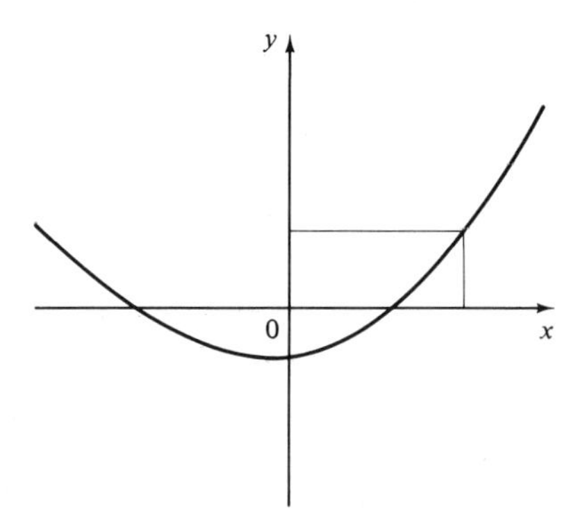

FIGURE 6-10. A function given graphically.

Example: The graph in Fig. 6-11 is *not* a function, because a single value of x can give *two* values of y. It is, however, a relation.

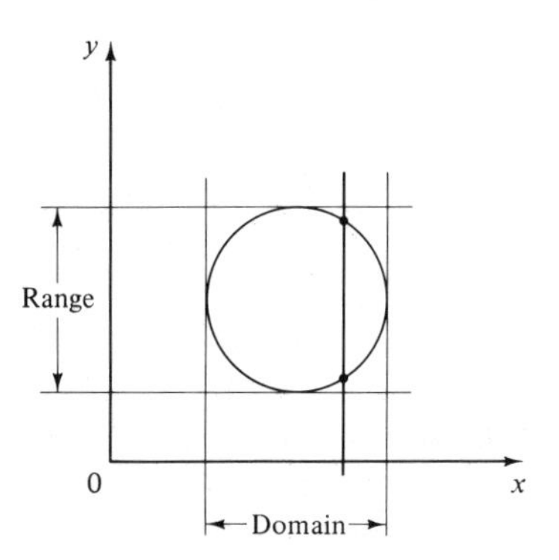

FIGURE 6-11. A graph of a relation showing domain and range.

Ordered Pairs of Numbers: In technical work we must often deal with *pairs* of numbers rather than with single values.

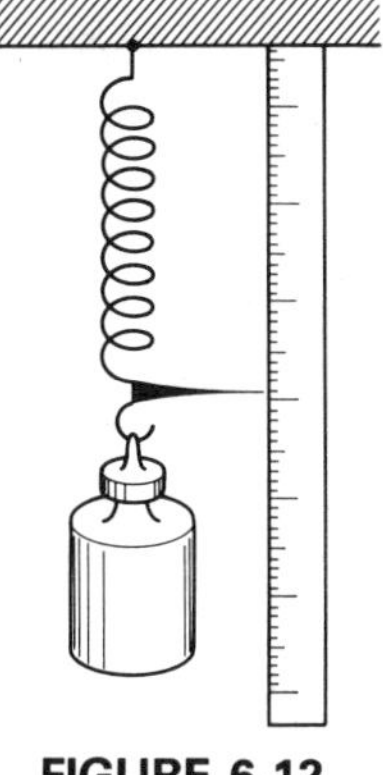

FIGURE 6-12

Example: In the experiment shown in Fig. 6-12 we change the load on the spring, and for each load, measure and record the distance that the spring has stretched from its unloaded position. If a load of 6 kg causes a stretch of 3 cm, then 6 kg and 3 cm are called *corresponding values*. The value 3 cm has no meaning *by itself,* only when it is *paired* with the load that produced it (6 kg). If we always write the pair of numbers in the *same order* (the load first, and the distance second, in this example), it is then called an *ordered pair* of numbers. It is written (6, 3).

A set of ordered pairs can be obtained by substituting values into an equation.

Example: Obtain a set of ordered pairs by substituting the integers from 0 to 5 for x in the function $y = x^2 - 1$.

Solution: Substituting,

$$y = 0^2 - 1 = -1$$
$$y = 1^2 - 1 = 0$$
$$y = 2^2 - 1 = 3$$
$$y = 3^2 - 1 = 8$$
$$y = 4^2 - 1 = 15$$
$$y = 5^2 - 1 = 24$$

We get the set of ordered pairs, (0, −1), (1, 0), (2, 3), (3, 8), (4, 15), and (5, 24).

Relation Defined in Terms of Sets: Consider a set D of elements x (called the *domain*), and another set R of elements y (called the *range*). If one or more of the elements in R are related to some of the elements in D (say, for example, those in R were the squares of those in D) we could form another set containing only the pairs of related elements. This new set is called a *relation*.

Example: Let D be a set of the integers from 1 to 10

$$D = \{x : x = 1, 2, 3, \ldots, 10\}$$

and let R be a set of integers from −10 to 10,

$$R = \{y : y = -10, -9, -8, \ldots, 0, 1, 2, \ldots, 10\}$$

and let x and y be related by the equation

$$y^2 = x$$

Our relation is then the set of those (x, y) pairs that satisfy the given equation, or

$$\{(1, 1), (4, 2), (4, -2), (9, 3), (9, -3)\}$$

Note that a single x (say 4) can be associated with more than one y (such as 2 and -2).

A Function Defined as a Set: A function, like a relation, can be regarded to be a set of ordered pairs (x, y), where each pair is an element of the set. However, each x is associated with *only one y*, whereas with a relation, one x could be associated with more than one y.

Although this definition is mathematically more correct, it is more useful in practical work for us to think of a function as being the rule (almost always an equation) from which the ordered pairs can be obtained.

A function is a set of ordered pairs of numbers (x, y) such that for each value of x there corresponds exactly one value of y.

Example: The set of ordered pairs (5, 9), (7, 11), (10, 18), and (16, 25) is a function.

The ordered pairs are often given in table form.

Example: The table of corresponding values

x	0	1	2	3
y	3	5	8	12

is a function.

We see, then, that a function or relation can be expressed in several different ways, as in Fig. 6-13. Starting with an equation or a verbal statement, you should be able to write the function in the four other ways. However, given a graph or a set or table of ordered pairs, it is difficult and sometimes impossible to obtain an equation. That operation is called *curve fitting* (see Sec. 15-8).

Domain and Range: For a function or a relation, the set of all the x values is called the *domain*, and the set of all y's is called the *range*.

The value of y corresponding to some value of x is sometimes called the image *of x. It is also said that the* function maps *a value of x in the domain to its image in the range.*

Example:

x	0	1	2	3	4	5	domain
y	-1	0	3	8	15	24	range

Example: For the circle in Fig. 6-11 the domain and range extend only over those regions in which the circle exists, as shown in the figure.

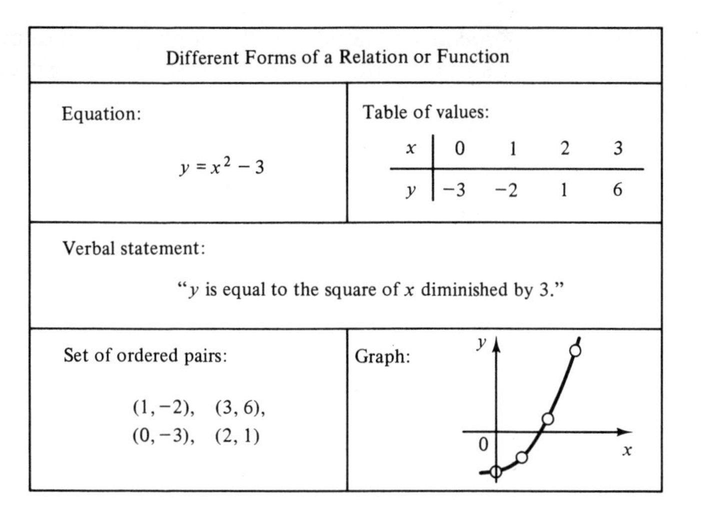

FIGURE 6-13

Example: For a set of ordered pairs:

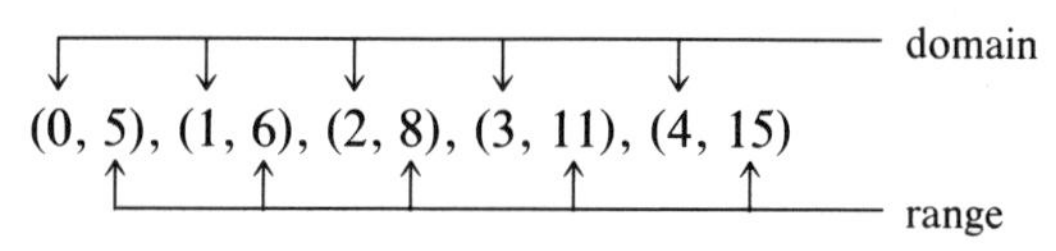

Determining Domain and Range: Sometimes the domain is given with the function, as in the equation for exponential growth,

$$y = ae^{nx} \qquad (x \geqslant 0)$$

That is, the domain is the set of all numbers greater than or equal to zero.

If the domain is *not* given, it is understood to be all the real numbers, except those that will make y imaginary or that will result in an illegal mathematical operation such as division by zero.

Example: Find the domain and range of the function

$$y = \sqrt{x - 2}$$

Solution: Any value of x less than 2 will make the quantity under the radical sign negative, resulting in an imaginary y. Thus, the domain of x is all the positive numbers equal to or greater than 2. The range of y will be all the positive numbers and zero. Can you explain why?

Example: Find the domain and range of the function

$$y = \frac{9}{\sqrt{4 - x}}$$

Solution: Since the denominator cannot be zero, x can not be 4. Also, any x greater than 4 will result in a negative quantity under the radical sign. The domain of x is then

$$x \leqslant 4$$

Restricted to these values, the quantity $4 - x$ is positive, and ranges from very small (when x is nearly 4) to very large (when x is large and negative). Thus the denominator is positive, since we allow only the principal (positive) root, and can vary from near zero to infinity. The range of y, then, includes all the values greater than zero.

Function Machine: It is possible to think of a function as a *machine*, as in Fig. 6-14. A value of x is put into the hopper and the machine delivers the single corresponding value of y at the chute. Values of x not within the domain of the function are rejected.

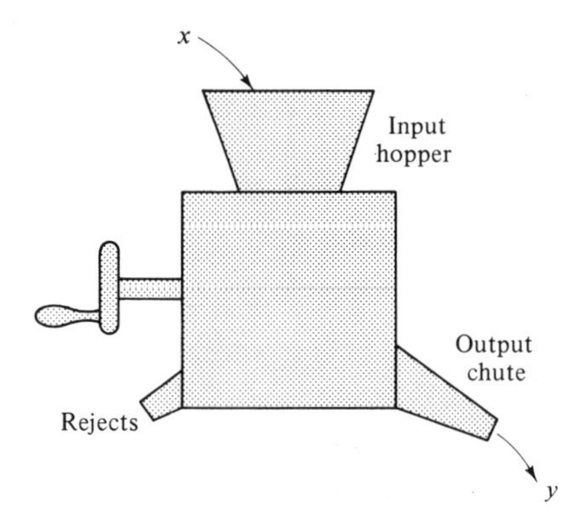

FIGURE 6-14. Function machine.

A flowchart *is a graphical way of representing a computation and is of great help to computer programmers. Steps in the computation are represented by boxes, which are connected by arrows showing the direction of "flow." You should have no trouble understanding the simple flowchart given here.*

Inside the machine can be anything that will return the proper y for a given x. We can represent the "insides" by a *flowchart*. For example, the flowchart for the function

$$y = \sqrt{x - 2}$$

is shown inside the function machine in Fig. 6-15.

EXERCISE 2

Relations vs. Functions

Which of the following relations are also functions?

1. $y = 3x^2 - 5$

2. $y = \sqrt{2x}$

3. $y = \pm\sqrt{2x}$

4. $y^2 = 3x - 5$

5. $2x^2 = 3y^2 - 4$

6. $x^2 - 2y^2 - 3 = 0$

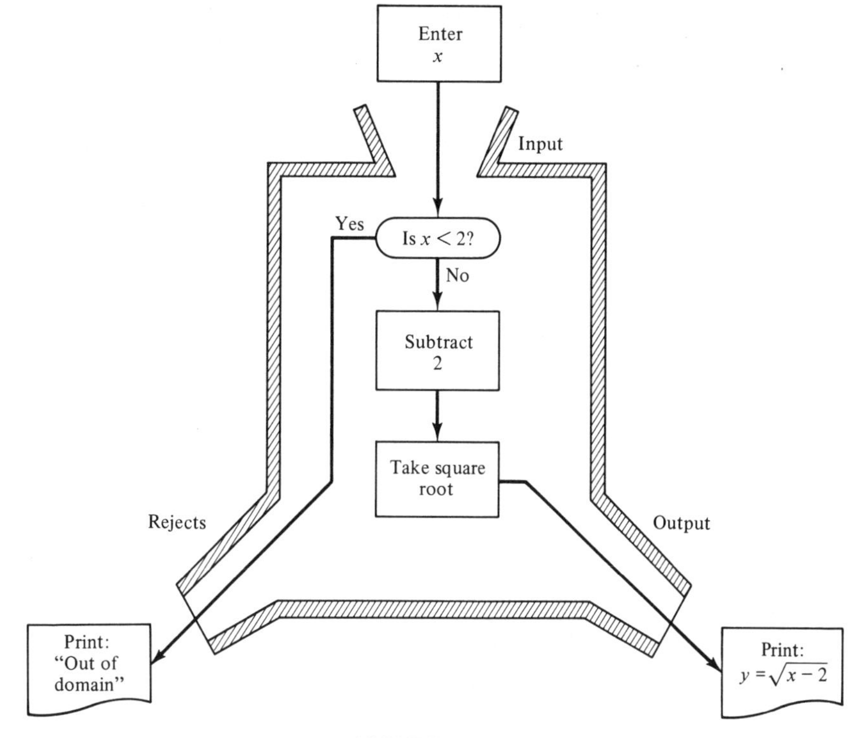

FIGURE 6-15

7. Is the set of ordered pairs (1, 3), (2, 5), (3, 8), (4, 12) a function? Explain.

8. Is the set of ordered pairs (0, 0), (1, 2), (1, −2), (2, 3), (2, −3), a function? Explain.

Functions in Verbal Form

For each expression, write y as a function of x, where the value of y is equal to:

9. The cube of x

10. The square root of x, diminished by 5

11. x increased by twice the square of x

12. The reciprocal of the cube of x

13. Two-thirds of the amount by which x exceeds 4

Replace each function by a verbal statement of the type given in Problems 9 to 13.

14. $y = 3x^2$ **15.** $y = 5 - x$ **16.** $y = \frac{1}{x} + x$

17. $y = 2\sqrt[3]{x}$ **18.** $y = 5(4 - x)$

Write the equation called for in the following statements. Refer to the Summary of Facts and Formulas (Appendix A).

19. Express the area A of a triangle as a function of its base b and altitude h.

20. Express the hypotenuse c of a right triangle as a function of its legs, a and b.

21. Express the volume V of a sphere as a function of the radius r.

22. Express the power P dissipated in a resistor as a function of its resistance R and the current I through the resistor.

Write the equations called for in each statement.

23. A car is traveling at a speed of 55 mi/h. Write its displacement d as a function of time t.

24. To ship their merchandise, a mail-order company charges 65 cents/lb plus $2.25 for handling and insurance. Express the total shipping charge s as a function of the item weight w.

25. A projectile is shot upward with an initial velocity of 125 m/s. Express the height H of the projectile as a function of time t.

Domain and Range

State the domain and range of each function.

26. (0, 2), (1, 4), (2, 8), (3, 16), (4, 32)

27. (−10, 20), (5, 7), (−7, 10), (10, 20), (0, 3)

28.

x	2	4	6	8	10
y	0	−2	−5	−9	−15

Find the largest possible domain and range for each function.

29. $y = \sqrt{x - 7}$

30. $y = \dfrac{3}{\sqrt{x - 2}}$

31. $y = x - \dfrac{1}{x}$

32. $y = \sqrt{x^2 - 25}$

6-3. FUNCTIONAL NOTATION

Implicit and Explicit Form: When one variable is isolated on one side of the equal sign, the equation is said to be in *explicit form.*

Example: The equations $y = 2x^3 + 5$, $z = ay + b$, and $x = 3z^2 + 2z - z$ are all in explicit form.

When a variable is *not* isolated, the equation is in *implicit form.*

Example: The equations $y = x^2 + 4y$, $x^2 + y^2 = 25$, and $w + x = y + z$ are all in implicit form.

Dependent and Independent Variables: In the equation

$$y = x + 5$$

y is called the *dependent variable,* because its value depends on the value of x, and x is called the *independent* variable. Of course, the same equation can be written $x = y - 5$, so that x becomes the dependent variable and y the independent variable.

The terms "dependent" and "independent" are used only for an equation in explicit form.

Example: In the implicit equation $y - x = 5$, neither x or y is called dependent or independent.

Functional Notation: Just as we use the symbol x to represent a *number,* without saying which number we are specifying, we use the notation

$$f(x)$$

to represent a *function* without having to specify which particular function we are talking about.

We can also use functional notation to designate a *particular* function, such as

$$f(x) = x^3 - 2x^2 - 3x + 4$$

A functional relation between two variables x and y, in explicit form, such as

$$y = 5x^2 - 6$$

could be written

$$y = f(x)$$

This is read "y is a function of x" or "y equals f of x." It does not mean "y equals f times x." The independent variable x is sometimes referred to as the argument.

Example: We may know that the horsepower P of an engine depends (somehow) on the engine displacement d. We can express this fact by

$$P = f(d)$$

We are saying that P is a function of d, even though we do not know (or perhaps even care) what the relationship is.

The letter f is usually used to represent a function, but *other letters* can, of course, be used (g and h being very common). Subscripts are also used to distinguish one function from another.

Examples:

$$y = f_1(x) \qquad y = g(x)$$
$$y = f_2(x) \qquad y = h(x)$$
$$y = f_3(x)$$

Implicit functions can also be represented in functional notation.

Example: The equation $x - 3xy + 2y = 0$ can be represented in functional notation by $f(x, y) = 0$.

Functions relating *more than two variables* can be represented in functional notation, as in the following examples.

Examples:

$y = 2x + 3z$	can be written	$y = f(x, z)$
$z = x^2 - 2y + w^2$	can be written	$z = f(w, x, y)$
$x^2 + y^2 + z^2 = 0$	can be written	$f(x, y, z) = 0$

Manipulating Functions: Functional notation provides us with a convenient way of indicating what is to be done with a function or functions. These include solving an equation for a different variable, changing from implicit to explicit form, or vice versa, combining two or more functions to make another function, or substituting numerical or literal values into an equation.

Example: Write the equation $y = 2x - 3$ in the form $x = f(y)$.

Solution: We are being asked to write the given equation with x, instead of y, as the dependent variable. Solving for x, we obtain

$$2x = y + 3$$
$$x = \frac{y + 3}{2}$$

Example: Write the equation $y = 3x^2 - 2x$ in the form $f(x, y) = 0$.

Solution: We are asked here to go from explicit to implicit form. Transposing,

$$3x^2 - 2x - y = 0$$

Composite Functions: A new function obtained by combining two or more given functions is called a *composite function.*

Example: If $y = 3w + 2$ and $w = x^2 + 5$, find $y = f(x)$.

Solution: To write y as a function of x, but not of w, we must eliminate w from the first given equation. We may do this by replacing it by $x^2 + 5$.

$$y = 3w + 2 = 3(x^2 + 5) + 2 = 3x^2 + 15 + 2$$
$$= 3x^2 + 17$$

Example: If

$$y = 3w$$

and $$w = z^2$$

and $$z = \frac{3x}{2}$$

write $$y = f(x)$$

Solution:

$$y = 3w = 3(z^2)$$
$$= 3\left(\frac{9x^2}{4}\right)$$
$$= \frac{27x^2}{4}$$

The *order* in which the functions are combined is important.

Example: If

$$g(w) = w^2$$

and $$f(x) = x + 1$$

find (a) $f[g(w)]$ and (b) $g[f(x)]$.

Solution:

(a) $$f[g(w)] = f(w^2)$$
$$= w^2 + 1$$

(b) $$g[f(x)] = g(x + 1)$$
$$= (x + 1)^2$$

Substituting into Functions

Example: Given $f(x) = x^3 - 5x$, find $f(2)$.

Solution: The notation $f(2)$ means that 2 is to be *substituted for x* in the given function. Substituting, we obtain

$$f(2) = (2)^3 - 5(2) = 8 - 10 = -2$$

Often, we must substitute *several values* into a function and combine them as indicated.

Example: If $f(x) = x^2 - 3x + 4$, find

$$\frac{f(5) - 3f(2)}{2f(3)}$$

Solution:

$$f(2) = 2^2 - 3(2) + 4 = 2$$

and

$$f(3) = 3^2 - 3(3) + 4 = 4$$

$$f(5) = 5^2 - 3(5) + 4 = 14$$

so

$$\frac{f(5) - 3f(2)}{2f(3)} = \frac{14 - 3(2)}{2(4)}$$

$$= \frac{8}{8} = 1$$

The substitution might involve *literal* quantities instead of numerical values.

Example: Given $f(x) = 3x^2 - 2x + 3$, find $f(5a)$.

Solution: We substitute $5a$ for x,

$$\begin{aligned} f(5a) &= 3(5a)^2 - 2(5a) + 3 \\ &= 3(25a^2) - 10a + 3 \\ &= 75a^2 - 10a + 3 \end{aligned}$$

There may be *more than one function* in a single problem.

Example: Given $f(x) = 3x$, $g(x) = x^2$, and $h(x) = \sqrt{x}$, find

$$\frac{2g(3) + 4h(9)}{f(5)}$$

Solution: First substitute into each function:

$$f(5) = 3(5) = 15 \qquad g(3) = 3^2 = 9 \qquad h(9) = \sqrt{9} = 3$$

Then combine these as indicated by the given equation:

$$\frac{2(9) + 4(3)}{15} = \frac{18 + 12}{15} = \frac{30}{15} = 2$$

Example: If $g(x) = 2x - 1$ and $f(u) = u^2$, find (a) $g[f(3)]$ and (b) $f[g(3)]$.

Solution:

$$g(3) = 2(3) - 1 = 5$$

and

$$f(3) = 3^2 = 9$$

so (a)

$$g[f(3)] = g(9) = 2(9) - 1 = 17$$

and (b)

$$f[g(3)] = f(5) = 5^2 = 25$$

Sometimes we must substitute into a function containing *more than one variable.*

Example: Given $f(x, y, z) = 2y - 3z + x$, find $f(3, 1, 2)$.

Solution: We substitute the given numerical values for the variables. Be sure that the numerical values are taken in the *same order* as the variable names in the functional notation.

$$\begin{array}{c} f(x,\ y,\ z) \\ \ \ \updownarrow\ \ \updownarrow\ \ \updownarrow \\ f(3,\ 1,\ 2) \end{array}$$

Substituting, we obtain

$$f(3, 1, 2) = 2(1) - 3(2) + 3 = 2 - 6 + 3$$
$$= -1$$

Inverse Functions: Suppose that we have a function in explicit form, say,

$$y = f(x) = x - 3$$

We can, of course, solve for x, so that x rather than y becomes the dependent variable.

$$x = g(y) = y + 3$$

The two functions $f(x)$ and $g(y)$ are said to be *inverse* functions.

Example: Find the inverse of the function $y = 3x + 5$.

Solution: Solving for x, we obtain

$$3x = y - 5$$

$$x = \frac{y - 5}{3}$$

Sometimes the inverse of a function gets a special name. The inverse of the sine function, for example, is the *arcsin* (see Sec. 12-2) and the inverse of an exponential function is a *logarithmic* function (Sec. 15-2).

Functional Notation on the Computer: Programming languages usually have a command for defining a function. BASIC, for example, uses the DEF FN statement, as in

```
50 DEF FND(X) = 3*X + 5
```

Here the *name* of the function is D, and the *independent variable or argument* is x.

If we have the line

```
80 PRINT FND(4)
```

later in the same program, the value 4 will be substituted into function D and the value

$$3(4) + 5 = 17$$

will be printed.

Functions of more than one variable can also be defined, as in

```
100  DEF FNB(X,Y,Z) = 3*X + Y/2 - Z
```

EXERCISE 3

Implicit and Explicit Form

Which equations are in explicit form and which in implicit form?

1. $y = 5x - 8$

2. $x = 2xy + y^2$

3. $3x^2 + 2y^2 = 0$

4. $y = wx + wz + xz$

Dependent and Independent Variables

Label the variables in each equation as dependent or independent.

5. $y = 3x^2 + 2x$

6. $x = 3y - 8$

7. $w = 3x + 2y$

8. $xy = x + y$

9. $x^2 + y^2 = r^2$

Manipulating Functions

10. If $y = 5x + 3$, write $x = f(y)$.

11. If $x = \dfrac{2}{y - 3}$, write $y = f(x)$.

12. If $y = \dfrac{1}{x} - \dfrac{1}{5}$, write $x = f(y)$.

13. If $x^2 + y = x - 2y + 3x^2$, write $y = f(x)$.

14. The power P dissipated in a resistor is given by Eq. A45, $P = I^2R$. Write $R = f(P, I)$.

15. The resistance R of two resistors wired in parallel can be found from Eq. A42,

$$\frac{1}{R} = \frac{1}{R_1} + \frac{1}{R_2}$$

Write $R_2 = f(R, R_1)$.

Composite Functions

16. If $x = 3y^2 - w$ and $y = 6 + 2z$, write $x = f(w, z)$.

17. If $x = 4z - y^2$ and $y = 2w + 5$, write $x = f(w, z)$.

18. If $y = 2x^2 - z^2$, $z = 2w + 3$, and $x = 3w^2 + z$, write $y = f(w)$.

19. If

$$Z = f(X, R) = \sqrt{R^2 - X^2}$$

$$X = f(X_L, X_C) = X_L - X_C$$

$$X_C = f(C, \omega) = \frac{1}{\omega C}$$

$$X_L = f(L, \omega) = \omega L$$

write

$$Z = f(\omega, L, C, R)$$

20. If the power P dissipated in a resistance wire is

$$P = VI$$

the current through the wire is

$$I = \frac{V}{R}$$

and the resistance of the wire is

$$R = \frac{\rho L}{A}$$

write

$$P = f(V, A, L, \rho)$$

21. If $f(u) = 3u + 2$ and $g(x) = 2x - 5$, find (a) $f[g(x)]$ and (b) $g[f(u)]$.

Substituting into Functions

Substitute the given numerical value(s) into each function.

22. If $f(x) = 2x^2 + 4$, find $f(3)$.

23. If $f(x) = 5x + 1$, find $f(1)$.

24. If $f(x) = 15x + 9$, find $f(3)$.

25. If $f(x) = 5 - 13x$, find $f(2)$.

26. If $g(x) = 9 - 3x^2$, find $g(-2)$.

27. If $h(x) = x^3 - 2x + 1$, find $h(2.55)$.

28. If $f(x) = 7 + 2x$, find $f(3)$.

29. If $f(x) = x^2 - 9$, find $f(-2)$.

30. If $f(x) = 2x + 7$, find $f(-1)$.

Substitute and combine as indicated.

31. Given $f(x) = 2x - x^2$, find $f(5) + 3f(2)$.

32. Given $f(x) = x^2 - 5x + 2$, find $f(2) + f(3) - f(1)$.

33. Given $f(x) = x^2 - 4$, find $\dfrac{f(2) + f(3)}{4}$.

34. Given $f(x) = 3x^2 + 1$, find $\dfrac{f(4) - f(1)}{f(2)}$.

Substitute the literal values into each function.

35. If $f(x) = 2x^2 + 4$, find $f(a)$.

36. If $f(x) = 2x - \dfrac{1}{x} + 4$, find $f(2a)$.

37. If $f(x) = 5x + 1$, find $f(a + b)$.

38. If $f(x) = 5 - 13x$, find $f(-2c)$.

Substitute in each function and combine.

39. If $f(x) = x^2$ and $g(x) = \dfrac{1}{x}$, find $f(3) + g(2)$.

40. If $f(x) = \dfrac{2}{x}$ and $g(x) = 3x$, find $f(4) - g\left(\dfrac{1}{3}\right)$.

41. If $f(x) = 8x^2 - 2x + 1$ and $g(x) = x - 5$, find $2f(3) - 4g(3)$.

42. If $g(x) = x^2 - 3$ and $h(x) = 2 + x$, find $3g(3) + 5h(2)$.

43. If $h(z) = 2z$ and $g(w) = w^2$, find

(a) $\dfrac{h(4) + g(1)}{5}$ (b) $h[g(3)]$ (c) $g[h(2)]$

Functions of More Than One Variable

Substitute the values into each function.

44. If $f(x, y) = 3x + 2y^2 - 4$, find $f(2, 3)$.

45. If $f(x, y, z) = 3z - 2x + y^2$, find $f(3, 1, 5)$.

46. If $g(a, b) = 2b - 3a^2$, find $g(4, -2)$.

47. If $h(x, y) = 2x^2 + 3y$, find $h(a, a^2)$.

48. If $f(x, y) = y - 3x$, find $3f(2, 1) + 2f(3, 2)$.

49. If $g(a, b, c) = 2b - a^2 + c$, find $\dfrac{3g(1, 1, 1) + 2g(1, 2, 3)}{g(2, 1, 3)}$.

Inverse Functions

Find the inverse of:

50. $y = 8 - 3x$

51. $x = 5(2y - 3) + 4y$

Applications

52. The distance traveled by a freely falling body is a function of the elapsed time t,

$$f(t) = V_0t + \tfrac{1}{2}gt^2$$

where V_0 is the initial velocity and g is the acceleration due to gravity (about 32 ft/s^2). If V_0 is 55 ft/s, find $f(10)$, $f(15)$, and $f(20)$.

53. The resistance R of a conductor is a function of temperature,

$$f(t) = R_0(1 + \alpha t)$$

where R_0 is the resistance at 0°C and α is the temperature coefficient of resistance (0.00427 for copper). If the resistance of a copper coil is 9800 Ω at 0°C, find $f(20)$, $f(25)$, and $f(30)$.

54. The maximum deflection of a certain cantilever beam, with a concentrated load applied r feet from the fixed end, is a function of r,

$$f(r) = 0.00003r^2(80 - r) \qquad \text{inches}$$

Find the deflections $f(10)$ and $f(15)$.

Functional Notation on the Computer

55. Using the DEFINE instructions on your machine, compute and print

$$y = \frac{4f_1(x)}{f_2(x) + 3}$$

for integer values of x from 0 to 10, where

$$f_1(x) = 2x - 5 \qquad \text{and} \qquad f_2(x) = 3x^2 + 2x - 3$$

6-4. GRAPHING RELATIONS AND FUNCTIONS

They are also called Cartesian *coordinates, after the French mathematician René Descartes (1596–1650). Another type of coordinate system we will use is called the polar coordinate system (Sec. 14-6).*

The Rectangular Coordinate System: In Chapter 1 we plotted numbers on the number line. Suppose, now, that we take a second number line and place it at right angles to the first one, so that each intersects the other at the zero mark, as in Fig. 6-16. We call this a *rectangular coordinate system.*

The horizontal number line is called the *x axis* and the vertical is called the *y axis*. They intersect at the *origin*. These two axes divide the plane into four *quadrants,* numbered counterclockwise, as in the figure.

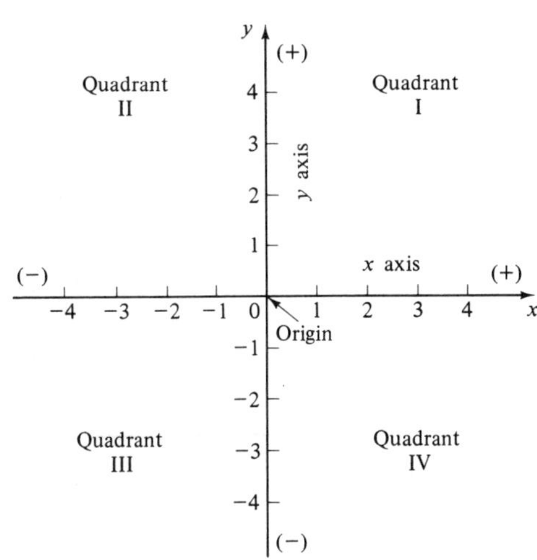

FIGURE 6-16. The rectangular coordinate system.

Graphing Ordered Pairs: Figure 6-17 shows a point P in the first quadrant. Its horizontal distance from the origin, called the *x coordinate* or *abscissa* of the point, is 3 units. Its vertical distance from the origin, called the *y coordinate* or *ordinate* of the point, is 2 units. The numbers in the *ordered pair* (3, 2) are called the *rectangular coordinates* (or simply, *coordinates*) of the point. They are always written in the same order, with the x coordinate first. The letter identifying the point is sometimes written before the coordinates, as $P(3, 2)$.

To plot any ordered pair (h, k), simply place a point at a distance h units from the y axis and k units from the x axis. Remember that negative values of x are located to the left of the origin and negative y's are below the origin.

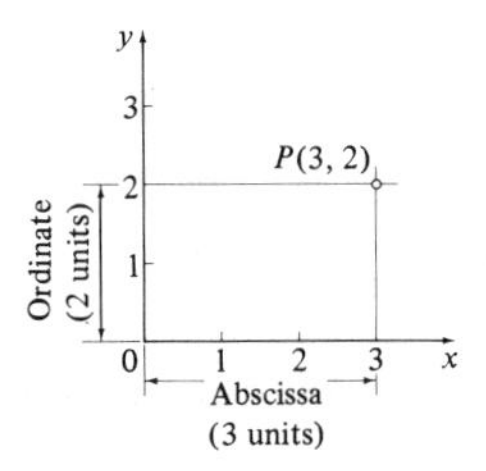

FIGURE 6-17. Is it clear from this figure why we call these rectangular coordinates?

Example: The points

$$P(4, 1) \quad Q(-2, 3) \quad R(-1, -2) \quad S(2, -3) \quad \text{and} \quad T(1.3, 2.7)$$

are shown plotted in Fig. 6-18.

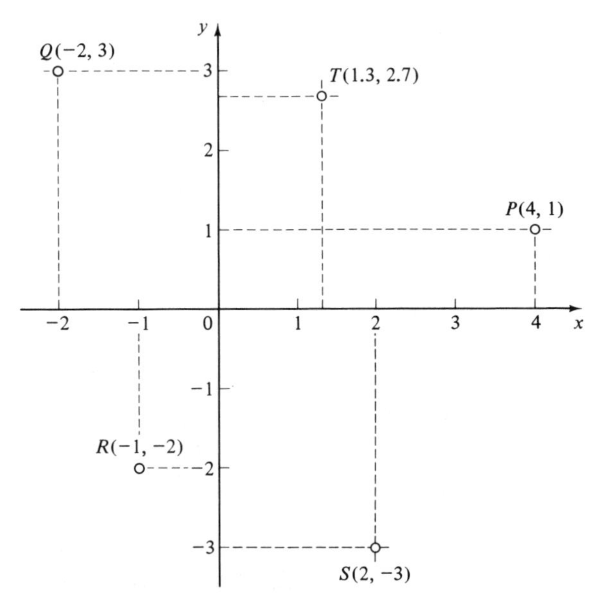

FIGURE 6-18

Graphing Relations or Functions: If the relation or function is given as a set of ordered pairs, simply plot each ordered pair. We usually connect the points with a smooth curve unless we have reason to believe that there are sharp corners, breaks, or gaps in the graph.

If the relation or function is in the form of an *equation*, obtain a table of ordered pairs by selecting values of x over the required domain, and compute corresponding values of y, as was done in Sec. 6-2. Then plot the set of ordered pairs.

Example: Graph the function $y = f(x) = 2x - 1$ for values of x from -2 to 2.

Solution: Substituting into the equation, we obtain

$$f(-2) = 2(-2) - 1 = -5$$
$$f(-1) = 2(-1) - 1 = -3$$
$$f(0) = 2(0) - 1 = -1$$
$$f(1) = 2(1) - 1 = 1$$
$$f(2) = 2(2) - 1 = 3$$

These points plot as a straight line (Fig. 6-19). Had we known in advance that the graph would be a straight line, we could have saved time by plotting just two points, with perhaps a third as a check.

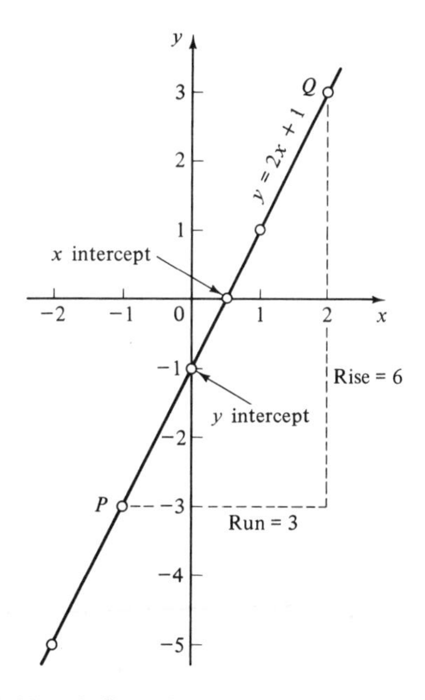

FIGURE 6-19. A first-degree equation will always plot as a straight line—hence the name *linear* equation. The points where the curve crosses the coordinate axes are called *intercepts*.

The horizontal distance between any two points on a straight line is called the *run* and the vertical distance between the same points is called the *rise*. For points P and Q (Fig. 6-19) the rise is 6 in a run of 3. The rise divided by the run is called the *slope* of the line, and is given the symbol m:

$$\text{slope} = m = \frac{\text{rise}}{\text{run}}$$

For the line in Fig. 6-19,

$$m = \frac{6}{3} = 2$$

Further, notice that in the equation of the line, the coefficient of x is equal to the slope and that the constant term is equal to the y intercept. For this reason, the equation is said to be in *slope-intercept* form.

$$y = 2x + 1$$

slope ↑ (2), y intercept ↑ (1)

Example: Graph the function $y = f(x) = x^2 - 4x - 3$ for values of x from -1 to 5.

Be especially careful when substituting negative values into an equation. It is easy to make an error.

Solution: Substituting into the equation, we obtain

$$f(-1) = (-1)^2 - 4(-1) - 3 = 1 + 4 - 3 = 2$$
$$f(0) = 0 - 0 - 3 = -3$$
$$f(1) = 1 - 4(1) - 3 = 1 - 4 - 3 = -6$$
$$f(2) = 2^2 - 4(2) - 3 = 4 - 8 - 3 = -7$$
$$f(3) = 3^2 - 4(3) - 3 = 9 - 12 - 3 = -6$$
$$f(4) = 4^2 - 4(4) - 3 = 16 - 16 - 3 = -3$$
$$f(5) = 5^2 - 4(5) - 3 = 25 - 20 - 3 = 2$$

These values are plotted in Fig. 6-20.

Solving Equations Graphically: We have said that a point where a graph of a function $y = f(x)$ crosses or touches the x axis is called an x intercept. Such a point is also called a *zero* of that function. Further, each such point is a *root* or *solution* to the equation $f(x) = 0$.

Example: Graphically find the approximate root(s) of the equation

$$4.1x^3 - 5.9x^2 - 3.8x + 7.5 = 0 \tag{1}$$

Solution: Let us represent the left side of the given equation by $f(x)$.

$$f(x) = 4.1x^3 - 5.9x^2 - 3.8x + 7.5$$

Any value of x for which $f(x) = 0$ will clearly be a solution to (1), so we simply plot $f(x)$ and look for the x intercepts. Not knowing the

If we had plotted many more points than these—say billions of them, they would be crowded so close together that they would seem to form a continuous line. The curve can be thought of as a collection *of all points that satisfy the equation. Such a curve (or the set of points) is called a* locus *of the equation.*

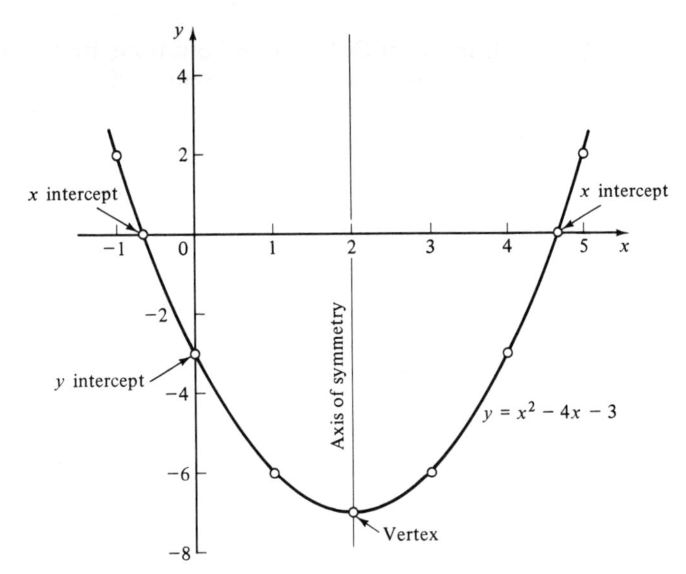

FIGURE 6-20. This curve is called a *parabola*, which we will meet again in Sec. 11-6. Note the *vertex* and the *axis of symmetry*.

shape of the curve, we compute $f(x)$ at various values of x until we are satisfied that we have located each region in which an x intercept is located. We then make a table of point pairs for this region.

x	-2	-1	0	1	2	3
$f(x)$	-41.3	1.3	7.5	1.9	9.1	53.7

We graph the function (Fig. 6-21) and read the approximate value of the x intercept,

$$x \cong -1.1$$

This, then, is an approximate solution to the given equation.

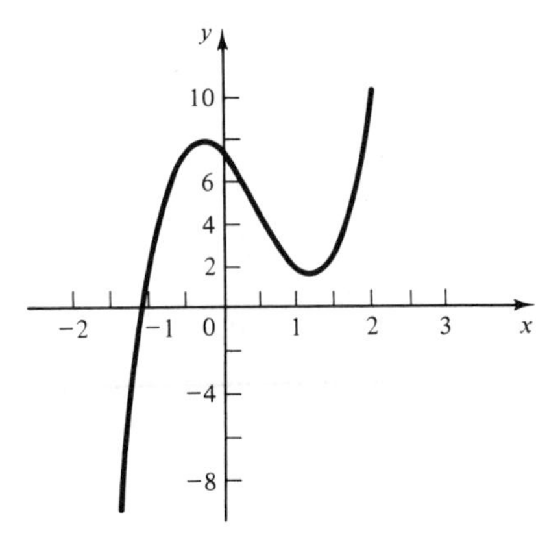

FIGURE 6-21. The shape of this curve is typical of a third-degree (cubic) function.

This method is also called the half-interval method *or the* bisection method.

Computer Solution of Equations by the Midpoint Method: There are several methods for finding the roots of equations by computer. In Chapter 11 we show the method of *simple iteration,* and when you study calculus you will learn the popular *Newton's method.* Here we cover the simple *midpoint* method.

Figure 6-22 shows a graph of some function $y = f(x)$. It is a smooth, continuous curve which crosses the x axis at R, which is the root of the equation $f(x) = 0$. We determine two values x_1 and x_2 which lie to the left and right of R, a fact that we verify by testing if y_1 and y_2 have opposite signs. We then locate x_3 midway between x_1 and x_2, and determine whether it is to the left or right of R by comparing the sign of y_3 with that of y_1. Thus x_3 becomes a new endpoint (either the left or the right as determined by the test), and we repeat the computation, halving the interval again and again until we get as close to R as we wish.

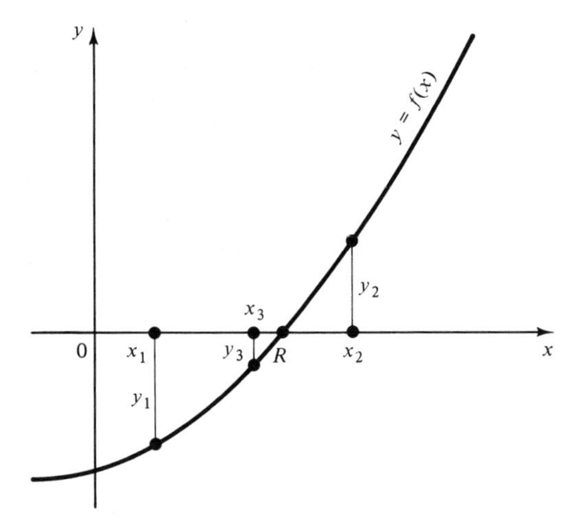

FIGURE 6-22. The midpoint method for finding roots.

Following is an algorithm for the midpoint method.

1. Make two initial guesses, x_1 and x_2, which must lie on either side of the root R.
2. Go to step 3 if y_1 and y_2 have opposite signs. Otherwise, go to step 1.
3. Compute the abscissa x_3 of the point midway between the two original points, with

$$x_3 = \frac{x_1 + x_2}{2}$$

4. Compute y_1 and y_3 by substituting x_1 and x_3 into the equation of the curve.

5. If the absolute value of y_3 is "small," then print the value of x_3 and end.
6. Determine if x_3 is to the right or left of the root by comparing the signs of y_1 and y_3.
7. If y_1 and y_3 have the same signs, replace x_1 with x_3. Otherwise, replace x_2 with x_3.
8. Go to step 2 and repeat the computation.

A technique such as this one, in which we repeat a computation many times, each time getting an approximate result that is (with luck) closer to the true value, is called iteration. *Iteration methods are used often on the computer, and we give several in this text.*

Using this algorithm as a guide, try to program the midpoint method on your computer. Save your program carefully because you can use it to check your manual solutions to the various equations in this text.

EXERCISE 4

Rectangular Coordinate System

If h and k are positive quantities, in which quadrants would the following points lie?

1. $(h, -k)$ **2.** (h, k) **3.** $(-h, k)$ **4.** $(-h, -k)$

5. Which quadrant contains points having a positive abscissa and a negative ordinate?

6. In what quadrants is the ordinate negative?

7. In which quadrants is the abscissa positive?

8. The ordinate of any point on a certain straight line is -5. Give the coordinates of the point of intersection of that line and the y axis.

9. Find the abscissa of any point on a vertical straight line that passes through the point (7, 5).

Graphing Ordered Pairs

10. Write the coordinates of points A, B, C, and D in Fig. 6-23.

11. Write the coordinates of points E, F, G, and H in Fig. 6-23.

12. Graph each point.

(a) (3, 5) (b) $(4, -2)$ (c) $(-2.4, -3.8)$

(d) $(-3.75, 1.42)$ (e) $(-4, 3)$ (f) $(-1, -3)$

Graph each set of points, connect them, and identify the geometric figure formed.

13. (0.7, 2.1), (2.3, 2.1), (2.3, 0.5), and (0.7, 0.5)

14. $(2, -\frac{1}{2})$, $(3, -1\frac{1}{2})$, $(1\frac{1}{2}, -3)$, and $(\frac{1}{2}, -2)$

15. $(-1\frac{1}{2}, 3)$, $(-2\frac{1}{2}, \frac{1}{2})$, and $(-\frac{1}{2}, \frac{1}{2})$

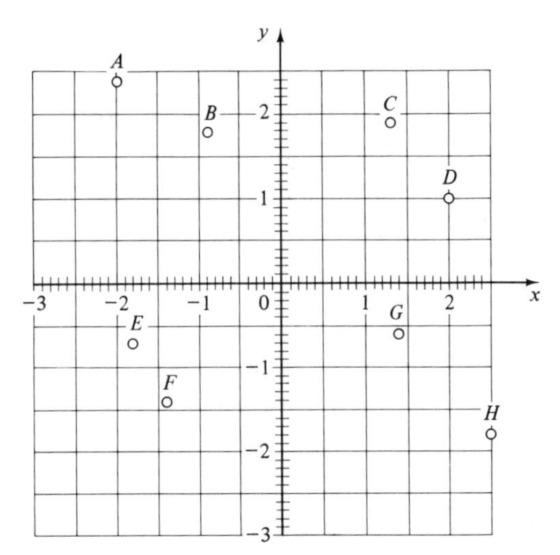

FIGURE 6-23

16. $(-3, -1)$, $(-1, -\frac{1}{2})$, $(-2, -3)$, and $(-4, -3\frac{1}{2})$

17. Three corners of a rectangle have the coordinates $(-4, 9)$, $(8, 3)$, and $(-8, 1)$. Graphically find the coordinate of the fourth corner.

Graphing Sets of Ordered Pairs

Graph each set of ordered pairs. Connect them with a curve that seems to you to fit the data best.

18. $(-3, -2)$, $(9, 6)$, $(3, 2)$, $(-6, -4)$

19. $(-7, 3)$, $(0, 3)$, $(4, 10)$, $(-6, 1)$, $(2, 6)$, $(-4, 0)$

20. $(-10, 9)$, $(-8, 7)$, $(-6, 5)$, $(-4, 3)$, $(-2, 4)$, $(0, 5)$, $(2, 6)$, $(4, 7)$

21. $(0, 4)$, $(3, 3.2)$, $(5, 2)$, $(6, 0)$, $(5, -2)$, $(3, -3.2)$, $(0, -4)$

Graphing Empirical Data

Data obtained by experiment or observation are called empirical data.

Graph the following experimental data. Label the graph completely. Take the first quantity in each table as the abscissa, and the second quantity as the ordinate. Connect the points with a smooth curve.

22. The melting point of a certain alloy, where T is the temperature (°C) and P is the percent lead in the alloy.

P	40	50	60	70	80	90
T	186	205	226	250	276	304

23. The current I(mA) through a tungsten lamp when a voltage V(V) is applied to the lamp.

V	10	20	30	40	50	60	70	80	90	100	110	120
I	158	243	306	367	420	470	517	559	598	639	676	710

24. A steel wire in tension, where σ is the stress (lb/in.^{2}) and ε is the strain (in./in.).

σ	5000	10,000	20,000	30,000	40,000	50,000	60,000	70,000
ε	0	0.00019	0.00057	0.00094	0.00134	0.00173	0.00216	0.00256

Graphing Equations

25. Which of the graphs in Fig. 6-24 are functions? Explain.

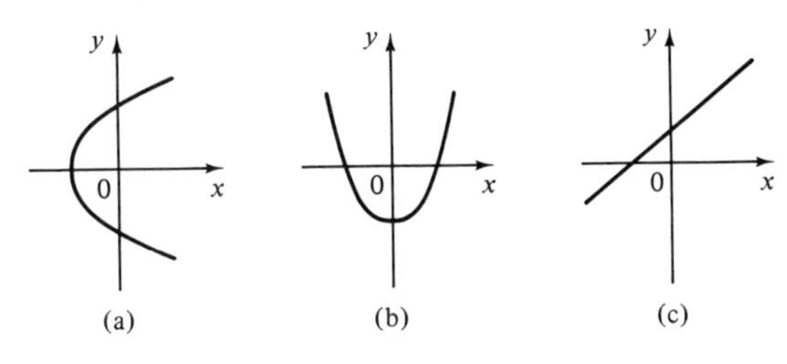

FIGURE 6-24

26. Which of the graphs in Fig. 6-25 are functions? Explain.

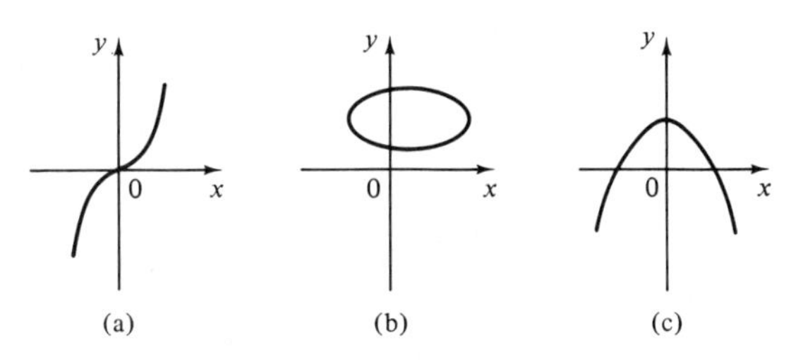

FIGURE 6-25

Fill in the missing values and graph the functions. Label any zeros or intercepts.

27. $y = x^2 - 1$

x	0	1	2	3	4	5
y						

28. $y = 5x - x^2$

x	-2	-1	0	1	2
y					

For each equation, make a table of ordered pairs, taking integer values of x from -3 to 3, and graph the function. Label any zeros or intercepts.

29. $y = 3x$

30. $y = -x + 2$

31. $y = x^2$

32. $y = 4 - 2x^2$

33. $y = \dfrac{x^2}{x + 3}$

34. $y = 5$

35. $y = 0$

36. $y = x$

37. $y = x^2 - 7x + 10$

38. State the domain and range of the relations in Fig. 6-26a and b.

39. State the domain and range of the relations in Fig. 6-26c and d.

40. Find the range of the function $y = 3x^2 - 5$ whose domain is $0 \leq x \leq 5$.

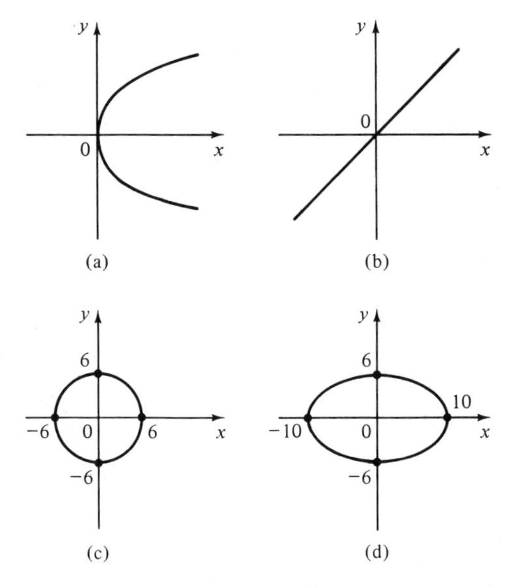

FIGURE 6-26

Graphical Solution of Equations

The following equations all have at least one root between $x = -10$ and $x = 10$. Put each equation into the form $y = f(x)$ and plot just enough of each to find the approximate value of the root(s).

41. $2.4x^3 - 7.2x^2 - 3.3 = 0$

42. $9.4x = 4.8x^3 - 7.2$

43. $25x^2 - 19 = 48x + x^3$

44. $1.2x + 3.4x^3 = 2.8$

45. $6.4x^4 - 3.8x = 5.5$

46. $621x^4 - 284x^3 - 25 = 0$

Graph each pair of equations on the same coordinate axes. Give the approximate coordinates of the point of intersection of the lines.

Can you guess the significance of the point of intersection? Peek ahead at Sec. 8-1.

47. $y = 2x + 1$
$y = -x + 2$

48. $x - y = 2$
$x + y = 6$

49. $y = -3x + 5$
$y = x - 4$

50. $2x + 3y = 9$
$3x - y = -4$

Computer

51. Write a program for solving equations by the midpoint method. Use it to solve any of the equations in Problems 41 to 46.

52. Write a program to generate a table of x, y point pairs for any function you enter. You should also be able to enter the domain of x, and the interval between your points. Use your program to obtain plotting points for any of the equations in Problems 29 to 37, and plot the points by hand.

CHAPTER TEST

Which of the following relations are also functions?

1. $y = 5x^3 - 2x^2$ **2.** $x^2 + y^2 = 25$ **3.** $y = \pm\sqrt{2x}$

If $A = \{p, q, r, s, t\}$, $B = \{a, b, c, s, t\}$, *and* $C = \{a, q, c, t\}$:

4. Find $A \cap (B \cup C)$. **5.** Find $(A \cup B) - C$. **6.** Find $(A \cap B) \cup (B \cap C)$.

7. Write y as a function of x if y is equal to half the cube of x, diminished by twice x.

8. Write an equation to express the surface area S of a sphere as a function of its radius r.

Find the largest possible domain and range for each function.

9. $y = \dfrac{5}{\sqrt{3 - x}}$ **10.** $y = \dfrac{1 + x}{1 - x}$

11. Are sets A and B equal if $A = \{x : 12 - 3x = 0\}$ and $B = \{4\}$?

Label each function as implicit or explicit. If explicit, name the dependent and independent variables.

12. $w = 3y - 7$ **13.** $x - 2y = 8$

14. Given $y = 3x - 5$, write $x = f(y)$. **15.** Given $x^2 + y^2 + 2w = 3$, write $w = f(x, y)$.

16. Graph the following points and connect them with a smooth curve.

x	0	1	2	3	4	5	6
y	2	$2\frac{1}{4}$	3	4	6	9	13

If $A = \{a, e, i, o, u\}$, $B = \{a, p, b, q, c, r\}$, *and* $U = \{a, b, c, d, \ldots, x, y, z\}$:

17. Find $A \cap B$. **18.** Find $A \cup B$. **19.** Find A^c.

20. Find B^c. **21.** Find $A - B$. **22.** Find $n(B)$.

23. If $y = 3x^2 + 2z$, and $z = 2x^2$, write $y = f(x)$.

24. If $f(x) = 5x^2 - 7x + 2$, find $f(3)$.

25. If $f(x) = 9 - 3x$, find $2f(3) + 3f(1) - 4f(2)$.

26. If $P = \{1, 3, 5, 7, \ldots, 13\}$ and $Q = \{x : x \in P, x > 5\}$, list the elements of set Q.

27. Plot the function $y = x^3 - 2x$ in the domain -3 to $+3$. Label any zeros or intercepts.

28. Graph the function $y = 3x - 2x^2$ from $x = -3$ to $x = 3$. Label any zeros or intercepts.

29. Graphically find the approximate value of the roots of the equation

$$(x + 3)^2 = x - 2x^2 + 4$$

30. If $f(x) = 5x$, $g(x) = 1/x$, and $h(x) = x^3$, find

$$\frac{5h(1) - 2g(3)}{3f(2)}$$

If $A = \{p, q, r, s, t\}$, $B = \{q, s, t\}$, $C = \{a, b, s, t\}$, and $U = \{a, b, c, d, \ldots, z\}$, which of the following statements are true, and which false?

31. $A \subset U$

32. $C \supset A$

33. $U \not\supset A$

34. $C \subset A$

35. $B \subset A$

36. $\emptyset \subset B$

37. $n(B) = n(C)$

38. $n(U) = 36$

39. $n(A) = n(A \cup B)$

40. Is the relation $x^2 + 3xy + y^2 = 1$ a function? Why?

41. Replace the function $y = 5x^3 - 7$ by a verbal statement.

42. What are the domain and range of the function (10, −8), (20, −5), (30, 0), (40, 3), and (50, 7)?

43. If $y = 6 - 3x^2$, write $x = f(y)$.

44. If $x = 6w - 5y$ and $y = 3z + 2w$, write $x = f(w, z)$.

45. If $f(x) = 8x + 3$ and $g(u) = u^2 - 4$, write $f[g(u)]$.

46. If $f(x) = 7x + 5$ and $g(x) = x^2$, find $3f(2) - 5g(3)$.

47. If $f(x, y, z) = x^2 + 3xy + z^3$, find $f(3, 2, 1)$.

48. Graphically find the point of intersection.

$$2.5x - 3.7y = 5.2$$
$$6.3x + 4.2y = 2.6$$

49. Write the inverse of the function $y = 9x - 5$.

50. Make a graph of the pressure p (lb/in.^{2}) in an engine cylinder, where v is the volume (in.^{3}) above the piston.

p	39.6	44.7	53.8	73.5	85.8	113.2	135.8	178.2
v	10.61	9.73	8.55	7.00	6.23	5.18	4.59	3.87

Factoring and Fractions

In Chapter 4 we learned how to multiply expressions together—here we do the reverse. We find those quantities (factors) which, when multiplied together, give the original product.

Why bother factoring? What is the point of changing an expression from one form to another? Because factoring is an essential step in simplifying expressions and in solving equations, especially the quadratic equations in Chapter 11. The value of factoring will be clear as we go on to use factoring to simplify fractions, and then to solve literal equations and to manipulate the formulas that are so important in technology.

Although the computer, the calculator, and the metric system (all of which use decimal notation) have somewhat reduced the importance of common fractions, they are still widely used. Algebraic fractions are, of course, as important as ever. We must be able to handle them in order to solve fractional equations and the word problems from which they arise.

7-1. COMMON FACTORS

Factors of an Expression: The *factors of an expression* are those quantities whose product is the original expression.

Although 1 and 10xyz are also factors of 10xyz, we would not usually state them.

Example: The factors of $10xyz$ are 2, 5, x, y, and z.

Example: The factors of $x^2 - 9$ are $x + 3$ and $x - 3$, because

$$(x + 3)(x - 3) = x^2 + 3x - 3x - 9$$
$$= x^2 - 9$$

Prime Factors: Many expressions have no factors other than 1 and themselves. Such expressions are called *prime*.

Example: The expressions x, $x - 3$, and $x^2 - 3x + 7$ are all prime.

Factoring: *Factoring* is the process of finding the factors of an expression. It is *the reverse of finding the product* of two or more quantities.

Multiplication (Finding the Product)
$x(x + 4) = x^2 + 4x$
Factoring (Finding the Factors)
$x^2 + 4x = x(x + 4)$

Common Factors: If each term of an expression contains the same quantity (called the common factor), that quantity may be *factored out*.

This is nothing but the distributive law *that we studied in Sec. 4-4.*

Common Factor	$ab + ac = a(b + c)$	**5**

Example: In the expression

$$2x + x^2$$

each term contains an x as a common factor. So we write

$$2x + x^2 = x(2 + x)$$

Most of the factoring we will do will be of this type.

Examples:

(a) $3x^3 - 2x + 5x^4 = x(3x^2 - 2 + 5x^3)$

(b) $3xy^2 - 9x^3y + 6x^2y^2 = 3xy(y - 3x^2 + 2xy)$

(c) $3x^3 - 6x^2y + 9x^4y^2 = 3x^2(x - 2y + 3x^2y^2)$

(d) $2x^2 + x = x(2x + 1)$

Common Error	Students are sometimes puzzled over the 1 in example (d) above. Why should it be there? After all, when you remove a chair from a room, it is *gone*; there is nothing (zero) remaining where the chair used to be. If you remove an x by factoring, shouldn't there be nothing (zero) remaining where the x used to be? $$2x^2 + x = x(2x + 0)?$$ Prove to yourself that this is not correct by multiplying the factors to see if you get back the original expression.

Factors in the Denominator: Common factors may appear in the denominators of the terms as well as in the numerators.

Examples:

(a)
$$\frac{1}{x} + \frac{2}{x^2} = \frac{1}{x}\left(1 + \frac{2}{x}\right)$$

(b)
$$\frac{x}{y^2} + \frac{x^2}{y} + \frac{2x}{3y} = \frac{x}{y}\left(\frac{1}{y} + x + \frac{2}{3}\right)$$

Checking: To check if factoring has been done correctly, simply multiply your factors together and see if you get back the original expression.

This check will tell us if we have factored correctly but, of course, not whether we have factored completely *(see Sec. 7 2).*

Example: Are $2xy$ and $x + 3 - y^2$ the factors of
$$2x^2y + 6xy - 2xy^3?$$

Solution: Multiplying the factors, we obtain
$$2xy(x + 3 - y^2) = 2x^2y + 6xy - 2xy^3 \qquad \text{checks}$$

EXERCISE 1

Common Factors

Factor each expression and check your result.

1. $3y^2 + y^3$

2. $6x - 3y$

3. $x^5 - 2x^4 + 3x^3$

4. $9y - 27xy$

5. $3a + a^2 - 3a^3$

6. $8xy^3 - 6x^2y^2 + 2x^3y$

7. $5(x + y) + 15(x + y)^2$

8. $\frac{a}{3} - \frac{a^2}{4} + \frac{a^3}{5}$

9. $\dfrac{3}{x} + \dfrac{2}{x^2} - \dfrac{5}{x^3}$

10. $\dfrac{3ab^2}{y^3} - \dfrac{6a^2b}{y^2} + \dfrac{12ab}{y}$

11. $5a^2b + 6a^2c$

12. $a^2c + b^2c + c^2d$

13. $4x^2y + cxy^2 + 3xy^3$

14. $4abx + 6a^2x^2 + 8ax$

15. $3a^3y - 6a^2y^2 + 9ay^3$

16. $2a^2c - 2a^2c^2 + 3ac$

17. $5acd - 2c^2d^2 + bcd$

18. $4b^2c^2 - 12abc - 9c^2$

19. $8x^2y^2 + 12x^2z^2$

20. $6xyz + 12x^2y^2z$

21. $3a^2b + abc - abd$

22. $5a^3x^2 - 5a^2x^3 + 10a^2x^2z$

Applications

23. When a bar of length L_0 is changed in temperature by an amount t, its new length L will be $L = L_0 + L_0\alpha t$, where α is the coefficient of thermal expansion. Factor the right side of this equation.

24. A sum of money a when invested for x years at an interest rate n will accumulate to an amount y, where $y = a + anx$. Factor the right side of this equation.

Compare your result with Eq. A48.

25. When a resistance R_1 is heated from a temperature t_1 to a new temperature t, it will increase in resistance by an amount $\alpha(t - t_1)R_1$, where α is the temperature coefficient of resistance. The final resistance will then be $R = R_1 + \alpha(t - t_1)R_1$. Factor the right side of this equation.

26. An item costing P dollars is reduced in price by 15%. The resulting price C is then $C = P - 0.15P$. Factor the right side of this equation.

27. The displacement of a uniformly accelerated body is given by Eq. A17.

$$s = v_0t + \frac{a}{2}t^2$$

Factor the right side of this equation.

28. The sum of the voltage drops across the resistors in Fig. 7-1 must equal the battery voltage E.

$$E = iR_1 + iR_2 + iR_3$$

Factor the right side of this equation.

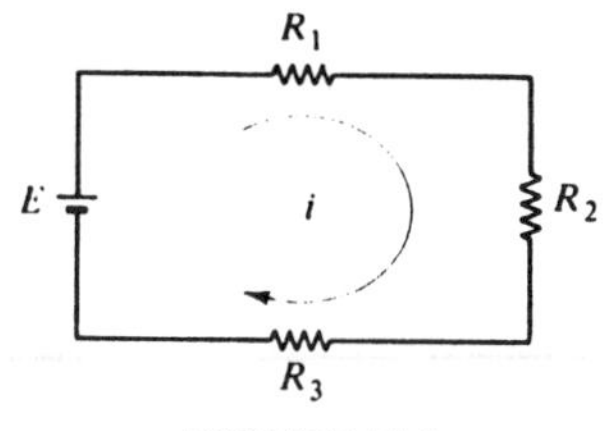

FIGURE 7-1

29. The weight of a spherical shell having an outside radius of r_2 and an inside radius r_1 is

$$\text{mass} = \tfrac{4}{3}\pi r_2^3 D - \tfrac{4}{3}\pi r_1^3 D$$

where D is the mass density of the material (see Eq. 128). Factor the right side of this equation.

7-2. DIFFERENCE OF TWO SQUARES

Form: An expression of the form

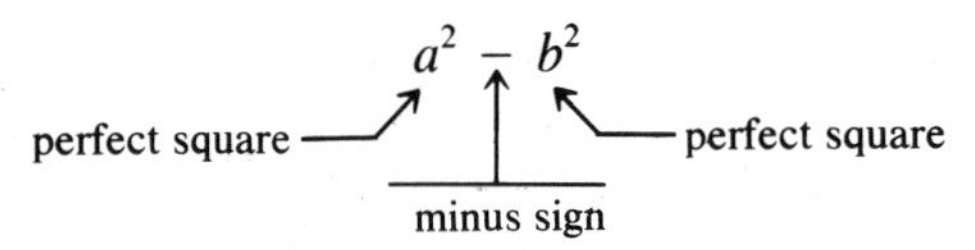

where one perfect square is subtracted from another, is called a *difference of two squares*. It arises when $(a - b)$ and $(a + b)$ are multiplied together.

Remember this from Sec. 4-4? There is no similar rule *for the* sum *of two squares* $(a^2 + b^2)$.

Difference of Two Squares	$a^2 - b^2 = (a + b)(a - b)$	**41**

Factoring the Difference of Two Squares: Once we recognize its form, the difference of two squares is easily factored.

Example:

$$4x^2 - 9 = (2x)^2 - (3)^2$$

$$(2x)^2 - (3)^2 = (2x \quad\ )(2x \quad\ )$$

square root of the first term

$$(2x)^2 - (3)^2 = (2x \quad 3)(2x \quad 3)$$

square root of the last term

$$4x^2 - 9 = (2x + 3)(2x - 3)$$

opposite signs

Example:

$$y^2 - 1 = (y + 1)(y - 1)$$

Example:

$$9a^2 - 16b^2 = (3a + 4b)(3a - 4b)$$

Factoring Completely: After factoring an expression, see if any of the factors themselves can be factored *again*.

We do not usually factor integers or powers. Thus, $6x^2$ would be left as is, not factored into $2 \cdot 3 \cdot x \cdot x$. Also, we would not usually factor x into $\sqrt{x} \cdot \sqrt{x}$.

Example: Factor $a - ab^2$.

Solution: Always remove any common factors first. Factoring, we obtain

$$a - ab^2 = a(1 - b^2)$$

Now factoring the difference of two squares, we get

$$a - ab^2 = a(1 + b)(1 - b)$$

Example:

$$x^4 - y^4 = (x^2 + y^2)(x + y)(x - y)$$

EXERCISE 2

Difference of Two Squares

Factor completely.

1. $4 - x^2$ **2.** $x^2 - 9$ **3.** $9a^2 - x^2$

4. $25 - x^2$ **5.** $4x^2 - 4y^2$ **6.** $9x^2 - y^2$

7. $x^2 - 9y^2$ **8.** $16x^2 - 16y^2$ **9.** $9c^2 - 16d^2$

10. $25a^2 - 9b^2$ **11.** $9y^2 - 1$ **12.** $4x^2 - 9y^2$

13. $x^2y^2 - 4y^2z^2$ **14.** $1 - x^2y^2$ **15.** $81x^2y^2 - 1$

16. $49a^2b^2 - 4$ **17.** $25x^2 - a^2$ **18.** $4a^2b^2 - c^2d^2$

19. $1 - 16b^2$ **20.** $5y^2 - 20x^2$ **21.** $49x^2y^2 - 16a^2b^2$

Applications

22. A body at temperature T will radiate an amount of heat kT^4 and will absorb from the surroundings an amount of heat kT_s^4, where T_s is the temperature of the surroundings. Write an expression for the net heat transfer by radiation (amount radiated minus amount absorbed) and factor this expression completely.

23. A spherical balloon shrinks from radius r_1 to radius r_2. The change in surface area is, from Eq. 129, $4\pi r_1^2 - 4\pi r_2^2$. Factor this expression.

24. When the distance between a lamp and a screen is increased from d_1 to d_2, the loss in brightness on the screen is

$$\frac{k}{d_1^2} - \frac{k}{d_2^2}$$

where k is a constant. Factor this expression.

25. When an object is released from rest, the distance fallen between time t_1 and time t_2 is, from Eq. A17, $\frac{1}{2}gt_2^2 - \frac{1}{2}gt_1^2$, where g is the acceleration due to gravity. Factor this expression.

26. When a body of mass m slows down from velocity v_1 to velocity v_2, the decrease in kinetic energy is $\frac{1}{2}mv_1^2 - \frac{1}{2}mv_2^2$. Factor this expression.

27. The work required to stretch a spring from a length x_1 (measured from the unstretched position) to a new length x_2 is $\frac{1}{2}kx_2^2 - \frac{1}{2}kx_1^2$. Factor this expression.

28. The reduction in power in a resistance R by lowering the voltage across the resistor from V_2 to V_1 is, from Eq. A44,

$$\frac{V_2^2}{R} - \frac{V_1^2}{R}$$

Factor this expression.

29. The formula for the volume of a cylinder of radius r and height h is $v = \pi r^2 h$. Write an expression for the volume of a hollow cylinder that has an inside radius r, an outside radius R, and a height h. Then factor the expression completely.

7-3. FACTORING TRINOMIALS

Trinomials: A *trinomial,* you recall, is an expression having *three terms.*

Example: $2x^3 - 3x + 4$ is a trinomial.

A *quadratic trinomial* has an x^2 term, an x term, and a constant term.

Example: $4x^2 + 3x - 5$ is a quadratic trinomial.

In this section we factor only quadratic trinomials.

Test for Factorability: Not all quadratic trinomials can be factored. To test for factorability,

$b^2 - 4ac$ is called the discriminant, Eq. 101. We use it in Sec. 11-4 to predict the nature of the roots of a quadratic equation.

Test for Factor-ability	The trinomial $ax^2 + bx + c$ (where a, b, and c are constants) is factorable if $b^2 - 4ac$ is a perfect square	**44**

Example: Can the trinomial $6x^2 + x - 12$ be factored?

Solution: Using the test for factorability, Eq. 44 with $a = 6$, $b = 1$, and $c = -12$,

$$\begin{aligned} b^2 - 4ac &= 1^2 - 4(6)(-12) \\ &= 1 + 288 = 289 \end{aligned}$$

By taking the square root of 289 on the calculator, we see that it is a perfect square ($289 = 17^2$), so the given trinomial is factorable. We see later that its factors are $(2x + 3)$ and $(3x - 4)$.

The coefficient of the x^2 term in a quadratic trinomial is called the *leading coefficient*. (It is the constant a in Eq. 44). We will first factor the easier type of trinomial, in which the leading coefficient is 1.

Trinomials with a Leading Coefficient of 1: We get such a trinomial when we multiply two binomials if the coefficients of the x terms in the binomials are also 1.

We wrote this equation back in Sec. 4-4.

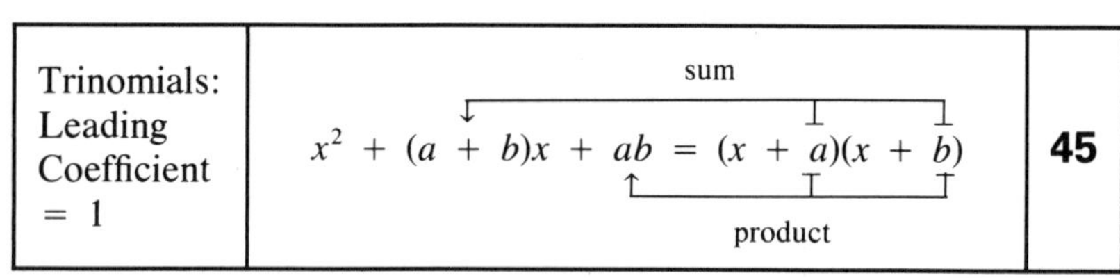

Note that the middle term of the trinomial equals the *sum* of a and b, and that the last term equals the *product* of a and b. Knowing this, we can reverse the process and find the factors when the trinomial is given.

Example: Factor the trinomial $x^2 + 8x + 15$.

Solution: From Eq. 45 this trinomial, if factorable, will factor as $(x + a)(x + b)$, where a and b have a sum of 8 and a product of 15. The integers 5 and 3 have a sum of 8 and a product of 15, so,

$$x^2 + 8x + 15 = (x + 5)(x + 3)$$

Using the Signs to Aid Factoring: The signs of the terms of the trinomial can tell you the signs of the factors.

1. The sign in one of the factors is the same as the sign of the middle term of the trinomial.
2. The sign in the other factor is the *product* of the signs of the middle and last terms of the trinomial.

Example:

same sign

$$x^2 - 2x - 8 = (x - 4)(x + 2)$$

product of signs

3. If the sign of the last term of the trinomial is *negative*, the sign of the middle term will tell which of the two numbers a and b is larger, the positive one or the negative one.

Example: Factor $x^2 + 2x - 15$.

Solution: We first determine the signs in the factors.

same sign

$$x^2 + 2x - 15 = (x + \quad)(x - \quad)$$

product of signs

We then find two numbers whose product is -15 and whose sum is 2. These two numbers must have opposite signs in order to have a negative product. Also, since the middle coefficient is $+2$, the positive number must be 2 greater than the negative number. Two numbers that meet these conditions are $+5$ and -3. So

$$x^2 + 2x - 15 = (x + 5)(x - 3)$$

Remember to remove any *common factors* before factoring a trinomial.

Example:

$$\begin{aligned} 2x^3 + 4x^2 - 30x &= 2x(x^2 + 2x - 15) \\ &= 2x(x - 3)(x + 5) \end{aligned}$$

EXERCISE 3

Test for Factorability

Test each trinomial for factorability.

1. $3x^2 - 5x - 9$ **2.** $3a^2 + 7a + 2$ **3.** $2x^2 - 12x + 18$

4. $5z^2 - 2z - 1$ **5.** $x^2 - 30x - 64$ **6.** $y^2 + 2y - 7$

Trinomials with a Leading Coefficient of 1

Factor completely.

7. $x^2 - 10x + 21$ **8.** $x^2 - 15x + 56$ **9.** $x^2 - 10x + 9$

10. $x^2 + 13x + 30$ **11.** $x^2 + 7x - 30$ **12.** $x^2 + 3x + 2$

13. $x^2 + 7x + 12$ **14.** $c^2 + 9c + 18$ **15.** $x^2 - 4x - 21$

16. $x^2 - x - 56$ **17.** $x^2 + 6x + 8$ **18.** $x^2 + 12x + 32$

19. $b^2 - 8b + 15$ **20.** $b^2 + b - 12$ **21.** $b^2 - b - 12$

22. $y^2 + y - 56$ **23.** $x^2 - 7x - 18$ **24.** $a^2 + 5a + 6$

25. $x^2 - 7x + 12$ **26.** $x^2 - 6x + 8$ **27.** $a^2 + 22a - 48$

28. $x^2 - 30x - 64$ **29.** $a^2 + 7a + 10$ **30.** $x^2 - 15x + 14$

7-4. TRINOMIALS WITH A LEADING COEFFICIENT OTHER THAN 1

General Quadratic Trinomial: When we multiply the two binomials $(ax + b)$ and $(cx + d)$, we get a trinomial with a leading coefficient of ac, a middle coefficient of $(ad + bc)$, and a constant term of bd.

We performed this multiplication in Sec. 4-4.

General Quadratic Trinomial	$(ax + b)(cx + d) = acx^2 + (ad + bc)x + bd$	**46**

The general quadratic trinomial may be factored by trial and error or by the grouping method.

Trial and Error: To factor this trinomial by trial and error, we look for four numbers, a, b, c, and d, such that

$$ac = \text{the leading coefficient}$$
$$ad + bc = \text{the middle coefficient}$$
$$bd = \text{the constant term}$$

Also, the *signs* in the factors are found in the same way as for trinomials with a leading coefficient of 1.

Example: Factor $2x^2 + 5x + 3$.

Solution: The leading coefficient $ac = 2$ and the constant term $bd = 3$.

Try

$$a = 1, \quad c = 2 \quad \text{and} \quad b = 3, \quad d = 1$$

Then $ad + bc = 1(1) + 3(2) = 7$. No good. It is supposed to be 5. We next try

$$a = 1, \quad c = 2, \quad b = 1, \quad \text{and} \quad d = 3$$

Then $ad + bc = 1(3) + 1(2) = 5$. This works. So

$$2x^2 + 5x + 3 = (x + 1)(2x + 3)$$

Some people have a knack for factoring and can quickly factor a trinomial by trial and error. Others rely on the longer but surer grouping method.

Grouping Method: The grouping method eliminates the need for trial and error.

Example: Factor $3x^2 - 16x - 12$.

Solution:

1. Multiply the leading coefficient and the constant term.

$$3(-12) = -36$$

2. Find two numbers whose product equals -36 and whose sum equals the middle coefficient, -16. Two such numbers are 2 and -18.

3. Rewrite the trinomial, splitting the middle term according to the selected factors ($-16x = 2x - 18x$)

$$3x^2 + 2x - 18x - 12$$

and group the first two terms together and the last two terms together.

$$(3x^2 + 2x) + (-18x - 12)$$

4. Remove common factors from each grouping.

$$x(3x + 2) - 6(3x + 2)$$

5. Remove the common factor $(3x + 2)$ from the entire expression:

$$(3x + 2)(x - 6)$$

which are the required factors.

EXERCISE 4

Trinomials with Leading Coefficient Other Than 1

Factor completely.

1. $4x^2 - 13x + 3$ **2.** $5a^2 - 8a + 3$ **3.** $5x^2 + 11x + 2$

4. $7x^2 + 23x + 6$ **5.** $12b^2 - b - 6$ **6.** $6x^2 - 7x + 2$

7. $2a^2 + a - 6$ **8.** $2x^2 + 3x - 2$ **9.** $5x^2 - 38x + 21$

10. $4x^2 + 7x - 15$ **11.** $3x^2 + 6x + 3$ **12.** $2x^2 + 11x + 12$

13. $3x^2 - x - 2$ **14.** $7x^2 + 123x - 54$ **15.** $4x^2 - 10x + 6$

16. $3x^2 + 11x - 20$ **17.** $4a^2 + 4a - 3$ **18.** $9x^2 - 27x + 18$

19. $9a^2 - 15a - 14$ **20.** $16c^2 - 48c + 35$

Applications

This group of problems includes trinomials with a leading coefficient of 1, which we studied in Sec. 7-3.

21. An object is thrown upward with an initial velocity of 32 ft/s from a building 128 ft above the ground. The height of the object above the ground, s, at any time t is given by (see Eq. A17)

$$s = 128 + 32t - 16t^2$$

Factor the right side of this equation.

22. To find the dimensions of a rectangular field having a perimeter of 70 ft and an area of 300 ft^2, we must solve the equation

$$x^2 - 35x + 300 = 0$$

Factor the left side of this equation.

23. To find the width $2m$ of a road that will give a sight distance of 1000 ft on a curve of radius 500 ft, we must solve the equation

$$m^2 - 1000m + 250{,}000 = 0$$

Factor the left side of this equation.

24. To find two resistors that will give an equivalent resistance of 400 ohms when wired in series and 75 Ω when wired in parallel, we must solve the equation

$$R^2 - 400R + 30{,}000 = 0$$

Factor the left side of this equation.

7-5. THE PERFECT SQUARE TRINOMIAL

The Square of a Binomial: In Sec. 4-4 we saw that the expression obtained when a binomial is squared is called a *perfect square trinomial.*

Example: Square the binomial $(2x + 3)$.

Solution: Squaring, we obtain

$$(2x + 3)^2 = 4x^2 + 6x + 6x + 9$$
$$= 4x^2 + 12x + 9$$

Note that in the perfect square trinomial obtained in the example, the first and last terms are the squares of the first and last terms of the binomial

$$(2x + 3)^2$$

square ↙ ↘ square

$$4x^2 + 12x + 9$$

and that the middle term is twice the product of the terms of the binomial.

$$(2x + 3)^2$$

product ↓

$$6x$$

twice the product

↓

$$4x^2 + 12x + 9$$

In general,

Any quadratic trinomial can be manipulated into the form of a perfect square trinomial by a procedure called completing the square. *We will use that method in Sec. 11-3 to solve quadratic equations.*

Perfect Square Trinomials	$(a + b)^2 = a^2 + 2ab + b^2$	**47**
	$(a - b)^2 = a^2 - 2ab + b^2$	**48**

Factoring a Perfect Square Trinomial: Write a binomial whose terms are the square roots of the first and last terms of the trinomial. The sign in the binomial will be the same as the sign of the middle term of the trinomial.

Example: Factor $a^2 - 4a + 4$.

Solution: The first and last terms are both perfect squares, and the middle term is twice the product of the square roots of the first and last terms. Thus, the trinomial is a perfect square. Factoring, we obtain

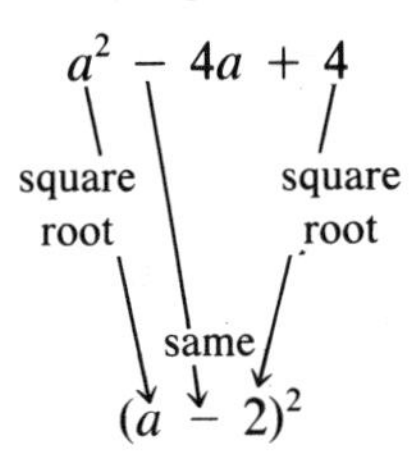

Example:

$$9x^2y^2 - 6xy + 1 = (3xy - 1)^2$$

EXERCISE 5

Perfect Square Trinomials

Factor completely.

1. $x^2 + 4x + 4$
2. $9 - 12a + 4a^2$
3. $2y^2 - 12y + 18$
4. $x^2 + 2x + 1$
5. $y^2 - 2y + 1$
6. $x^2 - 30x + 225$
7. $9 + 6x + x^2$
8. $4y^2 - 4y + 1$
9. $9x^2 + 6x + 1$
10. $16x^2 + 16x + 4$
11. $9y^2 - 18y + 9$
12. $16n^2 - 8n + 1$
13. $16 + 16a + 4a^2$
14. $1 + 20a + 100a^2$
15. $49a^2 - 28a + 4$

7-6. FRACTIONS

Parts of a Fraction: A fraction has a *numerator*, *denominator*, and a *fraction line*.

$$\text{fraction line} \rightarrow \frac{a}{b} \begin{matrix} \longleftarrow \text{numerator} \\ \longleftarrow \text{denominator} \end{matrix}$$

Quotient: A fraction is a way of indicating a *quotient* of two quantities. The fraction a/b can be read "a divided by b."

Ratio: The quotient of two numbers or quantities is also spoken of as the *ratio* of those quantities. Thus, the ratio of x to y is x/y.

Division by Zero: Since division by zero is not permitted, it should be understood in our work with fractions that *the denominator cannot be zero*.

Example: What values of x are not permitted in the fraction

$$\frac{3x}{x^2 + x - 6} \quad ?$$

Solution: Factoring the denominator, we get

$$\frac{3x}{x^2 + x - 6} = \frac{3x}{(x - 2)(x + 3)}$$

We see than an x equal to 2 or to -3 will result in division by zero, and so is not permitted.

Common Fraction: A fraction whose numerator and denominator are both integers is called a *common fraction.*

Examples:

$$\frac{2}{3}, \quad \frac{9}{5}, \quad \frac{124}{125}, \quad \text{and} \quad \frac{18}{11}$$

are common fractions.

Algebraic Fractions: An *algebraic fraction* is one whose numerator and/or denominator contain *literal* quantities.

Examples:

$$\frac{x}{y}, \quad \frac{\sqrt{x + 2}}{x}, \quad \frac{3}{y}, \quad \text{and} \quad \frac{x^2}{x - 3}$$

are algebraic fractions.

Rational Fractions: An algebraic fraction is called *rational* if the numerator and denominator are both *polynomials.*

Examples:

$$\frac{x}{y}, \quad \frac{3}{y}, \quad \text{and} \quad \frac{x^2}{x - 3}$$

are rational fractions but $\sqrt{x + 2}/x$ is not.

Proper and Improper Fractions: A *proper* common fraction is one whose numerator is smaller than its denominator.

Examples: $\frac{3}{5}$, $\frac{1}{3}$, and $\frac{9}{11}$ are proper fractions, whereas $\frac{8}{5}$, $\frac{3}{2}$, and $\frac{7}{4}$ are *improper* fractions.

Degree was defined in Sec. 4-1.

A proper *algebraic* fraction is a rational fraction whose numerator is of *lower degree* than the denominator.

Examples:

$$\frac{x}{x^2 + 2} \quad \text{and} \quad \frac{x^2 + 2x - 3}{x^3 + 9}$$

are proper fractions, while

$$\frac{x^3 - 2}{x^2 + x - 3} \quad \text{and} \quad \frac{x^2}{y}$$

are improper fractions.

Mixed Form: A *mixed number* is the sum of an integer and a fraction.

Examples:

$$2\frac{1}{2}, \quad 5\frac{3}{4}, \quad \text{and} \quad 3\frac{1}{3}$$

are mixed numbers.

A *mixed expression* is the sum of a polynomial and a rational algebraic expression.

Examples:

$$3x - 2 + \frac{1}{x} \quad \text{and} \quad y - \frac{y}{y^2 + 1}$$

are mixed expressions.

7-7. MANIPULATION OF FRACTIONS

Simplifying Fractions by Changing Signs: Recall from Sec. 4-5 that any two of the three signs of a fraction may be changed without changing the value of a fraction.

Rules of Signs		
Rules of Signs	$\frac{+a}{+b} = \frac{-a}{-b} = -\frac{-a}{+b} = -\frac{+a}{-b} = \frac{a}{b}$	**10**
	$\frac{+a}{-b} = \frac{-a}{+b} = -\frac{-a}{-b} = -\frac{a}{b}$	**11**

We can sometimes use this idea to simplify a fraction.

Example: Simplify the fraction

$$-\frac{3x - 2}{2 - 3x}$$

Solution: We change the sign of the denominator *and* the sign of the entire fraction;

$$-\frac{3x-2}{2-3x} = +\frac{3x-2}{-(2-3x)} = \frac{3x-2}{-2+3x} = \frac{3x-2}{3x-2} = 1$$

changed

Reducing a Fraction to Lowest Terms: Divide both numerator and denominator by any factor that is contained in both.

$\frac{ad}{bd} = \frac{a}{b}$	**50**

Examples:

(a) $$\frac{9}{12} = \frac{3(3)}{4(3)} = \frac{3}{4}$$

(b) $$\frac{3x^2yz}{9xy^2z^3} = \frac{3}{9} \cdot \frac{x^2}{x} \cdot \frac{y}{y^2} \cdot \frac{z}{z^3} = \frac{x}{3yz^2}$$

When possible, *factor* the numerator and denominator, and eliminate any factors common to both.

Examples:

(a) $$\frac{2x^2+x}{3x} = \frac{(2x+1)x}{3(x)} = \frac{2x+1}{3}$$

(b) $$\frac{ab+bc}{bc+bd} = \frac{b(a+c)}{b(c+d)} = \frac{a+c}{c+d}$$

Most students love canceling because they think they can cross out any term that stands in their way. If you use canceling, use it carefully!

The process of striking out the same factors from numerator and denominator is called *canceling*.

Common Error	If a factor is missing from *even one term* in the numerator or denominator, that factor *cannot* be canceled. $\frac{xy-z}{wx} \neq \frac{y-z}{w}$

Common Error	We may divide (or multiply) the numerator and denominator by the same quantity (Eq. 50), but we *may not add or subtract* the same quantity in the numerator and denominator, as this will change the value of the fraction. For example, $\frac{3}{5} \neq \frac{3+1}{5+1} = \frac{4}{6} = \frac{2}{3}$

Decimals and Fractions: To change a fraction to an equivalent decimal, simply divide the numerator by the denominator.

These are often written with a bar over the repeating part: $0.81\overline{81}$.

Example: To write $\frac{9}{11}$ as a decimal, we divide 9 by 11;

$$\frac{9}{11} = 0.8181818181\ldots$$

We get a *repeating decimal;* the dots following the number indicate that it repeats indefinitely.

To change a decimal number to a fraction, write a fraction with the decimal number in the numerator and 1 in the denominator. Multiply numerator and denominator by a multiple of 10 that will make the numerator a whole number. Finally, reduce to lowest terms.

Example:

$$0.875 = \frac{0.875}{1} = \frac{875}{1000} = \frac{7}{8}$$

To express a *repeating* decimal as a fraction, follow the steps in the next example.

Example: Change the repeating decimal $0.81\overline{81}$ to a fraction.

Solution: Let $x = 0.81\overline{81}$. Multiplying by 100,

$$100x = 81.\overline{81}$$

Subtracting the first equation from the second,

$$99x = 81 \quad \text{(exactly)}$$

Dividing by 99,

$$x = \frac{81}{99} = \frac{9}{11} \quad \text{(reduced)}$$

Changing Improper Fractions to Mixed Form: To write an improper fraction in mixed form, divide the numerator by the denominator and express the remainder as the numerator of a fraction whose denominator is the original denominator.

To change from mixed form to an improper fraction, see Sec. 7-9.

Example: Write $\frac{45}{7}$ as a mixed number.

Solution: Dividing 45 by 7, we get 6 with a remainder of 3, so

$$\frac{45}{7} = 6\frac{3}{7}$$

The procedure is the same for changing an *algebraic* fraction to mixed form. Divide the numerator by the denominator.

Example:

$$\frac{x^2 + 3}{x} = \frac{x^2}{x} + \frac{3}{x}$$
$$= x + \frac{3}{x}$$

EXERCISE 6

Division by Zero

In each fraction, what values of x, if any, are not permitted?

1. $\frac{12}{x}$ **2.** $\frac{x}{12}$ **3.** $\frac{18}{x - 5}$

Hint: *Factor the denominators in Problems 4, 5, and 6.*

4. $\frac{5x}{x^2 - 49}$ **5.** $\frac{7}{x^2 - 3x + 2}$ **6.** $\frac{3x}{8x^2 - 14x + 3}$

Simplifying Fractions

Simplify each fraction by manipulating the algebraic signs.

7. $\frac{a - b}{b - a}$ **8.** $-\frac{2x - y}{y - 2x}$

9. $\frac{(a - b)(c - d)}{b - a}$ **10.** $\frac{w(x - y - z)}{y - x + z}$

Reduce to lowest terms. Write your answers without negative exponents.

11. $\frac{2ab}{6b}$ **12.** $\frac{12m^2n}{15mn^2}$ **13.** $\frac{21m^2p^2}{28mp^4}$

14. $\frac{abx - bx^2}{acx - cx^2}$ **15.** $\frac{4a^2 - 9b^2}{4a^2 + 6ab}$ **16.** $\frac{3a^2 + 6a}{a^2 + 4a + 4}$

17. $\frac{x^2 + 5x}{x^2 + 4x - 5}$ **18.** $\frac{xy - 3y^2}{x^3 - 27y^3}$ **19.** $\frac{15x^2y^2z}{75xy^2z}$

20. $\frac{21x^2y^2z^2}{28x^2y^3z}$ **21.** $\frac{10abx^2y}{25abx^3y^2}$ **22.** $\frac{16xyz^3m}{24x^2y^2zm^2}$

23. $\frac{2a^2 - 2}{a^2 - 2a + 1}$ **24.** $\frac{3a^2 - 4ab + b^2}{a^2 - ab}$ **25.** $\frac{x^2 - z^2}{x^3 - z^3}$

Fractions and Decimals

Change each fraction to a decimal. Work to four digits.

26. $\frac{7}{12}$ **27.** $\frac{5}{9}$ **28.** $\frac{15}{16}$

29. $\frac{125}{155}$ **30.** $\frac{11}{3}$ **31.** $\frac{25}{9}$

Change each decimal to a fraction.

32. 0.7777 . . .	**33.** 0.636363 . . .	**34.** 0.6875
35. 0.28125	**36.** 0.4375	**37.** 0.390625

Improper Fractions and Mixed Form

Change each improper fraction to a mixed number.

38. $\frac{5}{3}$	**39.** $\frac{11}{5}$	**40.** $\frac{29}{12}$
41. $\frac{17}{3}$	**42.** $\frac{47}{5}$	**43.** $\frac{125}{12}$

Change each improper algebraic fraction to a mixed expression.

44. $\frac{x^2 + 1}{x}$ **45.** $\frac{1 - 4x}{2x}$

7-8. MULTIPLICATION AND DIVISION OF FRACTIONS

Multiplication: We multiply a fraction a/b by another fraction c/d as follows;

Multiplying Fractions	$\frac{a}{b} \cdot \frac{c}{d} = \frac{ac}{bd}$	**51**

The product of two or more fractions is a fraction whose numerator is the product of the numerators of the original fractions, and whose denominator is the product of the denominators of the original fractions.

Examples:

(a) $$\frac{2}{3} \cdot \frac{5}{7} = \frac{2(5)}{3(7)} = \frac{10}{21}$$

(b) $$\frac{1}{2} \cdot \frac{2}{3} \cdot \frac{3}{5} = \frac{1(2)(3)}{2(3)(5)} = \frac{1}{5}$$

(c) $$5\frac{2}{3} \cdot 3\frac{1}{2} = \frac{17}{3} \cdot \frac{7}{2} = \frac{119}{6}$$

(d) $$\frac{2a}{3b} \cdot \frac{5c}{4a} = \frac{10ac}{12ab} = \frac{5c}{6b}$$

Leave your product in factored form until after *you simplify.*

(e) $$\frac{x}{x + 2} \cdot \frac{x^2 - 4}{x^3} = \frac{x(x^2 - 4)}{(x + 2)x^3}$$

Factoring, $$= \frac{x(x + 2)(x - 2)}{(x + 2)x^3} = \frac{x - 2}{x^2}$$

Common Error	When multiplying mixed numbers, students sometimes try to multiply the whole parts and the fractional parts separately. $2\frac{2}{3} \times 4\frac{3}{5} \neq 8\frac{6}{15}$

Division: To divide one fraction, a/b, by another fraction, c/d:

$$\frac{\frac{a}{b}}{\frac{c}{d}}$$

Can you say why *it is permissible to multiply numerator and denominator by d/c?*

we multiply numerator and denominator by d/c:

$$\frac{\frac{a}{b}\cdot\frac{d}{c}}{\frac{c}{d}\cdot\frac{d}{c}} = \frac{\frac{a}{b}\cdot\frac{d}{c}}{\frac{cd}{cd}} = \frac{\frac{a}{b}\cdot\frac{d}{c}}{1} = \frac{a}{b}\cdot\frac{d}{c} = \frac{ad}{bc}$$

We see that dividing by a fraction is the same as *multiplying by the reciprocal* of that fraction.

Invert the divisor and multiply.

Division of Fractions	$\frac{a}{b} \div \frac{c}{d} = \frac{a}{b}\cdot\frac{d}{c} = \frac{ad}{bc}$	**52**

Examples:

(a) $\frac{2}{3} \div \frac{5}{7} = \frac{2}{3}\cdot\frac{7}{5} = \frac{14}{15}$

(b) $3\frac{2}{5} \div 2\frac{4}{15} = \frac{17}{5} \div \frac{34}{15}$

$$= \frac{17}{5} \times \frac{15}{34} = \frac{3}{2}$$

(c) $\frac{x}{y} \div \frac{x+2}{y-1} = \frac{x}{y}\cdot\frac{y-1}{x+2}$

$$= \frac{x(y-1)}{y(x+2)}$$

(d) $\frac{x^2+x-2}{x} \div \frac{x+2}{x^2} = \frac{x^2+x-2}{x}\cdot\frac{x^2}{x+2}$

Factoring, $= \frac{(x+2)(x-1)x^2}{x(x+2)} = x(x-1)$

EXERCISE 7

Multiplication of Fractions

Multiply the common fractions and mixed numbers and reduce.

1. $\frac{1}{3} \times \frac{2}{5}$

2. $\frac{3}{7} \times \frac{21}{24}$

3. $\frac{2}{3} \times \frac{9}{7}$

4. $\frac{3}{8} \times \frac{2}{9}$

5. $7 \times \frac{5}{8}$

6. $5 \times \frac{9}{20}$

7. $\frac{11}{3} \times 7$

8. $\frac{2}{3} \times 3\frac{1}{5}$

9. $3 \times 7\frac{2}{5}$

Multiply and simplify.

10. $\frac{15a^2}{7b^2} \cdot \frac{28ab}{9a^3c}$

11. $\frac{x + y}{x - y} \cdot \frac{x^2 - y^2}{(x + y)^2}$

12. $\frac{3ac}{4b} \cdot \frac{4x}{2ay}$

13. $\frac{x^4 - y^4}{a^2x^2} \cdot \frac{ax^3}{x^2 + y^2}$

14. $\frac{x + y}{10} \cdot \frac{ax}{3(x + y)}$

15. $\frac{5y^2}{a^2} \cdot \frac{3ax^2}{2y^2}$

16. $\frac{2a + 3b}{2x} \cdot \frac{2x}{4b}$

17. $\frac{a^4b^4}{2a^2y^n} \cdot \frac{a^2x}{xy^n}$

Division of Fractions

Divide and reduce.

18. $3\frac{7}{8} \div 2$

19. $\frac{7}{8} \div 4$

20. $8 \div \frac{3}{4}$

21. $\frac{3}{4} \div \frac{7}{10}$

22. $5\frac{3}{4} \div 8$

Divide and simplify.

23. $\frac{5abc^3}{3x^2} \div \frac{10ac^3}{6bx^2}$

24. $\frac{a^2 + 4a + 4}{d + c} \div (a + 2)$

25. $\frac{4a^3x}{6dy^2} \div \frac{2a^2x^2}{8a^2y}$

26. $\frac{5(x + y)^2}{x - y} \div (x + y)$

27. $\frac{ab + cd}{a + c} \div (a - c)$

28. $\frac{7x^2y}{3ad} \div \frac{2xy^2}{3a^2d}$

29. $\frac{3an + cm}{x^2 - y^2} \div (x^2 + y^2)$

30. $\frac{5x^2y^3z}{6a^2b^2c} \div \frac{10xy^3z^2}{8ab^2c^2}$

31. $\frac{4ayz}{8bcd} \div \frac{6a^2y^2z^2}{16bcd}$

32. $\frac{ac + ad + bc + bd}{c^2 - d^2} \div (a + b)$

Applications

33. The pressure on a surface is equal to the total force divided by the area, Eq. A26. Write and simplify an expression for the pressure on a circular surface of area $\pi d^2/4$ subjected to a distributed load F.

34. The stress on a bar in tension is equal to the load divided by the cross-sectional area, Eq. A30. Write and simplify an expression for the stress in a bar having a trapezoidal cross section of area $(a + b)h/2$, subject to a load P.

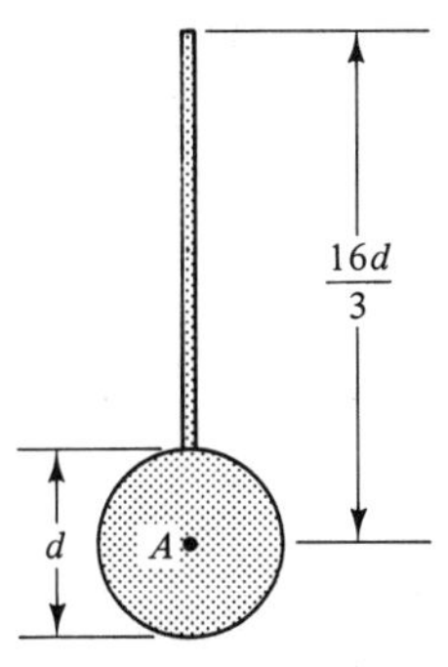

FIGURE 7-2

35. The acceleration on a body is equal to the force on the body divided by its mass, Eq. A19 and the mass equals the volume of the object times the density, Eq. A23. Write and simplify an expression for the acceleration of a sphere having a volume $4\pi r^3/3$ and a density D, subjected to a force F.

36. To find the moment of the area A in Fig. 7-2, we must multiply the area $\pi d^2/4$ by the distance of the pivot, $16d/3$. Multiply and simplify.

37. The circle in Fig. 7-3 has an area $\pi d^2/4$, and the sector has an area equal to $ds/4$. Find the ratio of the area of the circle to the area of the sector.

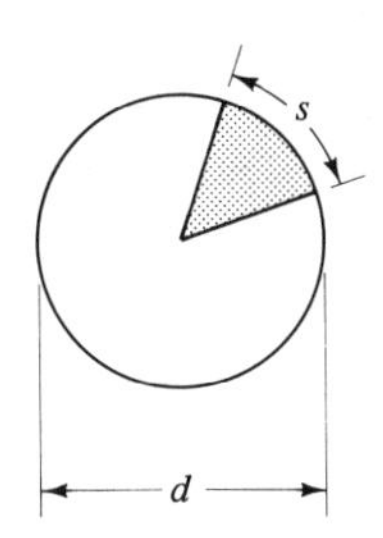

FIGURE 7-3

38. The mass density of an object is its mass divided by its volume (Eq. A23). Write and simplify an expression for the density of a sphere having a mass m and a volume equal to $4\pi r^3/3$ (Eq. 128).

7-9. ADDITION AND SUBTRACTION OF FRACTIONS

Similar Fractions: *Similar fractions* (also called *like* fractions) are those having the same (common) denominator. To add or subtract similar fractions, *combine the numerators and place them over the common denominator.*

Addition and Subtraction of Fractions	$\frac{a}{b} \pm \frac{c}{b} = \frac{a \pm c}{b}$	**53**

Examples:

(a) $$\frac{2}{3} + \frac{5}{3} = \frac{2 + 5}{3} = \frac{7}{3}$$

(b) $$\frac{1}{x} + \frac{3}{x} = \frac{1 + 3}{x} = \frac{4}{x}$$

(c) $$\frac{3x}{x + 1} - \frac{5}{x + 1} + \frac{x^2}{x + 1} = \frac{3x - 5 + x^2}{x + 1}$$

The least common denominator is also called the least common multiple (*LCM*) *of the denominators.*

Least Common Denominator: The *least common denominator,* LCD (also called the *lowest* common denominator) is the smallest expression that is exactly divisible by each of the denominators. Thus, the LCD must contain all the prime factors of each of the denominators. To find the LCD, form a product of all the factors of each denominator. Any factor that appears in more than one denominator is only included once.

Example: Find the LCD for the two fractions $\frac{3}{8}$ and $\frac{1}{18}$.

Solution: Factoring each denominator, we obtain

8		18
(2)(2)(2)		(2)(3)(3)
↖	duplicates; include only once in LCD	↗

Our LCD is then the product of these factors, dropping one of the 2's that appears in both sets of factors.

$$\text{LCD} = (2)(2)(2)(3)(3) = 8(9) = 72$$

For this simple problem you probably found the LCD by inspection in less time than it took to read this example, and are probably wondering what all the fuss is about. What we are really doing is developing a *method* that we can use later on complex algebraic fractions, where the LCD will *not* be obvious.

Example: Find the LCD for the fractions

$$\frac{5}{x^2 + x}, \quad \frac{x}{x^2 - 1}, \quad \text{and} \quad \frac{9}{x^3 - x^2}$$

Solution: The denominator $x^2 + x$ has the factors x and $x + 1$; the denominator $x^2 - 1$ has the factors $x + 1$ and $x - 1$; and the denominator $x^3 - x^2$ has the factors x, x, and $x - 1$. Our LCD is thus

$$(x)(x)(x + 1)(x - 1) \qquad \text{or} \qquad x^2(x^2 - 1)$$

Combining Fractions with Different Denominators: Find the LCD. Then multiply numerator and denominator of each fraction by that quantity that will make the denominator equal to the LCD. Finally, combine as shown above, and simplify.

Example:

$$\frac{1}{2} + \frac{2}{3} = \frac{1}{2}\left(\frac{3}{3}\right) + \frac{2}{3}\left(\frac{2}{2}\right)$$

$$= \frac{3}{6} + \frac{4}{6} = \frac{7}{6}$$

It is not necessary to write the fractions with the lowest common denominator; any common denominator will work as well. But your final result will then have to be reduced to lowest terms.

Example: Add $\frac{3}{8}$ and $\frac{1}{18}$.

Solution: The LCD, from a previous example, is 72. So

$$\frac{3}{8} + \frac{1}{18} = \frac{3}{8} \cdot \frac{9}{9} + \frac{1}{18} \cdot \frac{4}{4}$$

$$= \frac{27}{72} + \frac{4}{72} = \frac{31}{72}$$

Example: Combine: $x/2y - 5/x$.

Solution: The LCD will be the product of the two denominators, or $2xy$. So

$$\frac{x}{2y} - \frac{5}{x} = \frac{x}{2y}\left(\frac{x}{x}\right) - \frac{5}{x}\left(\frac{2y}{2y}\right)$$

$$= \frac{x^2}{2xy} - \frac{10y}{2xy}$$

$$= \frac{x^2 - 10y}{2xy}$$

The procedure is the same, of course, even when the denominators are more complicated.

Example: Combine the fractions

$$\frac{x + 2}{x - 3} + \frac{2x + 1}{3x - 2}$$

Solution: Our LCD is $(x - 3)(3x - 2)$, so

$$\frac{x + 2}{x - 3} + \frac{2x + 1}{3x - 2} = \frac{(x + 2)(3x - 2)}{(x - 3)(3x - 2)} + \frac{(2x + 1)(x - 3)}{(3x + 2)(x - 3)}$$

$$= \frac{(3x^2 + 4x - 4) + (2x^2 - 5x - 3)}{(x - 3)(3x + 2)}$$

$$= \frac{5x^2 - x - 7}{3x^2 - 7x - 6}$$

The method for adding and subtracting unlike fractions can be summarized by the formula

Combining Unlike Fractions	$\frac{a}{b} \pm \frac{c}{d} = \frac{ad}{bd} \pm \frac{bc}{bd} = \frac{ad \pm bc}{bd}$	**54**

Combining Integers and Fractions: Treat the integer as a fraction having 1 as a denominator, and combine as shown above.

Example:

$$3 + \frac{2}{9} = \frac{3}{1} + \frac{2}{9} = \frac{3}{1} \cdot \frac{9}{9} + \frac{2}{9}$$
$$= \frac{27}{9} + \frac{2}{9} = \frac{29}{9}$$

The same procedure may be used to change a *mixed number* to an *improper fraction.*

Example:

$$2\frac{1}{3} = 2 + \frac{1}{3} = \frac{6}{3} + \frac{1}{3}$$
$$= \frac{7}{3}$$

Changing a Mixed Algebraic Expression to an Improper Fraction: The procedure is no different from that in the preceding section. Write the nonfractional expression as a fraction with 1 as the denominator, find the LCD, and combine, as shown in the following example.

Example:

$$x^2 + \frac{5}{2x} = \frac{x^2}{1}\left(\frac{2x}{2x}\right) + \frac{5}{2x}$$

Adding,
$$= \frac{2x^3 + 5}{2x}$$

EXERCISE 8

Don't use your calculator for these numerical problems. The practice you get working with common fractions will help you when doing algebraic fractions.

Common Fractions and Mixed Numbers

Combine the common fractions and simplify.

1. $\frac{3}{5} + \frac{2}{5}$ **2.** $\frac{1}{8} - \frac{3}{8}$ **3.** $\frac{2}{7} + \frac{5}{7} - \frac{6}{7}$

4. $\frac{5}{9} + \frac{7}{9} - \frac{1}{9}$ **5.** $\frac{1}{3} - \frac{7}{3} + \frac{11}{3}$ **6.** $\frac{1}{5} - \frac{9}{5} + \frac{12}{5} - \frac{2}{5}$

Combine and simplify.

7. $\frac{1}{2} + \frac{2}{3}$ **8.** $\frac{3}{5} - \frac{1}{3}$ **9.** $\frac{3}{4} + \frac{7}{16}$

10. $\frac{2}{3} + \frac{3}{7}$ **11.** $\frac{5}{9} - \frac{1}{3} + \frac{3}{18}$ **12.** $\frac{1}{2} + \frac{1}{3} + \frac{1}{5}$

Combine the integers and fractions.

13. $2 + \frac{3}{5}$ **14.** $3 - \frac{2}{3} + \frac{1}{6}$ **15.** $\frac{1}{5} - 7 + \frac{2}{3}$

16. $\frac{1}{5} + 2 - \frac{1}{3} + 5$ **17.** $\frac{1}{7} + 2 - \frac{3}{7} + \frac{1}{5}$ **18.** $5 - \frac{3}{5} + \frac{2}{15} - 2$

Rewrite the mixed numbers as improper fractions.

19. $2\frac{2}{3}$ **20.** $1\frac{5}{8}$ **21.** $3\frac{1}{4}$

22. $2\frac{7}{8}$ **23.** $5\frac{11}{16}$ **24.** $9\frac{3}{4}$

25. $7\frac{2}{3}$ **26.** $5\frac{7}{8}$

Combining Algebraic Fractions

Combine the algebraic fractions and simplify.

27. $\frac{3}{x} + \frac{2}{x} - \frac{1}{x}$ **28.** $\frac{2a}{y} + \frac{3}{y} - \frac{a}{y}$ **29.** $\frac{x}{3a} - \frac{y}{3a} + \frac{z}{3a}$

30. $\frac{5x}{2} - \frac{3x}{2}$ **31.** $\frac{7}{x + 2} - \frac{5}{x + 2}$ **32.** $\frac{3x}{a - b} + \frac{2x}{b - a}$

33. $\frac{a}{x^2} - \frac{b}{x^2} + \frac{c}{x^2}$

Combine the algebraic fractions and simplify.

34. $\frac{1}{a} + \frac{1}{b}$ **35.** $\frac{3}{x} - \frac{2}{x^2}$ **36.** $\frac{3a}{2x} + \frac{2a}{5x}$

37. $\frac{1}{x} + \frac{1}{y} + \frac{1}{z}$ **38.** $\frac{a}{3} + \frac{2}{x} - \frac{1}{2}$ **39.** $\frac{1}{x + 3} + \frac{1}{x - 2}$

40. $\frac{2ab}{3xy} + \frac{5ad}{2xy}$ **41.** $\frac{3mn}{4y^2} + \frac{2mn}{4y}$ **42.** $\frac{a + b}{3} - \frac{a - b}{2}$

43. $\frac{3}{a + b} + \frac{2}{a - b}$ **44.** $\frac{4}{x - 1} - \frac{5}{x + 1}$ **45.** $\frac{x + 1}{x - 1} - \frac{x - 1}{x + 1}$

Mixed Form

Change the mixed expressions to improper fractions.

46. $x + \frac{1}{x}$ **47.** $\frac{1}{2} - x$

48. $\frac{x}{x - 1} - 1 - \frac{1}{x + 1}$ **49.** $a + \frac{1}{a^2 - b^2}$

50. $3 - a - \dfrac{2}{a}$

51. $\dfrac{5}{x} - 2x + \dfrac{3}{x^2}$

Applications

52. The resistance R (Fig. 7-4) is, from Eqs. A41 and A42,

$$\frac{R_1R_2}{R_1 + R_2} + R_3$$

Combine into a single term and simplify.

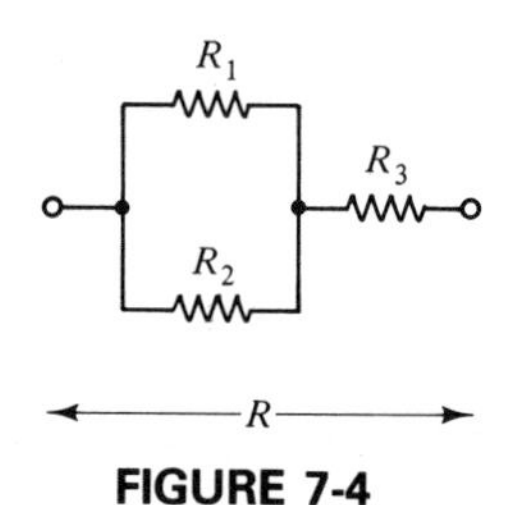

FIGURE 7-4

53. If a car travels a distance d at a rate V, the time required will be, by Eq. A16, d/V. The car then continues for a distance d_1 at a rate V_1, and for a third distance d_2 at rate V_2. Write an expression for the total travel time, then combine the three terms into a single term and simplify.

54. The acceleration a_1 of a body of mass m_1 subjected to a force F_1 is, by Eq. A19, $a_1 = F_1/m_1$. Write an expression for the difference in acceleration between that body and another having a mass m_2 and force F_2. Combine the terms of that expression and simplify.

7-10. COMPLEX FRACTIONS

So far, we have been dealing with *simple* fractions—fractions that have only *one* fraction line. Fractions that have *more than one* fraction line are called *complex* fractions.

Example:

$$\frac{\dfrac{a}{b}}{c} \quad \text{and} \quad \frac{x - \dfrac{1}{x}}{\dfrac{x}{y} - x}$$

are complex fractions.

Example: Simplify the complex fraction,

$$\frac{\dfrac{1}{2} + \dfrac{2}{3}}{3 + \dfrac{1}{4}}$$

Solution: We can simplify this fraction by multiplying numerator and denominator by the least common denominator for all the individual fractions. The denominators are 2, 3, and 4, so the LCD is 12. Multiplying, we obtain

$$\frac{\left(\frac{1}{2}+\frac{2}{3}\right)12}{\left(3+\frac{1}{4}\right)12}=\frac{6+8}{36+3}=\frac{14}{39}$$

Example: Simplify the complex fraction

$$\frac{1+\frac{a}{b}}{1-\frac{b}{a}}$$

Solution: The LCD for the two small fractions a/b and b/a is ab. Multiplying, we obtain

$$\frac{1+\frac{a}{b}}{1-\frac{b}{a}}\cdot\frac{ab}{ab}=\frac{ab+a^2}{ab-b^2}$$

or, in factored form,

$$\frac{a(b+a)}{b(a-b)}$$

EXERCISE 9

Complex Fractions

Simplify.

1. $\frac{\frac{2}{3}+\frac{3}{4}}{\frac{1}{5}}$ **2.** $\frac{\frac{3}{4}-\frac{1}{3}}{\frac{1}{2}+\frac{1}{6}}$ **3.** $\frac{\frac{1}{2}+\frac{1}{3}+\frac{1}{4}}{3-\frac{4}{5}}$

4. $\frac{\frac{4}{5}}{\frac{1}{5}+\frac{2}{3}}$ **5.** $\frac{5-\frac{2}{5}}{6+\frac{1}{3}}$ **6.** $\frac{1}{2}+\frac{3}{\frac{2}{5}+\frac{1}{3}}$

7. $\frac{x+\frac{y}{4}}{x-\frac{y}{3}}$ **8.** $\frac{\frac{a}{b}+\frac{x}{y}}{\frac{a}{z}-\frac{x}{c}}$ **9.** $\frac{1+\frac{x}{y}}{1+\frac{x^2}{y^2}}$

10. $\dfrac{\dfrac{x}{b}+\dfrac{y}{b}}{\dfrac{z}{b}}$ **11.** $\dfrac{\dfrac{1}{x}+\dfrac{1}{y}}{\dfrac{1}{x}-\dfrac{1}{y}}$ **12.** $1-\dfrac{1}{1+\dfrac{1}{a}}$

7-11. FRACTIONAL EQUATIONS

Solving Fractional Equations: An equation in which one or more terms is a fraction is called a *fractional equation*. To solve a fractional equation, first eliminate the fractions by multiplying both sides of the equation by the lowest common denominator (LCD) of every term. With the fractions thus eliminated, the equation is then solved like any nonfractional equation.

Example: Solve for x:

$$\frac{3x}{5}-\frac{x}{3}=\frac{2}{15}$$

Solution: Multiplying by the LCD (15), we obtain

$$15\left(\frac{3x}{5}-\frac{x}{3}\right)=15\left(\frac{2}{15}\right)$$
$$9x-5x=2$$
$$x=\frac{1}{2}$$

Equations with the Unknown in the Denominator: The procedure is the same when the unknown appears in the denominator of one or more terms. However, the LCD will now contain the unknown.

Example: Solve for x:

$$\frac{2}{3x}=\frac{5}{x}+\frac{1}{2}$$

Solution: The LCD is $6x$. Multiplying both sides of the equation yields

$$6x\left(\frac{2}{3x}\right)=6x\left(\frac{5}{x}+\frac{1}{2}\right)$$
$$6x\left(\frac{2}{3x}\right)=6x\left(\frac{5}{x}\right)+6x\left(\frac{1}{2}\right)$$
$$4=30+3x$$
$$3x=-26$$
$$x=-\frac{26}{3}$$

Be especially sure to check your answer after multiplying by an LCD containing the unknown, because this will often introduce an extraneous solution.

Check:

$$\frac{2}{3\left(-\frac{26}{3}\right)} \stackrel{?}{=} \frac{5}{-\frac{26}{3}} + \frac{1}{2}$$

$$-\frac{2}{26} \stackrel{?}{=} -\frac{15}{26} + \frac{13}{26}$$

$$-\frac{2}{26} = -\frac{2}{26} \quad \text{checks}$$

Example: Solve for x:

$$\frac{8x + 7}{5x + 4} = 2 - \frac{2x}{5x + 1}$$

Solution: Multiplying by the LCD $(5x + 4)(5x + 1)$ yields

$$(8x + 7)(5x + 1) = 2(5x + 4)(5x + 1) - 2x(5x + 4)$$
$$40x^2 + 35x + 8x + 7 = 2(25x^2 + 5x + 20x + 4) - 10x^2 - 8x$$
$$40x^2 + 43x + 7 = 50x^2 + 50x + 8 - 10x^2 - 8x$$
$$43x + 7 = 42x + 8$$
$$x = 1$$

Check:

$$\frac{8(1) + 7}{5(1) + 4} \stackrel{?}{=} 2 - \frac{2(1)}{5(1) + 1}$$

$$\frac{15}{9} \stackrel{?}{=} 2 - \frac{2}{6}$$

$$\frac{5}{3} = \frac{6}{3} - \frac{1}{3} \quad \text{checks}$$

Common Error	The technique of multiplying by the LCD in order to eliminate the denominators is valid *only when we have an equation.* Do not multiply through by the LCD when there is no equation!

EXERCISE 10

Fractional Equations

Solve for x.

1. $2x + \frac{x}{3} = 28$

2. $4x + \frac{x}{5} = 42$

3. $x + \frac{x}{5} = 24$

4. $\frac{x}{6} + x = 21$

5. $3x - \frac{x}{7} = 40$

6. $x - \frac{x}{6} = 25$

7. $\frac{4x}{6} - x = -24$

8. $\frac{3x}{5} + 7x = 38$

9. $\frac{x}{2} + \frac{x}{3} + \frac{x}{4} = 26$

10. $\frac{x-1}{2} = \frac{x+1}{3}$

11. $\frac{3x-1}{4} = \frac{2x+1}{3}$

12. $x + \frac{2x}{3} + \frac{3x}{4} = 29$

Equations with Unknown in Denominator

Solve for x.

13. $\frac{2}{3x} + 6 = 5$

14. $9 - \frac{4}{5x} = 7$

15. $4 + \frac{1}{x+3} = 8$

16. $\frac{x+5}{x-2} = 5$

17. $\frac{x-3}{x+2} = \frac{x+4}{x-5}$

18. $\frac{3}{x+2} = \frac{5}{x} + \frac{4}{x^2+2x}$

19. $\frac{3}{x+1} - \frac{x+1}{x-1} = \frac{x^2}{1-x^2}$

20. $\frac{3x+5}{2x-3} = \frac{3x-3}{2x-1}$

7-12. WORD PROBLEMS LEADING TO FRACTIONAL EQUATIONS

Solving Word Problems: The method for solving the word problems in this chapter (and in any other chapter) is no different from that given in Sec. 5-3. The resulting equations, however, will now contain fractions.

Example: One-fifth of a certain number is 3 greater than one-sixth of that same number. Find the number.

Solution: Let x = the number. Then from the problem statement,

$$\frac{1}{5}x = \frac{1}{6}x + 3$$

Multiplying by the LCD (30), we obtain

$$6x = 5x + 90$$

$$x = 90$$

Work Problems: One type of word problem not covered in Chapter 5 is the work problem. To tackle work problems, we need one simple idea:

(rate of work) × (time) = amount done	**A5**

We often use this equation to find the rate of work for a person or a machine.

Example: Find the work rate R for a technician who can assemble three amplifiers in 4 days.

Solution: From Eq. A5,

$$R(4) = 3$$

$$R = \frac{3}{4} \text{ amplifier per day}$$

In a typical work problem, there are two or more persons or machines doing work, each at a different rate. Find the work rate for each, and add them to get the combined rate for all the workers.

Example: If crew A can assemble 2 cars in 5 days, and crew B can assemble 3 cars in 7 days, how long will it take them to assemble 100 cars, working together?

Solution: We find the work rate for each crew, using Eq. A5:

Crew A:

$$\text{work rate} = \frac{2}{5} \text{ car per day}$$

Crew B:

$$\text{work rate} = \frac{3}{7} \text{ car per day}$$

Their combined rate is the sum of the individual rates:

$$\text{combined rate} = \frac{2}{5} + \frac{3}{7} = \frac{14}{35} + \frac{15}{35} = \frac{29}{35} \text{ car per day}$$

Letting x = the number of days needed to assemble 100 cars, we get, by Eq. A5

$$\left(\frac{29}{35}\right)x = 100$$

$$x = 100\left(\frac{35}{29}\right) = 121 \text{ days} \quad \text{(rounded)}$$

Example: A certain small hydroelectric generating station can produce 17 megawatthours (MWh) of energy per year. After 4 months of operation, another generator is added which, by itself, can produce 11 MWh in 5 months. How many additional months are needed for a total of 25 MWh to be produced?

Solution: Let x = additional months. The original generating station can produce 17/12 MWh per month for $4 + x$ months. The new generator can produce 11/5 MWh per month for x months.

$$\frac{17}{12}(4 + x) + \frac{11}{5}x = 25$$

Multiplying by the LCD, 60, we obtain

$$5(17)(4 + x) + 12(11)x = 25(60)$$
$$340 + 85x + 132x = 1500$$
$$217x = 1160$$
$$x = \frac{1160}{217} = 5.3 \text{ months}$$

EXERCISE 11

Number Problems

Depending on how you set them up, some of these problems may not result in fractional equations.

1. There is a number such that if $\frac{1}{5}$ of it is subtracted from 50 and this difference is multiplied by 4, the result will be 70 less than the number. What is the number?

2. Separate 100 into two parts such that if $\frac{1}{3}$ of one part is subtracted from $\frac{1}{4}$ of the other, the difference is 11.

3. The second of two numbers is 1 greater than the first, and $\frac{1}{2}$ of the first plus $\frac{1}{5}$ of the first is equal to the sum of $\frac{1}{3}$ of the second and $\frac{1}{4}$ of the second. What are the numbers?

4. Five years ago A's age was $2\frac{1}{3}$ times B's. One year hence it will be $1\frac{4}{9}$ times B's. How old is each now?

Work Problems

5. A laborer can do a certain job in 5 days, a second can do the job in 6 days, and a third in 8 days. In what time can the three together do the job?

6. A technician can assemble an instrument in 9.5 h. After working for 2 h he is joined by another technician who, alone, could do the job in 7.5 h. How many additional hours are needed to finish the job?

7. A certain screw machine can produce a box of parts in 3.3 h. A new machine is to be ordered having a speed such that both machines working together would produce a box of parts in 1.4 h. How long would it take the new machine alone to produce a box of parts?

Fluid Flow Problems

8. A cistern can be filled by a pipe in 3 h and emptied by another pipe in 4 h. How much time will be required to fill the cistern if both are running?

9. A tank holding 1200 liters has three pipes connected. The first, by itself, could fill the tank in $2\frac{1}{2}$ h, the second in 2 h, and the third in 1 h 40 min. In how many minutes will the tank be filled?

Motion Problems

10. A person sets out from Boston and walks toward Portland at the rate of 3 mi/h. Three hours afterward, a second person sets out from the same place and walks in the same direction at the rate of 4 mi/h. How far from Boston will the second overtake the first?

11. A courier who goes at the rate of $6\frac{1}{2}$ mi/h is followed, after 4 h, by another who goes at the rate of $7\frac{1}{2}$ mi/h. In how many hours will the second overtake the first?

12. A person walks to the top of a mountain at the rate of 2 mi/h and down the same way at the rate of 4 mi/h. If he is out 6 h, how far is it to the top of the mountain?

13. In going a certain distance, a train traveling at the rate of 40 mi/h takes 2 h less than a train traveling 30 mi/h. Find the distance.

Energy Flow Problems

In this group of problems, as in the previous ones, we assume that the rates remain constant.

14. A certain power plant consumes 1500 tons of coal in 4 weeks. There is a stockpile of 10,000 tons of coal available when the plant starts operating. After 3 weeks of operation, an additional boiler, capable of using 2300 tons in 3 weeks, is put on line with the first boiler. In how many more weeks will the stockpile of coal be consumed?

15. A certain array of solar cells can generate 2 megawatthours (MWh) in 5 months (under standard conditions). After operating for 3 months, another array of cells is added which alone can generate 5 MWh in 7 months. How many additional months, after the new array was added, is needed for the total energy generated from both arrays to be 10 MWh?

16. A wind generator can charge 20 storage batteries in 24 h. After charging for 6 h, another generator, which can charge the batteries in 36 h, is also connected to the batteries. How many additional hours are needed to charge the batteries?

17. A certain solar collector can absorb 9000 Btu in 7 h. Another panel is added and together they collect 35,000 Btu in 5 h. How long would it take the new panel alone to collect 35,000 Btu?

Financial Problems

18. After spending $\frac{1}{4}$ of my money, then $\frac{1}{5}$ of the remainder, I had remaining $66. How many dollars had I at first?

19. A person who had inherited money spent $\frac{3}{8}$ of it the first year, $\frac{4}{5}$ of the remainder the next year, and had $1420 left. What was the inheritance?

20. Three children were left an inheritance of which the eldest received $\frac{2}{3}$, the second $\frac{1}{5}$, and the third the rest, which was $200. How much did each receive?

21. A's capital was $\frac{3}{4}$ of B's. If A's had been \$500 less, it would have been only $\frac{1}{2}$ of B's. What was the capital of each?

7-13. LITERAL EQUATIONS AND FORMULAS

Literal Equations: A *literal equation* is one in which some or all of the constants are represented by letters.

Example:

$$a(x + b) = b(x + c)$$

is a literal equation, whereas

$$2(x + 5) = 3(x + 1)$$

is a numerical equation.

Formulas: As we said in Sec. 1-7 a *formula* is a literal equation that relates two or more mathematical or physical quantities. These are the equations, mentioned in the introduction of this chapter, that describe the workings of the physical world. A listing of some of the common formulas used in technology is given in the Summary of Facts and Formulas (Appendix A).

Example: Newton's second law,

$$F = ma$$

is a formula relating the force acting on a body with its mass and acceleration.

Solving Literal Equations and Formulas: When we solve a literal equation or formula, we cannot, of course, get a *numerical* answer, as we could with equations that had only one unknown. Our object here is to *isolate* one of the letters on one side of the equal sign. We "solve for" one of the literal quantities.

Example: Solve for x:

$$b\left(b + \frac{x}{a}\right) = d$$

Solution: Dividing both sides by b,

$$b + \frac{x}{a} = \frac{d}{b}$$

Subtracting b,

$$\frac{x}{a} = \frac{d}{b} - b$$

Multiplying both sides by a,

$$x = a\left(\frac{d}{b} - b\right)$$

Checking Literal Equations: As with numerical equations, we can substitute our expression for x back into the original equation and see if it checks, but this could get complicated. We can perform a simple "check" with numerical values, as in the following example.

Example: Check the results from the preceding example.

When choosing values, avoid 1 or 0 or values that will make a denominator zero.

Solution: Let us *choose* values for a, b, and d, say,

$$a = 2 \qquad b = 3 \qquad \text{and} \qquad d = 6$$

With these values,

$$x = a\left(\frac{d}{b} - b\right) = 2\left(\frac{6}{3} - 3\right) = -2$$

We now see if these values check in the original equation.

$$b\left(b + \frac{x}{a}\right) = d$$

$$3\left(3 + \frac{-2}{2}\right) \stackrel{?}{=} 6$$

$$3(3 - 1) \stackrel{?}{=} 6$$

$$3(2) = 6 \qquad \text{checks}$$

Realize that this is not a rigorous check, because we are testing the solution only for the specific values chosen. However, if it checks for those values, we can be *reasonably* sure that our solution is correct.

Example: Solve for x:

$$a(x + b) = b(x + c)$$

Solution: Removing parentheses,

$$ax + ab = bx + bc$$

Subtracting $bx + ab$,

$$ax - bx = bc - ab$$

Factoring,

$$x(a - b) = b(c - a)$$

Dividing by $(a - b)$,

$$x = \frac{b(c - a)}{a - b}$$

(Note that when $a = b$ this equation is not valid, for it would result in division by zero.)

Literal Fractional Equations: The procedure is the same as for other fractional equations; multiply by the LCD to eliminate the fractions.

Example: Solve for x in terms of a, b, and c:

$$\frac{x}{b} - \frac{a}{c} = \frac{x}{a}$$

Solution: Multiplying by the LCD (abc) yields

$$acx - a^2b = bcx$$

Transposing,

$$acx - bcx = a^2b$$

Factoring,

$$x(ac - bc) = a^2b$$

$$x = \frac{a^2b}{ac - bc}$$

EXERCISE 12

Literal Equations

Solve for x.

1. $2ax = bc$

2. $ax + dx = a - c$

3. $a(x + y) = b(x + z)$

4. $4x = 2x + ab$

5. $4acx - 3d^2 = a^2d - d^2x$

6. $a(2x - c) = a + c$

7. $a^2x - cd = b - ax + dx$

8. $3(x - r) = 2(x + p)$

9. $\frac{a}{2}(x - 3w) = z$

10. $cx - x = bc - b$

11. $3x + m = b$

12. $ax + m = cx + n$

13. $ax - bx = c + dx - m$

14. $3m + 2x - c = x + d$

15. $ax - ab = cx - bc$

16. $p(x - b) = qx + d$

17. $3(x - b) = 2(bx + c) - c(x - b)$

18. $a(bx + d) - cdx = c(dx + d)$

19. $(x + a)(x + b) = x^2 + c$

Literal Fractional Equations

Solve for x.

20. $\dfrac{w + x}{x} = w(w + y)$

21. $\dfrac{ax + b}{c} = bx - a$

22. $\dfrac{p - q}{x} = 3p$

23. $\dfrac{bx - c}{ax - c} = 5$

24. $\dfrac{a - x}{5} = \dfrac{b - x}{2}$

25. $\dfrac{x}{a} - b = \dfrac{c}{d} - x$

26. $\dfrac{x}{a} - a = \dfrac{a}{c} - \dfrac{x}{c - a}$

27. $\dfrac{x}{a - 1} - \dfrac{x}{a + 1} = b$

28. $ax - \dfrac{3a - bx}{2} = \dfrac{1}{4}$

Formulas

29. The correction for the sag in a surveyor's tape weighing w lb/ft and pulled with a force of P lb is

$$C = \frac{w^2L^3}{24P^2} \quad \text{feet}$$

Solve this equation for the distance measured, L.

30. When a bar of length L_0 having a coefficient of linear thermal expansion α is increased in temperature by an amount Δt, it will expand to a new length L, where

$$L = L_0(1 + \alpha\Delta t) \qquad \text{(A35)}$$

Solve this equation for Δt.

31. The formula for the equivalent resistance R for the parallel combination of two resistors, R_1 and R_2, is

$$\frac{1}{R} = \frac{1}{R_1} + \frac{1}{R_2} \qquad \text{(A42)}$$

Solve this formula for R_2.

32. A rod of cross-sectional area a and length L will stretch by an amount e when subject to a tensile load of P. The modulus of elasticity is given by

$$E = \frac{PL}{ae} \qquad \text{(A32)}$$

Solve this equation for a.

33. The formula for the displacement s of a freely falling body having an initial velocity V_0 and acceleration a is

$$s = v_0t + \tfrac{1}{2}at^2 \qquad \text{(A17)}$$

Solve this equation for a.

34. The formula for the amount of heat flowing through a wall by conduction is

$$q = \frac{kA(t_1 - t_2)}{L}$$

Solve this equation for t_2.

35. If the resistance of a conductor is R_1 at temperature t_1, the resistance will change to a value R when the temperature changes to t, where

$$R = R_1[1 + \alpha(t - t_1)] \qquad \text{(A48)}$$

and α is the temperature coefficient of resistance at temperature t_1. Solve this equation for t_1.

36. An amount a invested at a simple interest rate n for t years will accumulate to an amount y, where $y = a + ant$. Solve for a.

37. Applying Kirchhoff's law (Eq. A46) to loop 1 in Fig. 7-5 we get $E = I_1R_1 + I_1R_2 - I_2R_2$. Solve for I_1.

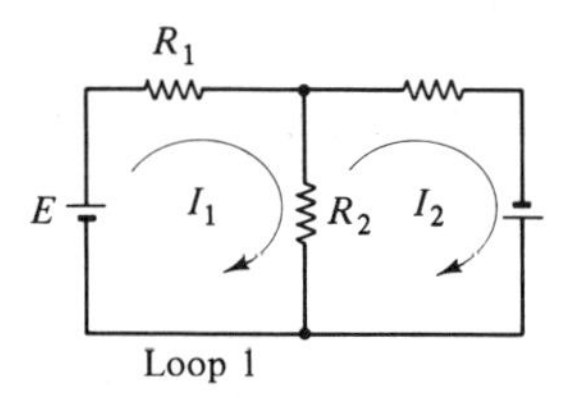

FIGURE 7-5

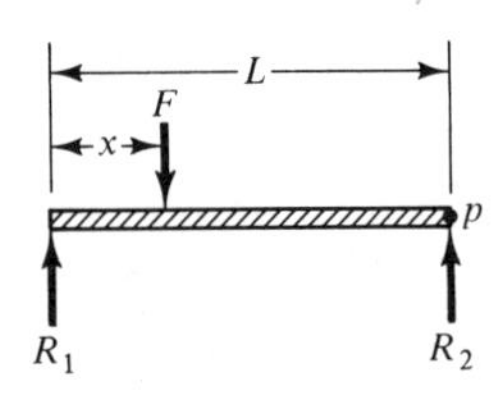

FIGURE 7-6

38. When a bar of length L_0 is changed in temperature by an amount Δt, it changes to a new length L, where

$$L = L_0 + L_0\alpha\Delta t$$

where α is the coefficient of thermal expansion. Solve for L_0.

39. Taking the moment M about point p in Fig. 7-6 we get $M - R_1L - F(L - x)$. Solve for L.

40. Three masses, m_1, m_2, and m_3, are attached together and accelerated by means of a force F. By Eq. A19,

$$F = m_1a + m_2a + m_3a$$

Solve for the acceleration a.

41. A resistance R_1, when changed in temperature from t_1 to t, will change in resistance to R, where

$$R = R_1 + R_1\alpha(t - t_1)$$

where α is the temperature coefficient of resistance. Solve for R_1.

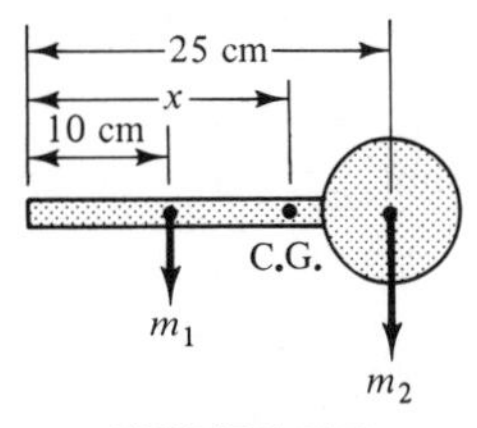

FIGURE 7-7

42. A bar of mass m_1 is attached to a sphere of mass m_2 (Fig. 7-7). The distance x to the center of gravity is

$$x = \frac{10m_1 + 25m_2}{m_1 + m_2}$$

Solve for m_1.

43. A ball of mass m is swung in a vertical circle (Fig. 7-8). At the top of its swing, the tension T in the cord plus the ball's weight mg is just balanced by the centrifugal force mv^2/R:

$$T + mg = \frac{mv^2}{R}$$

Solve for m.

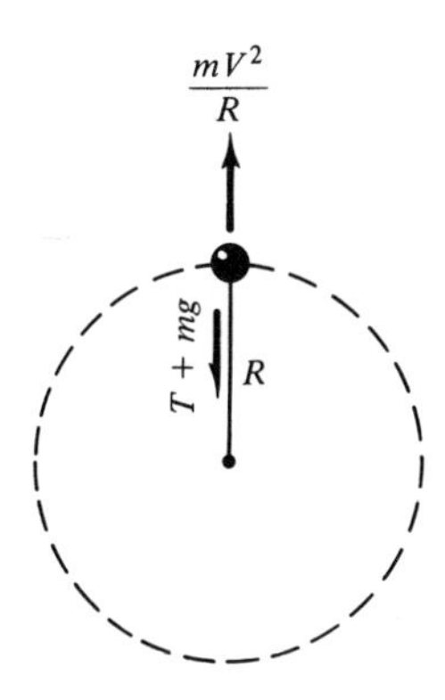

FIGURE 7-8

44. The total energy of a body of mass m, moving with velocity v and located at a height y above some datum, is the sum of the potential energy mgy and the kinetic energy $\frac{1}{2}mv^2$:

$$E = mgy + \tfrac{1}{2}mv^2$$

Solve for m.

CHAPTER TEST

Perform the indicated operations and simplify.

1. $\dfrac{a^2 + b^2}{a - b} - a + b$

2. $\dfrac{3}{5x^2} - \dfrac{2}{15xy} + \dfrac{1}{6y^2}$

3. $\dfrac{3m + 1}{3m - 1} + \dfrac{m - 4}{5 - 2m} - \dfrac{3m^2 - 2m - 4}{6m^2 - 17m + 5}$

4. $\dfrac{\dfrac{1}{1 - x} - \dfrac{1}{1 + x}}{\dfrac{1}{1 - x^2} - \dfrac{1}{1 + x^2}}$

5. $(a^2 + 1 + a)\left(1 - \dfrac{1}{a} + \dfrac{1}{a^2}\right)$

6. $\dfrac{\dfrac{a - 1}{6} - \dfrac{2a - 7}{2}}{\dfrac{3a}{4} - 3}$

7. $\dfrac{1 + \dfrac{a - c}{a + c}}{1 - \dfrac{a - c}{a + c}}$

8. $\left(1 + \dfrac{x + y}{x - y}\right)\left(1 - \dfrac{x - y}{x + y}\right)$

Factor completely.

9. $x^2 - 2x - 15$

10. $2a^2 + 3a - 2$

11. $x^6 - y^4$

12. $2x^2 + 3x - 2$

13. $8x^3 - \frac{y^3}{27}$

14. $\frac{x^2}{y} - \frac{x}{y}$

15. $2ax^2y^2 - 18a$

16. $\frac{2a^2}{12} - \frac{8b^2}{27}$

17. $3a^2 - 2a - 8$

18. $(y + 2)^2 - z^2$

19. $x^2 - 7x + 12$

20. $4a^2 - (3a - 1)^2$

21. $4a^6 - 4b^6$

22. $(x - y)^2 - z^2$

23. $x^2 - 2x - 3$

24. $1 - 16x^2$

25. $a^2 - 2a - 8$

26. $9x^4 - x^2$

27. $x^2 - 21x + 110$

Solve for x.

28. $cx - 5 = ax + b$

29. $a(x - 3) - b(x + 2) = c$

30. $\frac{5 - 3x}{4} + \frac{3 - 5x}{3} = \frac{3}{2} - \frac{5x}{3}$

31. $\frac{3x - 1}{11} - \frac{2 - x}{10} = \frac{6}{5}$

32. $\frac{x + 3}{2} + \frac{x + 4}{3} + \frac{x + 5}{4} = 16$

33. $\frac{2x + 1}{4} - \frac{4x - 1}{10} + \frac{5}{4} = 0$

34. $x^2 - (x - p)(x + q) = r$

35. $mx - n = \frac{nx - m}{p}$

36. $p = \frac{q - rx}{px - q}$

37. $m(x - a) + n(x - b) + p(x - c) = 0$

38. $\frac{x - 3}{4} - \frac{x - 1}{9} = \frac{x - 5}{6}$

39. $\frac{1}{x - 5} - \frac{1}{4} = \frac{1}{3}$

40. It is estimated that a bulldozer can prepare a certain site in 15 days and that a larger bulldozer can prepare the same site in 11 days. If we assume that they do not get in each other's way, how long will it take the two machines, working together, to prepare the site?

41. Multiply:

$$2\frac{3}{5} \times 7\frac{5}{9}$$

42. Divide:

$$\frac{8x^3y^2z}{5a^4b^3c} \text{ by } \frac{4xy^5z^3}{15abc^3}$$

43. Multiply:

$$\frac{3wx^2y^3}{7axyz} \text{ by } \frac{4a^3xz}{6aw^2y}$$

44. Divide:

$$9\frac{5}{7} \div 3\frac{4}{9}$$

Systems of Linear Equations

Usually, a physical situation can be described by a single equation. For example, the displacement of a freely falling body is described by Eq. A17, $s = v_0 t + \frac{1}{2}at^2$. Often, however, a situation can only be described by *more than one* equation. For example, to find the two currents I_1 and I_2 in the circuit of Fig. 8-1, we must solve the two equations

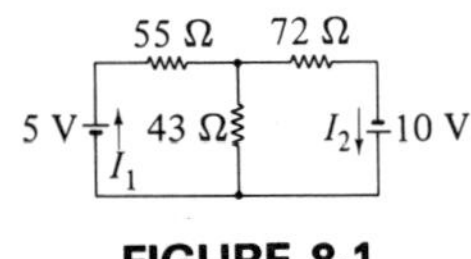

FIGURE 8-1

$$98I_1 - 43I_2 = 5$$

$$43I_1 - 115I_2 = 10$$

We must find values for I_1 and I_2 that satisfy *both* equations at the same time. In this chapter we learn how to solve such sets of two or three linear equations, and we also get practice in writing systems of equations to describe a variety of technical problems. In Chapter 9 we learn how to solve systems of equations by means of matrices and determinants.

8-1. SYSTEMS OF LINEAR EQUATIONS

Linear Equations: We have previously defined a *linear equation* as one of *first degree*.

Example: The equation

$$3x + 5 = 20$$

is a linear equation in *one unknown*. We learned how to solve these in Chapter 4.

Example: The equation

$$2x - 5y = 8$$

is a linear equation in *two* unknowns.

A linear equation can have *any number* of unknowns.

Example: $x - 3y + 2z = 5$ is a linear equation in three unknowns.

Example: $3x + 2y - 5z - w = 6$ is a linear equation in four unknowns.

Systems of equations are also called simultaneous *equations.*

Systems of Equations: A set of two or more equations that simultaneously impose conditions on all the variables is called a *system of equations*.

Examples:

(a)
$$\begin{aligned} 3x - 2y &= 5 \\ x + 4y &= 1 \end{aligned}$$

is a system of two linear equations in two unknowns.

(b)
$$\begin{aligned} x - 2y + 3z &= 4 \\ 3x + y - 2z &= 1 \\ 2x + 3y - z &= 3 \end{aligned}$$

is a system of three linear equations in three unknowns.

Note that some variables may have zero coefficients and may not appear in every equation.

(c)
$$\begin{aligned} 2x - y &= 5 \\ x + 2z &= 3 \\ 3y - z &= 1 \end{aligned}$$

is also a system of three linear equations in three unknowns.

(d)
$$\begin{aligned} y &= 2x^2 - 3 \\ y^2 &= x + 5 \end{aligned}$$

is a system of two quadratic equations in two unknowns.

Solution to a System of Equations: The solution to a *single* equation is the value(s) of the unknown(s) that will make the two sides of the equation equal to one another.

Example: The solution to $3x + 5 = 20$ is $x = 5$.

Example: A solution to $2x - 5y = 8$ is $x = 4$, $y = 0$. Another solution is the number pair $x = -1$, $y = -2$. There are, of course, an *infinite number* of solutions. If we were to plot the line $2x - 5y = 8$, *any point on the line* would be a solution to the equation.

The solution to a *system* of equations is a set of values of the unknowns that will *satisfy every equation* in the system.

Example: The system of equations

$$x + y = 5$$
$$x - y = 3$$

is satisfied *only* by the values $x = 4$, $y = 1$, and by *no other* set of values. Thus, the pair (4, 1) is the solution to the system, and the equations are said to be *independent.*

To get a numerical solution for all the unknowns in a system of linear equations, if one exists, *there must be as many independent equations as there are unknowns*. We first solve two equations in two unknowns, then later, three equations in three unknowns, and then larger systems, but the number of equations will always equal the number of unknowns.

Approximate Graphical Solution to a Pair of Equations: Since any point on a curve satisfies the equation of that curve the coordinates of the *points of intersection* of two curves will satisfy the equations of *both* curves. Thus we merely have to plot the two curves and find their points of intersection, which will be the solution to the pair of equations.

Example: Graphically find the approximate solution to the pair of linear equations

$$3x - y = 1$$
$$x + y = 3$$

This is also a good way to solve a pair of nonlinear *equations.*

Solution: We plot the lines as in Chapter 6, as shown in Fig. 8-2. The lines are seen to intersect at the point (1, 2).

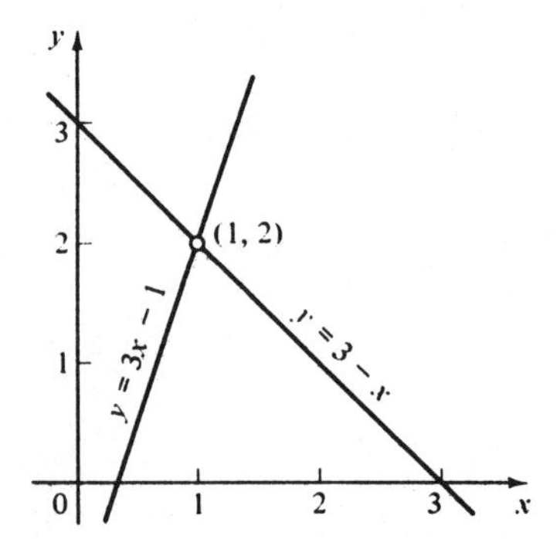

FIGURE 8-2 Graphical solution to a pair of equations.

Solving a Pair of Linear Equations by the Addition–Subtraction Method: The method of *addition–subtraction,* and the method of *substitution* that follows, have the object of *eliminating* one of the unknowns.

In the addition–subtraction method, this is accomplished by first (if necessary) multiplying each equation by such numbers that will make the coefficients of one unknown in both equations equal in absolute value. The two equations are then added or subtracted so as to eliminate that variable.

Example: Solve by the addition–subtraction method:

$$2x - 3y = -4$$
$$x + y = 3$$

Solution: Multiplying the second equation by 3,

$$\begin{array}{rcl} 2x - 3y &=& -4 \\ 3x + 3y &=& 9 \\ \hline 5x &=& 5 \end{array}$$

Adding,

We have thus reduced our two original equations to a single equation in one unknown. Solving for x gives

$$x = 1$$

Substituting into the second original equations,

$$1 + y = 3$$
$$y = 2$$

So the solution is $x = 1$, $y = 2$.

Check: Substituting into the first original equation,

$$2(1) - 3(2) \stackrel{?}{=} -4$$
$$2 - 6 = -4 \quad \text{checks}$$

and into the second original equation,

$$1 + 2 = 3 \quad \text{checks}$$

Example: Solve by addition or subtraction:

$$5x - 3y = 19$$
$$7x + 4y = 2$$

Solution: Multiplying the first equation by 4 and the second by 3, we get

$$\begin{array}{rcl} 20x - 12y &=& 76 \\ 21x + 12y &=& 6 \\ \hline 41x &=& 82 \\ x &=& 2 \end{array}$$

Adding,

Substituting $x = 2$ into the first equation,

$$\begin{aligned} 5(2) - 3y &= 19 \\ 3y &= -9 \\ y &= -3 \end{aligned}$$

So the solution is $x = 2$, $y = -3$.

The coefficients in the preceding examples were integers, but they may also be approximate numbers. If so, we must retain the proper number of significant digits, as in the following example.

Example: Solve for x and y.

$$\begin{aligned} 2.64x + 8.47y &= 3.72 \\ 1.93x + 2.61y &= 8.25 \end{aligned}$$

We are less likely to make a mistake adding rather than subtracting, so it is safer to multiply by a negative number and add the resulting equations, as we do here.

Solution: Let us eliminate y. We multiply the first equation by 2.61 and the second equation by -8.47. Let us carry at least three digits and/or two decimal places in our calculation, and round our answer to three digits at the end.

$$\begin{aligned} 6.89x + 22.11y &= 9.71 \\ -16.35x - 22.11y &= -69.88 \\ \hline -9.46x \qquad\qquad &= -60.17 \\ x &= 6.36 \end{aligned}$$

Substituting into the first equation,

$$\begin{aligned} 2.64(6.36) + 8.47y &= 3.72 \\ 8.47y = 3.72 - 16.79 &= -13.07 \\ y &= -1.54 \end{aligned}$$

Substitution Method: To use the *substitution method* to solve a pair of linear equations, first solve either original equation for one unknown in terms of the other unknown. Then substitute this expression into the other equation, thereby eliminating one unknown.

Example: Solve by substitution:

$$7x - 9y = 15 \qquad (1)$$

$$8y - 5x = -17 \qquad (2)$$

If one or both of the given equations is already in explicit form, then the substitution method is probably the easier. Otherwise, many students find the addition–subtraction method easier.

Solution: We choose to solve the second equation for y.

$$8y = 5x - 17$$

$$y = \frac{5x - 17}{8} \qquad (3)$$

Substituting this expression for y into the first equation,

$$7x - 9\left(\frac{5x - 17}{8}\right) = 15$$

We now have one equation in one unknown. Solving for x,

$$\begin{aligned} 56x - 9(5x - 17) &= 120 \\ 56x - 45x + 153 &= 120 \\ 11x &= -33 \\ x &= -3 \end{aligned}$$

Substituting $x = -3$ into (3),

$$y = \frac{5(-3) - 17}{8} = -4$$

So the solution is $x = -3$, $y = -4$.

Systems Having No Solution: Certain systems of equations have no unique solution. If you try to solve either of these types, *both variables will vanish.*

Example: Solve the system

$$\begin{aligned} 2x + 3y &= 5 \\ 6x + 9y &= 2 \end{aligned}$$

Solution: Multiplying the first equation by 3,

$$\begin{aligned} 6x + 9y &= 15 \\ 6x + 9y &= 9 \end{aligned}$$

Subtracting, $0 = 13$ no solution

It does not matter much whether a system is inconsistent or dependent; in either case we get no useful solution. But the practical problems we solve here will always have numerical solutions, so if your variables vanish, go back and check your work.

If both variables vanish and an *inequality* results, as in the preceding example, the system is *inconsistent*. The equations would plot as two *parallel lines*. There is no point of intersection and hence no solution.

If both variables vanish and an *equality* results (such as $4 = 4$), the system is *dependent*. The two equations would plot as a *single line*. This indicates that there are infinitely many solutions.

A Computer Technique: The Gauss–Seidel Method: The *Gauss–Seidel method* is a simple computer technique for solving systems of equations. Suppose that we wish to solve the set of equations

$$x - 2y + 2 = 0 \qquad (1)$$

$$3x - 2y - 6 = 0 \qquad (2)$$

We first solve the first equation for x in terms of y and the second equation for y in terms of x.

$$x = 2y - 2 \tag{3}$$

$$y = \frac{3x - 6}{2} \tag{4}$$

A method of this type is called iteration *or the method of* successive approximations.

We then *guess* at the value of y, substitute this value into (3), and obtain a value for x. This x is substituted into (4) and a value of y is obtained. These values for x and y will not be the correct values; they are only our *first approximation* to the true values.

The latest value for y is then put into (3), producing a new x; which is then put into (4), producing a new y; which is then put into (3); and so on. We repeat the computation until the values no longer change (we say that they *converge* on the true values). If the values of x and y get very large (*diverge* instead of converge), we solve the first equation for y and the second equation for x, and the computation will converge.

Example: Calculate x and y for the equations above using the Gauss–Seidel method. Take $y = 0$ for the first guess.

Solution: From equation (3) we get

$$x = 2(0) - 2 = -2$$

and from (4),

$$y = \frac{3(-2) - 6}{2} = -6$$

With y equal to -6, we then use (3) to find a new x, and so on. We get the values

x	y
−2	−6
−14	−24
−50	−78
−158	−240
−482	−726
−1,454	−2,184
−4,370	−6,558
−13,118	−19,680
−39,362	−59,046
.	.
.	.
.	.

Note that the values are *diverging*. We try again, this time solving equation (1) for y,

$$y = \frac{x + 2}{2}$$

and solving (2) for x,

$$x = \frac{2y + 6}{3}$$

Repeating the computation gives the values

x	y
2.666667	1
3.555556	2.333334
3.851852	2.777778
3.950617	2.925926
3.983539	2.975309
3.994513	2.991769
3.998171	2.997256
3.99939	2.999086
3.999797	2.999695
3.999932	2.999899
3.999977	2.999966
3.999992	2.999989
3.999998	2.999996
3.999999	2.999999
4	3
4	3
.	.
.	.
.	.

which converge on

$$x = 4 \quad \text{and} \quad y = 3$$

EXERCISE 1

Graphical Solution

Graphically find the approximate solution to each system of equations.

There are no applications given here. They are located in Exercise 3.

1. $2x - y = 5$
$x - 3y = 5$

2. $x + 2y = -7$
$5x - y = 9$

3. $x - 2y = -3$
$3x + y = 5$

4. $4x + y = 8$
$2x - y = 7$

5. $3x + 2y = 4$
$4x - 3y = 11$

6. $2x - 3y = -5$
$x + 6y = 2$

7. $2x + 5y = 4$
$5x - 2y = -3$

8. $x - 2y + 2 = 0$
$3x - 6y + 2 = 0$

Algebraic Solution

Solve each system of equations by addition–subtraction, by substitution, or by computer, as directed by your instructor.

9. $2x + y = 11$
$3x - y = 4$

10. $5x + 7y = 101$
$y = 7x - 55$

11. $3x - 2y = -15$
$5x + 6y = 3$

12. $7x + 6y = 20$
$2x + 5y = 9$

13. $x + 5y = 11$
$3x + 2y = 7$

14. $4x - 5y = -34$
$2x - 3y = -22$

15. $x = 11 - 4y$
$5x - 2y = 11$

16. $2x - 3y = 3$
$4x + 5y = 39$

17. $7x - 4y = 81$
$5x - 3y = 57$

18. $3x + 4y = 85$
$5x + 4y = 107$

19. $3x - 2y = 1$
$2x + y = 10$

20. $5x - 2y = 3$
$2x + 3y = 5$

21. $y = 9 - 3x$
$x = 8 - 2y$

22. $y = 2x - 3$
$x = 19 - 3y$

23. $29.1x - 47.6y = 42.8$
$11.5x + 72.7y = 25.8$

24. $4.92x - 8.27y = 2.58$
$6.93x + 2.84y = 8.36$

25. $3x + 4y = 18$
$x = 8 - 2y$

26. $5x + 4y = 14$
$17x - 3y = 31$

27. $3x - 5y = 13$
$4x - 7y = 17$

28. $3x - 2y = 5$
$2x + 5y = 16$

Computer

29. Write a program for the Gauss–Seidel method described in the preceding section. Use it to solve any of the equations in this exercise.

8-2. SYSTEMS OF FRACTIONAL, NONLINEAR, AND LITERAL EQUATIONS

The techniques in this section apply not only to two equations in two unknowns, but to larger systems as well.

Systems with Fractional Coefficients: When one or more of the equations in our system has fractional coefficients, simply multiply the entire equation by the LCD, and proceed as before.

Example: Solve for x and y:

$$\frac{x}{2} + \frac{y}{3} = \frac{5}{6} \qquad (1)$$

$$\frac{x}{4} - \frac{y}{2} = \frac{7}{4} \qquad (2)$$

Solution: Multiplying (1) by 6 and (2) by 4,

$$3x + 2y = 5 \quad (3)$$
$$x - 2y = 7 \quad (4)$$

Adding (3) and (4),

$$4x = 12$$
$$x = 3$$

Substituting $x = 3$ into (4),

$$3 - 2y = 7$$
$$2y = 3 - 7 = -4$$
$$y = -2$$

Fractional Equations with Unknowns in the Denominator: The same method (multiplying by the LCD) can be used to clear fractions when the unknowns appear in the denominators. Note that such equations are not linear (that is, not of first degree) as were the equations we have solved so far, but we are able to solve them by the same methods.

Example: Solve the system

$$\frac{10}{x} - \frac{9}{y} = 8 \quad (1)$$
$$\frac{8}{x} + \frac{15}{y} = -1 \quad (2)$$

Solution: Multiplying (1) by xy and (2) by xy,

$$10y - 9x = 8xy \quad (3)$$
$$8y + 15x = -xy \quad (4)$$

Multiplying (3) by 5, and (4) by 3,

$$50y - 45x = 40xy$$
$$24y + 45x = -3xy$$

Adding,

$$74y = 37xy$$
$$2 = x$$

Substituting $x = 2$ into (1),

$$\frac{10}{2} - \frac{9}{y} = 8$$
$$\frac{9}{y} = 5 - 8 = -3$$
$$y = -3$$

A convenient way to solve nonlinear systems such as these is to *substitute new variables* such as to make the equations linear. Solve in the usual way and then substitute back.

Example: Solve the nonlinear system of equations

$$\frac{2}{3x} + \frac{3}{5y} = 17 \tag{1}$$

$$\frac{3}{4x} + \frac{2}{3y} = 19 \tag{2}$$

Solution: We substitute $m = 1/x$ and $n = 1/y$ and get the linear system

$$\frac{2m}{3} + \frac{3n}{5} = 17 \tag{3}$$

$$\frac{3m}{4} + \frac{2n}{3} = 19 \tag{4}$$

This technique of substitution is very useful, and we will use it again to reduce certain equations to quadratic form (Sec. 11-7). Do not confuse this kind of substitution with the method of substitution we studied in Sec. 8-1.

Again, we clear fractions by multiplying each equation by its LCD. Multiplying (3) by 15 gives

$$10m + 9n = 255 \tag{5}$$

and multiplying (4) by 12 gives

$$9m + 8n = 228 \tag{6}$$

Using the addition–subtraction method, we multiply (5) by 8 and multiply (6) by -9.

$$\begin{aligned} 80m + 72n &= 2040 \\ -81m - 72n &= -2052 \\ \hline \end{aligned}$$

Adding,

$$\begin{aligned} -m &= -12 \\ m &= 12 \end{aligned}$$

Substituting $m = 12$ into (5),

$$\begin{aligned} 120 + 9n &= 255 \\ n &= 15 \end{aligned}$$

Finally, we substitute back to get x and y.

$$x = \frac{1}{m} = \frac{1}{12} \quad \text{and} \quad y = \frac{1}{n} = \frac{1}{15}$$

Common Error	Students often forget that last step. We are not solving for m and n, but for x and y.

Literal Equations: We use the method of addition–subtraction or the method of substitution for solving systems of equations with literal coefficients, treating the literals as if they were numbers.

Example: Solve for x and y in terms of the other literals:

$$2mx + ny = 3 \tag{1}$$

$$mx + 3ny = 2 \tag{2}$$

Solution: We will use the addition–subtraction method. Multiplying the second equation by -2,

$$\begin{array}{r} 2mx + ny = 3 \\ -2mx - 6ny = -4 \\ \hline \end{array}$$

Adding,

$$-5ny = -1$$

$$y = \frac{1}{5n}$$

Substituting back into the second original equation, we obtain

$$mx + 3n\left(\frac{1}{5n}\right) = 2$$

$$mx + \frac{3}{5} = 2$$

$$mx = 2 - \frac{3}{5} = \frac{7}{5}$$

$$x = \frac{7}{5m}$$

Example: Solve by the addition–subtraction method:

$$a_1x + b_1y = c_1$$

$$a_2x + b_2y = c_2$$

Solution: Multiplying the first equation by b_2 and the second equation by b_1, we obtain

$$\begin{array}{r} a_1b_2x + b_1b_2y = c_1b_2 \\ a_2b_1x + b_1b_2y = c_2b_1 \\ \hline \end{array}$$

Subtracting,

$$(a_1b_2 - a_2b_1)x = c_1b_2 - c_2b_1$$

Dividing by $a_1b_2 - a_2b_1$,

$$x = \frac{c_1b_2 - c_2b_1}{a_1b_2 - a_2b_1}$$

Now solving for y, we multiply the first equation by a_2 and the second equation by a_1. Writing the second equation above the first, we get

$$\begin{array}{r} a_1a_2x + a_1b_2y = a_1c_2 \\ a_1a_2x + a_2b_1y = a_2c_1 \\ \hline \end{array}$$

Subtracting,

$$(a_1b_2 - a_2b_1)y = a_1c_2 - a_2c_1$$

Dividing by $a_1b_2 - a_2b_1$,

$$y = \frac{a_1c_2 - a_2c_1}{a_1b_2 - a_2b_1}$$

This result may be summarized as follows:

This provides us with a formula *for solving a pair of linear equations. We are going to need this formula again when we discuss determinants in Sec. 9-2.*

Two Equations in Two Unknowns	The solution to the set of equations $a_1x + b_1y = c_1$, $a_2x + b_2y = c_2$ is $x = \dfrac{c_1b_2 - c_2b_1}{a_1b_2 - a_2b_1}$ $\quad y = \dfrac{a_1c_2 - a_2c_1}{a_1b_2 - a_2b_1}$ $(a_1b_2 - a_2b_1 \neq 0)$	63

EXERCISE 2

Fractional Coefficients

Solve for x and y.

1. $\dfrac{x}{5} + \dfrac{y}{6} = 18$, $\dfrac{x}{2} - \dfrac{y}{4} = 21$

2. $\dfrac{x}{2} + \dfrac{y}{3} = 7$, $\dfrac{x}{3} + \dfrac{y}{4} = 5$

3. $\dfrac{x}{3} + \dfrac{y}{4} = 8$, $x - y = -3$

4. $\dfrac{x}{2} + \dfrac{y}{3} = 5$, $\dfrac{x}{3} + \dfrac{y}{2} = 5$

5. $\dfrac{3x}{5} + \dfrac{2y}{3} = 17$, $\dfrac{2x}{3} + \dfrac{3y}{4} = 19$

6. $\dfrac{x}{7} + 7y = 251$, $\dfrac{y}{7} + 7x = 299$

7. $\dfrac{x}{2} + \dfrac{y}{3} = 3$, $\dfrac{x}{5} + \dfrac{y}{2} = \dfrac{23}{10}$

8. $\dfrac{x}{6} - \dfrac{y}{3} = -\dfrac{1}{3}$, $\dfrac{2x}{3} - \dfrac{3y}{4} = 1$

Unknowns in the Denominator

Solve for x and y.

9. $\dfrac{8}{x} + \dfrac{6}{y} = 3$, $\dfrac{6}{x} + \dfrac{15}{y} = 4$

10. $\dfrac{1}{x} + \dfrac{3}{y} = 11$, $\dfrac{5}{x} + \dfrac{4}{y} = 22$

11. $\dfrac{5}{x} + \dfrac{6}{y} = 7$, $\dfrac{7}{x} + \dfrac{9}{y} = 10$

12. $\dfrac{2}{x} + \dfrac{4}{y} = 14$, $\dfrac{6}{x} - \dfrac{2}{y} = 14$

13. $\dfrac{6}{x} + \dfrac{8}{y} = 1$, $\dfrac{7}{x} - \dfrac{11}{y} = -9$

14. $\dfrac{2}{5x} + \dfrac{5}{6y} = 14$, $\dfrac{2}{5x} - \dfrac{3}{4y} = -5$

15. $\dfrac{5}{3x} + \dfrac{2}{5y} = 7$, $\dfrac{7}{6x} - \dfrac{1}{10y} = 3$

16. $\dfrac{1}{6x} + \dfrac{1}{5y} = 18$, $\dfrac{1}{4x} - \dfrac{1}{2y} = -21$

Literal Equations

Solve for x and y in terms of the other literal quantities.

17. $3x - 2y = a$
$2x + y = b$

18. $ax + by = r$
$ax + cy = s$

19. $ax - dy = c$
$mx - ny = c$

20. $px - qy + pq = 0$
$2px - 3qy = 0$

Computer

21. Program Eqs. 63 into your computer so that x and y will be computed and printed whenever the six coefficients

$$a_1, \quad b_1, \quad c_1 \qquad \text{and} \qquad a_2, \quad b_2, \quad c_2$$

are entered into the program. Test the program on a system of equations that you have already solved algebraically.

8-3. WORD PROBLEMS WITH TWO UNKNOWNS

In many problems there are two or more unknowns that must be found. To solve such problems, we must write *as many independent equations as there are unknowns.* Otherwise, it is not possible to obtain numerical answers, although one unknown could be expressed in terms of another.

Set up these problems as we did in Chapter 5 and solve the resulting system of equations by any of the methods of this chapter.

Example: During a certain day, two computer printers are observed to process 1705 form letters, with the slower printer in use for 5.5 h and the faster for 4.0 h. On another day the slower printer works for 6.0 h and the faster for 6.5 h, and together they print 2330 form letters. How many letters can each print in an hour, working alone?

Solution: We let

x = rate of slow printer, letters/h
y = rate of fast printer, letters/h

On the first day, the slow printer produces $5.5x$ letters while the fast printer produces $4.0y$ letters. Together they produce

$$5.5x + 4.0y = 1705$$

Similarly, for the second day,

$$6.0x + 6.5y = 2330$$

Using the addition–subtraction method, we multiply the first equation by 6.5 and the second by -4.0.

$$\begin{aligned} 35.75x + 26y &= 11{,}083 \\ -24x - 26y &= 9{,}320 \\ \hline \end{aligned}$$

Adding,

$$\begin{aligned} 11.75x \qquad &= 1{,}763 \\ x &= 150 \text{ letters/h} \end{aligned}$$

Substituting back,

$$5.5(150) + 4y = 1705$$
$$y = 220 \text{ letters/h}$$

EXERCISE 3

Number Problems

1. The sum of two numbers is 24 and their difference is 8. What are the numbers?

2. The sum of the two digits of a number is 9, and if 9 is added to the number, the digits will be reversed. Find the number.

3. There is a fraction such that if 3 is added to the numerator its value will be $\frac{1}{3}$, and if 1 is subtracted from the denominator its value will be $\frac{1}{5}$. What is the fraction?

Some of these problems can be set up using just one unknown, but for now use two unknowns for the practice.

4. The sum of two numbers is 29 and their difference is 5. What are the numbers?

5. The sum of two numbers divided by 2 gives a quotient of 24, and their difference divided by 2 gives a quotient of 17. What are the numbers?

Motion Problems

6. A gives B 100 ft start and overtakes him in 4 min. In a second race he gains 750 ft on B when B runs 9000 ft. Find the rate at which each runs.

7. A and B run two races of 280 ft. In the first race, A gives B a start of 70 ft and neither wins the race. In the second race, A gives B a start of 35 ft and beats him by $6\frac{2}{3}$s. How many feet can each run in a second?

8. A and B run two races from P to Q and back, the distance from P to Q being 108 m. In the first race, A reaches Q first and meets B on his return at a point 12 m from Q. In the second race, A increases his speed by 2 m/s and B by 1 m/s. Now A meets B 18 m from Q. How fast did each run during the first race?

Financial Problems

9. A sum of money, at simple interest, amounted in 5 years to \$3000 and in 6 years to \$3100. Find the sum and the rate of interest (see Eq. A8).

10. If 4 yd of velvet and 3 yd of silk are sold for \$53, and 5 yd of velvet and 6 yd of silk for \$72, what is the price per yard of the velvet and of the silk?

11. A sum of \$10,000 is invested, part at 4% and the remainder at 5%. The annual income from the 4% investment is \$40 more than from the 5% investment. Find the amounts invested at 4% and at 5%.

12. A person invested \$4400, part of it in railroad bonds bearing 6.2% interest and the remainder in state bonds bearing 9.7% interest, and received the same income from each. How much was invested in each?

Mixture Problems

13. A certain brass alloy contains 62% copper, 35.0% zinc, and 3.0% lead. How many kilograms of lead and of zinc must be added to 200 kg of this alloy to produce an alloy that is 40% zinc and 4.0% lead?

14. A certain concrete mixture contains 5.0% cement and 8.0% sand. How many pounds of this mixture and how many pounds of sand should be combined with 255 lb of cement to make a batch that is 12% cement and 15% sand?

Work Problems

15. A carpenter and a helper can do a certain job in 15 days. If the carpenter works 1.5 times as fast as the helper, how long would it take each to do the job, working alone?

16. During 1 week two machines produce a total of 27,210 parts, with the faster machine working for 37.5 h and the slower for 28.2 h. During another week, they produce 59,830 parts, with the faster machine working 66.5 h and the slower machine working for 88.6 h. How many parts can each produce in an hour working alone?

Flow Problems

17. Two pipelines lead from a dockside to a group of oil storage tanks. On one occasion the two pipes are seen to deliver 117,000 gal, with the larger pipe in use for 3.5 h and the smaller pipe for 4.5 h. On another day the two pipes deliver 151,200 gal, with the larger pipe operating for 5.2 h and the smaller for 4.8 h. Find the hourly flow rate through each pipe.

18. Two conveyors can fill a certain bin in 6 h, working together. If one conveyor works 1.8 times as fast as the other, how long would it take each to fill the bin working alone?

Energy Flow

19. A hydroelectric generating plant and a coal-fired generating plant together supply a city of 255,000 people, with the hydro plant producing 1.75 times the power of the coal plant. How many people could each service alone?

20. During a certain week a small wind generator and a small hydro unit together produce 5880 kWh, with the wind generator operating only 85% of the time. During another week the two units produce 6240 kWh, with the wind generator working 95% of the time and the hydro unit down 7.5 h for repairs. Assuming that each unit has a constant output when operating, find the number of kilowatts produced by each in 1 h.

Electrical

21. To find the currents I_1 and I_2 in Fig. 8-3 we use Eq. A46 in each loop and get the pair of equations

$$6.00 - R_1I_1 - R_2I_1 + R_2I_2 = 0$$

$$12.0 - R_3I_2 - R_2I_2 + R_2I_1 = 0$$

Solve for I_1 and I_2 if $R_1 = 736\ \Omega$, $R_2 = 386\ \Omega$, and $R_3 = 375\ \Omega$.

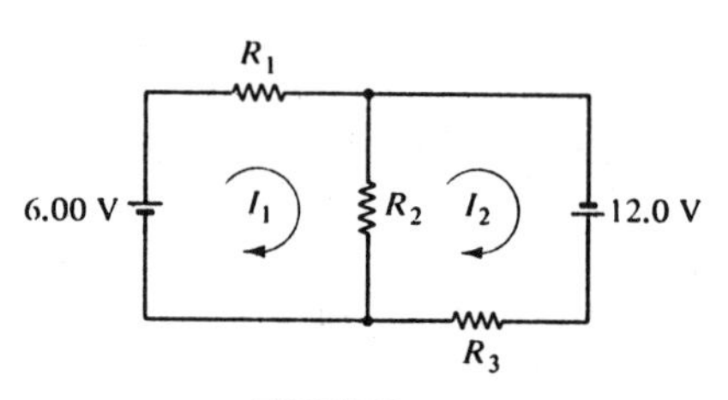

FIGURE 8-3

22. Use Eq. A46 to write a pair of equations for the circuit of Fig. 8-4, as in Problem 21. Solve these equations for I_1 and I_2.

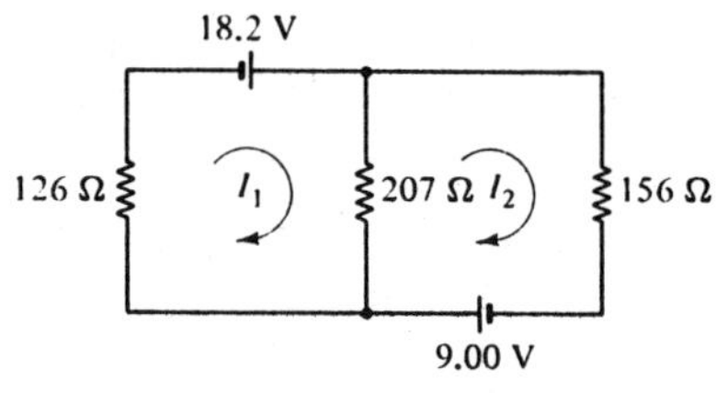

FIGURE 8-4

8-4. SYSTEMS OF THREE EQUATIONS

Our strategy here is to reduce a given system of three equations in three unknowns to a system of two equations in two unknowns, which we already know how to solve.

We take any two of the given equations and, by addition–subtraction or substitution, eliminate one variable, obtaining a single equation in two unknowns. We then take another pair of equations (which must include the one not yet used, as well as one of those already used), and similarly obtain a second equation in the *same two* unknowns. This pair of equations can then be solved simultaneously, and the values obtained are substituted back to obtain the third variable.

In the following chapter we will use determinants *to solve sets of three or more equations.*

Example: Solve

$$
\begin{aligned}
6x - 4y - 7z &= 17 && (1)\\
9x - 7y - 16z &= 29 && (2)\\
10x - 5y - 3z &= 23 && (3)
\end{aligned}
$$

Solution:

Multiplying (1) by 3, $\qquad 18x - 12y - 21z = 51 \qquad (4)$

and multiplying (2) by 2, $\qquad 18x - 14y - 32z = 58 \qquad (5)$

Subtracting, $\qquad 2y + 11z = -7 \qquad (6)$

We now take another pair of equations (1) and (3).

Multiplying (1) by 5, $\qquad 30x - 20y - 35z = 85 \qquad (7)$

and multiplying (3) by 3, $\qquad 30x - 15y - 9z = 69 \qquad (8)$

Subtracting (7) from (8), $\qquad 5y + 26z = -16 \qquad (9)$

Now we solve (6) and (9) simultaneously.

Multiplying (6) by 5, $10y + 55z = -35$
and (9) by 2, $10y + 52z = -32$

Subtracting, $3z = -3$

$$z = -1$$

Substituting $z = -1$ into (6),

$$2y + 11(-1) = -7$$

$$y = 2$$

Substituting $y = 2$ and $z = -1$ into (1),

$$6x - 4(2) - 7(-1) = 17$$

$$x = 3$$

Fractional Equations: We use the same techniques for solving a set of three fractional equations that we did for two equations; (1) multiply each equation by its LCD to eliminate fractions; and (2) if the unknowns are in the denominators, substitute new variables which are the reciprocals of the originals.

Example: Solve for x, y, and z.

$$\frac{4}{x} + \frac{9}{y} - \frac{8}{z} = 3 \quad (1)$$

$$\frac{8}{x} - \frac{6}{y} + \frac{4}{z} = 3 \quad (2)$$

$$\frac{5}{3x} + \frac{7}{2y} - \frac{2}{z} = \frac{3}{2} \quad (3)$$

Solution: We make the substitution

$$p = \frac{1}{x} \qquad q = \frac{1}{y} \qquad \text{and} \qquad r = \frac{1}{z}$$

and also multiply (3) by its LCD, 6, to clear fractions.

$$4p + 9q - 8r = 3 \quad (4)$$

$$8p - 6q + 4r = 3 \quad (5)$$

$$10p + 21q - 12r = 9 \quad (6)$$

We multiply (5) by 2 and add it to (4),

$$20p - 3q = 9 \quad (7)$$

Then we multiply (5) by 3 and add it to (6),

$$34p + 3q = 18 \quad (8)$$

Adding (7) and (8) gives

$$54p = 27$$

$$p = \frac{1}{2}$$

Then from (8),

$$3q = 18 - 17 = 1$$

$$q = \frac{1}{3}$$

and from (5),

$$4r = 3 - 4 + 2 = 1$$

$$r = \frac{1}{4}$$

Returning to our original variables,

$$x = 2 \qquad y = 3 \qquad \text{and} \qquad z = 4$$

EXERCISE 4

Solve for x, y, and z.

1. $x + y = 35$
$x + z = 40$
$y + z = 45$

2. $x + y + z = 12$
$x - y = 2$
$x - z = 4$

3. $3x + y = 5$
$2y - 3z = -5$
$x + 2z = 7$

4. $x - y = 5$
$y - z = -6$
$2x \quad z = 2$

5. $x + y + z - 18$
$x - y + z = 6$
$x + y - z = 4$

6. $x + y + z = 90$
$2x - 3y = -20$
$2x + 3z = 145$

7. $x + 2y + 3z = 14$
$2x + y + 2z = 10$
$3x + 4y - 3z = 2$

8. $x + y + z = 35$
$x - 2y + 3z = 15$
$y - x + z = -5$

9. $x - 2y + 2z = 5$
$5x + 3y + 6z = 57$
$x + 2y + 2z = 21$

10. $x + y + z = 6$
$5x + 4y + 3z = 22$
$3x + 4y - 3z = 2$

11. $2x - 4y + 3z = 10$
$3x + y - 2z = 6$
$x + 3y - z = 20$

12. $5x + y - 4z = -5$
$3x - 5y - 6z = -20$
$x - 3y + 8z = -27$

13. $x + 3y - z = 10$
$5x - 2y + 2z = 6$
$3x + 2y + z = 13$

Fractional Equations

Solve for x, y, and z.

14. $x + \frac{y}{3} = 5$

$x + \frac{z}{3} = 6$

$y + \frac{z}{3} = 9$

15. $\frac{1}{x} + \frac{1}{y} = 5$

$\frac{1}{y} + \frac{1}{z} = 7$

$\frac{1}{x} + \frac{1}{z} = 6$

16. $\dfrac{x}{10} + \dfrac{y}{5} + \dfrac{z}{20} = \dfrac{1}{4}$

$x + y + z = 6$

$\dfrac{x}{3} + \dfrac{y}{2} + \dfrac{z}{6} = 1$

17. $\dfrac{1}{x} + \dfrac{2}{y} - \dfrac{1}{z} = -3$

$\dfrac{3}{x} + \dfrac{1}{y} + \dfrac{1}{z} = 4$

$\dfrac{1}{x} - \dfrac{1}{y} + \dfrac{2}{z} = 6$

Number Problems

18. There are two fractions which have the same denominator. If 1 is subtracted from the numerator of the smaller fraction, its value will be $\frac{1}{3}$ of the larger fraction; but if 1 is subtracted from the numerator of the larger, its value will be twice that of the smaller fraction. The difference between the fractions is $\frac{1}{3}$. What are the fractions?

19. A certain number is expressed by three digits whose sum is 10. The sum of the first and last digits is $\frac{2}{3}$ of the second digit; and if 198 is subtracted from the number, the digits will be reversed. What is the number?

Financial Problems

20. Three persons, A, B, and C, jointly purchased a quantity of lumber for \$900. One-half of what A paid added to $\frac{1}{4}$ of what B paid and $\frac{1}{5}$ of what C paid will make \$279; and the sum that A paid increased by $\frac{2}{3}$ of what B paid and diminished by $\frac{1}{2}$ of what C paid will make \$320. How much did each pay?

21. A merchant found, on counting her cash, that she had 84 coins, in dollars, half-dollars, and quarters, worth \$42. She also found that $\frac{1}{3}$ of her half-dollars and $\frac{1}{4}$ of her quarters were worth \$6.50. How many pieces of each kind had she?

Electrical Problems

22. When writing Kirchhoff's law (Eq. A46) for a certain three-loop network, we get the set of equations

$$3I_1 + 2I_2 - 4I_3 = 4$$
$$I_1 - 3I_2 + 2I_3 = -5$$
$$2I_1 + I_2 - I_3 = 3$$

where I_1, I_2, and I_3 are the loop currents in amperes. Solve for these currents.

23. Using Kirchhoff's laws, Eq. A46, write a set of three equations for the circuit of Fig. 8-5 and solve for the three currents I_1, I_2, and I_3.

24. The Gauss–Seidel method described in Sec. 8-1 works for any number of equations. For three equations in three unknowns, for example, write the equations in the form

$$x = f(y, z)$$
$$y = g(x, z)$$
$$z = h(x, y)$$

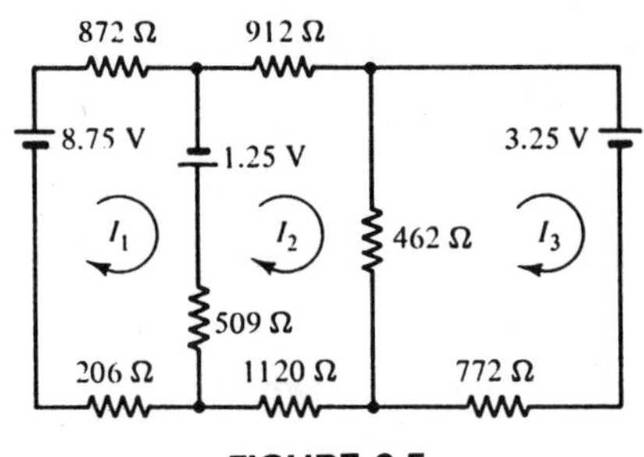

FIGURE 8-5

Try it on one of the systems given in this exercise. If the computation diverges, then solve each equation for a different variable than before and try again. It may take two or three tries to find a combination that works.

CHAPTER TEST

Solve each system of equations by any method.

1. $4x + 3y = 27$
$2x - 5y = -19$

2. $\dfrac{x}{3} + \dfrac{y}{2} = \dfrac{4}{3}$
$\dfrac{x}{2} + \dfrac{y}{3} = \dfrac{7}{6}$

3. $\dfrac{15}{x} + \dfrac{4}{y} = 1$
$\dfrac{5}{x} - \dfrac{12}{y} = 7$

4. $2x + 4y - 3z = 22$
$4x - 2y + 5z = 18$
$6x + 7y - z = 63$

5. $5x + 3y - 2z = 5$
$3x - 4y + 3z = 13$
$x + 6y - 4z = -8$

6. $x + y + 2z = 2(b + c)$
$x + 2y + z = 2(a + c)$
$2x + y + z = 2(a + b)$

7. $\dfrac{3}{x + y} + \dfrac{4}{x - z} = 2$
$\dfrac{6}{x + y} + \dfrac{5}{y - z} = 1$
$\dfrac{4}{x - z} + \dfrac{5}{y - z} = 2$

8. $4x + 2y = 26$
$3x + 4y = 39$

9. $2x - 3y = -14$
$3x + 2y = 44$

10. $\dfrac{2}{x} + \dfrac{1}{y} = \dfrac{4}{3}$
$\dfrac{3}{x} + \dfrac{5}{y} = \dfrac{19}{6}$

11. $\dfrac{9}{x} + \dfrac{8}{y} = \dfrac{43}{6}$
$\dfrac{3}{x} + \dfrac{10}{y} = \dfrac{29}{6}$

12. $\dfrac{x}{a} + \dfrac{y}{b} = p$
$\dfrac{x}{b} + \dfrac{y}{a} = q$

13. $x + y = a$
$x + z = b$
$y + z = c$

14. $\dfrac{5x}{6} + \dfrac{2y}{5} = 14$
$\dfrac{3x}{4} - \dfrac{2y}{5} = 5$

15. $\dfrac{2x}{7} + \dfrac{2y}{3} = \dfrac{16}{3}$
$\dfrac{3x}{5} + \dfrac{3y}{5} = \dfrac{36}{5}$

16. A sum of money was divided between *A* and *B* so that the share of *A* was to that of *B* as 5 to 3. The share of *A* also exceeded $\frac{5}{9}$ of the whole sum by \$50. What was the share of each?

17. A and B can do a job in 12 days, and B and C can do the same job in 16 days. How long would it take them all working together to do the job if A does $1\frac{1}{2}$ times as much as C?

18. If the numerator of a certain fraction is increased by 2 and its denominator diminished by 2, its value will be 1. If the numerator is increased by the denominator and the denominator is diminished by 5, its value will be 5. Find the fraction.

19. A farmer found that the number of his sheep was 26 more than the number of his cows and horses together; that $\frac{1}{5}$ of the number of sheep was equal to the number of horses together with $\frac{1}{4}$ of his cows; and that $\frac{1}{8}$ of his cows, $\frac{1}{2}$ of his horses, and $\frac{1}{5}$ of his sheep amounted to 12. How many had he of each?

Matrices and Determinants

In Chapter 8 we learned how to solve sets of two or three equations by graphing, addition–subtraction, substitution, and by computer using the Gauss–Seidel method. These methods, however, become unwieldy when the number of equations to be solved is greater than three.

In this chapter we introduce the idea of a *matrix*, and then show how the *determinant* of a matrix allows us to solve large systems of equations, using Cramer's rule.

We then apply our matrix algebra to the solution of sets of linear equations by two methods that are easier to program for the computer than are determinants. These methods are called *Gauss elimination* and *matrix inversion*.

9-1. MATRICES

Arrays: A set of numbers, called *elements*, arranged in a pattern, is called an *array*. Arrays are named for the shape of the pattern made by the elements.

Example: The array

$$\begin{pmatrix} 7 & 3 & 9 & 1 \\ & 8 & 3 & 2 \\ & & 9 & 1 \\ & & & 5 \end{pmatrix}$$

is called a *triangular array*.

Matrices: A *matrix* is a *rectangular* array.

Examples: The rectangular arrays

(a) $$\begin{pmatrix} 2 & 5 & 7 \\ 6 & 3 & 1 \end{pmatrix}$$

(b) $$\begin{pmatrix} 7 & 3 & 5 \\ 1 & 9 & 3 \\ 8 & 3 & 2 \end{pmatrix}$$

(c) $$\begin{pmatrix} 5 \\ 0 \\ 4 \\ 1 \end{pmatrix}$$

(d) $$(3 \quad 2 \quad 9 \quad 6)$$

are matrices. Further, (b) is a *square* matrix, (c) is called a *column vector*, and (d) is a *row vector*.

We will come across the word vector *again in Chapter 12 when we study force and velocity vectors.*

In everyday language, an array is a *table* and a vector is a *list*.

Subscripts: Each element in an array is located in a horizontal *row* and a vertical *column*. We indicate the row and column by means of *subscripts*.

Example: The element

$$a_{25}$$

is located in row 2 and column 5.

Thus, an element in an array needs *double subscripts* to give its location. An element in a list, of course, needs only a *single subscript*, such as b_7.

Subscripted Variables on the Computer: In many computer languages, including BASIC, FORTRAN, and Pascal, subscripts for a variable are enclosed in parentheses.

Example: The subscripted variables b_7 and a_{25} would be written

B(7) and A(2,5)

These subscripted variables are essential for manipulating arrays, as we do in this chapter.

Dimensions: A matrix will have, in general, m rows and n columns. The numbers m and n are the *dimensions* of the matrix, as, for example, a 4×5 matrix.

Scalars: A single number, as opposed to an array of numbers, is called a *scalar.*

Example: Some scalars are

$$5 \qquad 693.6 \qquad -24.3$$

A scalar can also be thought of as an array having just one row and one column.

Naming a Matrix: We will often denote or *name* a matrix with a single letter.

Example: We can let

$$\mathbf{A} = \begin{pmatrix} 2 & 5 & 1 \\ 0 & 4 & 3 \end{pmatrix}$$

Thus we can represent an entire array by a single symbol.

The Null or Zero Matrix: A matrix in which all elements are zero is called the *null* or *zero* matrix.

Example: The matrix

$$\mathbf{O} = \begin{pmatrix} 0 & 0 & 0 \\ 0 & 0 & 0 \end{pmatrix}$$

is a 2×3 zero matrix. We will denote the zero matrix by the letter **O**.

Square Matrices: A *square matrix* is one that has the same number of rows as columns. If there are n rows and n columns, the matrix is said to be of *order n*, and is also called an *n-square matrix*. The *main diagonal* of a square matrix runs from the upper left to the lower right corners. The other diagonal is called the *secondary diagonal*.

Example: The matrix

$$\begin{pmatrix} 3 & 6 & 1 \\ 5 & 8 & 2 \\ 0 & 4 & 7 \end{pmatrix}$$

is a square matrix of order 3. The elements 3, 8, and 7 are on the main diagonal and the elements 1, 8, and 0 are on the secondary diagonal.

The Unit Matrix: If all the elements *not* on the main diagonal of a square matrix are 0, we have a *diagonal matrix*. If, in addition, the diagonal elements are equal, we have a *scalar matrix*. If the diagonal elements of a scalar matrix are 1's, we have a *unit matrix* or *identity matrix*.

Example: The matrices

$$\mathbf{A} = \begin{pmatrix} 3 & 0 & 0 \\ 0 & 5 & 0 \\ 0 & 0 & 7 \end{pmatrix} \qquad \mathbf{B} = \begin{pmatrix} 6 & 0 & 0 \\ 0 & 6 & 0 \\ 0 & 0 & 6 \end{pmatrix} \qquad \mathbf{I} = \begin{pmatrix} 1 & 0 & 0 \\ 0 & 1 & 0 \\ 0 & 0 & 1 \end{pmatrix}$$

are all diagonal matrices. **B** and **I** are, in addition, scalar matrices, and **I** is a unit matrix. We denote the unit matrix by the letter **I**.

Equality of Matrices: When comparing two matrices *of the same shape*, elements in each matrix having the same row and column subscripts are called *corresponding elements*. Two matrices of the same shape are equal if their corresponding elements are equal.

Example: The matrices

$$\begin{pmatrix} x & 2 & 5 \\ 3 & 1 & 7 \end{pmatrix} \quad \text{and} \quad \begin{pmatrix} 4 & 2 & 5 \\ 3 & 1 & y \end{pmatrix}$$

are equal only if $x = 4$ and $y = 7$.

Transpose of a Matrix: The *transpose* $\mathbf{A}'$ of a matrix **A** is obtained by changing the rows to columns and changing the columns to rows.

Example: The transpose of the matrix

$$\begin{pmatrix} 1 & 2 & 3 & 4 \\ 5 & 6 & 7 & 8 \end{pmatrix}$$

is

$$\begin{pmatrix} 1 & 5 \\ 2 & 6 \\ 3 & 7 \\ 4 & 8 \end{pmatrix}$$

EXERCISE 1

$$\mathbf{A} = \begin{pmatrix} 2 & 5 & 1 \\ 6 & 3 & 7 \\ 1 & 6 & 9 \\ 7 & 4 & 2 \end{pmatrix} \qquad \mathbf{B} = \begin{pmatrix} 7 \\ 3 \\ 9 \\ 2 \end{pmatrix} \qquad \mathbf{C} = \begin{pmatrix} f & i & q & w \\ & g & w & k \\ & & c & z \\ & & & b \end{pmatrix}$$

$$\mathbf{D} = \begin{pmatrix} 6 & 2 & 0 & 1 \\ 2 & 8 & 3 & 9 \end{pmatrix} \qquad \mathbf{E} = \begin{pmatrix} x & y \\ z & w \end{pmatrix} \qquad \mathbf{F} = \begin{pmatrix} 0 & 0 & 0 & 0 \\ 0 & 0 & 0 & 0 \end{pmatrix}$$

$$\mathbf{G} = 7 \qquad \mathbf{H} = \begin{pmatrix} 3 & 8 \\ & 5 \end{pmatrix} \qquad \mathbf{I} = (3 \quad 8 \quad 4 \quad 6)$$

Which of the nine arrays shown above is:

1. a rectangular array?

2. a square array?

3. a triangular array?

4. a column vector?

5. a row vector?

6. a table?

7. a list?

8. a scalar?

9. a null matrix?

10. a matrix?

For the matrix **A** above, find the element:

11. a_{32}

12. a_{41}

Give the dimensions of:

13. matrix **A**

14. matrix **B**

15. matrix **D**

16. matrix **E**

Write the transpose of:

17. matrix **F**

18. matrix **D**

19. Under what conditions will the matrices

$$\begin{pmatrix} 6 & x \\ w & 8 \end{pmatrix} \quad \text{and} \quad \begin{pmatrix} y & 2 \\ 7 & z \end{pmatrix}$$

be equal?

Computer

20. Using subscripted variables, write a computer program to find and print the transpose of any matrix. Use DATA statements to enter the number of rows and of columns, and then the elements of the original matrix, row by row, into your program.

9-2. DETERMINANTS

Definitions: Given a square matrix of order 2, such as

$$\begin{pmatrix} a_1 & b_1 \\ a_2 & b_2 \end{pmatrix}$$

we define a *determinant* of this matrix to be the number

$$a_1b_2 - a_2b_1$$

This determinant can be denoted either by writing "det" before the matrix, or by enclosing the array between vertical bars. Thus,

$$\det \begin{pmatrix} a_1 & b_1 \\ a_2 & b_2 \end{pmatrix} = \begin{vmatrix} a_1 & b_1 \\ a_2 & b_2 \end{vmatrix}$$

Why is this number so important that we give it a special name? We will see this soon, when we use determinants to solve sets of equations.

We can easily tell a matrix from a determinant because a matrix is enclosed in parentheses while a determinant is enclosed between vertical bars. Note that the determinant is not a matrix, although it looks something like one. Rather, it is a *number* that is obtained from the matrix.

Order: The *order* of a determinant is equal to the number of rows or of columns it contains. The determinants in the examples above are all of *second order*. We first study second-order determinants, and later cover determinants of higher order.

Value of a Second-Order Determinant: To evaluate a second-order determinant, take the product of the elements along the principal diagonal, minus the product of the elements along the secondary diagonal.

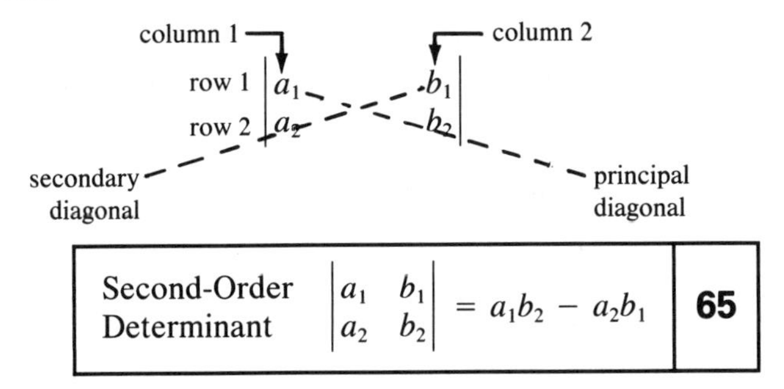

Second-Order Determinant	$\begin{vmatrix} a_1 & b_1 \\ a_2 & b_2 \end{vmatrix} = a_1b_2 - a_2b_1$	**65**

Examples:

(a) $$\begin{vmatrix} 2 & -1 \\ 3 & 5 \end{vmatrix} = 2(5) - 3(-1) = 13$$

(b) $$\begin{vmatrix} 4 & a \\ x & 3b \end{vmatrix} = 4(3b) - x(a) = 12b - ax$$

Example: Write the determinant of the matrix

$$\begin{pmatrix} 2 & 3 \\ 4 & 1 \end{pmatrix}$$

Solution:

$$\det \begin{pmatrix} 2 & 3 \\ 4 & 1 \end{pmatrix} = \begin{vmatrix} 2 & 3 \\ 4 & 1 \end{vmatrix} = (2)(1) - (3)(4) = -10$$

Solving a Pair of Linear Equations by Determinants: In Sec. 8-2 we solved the pair of literal equations

$$a_1x + b_1y = c_1$$
$$a_2x + b_2y = c_2$$

and got the result

$$x = \frac{c_1b_2 - c_2b_1}{a_1b_2 - a_2b_1} \quad \text{and} \quad y = \frac{a_1c_2 - a_2c_1}{a_1b_2 - a_2b_1}$$	**63**

This is called the determinant of the coefficients, *or sometimes the* determinant of the system. *It is usually given the Greek capital letter delta (Δ). If this determinant equals zero, there is no solution to the set of equations.*

Does the denominator of each of these expressions look familiar? It is nothing but the value of the determinant

$$\Delta = \begin{vmatrix} a_1 & b_1 \\ a_2 & b_2 \end{vmatrix} = a_1b_2 - a_2b_1$$

and the numerators are the same as the values of the two determinants

$$\begin{vmatrix} c_1 & b_1 \\ c_2 & b_2 \end{vmatrix} = c_1b_2 - c_2b_1 \quad \text{and} \quad \begin{vmatrix} a_1 & c_1 \\ a_2 & c_2 \end{vmatrix} = a_1c_2 - a_2c_1$$

Writing the solution to the pair of equations in terms of determinants,

After the Swiss Mathematician Gabriel Cramer, 1704–1752

Cramer's Rule	$$x = \frac{\begin{vmatrix} c_1 & b_1 \\ c_2 & b_2 \end{vmatrix}}{\begin{vmatrix} a_1 & b_1 \\ a_2 & b_2 \end{vmatrix}} \quad \text{and} \quad y = \frac{\begin{vmatrix} a_1 & c_1 \\ a_2 & c_2 \end{vmatrix}}{\begin{vmatrix} a_1 & b_1 \\ a_2 & b_2 \end{vmatrix}}$$	**70**

The solution for any variable is a fraction whose denominator is the determinant of the coefficients, and whose numerator is that same determinant, except that the column of coefficients for the variable for which we are solving is replaced by the column of constants.

Of course, the denominator Δ cannot be zero or we get division by zero. This is an indication that the set of equations has no unique solution.

Example: Solve by determinants:

$$2x - 3y = 1$$
$$x + 4y = -5$$

Cramer's rule, as we will see, is valid for systems with any number of equations.

Solution: We first evaluate the determinant of the coefficients, because, if $\Delta = 0$ there is no unique solution, and we have no need to proceed further.

$$\Delta = \begin{vmatrix} 2 & -3 \\ 1 & 4 \end{vmatrix} = 2(4) - 1(-3) = 8 + 3 = 11$$

Solving for x,

column of constants

$$x = \frac{\begin{vmatrix} 1 & -3 \\ -5 & 4 \end{vmatrix}}{\Delta} = \frac{1(4) - (-5)(-3)}{11} = \frac{4 - 15}{11} = -1$$

It's easier to find y by substituting x back into one of the previous equations, but we use determinants instead, to show how it is done.

Solving for y,

column of constants

$$y = \frac{\begin{vmatrix} 2 & 1 \\ 1 & -5 \end{vmatrix}}{\Delta} = \frac{2(-5) - 1(1)}{11} = \frac{-10 - 1}{11} = -1$$

Example: Solve for x and y by determinants:

$$2ax - by + 3a = 0$$
$$4y = 5x - a$$

Solution: We first rearrange our equation into the form $Ax + By = C$.

$$2ax - by = -3a$$
$$5x - 4y = \quad a$$

The determinant of the coefficients is

$$\Delta = \begin{vmatrix} 2a & -b \\ 5 & -4 \end{vmatrix} = 2a(-4) - 5(-b) = -8a + 5b$$

So

$$x = \frac{\begin{vmatrix} -3a & -b \\ a & -4 \end{vmatrix}}{\Delta} = \frac{(-3a)(-4) - a(-b)}{5b - 8a}$$
$$= \frac{12a + ab}{5b - 8a}$$

and

$$y = \frac{\begin{vmatrix} 2a & -3a \\ 5 & a \end{vmatrix}}{\Delta} = \frac{2a(a) - 5(-3a)}{5b - 8a} = \frac{2a^2 + 15a}{5b - 8a}$$

EXERCISE 2

Second-Order Determinants

Find the value of each second-order determinant.

1. $\begin{vmatrix} 3 & 2 \\ 1 & -4 \end{vmatrix}$ **2.** $\begin{vmatrix} -2 & 5 \\ 3 & -3 \end{vmatrix}$ **3.** $\begin{vmatrix} 0 & 5 \\ -3 & 4 \end{vmatrix}$

4. $\begin{vmatrix} 1 & 1 \\ 1 & 1 \end{vmatrix}$ **5.** $\begin{vmatrix} 5 & 5 \\ 5 & 5 \end{vmatrix}$ **6.** $\begin{vmatrix} 3 & 0 \\ 7 & 0 \end{vmatrix}$

7. $\begin{vmatrix} a & b \\ c & d \end{vmatrix}$ **8.** $\begin{vmatrix} 2m & 3n \\ -m & 4n \end{vmatrix}$ **9.** $\begin{vmatrix} \frac{1}{2} & \frac{1}{4} \\ -\frac{1}{2} & -\frac{1}{4} \end{vmatrix}$

10. $\begin{vmatrix} \sin\theta & 3 \\ \tan\theta & \sin\theta \end{vmatrix}$ **11.** $\begin{vmatrix} 2i_1 & i_2 \\ -3i_1 & 4i_2 \end{vmatrix}$ **12.** $\begin{vmatrix} 3i_1 & i_2 \\ \sin\theta & \cos\theta \end{vmatrix}$

Solving Equations by Determinants

Solve each pair of equations by determinants.

13. $2x - 3y = 7$
$5x + 2y = 27$

14. $7x - 5y = 13$
$3x + 3y = 21$

15. $6x + y = 60$
$3x + 2y = 39$

16. $2x + 5y = 29$
$2x - 5y = -21$

17. $4x + 3y = 7$
$2x - 3y = -1$

18. $4x - 5y = 3$
$3x + 5y = 11$

9-3. THIRD-ORDER DETERMINANTS

There is a method for evaluating a third-order determinant by multiplying along diagonals. We omit it here because it works only *for third-order determinants, and also is not suitable for computer solution.*

Minors: The technique we used for finding the value of a second-order determinant *cannot* be used for higher-order determinants. Instead, we use a technique called *development by minors* to reduce the order of a determinant to second order so that it may be evaluated by the method described in Sec. 9-2.

The *minor of an element in a determinant* is a determinant of next lower order, obtained by deleting the row and the column in which that element lies.

Example: The minor of element b in the determinant

$$\begin{vmatrix} a & b & c \\ d & e & f \\ g & h & i \end{vmatrix}$$

is found by striking out the first row and the second column

$$\begin{vmatrix} \not{a} & \textcircled{b} & \not{c} \\ d & \not{e} & f \\ g & \not{h} & i \end{vmatrix} \quad \text{or} \quad \begin{vmatrix} d & f \\ g & i \end{vmatrix}$$

A signed minor of an element is also called a cofactor *of that element.*

A *signed minor* is a minor which is given an algebraic sign depending on the position of the element in the determinant. The signed minor for the element in the upper left-hand corner of the determinant is positive and the signs alternate according to the pattern

$$\begin{vmatrix} + & - & + & - & + & - & \cdot \\ - & + & - & + & - & \cdot & \cdot \\ + & - & + & - & \cdot & \cdot & \cdot \\ - & + & - & + & \cdot & \cdot & \cdot \\ + & - & \cdot & \cdot & \cdot & \cdot & \cdot \\ - & \cdot & \cdot & \cdot & \cdot & \cdot & \cdot \\ \cdot & \cdot & \cdot & \cdot & \cdot & \cdot & \cdot \end{vmatrix}$$

so that the sign may be found simply by *counting off* from the upper left corner.

Example: In the preceding example, the signed minor of element b is

$$-\begin{vmatrix} d & f \\ g & i \end{vmatrix}$$

and the signed minor of element e is

$$+\begin{vmatrix} a & c \\ g & i \end{vmatrix}$$

Finally, the quantity we will need to evaluate a determinant is the *product of an element and its signed minor.*

Examples:

(a) The product of element b and its signed minor is

$$-b\begin{vmatrix} d & f \\ g & i \end{vmatrix}$$

(b) The product of element e and its signed minor is

$$+e\begin{vmatrix} a & c \\ g & i \end{vmatrix}$$

Evaluating a Determinant by Minors:

To find the value of a determinant: 1. Choose any row or any column to develop by minors. 2. Write the product of every element in that row or column and its signed minor. 3. Add these products to get the value of the determinant.	**68**

Example: Develop by minors:

$$\begin{vmatrix} a_1 & b_1 & c_1 \\ a_2 & b_2 & c_2 \\ a_3 & b_3 & c_3 \end{vmatrix}$$

Solution: Choosing, say, the first row for development, we get

$$a_1\begin{vmatrix} b_2 & c_2 \\ b_3 & c_3 \end{vmatrix} - b_1\begin{vmatrix} a_2 & c_2 \\ a_3 & c_3 \end{vmatrix} + c_1\begin{vmatrix} a_2 & b_2 \\ a_3 & b_3 \end{vmatrix}$$

$$= a_1(b_2c_3 - b_3c_2) - b_1(a_2c_3 - a_3c_2) + c_1(a_2b_3 - a_3b_2)$$

$$= a_1b_2c_3 - a_1b_3c_2 - a_2b_1c_3 + a_3b_1c_2 + a_2b_3c_1 - a_3b_2c_1$$

$$= a_1b_2c_3 + a_3b_1c_2 + a_2b_3c_1 - a_3b_2c_1 - a_1b_3c_2 - a_2b_1c_3$$

The value of the determinant is then

Third-Order Determinant	$\begin{vmatrix} a_1 & b_1 & c_1 \\ a_2 & b_2 & c_2 \\ a_3 & b_3 & c_3 \end{vmatrix} = a_1b_2c_3 + a_3b_1c_2 + a_2b_3c_1 - a_3b_2c_1 - a_1b_3c_2 - a_2b_1c_3$	**66**

Example: Evaluate by minors:

$$\begin{vmatrix} 3 & 2 & 4 \\ 1 & 6 & -2 \\ -1 & 5 & -3 \end{vmatrix}$$

Solution: If we choose the first column for development, we get

$$\begin{vmatrix} 3 & 2 & 4 \\ 1 & 6 & -2 \\ -1 & 5 & -3 \end{vmatrix} = 3\begin{vmatrix} 6 & -2 \\ 5 & -3 \end{vmatrix} - 1\begin{vmatrix} 2 & 4 \\ 5 & -3 \end{vmatrix} + (-1)\begin{vmatrix} 2 & 4 \\ 6 & -2 \end{vmatrix}$$

$$= 3[(6)(-3) - (-2)(5)] - 1[(2)(-3) - (4)(5)] - 1[(2)(-2) - 4(6)]$$

$$= -24 + 26 + 28$$

$$= 30$$

Common Error	Remember that when we say that a signed minor is negative, what we mean is to *reverse the sign* of the element. If the element was initially negative, the signed minor will be positive.

Solving a System of Three Equations by Determinants: If we were to *algebraically* solve the system of equations

$$a_1x + b_1y + c_1z = k_1$$
$$a_2x + b_2y + c_2z = k_2$$
$$a_3x + b_3y + c_3z = k_3$$

we would get the solution

Three Equations in Three Unknowns	$x = \dfrac{b_2c_3k_1 + b_1c_2k_3 + b_3c_1k_2 - b_2c_1k_3 - b_3c_2k_1 - b_1c_3k_2}{a_1b_2c_3 + a_3b_1c_2 + a_2b_3c_1 - a_3b_2c_1 - a_1b_3c_2 - a_2b_1c_3}$ $y = \dfrac{a_1c_3k_2 + a_3c_2k_1 + a_2c_1k_3 - a_3c_1k_2 - a_1c_2k_3 - a_2c_3k_1}{a_1b_2c_3 + a_3b_1c_2 + a_2b_3c_1 - a_3b_2c_1 - a_1b_3c_2 - a_2b_1c_3}$ $z = \dfrac{a_1b_2k_3 + a_3b_1k_2 + a_2b_3k_1 - a_3b_2k_1 - a_1b_3k_2 - a_2b_1k_3}{a_1b_2c_3 + a_3b_1c_2 + a_2b_3c_1 - a_3b_2c_1 - a_1b_3c_2 - a_2b_1c_3}$	**64**

Notice that the three denominators are identical and are equal to the value of the determinant formed from the coefficients of the unknowns (Eq. 66). As before, we call it the *determinant of the coefficients*.

Furthermore, the numerator of each can be obtained from the

determinant of the system by replacing the coefficients of the variable in question with the constants k_1, k_2, and k_3. The solution to our set of equations

$$a_1x + b_1y + c_1z = k_1$$
$$a_2x + b_2y + c_2z = k_2$$
$$a_3x + b_3y + c_3z = k_3$$

is then

Cramer's Rule	$x = \dfrac{\begin{vmatrix} k_1 & b_1 & c_1 \\ k_2 & b_2 & c_2 \\ k_3 & b_3 & c_3 \end{vmatrix}}{\Delta} \quad y = \dfrac{\begin{vmatrix} a_1 & k_1 & c_1 \\ a_2 & k_2 & c_2 \\ a_3 & k_3 & c_3 \end{vmatrix}}{\Delta} \quad z = \dfrac{\begin{vmatrix} a_1 & b_1 & k_1 \\ a_2 & b_2 & k_2 \\ a_3 & b_3 & k_3 \end{vmatrix}}{\Delta}$ where $\Delta = \begin{vmatrix} a_1 & b_1 & c_1 \\ a_2 & b_2 & c_2 \\ a_3 & b_3 & c_3 \end{vmatrix} \neq 0$	**71**

Cramer's rule, although we do not prove it, works for higher-order systems as well. We restate it now in words.

Cramer's Rule	The solution for any variable is a fraction whose denominator is the determinant of the coefficients, and whose numerator is the same determinant, except that the column of coefficients for the variable for which we are solving is replaced by the column of constants.	**69**

Example: Solve by determinants:

$$x + 2y + 3z = 14$$
$$2x + y + 2z = 10$$
$$3x + 4y - 3z = 2$$

Solution: We first write the determinant of the coefficients

$$\Delta = \begin{vmatrix} 1 & 2 & 3 \\ 2 & 1 & 2 \\ 3 & 4 & -3 \end{vmatrix}$$

and expand it by minors. Choosing the first row for expansion gives

$$\Delta = \begin{vmatrix} 1 & 2 & 3 \\ 2 & 1 & 2 \\ 3 & 4 & -3 \end{vmatrix} = 1\begin{vmatrix} 1 & 2 \\ 4 & -3 \end{vmatrix} - 2\begin{vmatrix} 2 & 2 \\ 3 & -3 \end{vmatrix} + 3\begin{vmatrix} 2 & 1 \\ 3 & 4 \end{vmatrix}$$
$$= 1(-3 - 8) - 2(-6 - 6) + 3(8 - 3)$$
$$= -11 + 24 + 15 = 28$$

As when solving a set of two equations, a Δ of zero would mean that the system had no unique solution, and we would stop here.

We now make a new determinant by replacing the coefficients of x, $\begin{vmatrix} 1 \\ 2 \\ 3 \end{vmatrix}$, by the column of constants $\begin{vmatrix} 14 \\ 10 \\ 2 \end{vmatrix}$, and expand it by, say, the second column.

$$\begin{vmatrix} 14 & 2 & 3 \\ 10 & 1 & 2 \\ 2 & 4 & -3 \end{vmatrix} = -2\begin{vmatrix} 10 & 2 \\ 2 & -3 \end{vmatrix} + 1\begin{vmatrix} 14 & 3 \\ 2 & -3 \end{vmatrix} - 4\begin{vmatrix} 14 & 3 \\ 10 & 2 \end{vmatrix}$$

$$= -2(-30 - 4) + 1(-42 - 6) - 4(28 - 30)$$

$$= 68 - 48 + 8 = 28$$

Dividing this value by Δ gives x.

$$x = \frac{28}{\Delta} = \frac{28}{28} = 1$$

Next we replace the coefficients of y with the column of constants and expand the first column by minors.

$$\begin{vmatrix} 1 & 14 & 3 \\ 2 & 10 & 2 \\ 3 & 2 & -3 \end{vmatrix} = 1\begin{vmatrix} 10 & 2 \\ 2 & -3 \end{vmatrix} - 2\begin{vmatrix} 14 & 3 \\ 2 & -3 \end{vmatrix} + 3\begin{vmatrix} 14 & 3 \\ 10 & 2 \end{vmatrix}$$

$$= 1(-30 - 4) - 2(-42 - 6) + 3(28 - 30)$$

$$= -34 + 96 - 6 = 56$$

Dividing by Δ gives the value of y.

$$y = \frac{56}{\Delta} = \frac{56}{28} = 2$$

We can get z by using Eq. 71 or, more easily, by back substitution. Substituting $x = 1$ and $y = 2$ into the first equation,

$$1 + 2(2) + 3z = 14$$

$$3z = 9$$

$$z = 3$$

Often, some terms will have zero coefficients or decimal coefficients. Further, the terms may be out of order, as in the following example.

Example: Solve for x, y, and z.

$$23.7y + 72.4x = 82.4 - 11.3x$$

$$25.5x - 28.4z + 19.3 = 48.2y$$

$$13.4 + 66.3z = 39.2x - 10.5$$

Solution: We rewrite each equation in the order $ax + by + cz = k$, combining like terms as we go, and putting in the missing terms with zero coefficients. We do this *before* writing the determinant.

$$\begin{aligned} 83.7x + 23.7y + \quad 0z &= \quad 82.4 \\ 25.5x - 48.2y - 28.4z &= -19.3 \\ -39.2x + \quad 0y + 66.3z &= -23.9 \end{aligned}$$

The determinant of the system is then

$$\Delta = \begin{vmatrix} 83.7 & 23.7 & 0 \\ 25.5 & -48.2 & -28.4 \\ -39.2 & 0 & 66.3 \end{vmatrix}$$

Let us develop the first row by minors.

$$\Delta = 83.7\begin{vmatrix} -48.2 & -28.4 \\ 0 & 66.3 \end{vmatrix} - 23.7\begin{vmatrix} 25.5 & -28.4 \\ -39.2 & 66.3 \end{vmatrix}$$

$$= 83.7\,(-48.2)\,(66.3) - 23.7[25.5(66.3) - (-28.4)\,(-39.2)]$$

$$= -281{,}000 \qquad \text{(to three significant digits)}$$

Now solving for x,

$$x = \frac{\begin{vmatrix} 82.4 & 23.7 & 0 \\ -19.3 & -48.2 & -28.4 \\ -23.9 & 0 & 66.3 \end{vmatrix}}{\Delta}$$

Let us develop the first row of thc determinant by minors. We get

$$82.4\begin{vmatrix} -48.2 & -28.4 \\ 0 & 66.3 \end{vmatrix} - 23.7\begin{vmatrix} -19.3 & -28.4 \\ -23.7 & 66.3 \end{vmatrix}$$

$$= 82.4(-48.2)\,(66.3) - 23.7[(-19.3)\,(66.3) - (-28.4)\,(-23.7)]$$

$$= -217{,}000$$

Dividing by Δ,

$$x = \frac{-217{,}000}{-281{,}000} = 0.772$$

We solve for y and z by back substitution.

$$\begin{aligned} 23.7y &= 82.4 - 83.7(0.772) = 17.8 \\ y &= 0.751 \end{aligned}$$

and

$$\begin{aligned} 66.3z &= -23.9 + 39.2(0.772) = 6.34 \\ z &= 0.0960 \end{aligned}$$

EXERCISE 3

Evaluate each determinant.

1. $\begin{vmatrix} 1 & 0 & 2 \\ 3 & 1 & 0 \\ 1 & 2 & 1 \end{vmatrix}$

2. $\begin{vmatrix} 2 & -1 & 3 \\ 0 & 2 & 1 \\ 3 & -2 & 4 \end{vmatrix}$

3. $\begin{vmatrix} -3 & 1 & 2 \\ 0 & -1 & 5 \\ 6 & 0 & 1 \end{vmatrix}$

4. $\begin{vmatrix} -1 & 0 & 3 \\ 2 & 0 & -2 \\ 1 & -3 & 4 \end{vmatrix}$

5. $\begin{vmatrix} 5 & 1 & 2 \\ -3 & 2 & -1 \\ 4 & -3 & 5 \end{vmatrix}$

6. $\begin{vmatrix} 1.0 & 2.4 & -1.5 \\ -2.6 & 0 & 3.2 \\ -2.9 & 1.0 & 4.1 \end{vmatrix}$

7. $\begin{vmatrix} 2 & 1 & 3 \\ 0 & -2 & 4 \\ 0 & 1 & 5 \end{vmatrix}$

8. $\begin{vmatrix} 1 & 5 & 4 \\ -3 & 6 & -2 \\ -1 & 5 & 3 \end{vmatrix}$

Three Equations in Three Unknowns

Solve by determinants.

Problems 9 to 18 are identical to those given in Chapter 8, Exercise 4.

9. $x + y + z = 18$
$x - y + z = 6$
$x + y - z = 4$

10. $x + y + z = 12$
$x - y = 2$
$x - z = 4$

11. $x + y = 35$
$x + z = 40$
$y + z = 45$

12. $x + y + z = 35$
$x - 2y + 3z = 15$
$y - x + z = -5$

13. $x + 2y + 3z = 14$
$2x + y + 2z = 10$
$3x + 4y - 3z = 2$

14. $x + y + z = 90$
$2x - 3y = -20$
$2x + 3z = 145$

15. $x - 2y + 2z = 5$
$5x + 3y + 6z = 57$
$x + 2y + 2z = 21$

16. $3x + y = 5$
$2y - 3z = -5$
$x + 2z = 7$

17. $2x - 4y + 3z = 10$
$3x + y - 2z = 6$
$x + 3y - z = 20$

18. $x - y = 5$
$y - z = -6$
$2x - z = 2$

Fractional Equations

Solve by determinants.

19. $x + \frac{y}{3} = 5$
$x + \frac{z}{3} = 6$
$y + \frac{z}{3} = 9$

20. $\frac{1}{x} + \frac{1}{y} = 5$
$\frac{1}{y} + \frac{1}{z} = 7$
$\frac{1}{x} + \frac{1}{z} = 6$

21. $\frac{x}{10} + \frac{y}{5} + \frac{z}{20} = \frac{1}{4}$
$x + y + z = 6$
$\frac{x}{3} + \frac{y}{2} + \frac{z}{6} = 1$

22. $\frac{1}{x} + \frac{2}{y} - \frac{1}{z} = -3$
$\frac{3}{x} + \frac{1}{y} + \frac{1}{z} = 4$
$\frac{1}{x} - \frac{1}{y} + \frac{2}{z} = 6$

9-4. HIGHER-ORDER DETERMINANTS

We can evaluate a determinant of any order by repeated use of the method of minors. Thus any row or column of a fifth-order determinant can be developed, thus reducing the determinant to five fourth-order determinants. Each of these can then be developed into four third-order determinants, and so on, until only second-order determinants remain.

Obviously, this is lots of work. To avoid this, we first *simplify* the determinant, using the properties of determinants in the following section, and then apply a systematic method for reducing the determinants by minors shown after that.

Properties of Determinants: A determinant larger than third order can usually be evaluated faster if we first reduce its order before expanding by minors. We do this by applying the various *properties of determinants* as shown in the following examples.

Zero Row or Column:

If all elements in a row (or column) are zero, the value of the determinant is zero.	**72**

Example:

$$\begin{vmatrix} 4 & 8 & 0 \\ 5 & 2 & 0 \\ 3 & 9 & 0 \end{vmatrix} = 0$$

Identical Rows or Columns:

The value of a determinant is zero if two rows (or columns) are identical.	**73**

Example: The value of the determinant

$$\begin{vmatrix} 2 & 1 & 3 \\ 4 & 2 & 5 \\ 4 & 2 & 5 \end{vmatrix}$$

is found by expanding by minors using the first row,

$$2\begin{vmatrix} 2 & 5 \\ 2 & 5 \end{vmatrix} - 1\begin{vmatrix} 4 & 5 \\ 4 & 5 \end{vmatrix} + 3\begin{vmatrix} 4 & 2 \\ 4 & 2 \end{vmatrix}$$

$$= 2(10 - 10) - 1(20 - 20) + 3(8 - 8)$$

$$= 0$$

Zeros Below the Principal Diagonal:

If all elements below the principal diagonal are zeros, then the value of the determinant is the product of the elements along the principal diagonal.	**74**

Example: The determinant

$$\begin{vmatrix} 2 & 9 & 7 & 8 \\ 0 & 1 & 5 & 9 \\ 0 & 0 & 3 & 4 \\ 0 & 0 & 0 & 1 \end{vmatrix}$$

Expand this determinant by minors and see for yourself that its value is indeed 6.

has the value

$$(2)(1)(3)(1) = 6$$

Interchanging Rows with Columns:

The value of a determinant is unchanged if we change the rows to columns and the columns to rows.	**75**

In other words, the determinant of a matrix, and the determinant of the transpose of that matrix, are equal.

Example: The value of the determinant

$$\begin{vmatrix} 5 & 2 \\ 4 & 3 \end{vmatrix}$$

is

$$(5)(3) - (2)(4) = 7$$

If the first column now becomes the first row, and the second column becomes the second row, we get the determinant

$$\begin{vmatrix} 5 & 4 \\ 2 & 3 \end{vmatrix}$$

which has the value

$$(5)(3) - (4)(2) = 7$$

as before.

Interchange of Rows or Columns:

A determinant will change sign when we interchange two rows (or columns).	**76**

Example: The value of the determinant

$$\begin{vmatrix} 3 & 0 & 2 \\ 1 & 4 & 3 \\ 2 & 1 & 2 \end{vmatrix}$$

is found by expanding by minors using the first row.

$$3\begin{vmatrix} 4 & 3 \\ 1 & 2 \end{vmatrix} - 0 + 2\begin{vmatrix} 1 & 4 \\ 2 & 1 \end{vmatrix} = 3(8 - 3) + 2(1 - 8) = 1$$

Let us now interchange, say, the first and second columns.

$$\begin{vmatrix} 0 & 3 & 2 \\ 4 & 1 & 3 \\ 1 & 2 & 2 \end{vmatrix}$$

Its value is again found by expanding using the first row,

$$0 - 3\begin{vmatrix} 4 & 3 \\ 1 & 2 \end{vmatrix} + 2\begin{vmatrix} 4 & 1 \\ 1 & 2 \end{vmatrix} = -3(8 - 3) + 2(8 - 1) = -1$$

So interchanging two columns has reversed the sign of the determinant.

Multiplying by a Constant:

If each element in a row (or column) is multiplied by some constant, the value of the determinant is multiplied by that constant.	**77**

Example: We saw that the value of the determinant in the preceding example was equal to 1.

$$\begin{vmatrix} 3 & 0 & 2 \\ 1 & 4 & 3 \\ 2 & 1 & 2 \end{vmatrix} = 1$$

Let us now multiply the elements in the third row by some constant, say, 3.

$$\begin{vmatrix} 3 & 0 & 2 \\ 1 & 4 & 3 \\ 6 & 3 & 6 \end{vmatrix}$$

This new determinant has the value

$$3(24 - 9) + 2(3 - 24) = 45 - 42 = 3$$

or three times its previous value.

We can also use this rule *in reverse*, to *remove a factor* from a row or column.

Example: The determinant

$$\begin{vmatrix} 3 & 1 & 0 \\ 30 & 10 & 15 \\ 3 & 2 & 1 \end{vmatrix}$$

could be evaluated as it is, but let us first factor a 3 from the first column,

$$\begin{vmatrix} 3 & 1 & 0 \\ 30 & 10 & 15 \\ 3 & 2 & 1 \end{vmatrix} = 3 \begin{vmatrix} 1 & 1 & 0 \\ 10 & 10 & 15 \\ 1 & 2 & 1 \end{vmatrix}$$

and then factor a 5 from the second row,

$$3 \begin{vmatrix} 1 & 1 & 0 \\ 10 & 10 & 15 \\ 1 & 2 & 1 \end{vmatrix} = 3(5) \begin{vmatrix} 1 & 1 & 0 \\ 2 & 2 & 3 \\ 1 & 2 & 1 \end{vmatrix}$$

$$= 15[(2 - 6) - (2 - 3)] = -45$$

Multiples of One Row or Column Added to Another:

The value of a determinant is unchanged when the elements of a row or column are multiplied by some factor, and then added to the corresponding elements of another row or column.	**78**

Example: Again using the determinant from a previous example,

$$\begin{vmatrix} 3 & 0 & 2 \\ 1 & 4 & 3 \\ 2 & 1 & 2 \end{vmatrix} = 1$$

let us get a new second row by multiplying the third row by -4 and adding those elements to the second row. The new elements in the second row are then

$$1 + 2(-4) = -7, \qquad 4 + 1(-4) = 0, \qquad \text{and} \qquad 3 + 2(-4) = -5$$

giving a new determinant

$$\begin{vmatrix} 3 & 0 & 2 \\ -7 & 0 & -5 \\ 2 & 1 & 2 \end{vmatrix}$$

whose value we find by developing the second column by minors,

$$-1 \begin{vmatrix} 3 & 2 \\ -7 & -5 \end{vmatrix} = -1[3(-5) - 2(-7)] = -[-15 + 14] = 1$$

as before. Notice, however, that we have *introduced another zero* into the determinant, making it easier to evaluate. This, of course, is the point of the whole operation.

A Systematic Method for Evaluating a Determinant: In one method for evaluating a determinant, you first select the row or column having the most zeros, and then use the properties of determinants to introduce zeros into all but one element of that row or column, and then develop that row or column by minors. That method is fine for manual computation, but not as good for a computer. Although we could program a computer to select the best row for development, the logic gets complicated.

The following algorithm, based on the properties of determinants in Eqs. 77 and 78, gives a systematic method for manual or computer evaluation of a determinant of any size. It is sometimes called the *method of Chió.*

1. Factor out the element in row 1, column 1.
2. Multiply each element in row 1 by the first element in row 2. Subtract these products from the corresponding elements in row 2.
3. Repeat step 2 for each row until all but the first element in the first column are zeros.
4. Delete the first row and first column, thus reducing the order of the determinant by 1.
5. Repeat steps 1 through 4 until the determinant is reduced to a single element. The value of the determinant is then the product of that element and all the factors removed in step 1.

Example: Evaluate the following determinant by the method of Chió. Work to three significant digits.

$$\begin{vmatrix} 5 & 3 & 8 \\ 2 & 6 & 1 \\ 4 & 7 & 9 \end{vmatrix}$$

Solution: For the first pass:

Step 1: We factor out the 5.

$$5\begin{vmatrix} 1 & 0.6 & 1.6 \\ 2 & 6 & 1 \\ 4 & 7 & 9 \end{vmatrix}$$

Step 2: Multiply row 1 by 2 and subtract it from row 2.

$$5\begin{vmatrix} 1 & 0.6 & 1.6 \\ 0 & 4.8 & -2.2 \\ 4 & 7 & 9 \end{vmatrix}$$

Step 3: Multiply row 1 by 4 and subtract it from row 3.

$$5\begin{vmatrix} 1 & 0.6 & 1.6 \\ 0 & 4.8 & -2.2 \\ 0 & 4.6 & 2.6 \end{vmatrix}$$

Step 4: We develop the first column by minors. Since the first element is a 1 and the others are zeros, we simply delete the first row and first column.

$$5(1)\begin{vmatrix} 4.8 & -2.2 \\ 4.6 & 2.6 \end{vmatrix}$$

Thus we have reduced the order of the determinant from three to two. We then repeat the whole process.

For the second pass:

Step 1: Factor out 4.8.

If, during step 1, the element in row 1, column 1 happens to be zero, we will get division by zero. If this happens, interchange two rows or two columns and try again. Remember to change the sign on the determinant when interchanging rows or columns.

$$5(4.8)\begin{vmatrix} 1 & -0.458 \\ 4.6 & 2.6 \end{vmatrix}$$

Step 2:

$$5(4.8)\begin{vmatrix} 1 & -0.458 \\ 0 & 0.492 \end{vmatrix}$$

Step 4: Delete the first row and first column. The value of the determinant is then

$$5(4.8)\,[0.492] = 11.8$$

Example: Evaluate the following fourth-order determinant by the method of Chió. Work to three significant digits.

$$\begin{vmatrix} 5 & 2 & 6 & 1 \\ 4 & 2 & 9 & 4 \\ 3 & 6 & 1 & 0 \\ 7 & 1 & 3 & 4 \end{vmatrix}$$

Solution: For the first pass:

Step 1: We factor out the 5.

$$5\begin{vmatrix} 1 & 0.4 & 1.2 & 0.2 \\ 4 & 2 & 9 & 4 \\ 3 & 6 & 1 & 0 \\ 7 & 1 & 3 & 4 \end{vmatrix}$$

Step 2: Multiply row 1 by 4 and subtract it from row 2.

$$5\begin{vmatrix} 1 & 0.4 & 1.2 & 0.2 \\ 0 & 0.4 & 4.2 & 3.2 \\ 3 & 6 & 1 & 0 \\ 7 & 1 & 3 & 4 \end{vmatrix}$$

Step 3: Multiply row 1 by 3 and subtract it from row 3. Multiply row 1 by 7 and subtract it from row 4.

$$5\begin{vmatrix} 1 & 0.4 & 1.2 & 0.2 \\ 0 & 0.4 & 4.2 & 3.2 \\ 0 & 4.8 & -2.6 & -0.6 \\ 0 & -1.8 & -5.4 & 2.6 \end{vmatrix}$$

Step 4: Delete the first row and first column.

$$5\begin{vmatrix} 0.4 & 4.2 & 3.2 \\ 4.8 & -2.6 & -0.6 \\ -1.8 & -5.4 & 2.6 \end{vmatrix}$$

For the second pass:

Step 1: Factor out 0.4.

$$5(0.4)\begin{vmatrix} 1 & 10.5 & 8 \\ 4.8 & -2.6 & -0.6 \\ -1.8 & -5.4 & 2.6 \end{vmatrix}$$

Steps 2 and 3:

$$5(0.4)\begin{vmatrix} 1 & 10.5 & 8 \\ 0 & -53 & -39 \\ 0 & 13.5 & 17 \end{vmatrix}$$

Step 4: Delete the first row and first column.

$$5(0.4)\begin{vmatrix} -53 & -39 \\ 13.5 & 17 \end{vmatrix}$$

For the third pass:

Step 1: Factor out -53.

$$5(0.4)(-53)\begin{vmatrix} 1 & 0.736 \\ 13.5 & 17 \end{vmatrix}$$

Step 2: Multiply row 1 by 13.5 and subtract it from row 2.

$$5(0.4)(-53)\begin{vmatrix} 1 & 0.736 \\ 0 & 6.70 \end{vmatrix}$$

Step 4: Delete the first row and first column. The value of the determinant is, then,

$$5(0.4)(-53)(6.70) = -710$$

EXERCISE 4

Evaluate each determinant.

1. $\begin{vmatrix} 4 & 3 & 1 & 0 \\ -1 & 2 & -3 & 5 \\ 0 & 1 & -1 & 2 \\ 0 & 2 & -3 & 5 \end{vmatrix}$

2. $\begin{vmatrix} -1 & 3 & 0 & 2 \\ 2 & -1 & 1 & 0 \\ 5 & 2 & -2 & 0 \\ 1 & -1 & 3 & 1 \end{vmatrix}$

3. $\begin{vmatrix} 2 & 0 & -1 & 0 \\ 0 & 0 & 2 & -1 \\ 1 & 3 & 2 & 1 \\ 3 & 1 & 1 & -2 \end{vmatrix}$

4. $\begin{vmatrix} 1 & 2 & -1 & 1 \\ -1 & 1 & 2 & 3 \\ 3 & -1 & 1 & 2 \\ 1 & 2 & -1 & 1 \end{vmatrix}$

5. $\begin{vmatrix} 3 & 1 & 0 & 2 & 4 \\ 1 & 2 & 4 & 0 & 1 \\ 2 & 3 & 1 & 4 & 2 \\ 1 & 2 & 0 & 2 & 1 \\ 3 & 4 & 1 & 3 & 1 \end{vmatrix}$

6. $\begin{vmatrix} 2 & 1 & 5 & 3 & 6 \\ 1 & 4 & 2 & 4 & 3 \\ 3 & 1 & 2 & 4 & 1 \\ 5 & 2 & 3 & 1 & 4 \\ 4 & 5 & 2 & 3 & 1 \end{vmatrix}$

Four Equations in Four Unknowns

Solve by determinants.

7. $\begin{aligned} x + y + 2z + w &= 18 \\ x + 2y + z + w &= 17 \\ x + y + z + 2w &= 19 \\ 2x + y + z + w &= 16 \end{aligned}$

8. $\begin{aligned} 2x - y - z - w &= 0 \\ x - 3y + z + w &= 0 \\ x + y - 4z + w &= 0 \\ x + y + w &= 36 \end{aligned}$

9. $\begin{aligned} 3x - 2y - z + w &= -3 \\ -x - y + 3z + 2w &= 23 \\ x + 3y - 2z + w &= -12 \\ 2x - y - z - 3w &= -22 \end{aligned}$

10. $\begin{aligned} x + 2y &= 5 \\ y + 2z &= 8 \\ z + 2u &= 11 \\ u + 2x &= 6 \end{aligned}$

11. $\begin{aligned} x + y &= a + b \\ y + z &= b + c \\ z + w &= a - b \\ w - x &= c - b \end{aligned}$

12. $\begin{aligned} 2x - 3y + z - w &= -6 \\ x + 2y - z &= 8 \\ 3y + z + 3w &= 0 \\ 3x - y + w &= 0 \end{aligned}$

Five Equations in Five Unknowns

Solve by determinants.

13. $\begin{aligned} x + y &= 9 \\ y + z &= 11 \\ z + w &= 13 \\ w + u &= 15 \\ u + x &= 12 \end{aligned}$

14. $\begin{aligned} 3x + 4y + z &= 35 \\ 3z + 2y - 3w &= 4 \\ 2x - y + 2w &= 17 \\ 3z - 2w + v &= 9 \\ w + y &= 13 \end{aligned}$

15. $\begin{aligned} w + v + x + y &= 14 \\ w + v + x + z &= 15 \\ w + v + y + z &= 16 \\ w + x + y + z &= 17 \\ v + x + y + z &= 18 \end{aligned}$

16. Separate 125 into four parts such that if the first is increased by 4, the second diminished by 4, the third multiplied by 4, and the fourth divided by 4, these four quantities will all be equal.

17. A person divided a sum of money among four children so that the share of the eldest was $\frac{1}{2}$ of the sum of the shares of the other three, the share of the second was $\frac{1}{3}$ of the sum of the shares of the other three, and the share of the third was $\frac{1}{4}$ of the sum of the shares of the other three. The eldest has $14,000 more than the youngest. What was the share of each?

18. Applying Kirchhoff's law to a certain four-loop network gives the equations

$$57.2I_1 + 92.5I_2 - 23.0I_3 - 11.4I_4 = 38.2$$
$$95.3I_1 - 14.9I_2 + 39.0I_3 + 59.9I_4 = 29.3$$
$$66.3I_1 + 81.4I_2 - 91.5I_3 + 33.4I_4 = -73.6$$
$$38.2I_1 - 46.6I_2 + 30.1I_3 + 93.2I_4 = 55.7$$

Solve for the four loop currents, by any method.

19. We have available four bronze alloys containing the following percentages of copper, zinc, and lead.

	Alloy 1	*Alloy 2*	*Alloy 3*	*Alloy 4*
Copper	52	48	61	55
Zinc	30	38	20	38
Lead	3	2	4	3

How many pounds of each alloy should be taken to produce 600 kg of a new alloy that is 53.8% copper, 30.1% zinc, and 3.2% lead?

Computer

20. Write a program to evaluate a determinant of any order. Use the algorithm given for the method of Chió. Use DATA statements to input the order of the determinant, and then the elements, row by row. Use it to evaluate any of the determinants in this chapter.

9-5. OPERATIONS WITH MATRICES

The method of determinants is fine for solving sets of equations by hand, but it is not easy to program for the computer. Two other methods that are frequently used on the computer are called *Gauss elimination* and *matrix inversion*. We cover these methods after first learning how to add, subtract, and multiply matrices.

Sum of Two Matrices: To add two matrices *of the same dimensions*, simply combine corresponding elements.

Example:

$$\begin{pmatrix} 2 & -3 & 4 & 1 \\ 0 & 5 & 6 & -8 \end{pmatrix} + \begin{pmatrix} 5 & 8 & 1 & 2 \\ 9 & -3 & 0 & 1 \end{pmatrix} = \begin{pmatrix} 7 & 5 & 5 & 3 \\ 9 & 2 & 6 & -7 \end{pmatrix}$$

Subtraction is done in a similar way.

Addition and subtraction is not defined *for two matrices of* different *dimensions.*

Example:

$$\begin{pmatrix} 2 & -3 & 4 & 1 \\ 0 & 5 & 6 & -8 \end{pmatrix} - \begin{pmatrix} 5 & 8 & 1 & 2 \\ 9 & -3 & 0 & 1 \end{pmatrix} = \begin{pmatrix} -3 & -11 & 3 & -1 \\ -9 & 8 & 6 & -9 \end{pmatrix}$$

As with ordinary addition, the *commutative law* (Eq. 1) and the *associative law* (Eq. 3) apply.

Commutative Law	$\mathbf{A} + \mathbf{B} = \mathbf{B} + \mathbf{A}$	**79**

Associative Law	$\mathbf{A} + (\mathbf{B} + \mathbf{C}) = (\mathbf{A} + \mathbf{B}) + \mathbf{C} = (\mathbf{A} + \mathbf{C}) + \mathbf{B}$	**80**

Product of a Scalar and a Matrix: To multiply a matrix by a scalar, multiply each element in the matrix by the scalar.

Worked in reverse, we can factor *a scalar from a matrix. Thus, in the example shown, we can think of the number 3 as being* factored out.

Example:

$$3\begin{pmatrix} 2 & 4 & 1 \\ 5 & 7 & 3 \end{pmatrix} = \begin{pmatrix} 6 & 12 & 3 \\ 15 & 21 & 9 \end{pmatrix}$$

In symbols

Product of a Scalar and a Matrix	$k\begin{pmatrix} a & b \\ c & d \end{pmatrix} = \begin{pmatrix} ka & kb \\ kc & kd \end{pmatrix}$	**87**

Product of Two Matrices: Not all matrices can be multiplied. Further, for matrices **A** and **B**, it may be possible to find the product **AB** but *not* the product **BA**.

Conformable Matrices	The product **AB** of two matrices **A** and **B** is defined only when the number of columns in **A** equals the number of rows in **B**.	**82**

Two matrices for which multiplication is defined are called conformable matrices.

Example: If

$$\mathbf{A} = \begin{pmatrix} 3 & 5 \\ 2 & 1 \\ 7 & 4 \\ 9 & 0 \end{pmatrix} \quad \text{and} \quad \mathbf{B} = \begin{pmatrix} 7 & 3 & 8 \\ 4 & 7 & 1 \end{pmatrix}$$

we can find the product **AB** but *not* **BA**.

For two or more conformable matrices, the *associative* law and the *distributive* law for multiplication applies.

Associative Law	$\mathbf{A(BC)} = \mathbf{(AB)C} = \mathbf{ABC}$	**84**

Distributive Law	$\mathbf{A(B + C)} = \mathbf{AB} + \mathbf{AC}$	**85**

These are similar to Eqs. 4 and 5 for ordinary multiplication.

The *dimensions* of the matrix obtained by multiplying an $m \times p$ matrix with a $p \times n$ matrix is equal to $m \times n$. In other words, the product matrix **AB** will have the same number of *rows* as **A** and the same number of *columns* as **B**.

Dimensions of the Product	$(m \times p)(p \times n) = (m \times n)$	**86**

Example: The product **AB** of a 3×5 matrix **A** and a 5×7 matrix **B** is a 3×7 matrix.

Whatever we said about matrices applies, of course, to vectors as well.

Scalar Product of Two Vectors: We will show matrix multiplication first with vectors, and then extend it to other matrices.

The *scalar product* (also called the *dot product*) of a row vector **A** and a column vector **B** is defined only when the number of columns in **A** is equal to the number of rows in **B**.

If each vector contains m elements, the dimensions of the product will be, by Eq. 86,

$$(1 \times m)(m \times 1) = 1 \times 1$$

Thus, the product is a scalar, hence the name *scalar product*.

To multiply a row vector by a column vector, multiply the first elements of each, then the second, and so on, and add the products obtained.

Example:

$$(2 \quad 6 \quad 4 \quad 8)\begin{pmatrix}3\\5\\1\\7\end{pmatrix} = [(2)(3) + (6)(5) + (4)(1) + (8)(7)]$$

$$= (6 + 30 + 4 + 42) = 82$$

Example: A certain store has four types of radios, priced at $71, $62, $83, and $49. On a particular day, they sell quantities of 8, 3, 7, and 6, respectively. If we represent the prices by a row vector and the quantities by a column vector, and multiply, we get

$$(71 \quad 62 \quad 83 \quad 49)\begin{pmatrix}8\\3\\7\\6\end{pmatrix} = [71(8) + 62(3) + 83(7) + 49(6)]$$

$$= 1629$$

The interpretation of this number is obviously the total dollar income from the sale of the four radios on that day.

Product of a Row Vector and a Matrix: By Eq. 86, the product of a row vector and a matrix having n columns will be a row vector having n elements.

$$(1 \times p)\,(p \times n) = (1 \times n)$$

We saw earlier that the product of a row vector and a column vector is a single number, or scalar. Thus, the product of the given row vector and the first column of the matrix will give a scalar. This scalar will be *the first element in the product vector.*

The *second* element in the product vector is found by multiplying the second column in the matrix by the row vector, and so on.

Example: Multiply

$$(1 \quad 3 \quad 0)\begin{pmatrix}7 & 5\\1 & 0\\4 & 9\end{pmatrix}$$

Solution: The dimensions of the product will be

$$(1 \times 3)(3 \times 2) = (1 \times 2)$$

Multiplying the vector and the first column of the matrix gives

$$(1 \quad 3 \quad 0)\begin{pmatrix}\mathbf{7} & 5\\\mathbf{1} & 0\\\mathbf{4} & 9\end{pmatrix} = [\mathbf{1(7) + 3(1) + 0(4)} \qquad * \qquad]$$

$$= (\mathbf{10} \qquad *)$$

Multiplying the vector and the second column of the matrix,

$$(1 \quad 3 \quad 0)\begin{pmatrix} 7 & \mathbf{5} \\ 1 & \mathbf{0} \\ 4 & \mathbf{9} \end{pmatrix} = [10 \qquad \mathbf{1(5) + 3(0) + 0(9)}]$$

$$= (10 \quad \mathbf{5})$$

Product of Two Matrices: We now multiply two matrices **A** and **B**. We already know how to multiply a row vector and a matrix. So let us think of the matrix **A** as several row vectors. Each of these row vectors, multiplied by the matrix **B**, will produce a row vector in the product **AB**.

Example: Find the product **AB** of the matrices

$$\mathbf{A} = \begin{pmatrix} 1 & 0 \\ 3 & 2 \end{pmatrix} \qquad \mathbf{B} = \begin{pmatrix} 5 & 4 & 8 \\ 6 & 7 & 9 \end{pmatrix}$$

Solution: The dimensions of the product will be

$$(2 \times 2)(2 \times 3) = (2 \times 3)$$

We think of the matrix **A** as being two row vectors, each of which, when multiplied by matrix **B** will produce a row vector in the product matrix.

$$\underset{\mathbf{A}}{\begin{matrix} (1 & 0) \\ (3 & 2) \end{matrix}} \underset{\mathbf{B}}{\begin{pmatrix} 5 & 4 & 8 \\ 6 & 7 & 9 \end{pmatrix}} = \underset{\mathbf{AB}}{\begin{matrix} (& * & * & * &) \\ (& * & * & * &) \end{matrix}}$$

Multiplying by the first row vector in **A**,

$$\underset{\mathbf{A}}{\begin{matrix} \mathbf{(1} & \mathbf{0)} \\ (3 & 2) \end{matrix}} \underset{\mathbf{B}}{\begin{pmatrix} 5 & 4 & 8 \\ 6 & 7 & 9 \end{pmatrix}} = \underset{\mathbf{AB}}{\begin{matrix} [\mathbf{1(5) + 0(6)} & \mathbf{1(4) + 0(7)} & \mathbf{1(8) + 0(9)}] \\ [\quad * & * & * \quad] \end{matrix}}$$

$$= \begin{matrix} [\mathbf{5} & \mathbf{4} & \mathbf{8}] \\ [* & * & *] \end{matrix}$$

Then multiplying by the second row vector in **A**,

$$\underset{\mathbf{A}}{\begin{matrix} (1 & 0) \\ \mathbf{(3} & \mathbf{2)} \end{matrix}} \underset{\mathbf{B}}{\begin{pmatrix} 5 & 4 & 8 \\ 6 & 7 & 9 \end{pmatrix}} = \underset{\mathbf{AB}}{\begin{matrix} [\quad 5 & 4 & 8 \quad] \\ [\mathbf{3(5) + 2(6)} & \mathbf{3(4) + 2(7)} & \mathbf{3(8) + 2(9)}] \end{matrix}}$$

giving the final product

$$\underset{\mathbf{A}}{\begin{pmatrix} 1 & 0 \\ 3 & 2 \end{pmatrix}} \underset{\mathbf{B}}{\begin{pmatrix} 5 & 4 & 8 \\ 6 & 7 & 9 \end{pmatrix}} = \underset{\mathbf{AB}}{\begin{pmatrix} 5 & 4 & 8 \\ 27 & 26 & 42 \end{pmatrix}}$$

Example: Find the product **AB** of the matrices

$$\underset{\mathbf{A}}{\begin{pmatrix} 1 & 0 & 2 \\ 0 & 1 & 1 \\ 3 & 0 & 2 \\ 1 & 1 & 0 \\ 2 & 1 & 1 \end{pmatrix}} \underset{\mathbf{B}}{\begin{pmatrix} 3 & 0 & 1 & 1 \\ 1 & 2 & 3 & 0 \\ 0 & 1 & 2 & 1 \end{pmatrix}}$$

Solution: We think of the matrix **A** as five row vectors. Each of these five row vectors, multiplied by the matrix **B**, will produce a row vector in the product **AB**.

$$\underset{\substack{\mathbf{A}\\(5 \times 3)}}{\begin{array}{c} (1 \;\; 0 \;\; 2) \\ (0 \;\; 1 \;\; 1) \\ (3 \;\; 0 \;\; 2) \\ (1 \;\; 1 \;\; 0) \\ (2 \;\; 1 \;\; 1) \end{array}} \underset{\substack{\mathbf{B}\\(3 \times 4)}}{\begin{pmatrix} 3 & 0 & 1 & 1 \\ 1 & 2 & 3 & 0 \\ 0 & 1 & 2 & 1 \end{pmatrix}} = \underset{\substack{\mathbf{AB}\\(5 \times 4)}}{\begin{array}{c} (\qquad\qquad) \\ (\qquad\qquad) \\ (\qquad\qquad) \\ (\qquad\qquad) \\ (\qquad\qquad) \end{array}}$$

Multiplying the first row vector $(1 \;\; 0 \;\; 2)$ in **A** by the matrix **B** gives the first row vector in **AB**.

$$\underset{\mathbf{A}}{\begin{array}{c} (\mathbf{1} \;\; \mathbf{0} \;\; \mathbf{2}) \\ (0 \;\; 1 \;\; 1) \\ (3 \;\; 0 \;\; 2) \\ (1 \;\; 1 \;\; 0) \\ (2 \;\; 1 \;\; 1) \end{array}} \underset{\mathbf{B}}{\begin{pmatrix} 3 & 0 & 1 & 1 \\ 1 & 2 & 3 & 0 \\ 0 & 1 & 2 & 1 \end{pmatrix}} = \underset{\mathbf{AB}}{\begin{array}{c} (\mathbf{3} \;\; \mathbf{2} \;\; \mathbf{5} \;\; \mathbf{3}) \\ (\qquad\qquad) \\ (\qquad\qquad) \\ (\qquad\qquad) \\ (\qquad\qquad) \end{array}}$$

Multiplying the *second* row vector in **A** and the matrix **B** gives a second row vector in **AB**.

$$\underset{\mathbf{A}}{\begin{array}{c} (1 \;\; 0 \;\; 2) \\ (\mathbf{0} \;\; \mathbf{1} \;\; \mathbf{1}) \\ (3 \;\; 0 \;\; 2) \\ (1 \;\; 1 \;\; 0) \\ (2 \;\; 1 \;\; 1) \end{array}} \underset{\mathbf{B}}{\begin{pmatrix} 3 & 0 & 1 & 1 \\ 1 & 2 & 3 & 0 \\ 0 & 1 & 2 & 1 \end{pmatrix}} = \underset{\mathbf{AB}}{\begin{array}{c} (3 \;\; 2 \;\; 5 \;\; 3) \\ (\mathbf{1} \;\; \mathbf{3} \;\; \mathbf{5} \;\; \mathbf{1}) \\ (\qquad\qquad) \\ (\qquad\qquad) \\ (\qquad\qquad) \end{array}}$$

The *third* row vector in **A** produces a third row vector in **AB**.

$$\underset{\mathbf{A}}{\begin{array}{c} (1 \;\; 0 \;\; 2) \\ (0 \;\; 1 \;\; 1) \\ (\mathbf{3} \;\; \mathbf{0} \;\; \mathbf{2}) \\ (1 \;\; 1 \;\; 0) \\ (2 \;\; 1 \;\; 1) \end{array}} \underset{\mathbf{B}}{\begin{pmatrix} 3 & 0 & 1 & 1 \\ 1 & 2 & 3 & 0 \\ 0 & 1 & 2 & 1 \end{pmatrix}} = \underset{\mathbf{AB}}{\begin{array}{c} (3 \;\; 2 \;\; 5 \;\; 3) \\ (1 \;\; 3 \;\; 5 \;\; 1) \\ (\mathbf{9} \;\; \mathbf{2} \;\; \mathbf{7} \;\; \mathbf{5}) \\ (\qquad\qquad) \\ (\qquad\qquad) \end{array}}$$

And so on, until we get the final product

$$\underset{\mathbf{A}}{\begin{pmatrix}1 & 0 & 2\\0 & 1 & 1\\3 & 0 & 2\\1 & 1 & 0\\2 & 1 & 1\end{pmatrix}}\underset{\mathbf{B}}{\begin{pmatrix}3 & 0 & 1 & 1\\1 & 2 & 3 & 0\\0 & 1 & 2 & 1\end{pmatrix}} = \underset{\mathbf{AB}}{\begin{pmatrix}3 & 2 & 5 & 3\\1 & 3 & 5 & 1\\9 & 2 & 7 & 5\\4 & 2 & 4 & 1\\7 & 3 & 7 & 3\end{pmatrix}}$$

We can state the product of two matrices as a formula as follows:

Product of Two Matrices	$\begin{pmatrix}a & b & c\\d & e & f\end{pmatrix}\begin{pmatrix}u & x\\v & y\\w & z\end{pmatrix} = \begin{pmatrix}au + bv + cw & ax + by + cz\\du + ev + fw & dx + ey + fz\end{pmatrix}$	**92**

Product of Square Matrices: We multiply square matrices **A** and **B** just as we do other matrices. But unlike for a rectangular matrix, both the products **AB** and **BA** will be defined, although they will not usually be equal.

Example: Find the products **AB** and **BA** for the square matrices

$$\mathbf{A} = \begin{pmatrix}0 & 1 & 2\\1 & 1 & 0\\2 & 3 & 1\end{pmatrix} \qquad \mathbf{B} = \begin{pmatrix}2 & 1 & 0\\1 & 0 & 3\\3 & 1 & 1\end{pmatrix}$$

Solution: Multiplying as shown earlier,

$$\mathbf{AB} = \underset{(3 \times 3)}{\begin{pmatrix}40 & 1 & 2\\1 & 1 & 0\\2 & 3 & 1\end{pmatrix}}\underset{(3 \times 3)}{\begin{pmatrix}2 & 1 & 0\\1 & 0 & 3\\3 & 1 & 1\end{pmatrix}} = \underset{(3 \times 3)}{\begin{pmatrix}7 & 2 & 5\\3 & 1 & 3\\10 & 3 & 10\end{pmatrix}}$$

and

$$\mathbf{BA} = \begin{pmatrix}2 & 1 & 0\\1 & 0 & 3\\3 & 1 & 1\end{pmatrix}\begin{pmatrix}0 & 1 & 2\\1 & 1 & 0\\2 & 3 & 1\end{pmatrix} = \begin{pmatrix}1 & 3 & 4\\6 & 10 & 5\\3 & 7 & 7\end{pmatrix}$$

We see that multiplying the matrices in the reverse order does *not* give the same result.

$\mathbf{AB} \neq \mathbf{BA}$	**83**

Matrix multiplication is not commutative.

Product of a Matrix and a Column Vector: We have already multiplied a row vector **A** by a matrix **B**. Now we multiply a matrix **A** by a column vector **B**.

Example: Multiply

$$\begin{pmatrix} 3 & 1 & 2 & 0 \\ 1 & 3 & 5 & 2 \end{pmatrix}\begin{pmatrix} 2 \\ 1 \\ 3 \\ 4 \end{pmatrix}$$

Solution: The dimensions of the product will be

$$(2 \times 4)\,(4 \times 1) = (2 \times 1)$$

Multiplying the column vector by the first row in the matrix,

$$\begin{pmatrix} \mathbf{3} & \mathbf{1} & \mathbf{2} & \mathbf{0} \\ 1 & 3 & 5 & 2 \end{pmatrix}\begin{pmatrix} 2 \\ 1 \\ 3 \\ 4 \end{pmatrix} = \begin{pmatrix} \mathbf{3(2) + 1(1) + 2(3) + 0(4)} \\ \end{pmatrix} = \begin{pmatrix} \mathbf{13} \\ \end{pmatrix}$$

and then by the second row,

$$\begin{pmatrix} 3 & 1 & 2 & 0 \\ \mathbf{1} & \mathbf{3} & \mathbf{5} & \mathbf{2} \end{pmatrix}\begin{pmatrix} 2 \\ 1 \\ 3 \\ 4 \end{pmatrix} = \begin{pmatrix} 13 \\ \mathbf{1(2) + 3(1) + 5(3) + 2(4)} \end{pmatrix} = \begin{pmatrix} 13 \\ \mathbf{28} \end{pmatrix}$$

Most of the matrix multiplication we do later will be of a matrix and a column vector.

Note that the product of a matrix and a column vector is a column vector.

Example: During a three-night beer party, five students drank the following amounts.

		Night		
		1	*2*	*3*
Student	*1*	8	2	6
	2	3	5	2
	3	6	5	4
	4	7	3	1
	5	2	3	4

and the cost per bottle on the three nights was $0.75, $0.60, and $0.55. Find the total cost for each student for the three-night party.

Solution: To get the total costs, we know we must multiply the matrix **B** representing number of bottles drunk,

$$\mathbf{B} = \begin{pmatrix} 8 & 2 & 6 \\ 3 & 5 & 2 \\ 6 & 5 & 4 \\ 7 & 3 & 1 \\ 2 & 3 & 4 \end{pmatrix}$$

by a vector **C** representing the costs per bottle. But should **C** be a row vector or a column vector? And should we find the product **BC** or the product **CB**?

If we let **C** be a 1×3 row vector and try to multiply it by the 5×3 matrix **B**, we see that we cannot, because these matrices are not conformable for either product **BC** or **CB**.

On the other hand, if we let **C** be a 3×1 column vector,

$$\mathbf{C} = \begin{pmatrix} 0.75 \\ 0.60 \\ 0.55 \end{pmatrix}$$

the product **CB** is not defined, but the product **BC** gives a column vector

$$\mathbf{BC} = \begin{pmatrix} 8 & 2 & 6 \\ 3 & 5 & 2 \\ 6 & 5 & 4 \\ 7 & 3 & 1 \\ 2 & 3 & 4 \end{pmatrix} \begin{pmatrix} 0.75 \\ 0.60 \\ 0.55 \end{pmatrix} = \begin{pmatrix} 10.50 \\ 6.35 \\ 9.70 \\ 7.60 \\ 5.50 \end{pmatrix}$$

which represents the total cost for each student.

Tensor Product of Two Vectors: We have already found the product **AB** of a row vector **A** and a column vector **B**. We found it to be a scalar, and we called it the scalar or dot product. We now find the product **BA**, and find that the product is a matrix, rather than a scalar. It is called the *tensor product.*

Example:

$$\begin{pmatrix} 2 \\ 0 \\ 1 \end{pmatrix} \begin{pmatrix} 1 & 2 & 0 & 3 \end{pmatrix} = \begin{pmatrix} 2(1) & 2(2) & 2(0) & 2(3) \\ 0(1) & 0(2) & 0(0) & 0(3) \\ 1(1) & 1(2) & 1(0) & 1(3) \end{pmatrix}$$

$$= \begin{pmatrix} 2 & 4 & 0 & 6 \\ 0 & 0 & 0 & 0 \\ 1 & 2 & 0 & 3 \end{pmatrix}$$

$$(3 \times 1) \quad (1 \times 4) \qquad\qquad (3 \times 4)$$

EXERCISE 5

Addition and Subtraction of Matrices

Combine the following vectors and matrices.

1. $(3 \quad 8 \quad 2 \quad 1) + (9 \quad 5 \quad 3 \quad 7)$

2. $\begin{pmatrix} 6 & 3 & 9 \\ 2 & 1 & 3 \end{pmatrix} + \begin{pmatrix} 8 & 2 & 7 \\ 1 & 6 & 3 \end{pmatrix}$

3. $\begin{pmatrix} 5 \\ -3 \\ 5 \\ 8 \end{pmatrix} + \begin{pmatrix} 7 \\ 0 \\ -4 \\ 8 \end{pmatrix} - \begin{pmatrix} -2 \\ 7 \\ -5 \\ 2 \end{pmatrix}$

4. $\begin{pmatrix} 6 & 2 & -7 \\ -3 & 5 & 9 \\ 2 & 5 & 3 \end{pmatrix} - \begin{pmatrix} 5 & 2 & 9 \\ 1 & 4 & 8 \\ 7 & -3 & 4 \end{pmatrix} + \begin{pmatrix} 1 & -6 & 7 \\ 2 & 4 & -9 \\ -6 & 2 & 7 \end{pmatrix}$

5. $\begin{pmatrix} 1 & 5 & -2 & 1 \\ 0 & -2 & 3 & 5 \\ -2 & 9 & 3 & -2 \\ 5 & -1 & 2 & 6 \end{pmatrix} - \begin{pmatrix} -6 & 4 & 2 & 8 \\ 9 & 2 & -6 & 3 \\ -5 & 2 & 6 & 7 \\ 6 & 3 & 7 & 0 \end{pmatrix}$

Product of a Scalar and a Matrix.

Multiply each scalar and vector.

6. $6 \times \begin{pmatrix} 9 \\ -2 \\ 6 \\ -3 \end{pmatrix}$

7. $-3 \times (5 \quad 9 \quad -2 \quad 7 \quad 3)$

Multiply each scalar and matrix.

8. $7 \times \begin{pmatrix} 7 & -3 & 7 \\ -2 & 9 & 5 \\ 9 & 1 & 0 \end{pmatrix}$

9. $-3 \times \begin{pmatrix} 4 & 2 & -3 & 0 \\ -1 & 6 & 3 & -4 \end{pmatrix}$

Remove any factors from the vector or matrix.

10. $\begin{pmatrix} 25 \\ -15 \\ 30 \\ -5 \end{pmatrix}$

11. $\begin{pmatrix} 21 & 7 & -14 \\ 49 & 63 & 28 \\ 14 & -21 & 56 \\ -42 & 70 & 84 \end{pmatrix}$

If

$$\mathbf{A} = \begin{pmatrix} 2 & 5 & -1 \\ -3 & 2 & 6 \\ 9 & -4 & 5 \end{pmatrix} \quad \text{and} \quad \mathbf{B} = \begin{pmatrix} 0 & 1 & 6 \\ -7 & 2 & 1 \\ 4 & -2 & 0 \end{pmatrix}$$

find:

12. $\mathbf{A} - 4\mathbf{B}$

13. $3\mathbf{A} + \mathbf{B}$

14. $2\mathbf{A} + 2\mathbf{B}$

15. Show that $2(\mathbf{A} + \mathbf{B}) = 2\mathbf{A} + 2\mathbf{B}$, thus demonstrating the distributive law (Eq. 85)

Scalar Product of Two Vectors

Multiply each pair of vectors.

16. $(6 \quad -2)\begin{pmatrix} -1 \\ 7 \end{pmatrix}$

17. $(5 \quad 3 \quad 7)\begin{pmatrix} 2 \\ 6 \\ 1 \end{pmatrix}$

18. $(-3 \quad 6 \quad -9)\begin{pmatrix} 8 \\ -3 \\ 4 \end{pmatrix}$

19. $(-4 \quad 3 \quad 8 \quad -5)\begin{pmatrix} 7 \\ -6 \\ 0 \\ 3 \end{pmatrix}$

Product of a Row Vector and a Matrix

Multiply each row vector and matrix.

20. $(4 \quad -2)\begin{pmatrix} 3 & 6 \\ -1 & 5 \end{pmatrix}$

21. $(3 \quad -1)\begin{pmatrix} 4 & 7 & -2 & 4 & 0 \\ 2 & -6 & 8 & -3 & 7 \end{pmatrix}$

22. $(2 \quad 8 \quad 1)\begin{pmatrix} 5 & 6 \\ 3 & 1 \\ 1 & 0 \end{pmatrix}$

23. $(-3 \quad 6 \quad -1)\begin{pmatrix} 5 & -2 & 7 \\ -6 & 2 & 5 \\ 9 & 0 & -1 \end{pmatrix}$

24. $(9 \quad 1 \quad 4)\begin{pmatrix} 6 & 2 & 1 & 5 & -8 \\ 7 & 1 & -3 & 0 & 3 \\ 1 & -5 & 0 & 3 & 2 \end{pmatrix}$

25. $(2 \quad 4 \quad 3 \quad -1 \quad 0)\begin{pmatrix} 4 & 3 & 5 & -1 & 0 \\ 1 & 3 & 5 & -2 & 6 \\ 2 & -4 & 3 & 6 & 5 \\ 3 & 4 & -1 & 6 & 5 \\ 2 & 3 & 1 & 4 & -6 \end{pmatrix}$

Product of Two Matrices

Multiply each pair of matrices.

26. $\begin{pmatrix} 7 & 2 \\ 1 & 3 \end{pmatrix}\begin{pmatrix} 4 & 1 & 2 \\ 3 & 2 & 5 \end{pmatrix}$

27. $\begin{pmatrix} 3 & 1 & 4 \\ 1 & -3 & 5 \end{pmatrix}\begin{pmatrix} 4 & 2 \\ -1 & 5 \\ 2 & -6 \end{pmatrix}$

28. $\begin{pmatrix} 4 & 2 & -1 \\ -2 & 1 & 6 \end{pmatrix}\begin{pmatrix} 5 & -1 & 2 & 3 \\ 1 & 3 & 5 & 2 \\ 0 & 1 & -2 & 4 \end{pmatrix}$

29. $\begin{pmatrix} 5 & 1 \\ -2 & 0 \\ 1 & 3 \end{pmatrix}\begin{pmatrix} -4 & 1 \\ 0 & 5 \end{pmatrix}$

30. $\begin{pmatrix} 2 & 4 \\ 0 & -1 \\ 2 & 5 \end{pmatrix}\begin{pmatrix} 0 & -2 & 3 & -2 \\ 1 & 3 & -4 & 1 \end{pmatrix}$

31. $\begin{pmatrix} 4 & 1 & -2 & 3 \\ 0 & 2 & 1 & 5 \end{pmatrix}\begin{pmatrix} 3 & -1 \\ 0 & 2 \\ 3 & 1 \\ 0 & -2 \end{pmatrix}$

32. $\begin{pmatrix} -1 & 0 & 3 & 1 \\ 2 & 1 & 0 & -3 \end{pmatrix}\begin{pmatrix} 1 & 2 & 3 & 0 \\ 3 & 0 & -1 & 2 \\ -1 & 4 & 3 & 0 \\ 2 & 0 & -2 & 1 \end{pmatrix}$

33. $\begin{pmatrix} 3 & 1 \\ -2 & 0 \\ 1 & -3 \\ 5 & 0 \end{pmatrix}\begin{pmatrix} -1 & 0 \\ 3 & 2 \end{pmatrix}$

34. $\begin{pmatrix} 1 & 0 & -2 \\ 2 & 4 & 1 \\ 0 & -1 & 3 \end{pmatrix}\begin{pmatrix} 4 & 0 & 1 & -2 \\ 0 & 2 & 1 & 3 \\ 1 & 0 & -3 & 2 \end{pmatrix}$

35. $\begin{pmatrix} 2 & 1 & 0 & -1 \\ 1 & 2 & -1 & 1 \\ 0 & -2 & 4 & 0 \end{pmatrix}\begin{pmatrix} 1 & 2 & 0 & -1 \\ 3 & 1 & -1 & 3 \\ 2 & 1 & 0 & 3 \\ 3 & 5 & 1 & 0 \end{pmatrix}$

Product of Square Matrices

Find the product **AB** and the product **BA** for each pair of square matrices.

36. $\mathbf{A} = \begin{pmatrix} 2 & 3 \\ 5 & 1 \end{pmatrix} \qquad \mathbf{B} = \begin{pmatrix} 4 & 6 \\ 7 & 9 \end{pmatrix}$

37. $\mathbf{A} = \begin{pmatrix} 2 & -1 & 0 \\ 1 & 0 & 2 \\ -2 & 1 & 0 \end{pmatrix} \qquad \mathbf{B} = \begin{pmatrix} 2 & 1 & 0 \\ 3 & 0 & 1 \\ -2 & 4 & 1 \end{pmatrix}$

Using the matrices in Problem 36, find

38. $\mathbf{A}^2$

39. $\mathbf{B}^3$

Product of a Matrix and a Column Vector

Multiply each matrix and column vector.

40. $\begin{pmatrix} 2 & 1 \\ 0 & 3 \end{pmatrix}\begin{pmatrix} 3 \\ 1 \end{pmatrix}$

41. $\begin{pmatrix} 2 & 0 & 1 \\ 1 & 5 & 2 \end{pmatrix}\begin{pmatrix} 3 \\ 0 \\ -1 \end{pmatrix}$

42. $\begin{pmatrix} 2 & 1 & 0 \\ 1 & 0 & -2 \\ 3 & 5 & 0 \\ 0 & -1 & 3 \end{pmatrix} \begin{pmatrix} 3 \\ 0 \\ -1 \end{pmatrix}$

43. $\begin{pmatrix} 0 & 2 & -1 & 4 \\ 1 & 5 & 3 & 0 \\ -2 & 0 & 1 & 5 \end{pmatrix} \begin{pmatrix} 2 \\ 0 \\ -1 \\ 3 \end{pmatrix}$

Tensor Product of a Column Vector and a Row Vector

Find the tensor product.

44. $\begin{pmatrix} 3 \\ 1 \end{pmatrix} (4 \quad 2 \quad 1)$

45. $\begin{pmatrix} 2 \\ 0 \\ 3 \\ -1 \end{pmatrix} (3 \quad 2 \quad -4)$

46. $\begin{pmatrix} 1 \\ 0 \\ 3 \end{pmatrix} (5 \quad -1 \quad 2 \quad 3)$

47. The unit matrix **I** is a square matrix that has 1's on the main diagonal and 0's elsewhere. Show that the product of **I** and another square matrix **A** is simply **A**.

$\mathbf{AI} = \mathbf{IA} = \mathbf{A}$	**94**

Computer

Write computer programs to:

48. Add two matrices

49. Multiply a matrix by a scalar

50. Multiply two matrices

Test your programs on any of the problems in this exercise.

9-6. SOLVING SYSTEMS OF LINEAR EQUATIONS BY MATRICES

Representing a System of Equations by Matrices: Suppose that we have the set of equations

$$\begin{aligned} 2x + y + 2z &= 10 \\ x + 2y + 3z &= 14 \\ 3x + 4y - 3z &= 2 \end{aligned}$$

The left sides of these equations can be represented by the product of a square matrix **A** made up of the coefficients, called the *coefficient matrix*, and a column vector **X** made up of the unknowns. If

$$\mathbf{A} = \begin{pmatrix} 2 & 1 & 2 \\ 1 & 2 & 3 \\ 3 & 4 & -3 \end{pmatrix} \quad \text{and} \quad \mathbf{X} = \begin{pmatrix} x \\ y \\ z \end{pmatrix}$$

then

$$\mathbf{AX} = \begin{pmatrix} 2 & 1 & 2 \\ 1 & 2 & 3 \\ 3 & 4 & -3 \end{pmatrix}\begin{pmatrix} x \\ y \\ z \end{pmatrix} = \begin{pmatrix} 2x + y + 2z \\ x + 2y + 3z \\ 3x + 4y - 3z \end{pmatrix}$$

Note that our product is a column vector having *three* elements. The expression $(2x + y + 2z)$ is a *single element* in that vector.

If we now represent the constants in our set of equations by the column vector

$$\mathbf{B} = \begin{pmatrix} 10 \\ 14 \\ 2 \end{pmatrix}$$

our system of equations can be written

$$\begin{pmatrix} 2 & 1 & 2 \\ 1 & 2 & 3 \\ 3 & 4 & -3 \end{pmatrix}\begin{pmatrix} x \\ y \\ z \end{pmatrix} = \begin{pmatrix} 10 \\ 14 \\ 2 \end{pmatrix}$$

or in the more compact form

$\mathbf{AX} = \mathbf{B}$	**95**

Gauss Elimination: When we solved sets of linear equations in Chapter 8, we were able to:

1. Interchange two equations
2. Multiply an equation by a nonzero constant
3. Add a constant multiple of one equation to another equation

Thus, for a matrix that represents such a system of equations, we may:

Elementary Transformations of a Matrix	1. Interchange any rows 2. Multiply a row by a nonzero constant 3. Add a constant multiple of one row to another row	**96**

This is just one of many mathematical ideas named for Carl Friedrich Gauss (1777–1855). It is also called the Gauss–Jordan *method.*

In the *Gauss elimination* method we apply the elementary transformations to transform the coefficient matrix for a system of equations into the form

$$\begin{pmatrix} 1 & k_1 & k_2 \\ 0 & 1 & k_3 \\ 0 & 0 & 1 \end{pmatrix} \begin{pmatrix} x \\ y \\ z \end{pmatrix} = \begin{pmatrix} r_1 \\ r_2 \\ r_3 \end{pmatrix}$$

where the k's and r's are constants. The coefficient matrix is now said to be in *triangular form* (all 1's on the main diagonal, and 0's below it). The advantage to this form becomes clear when we multiply the coefficient matrix by the column of unknowns. We get

$$\begin{aligned} x + k_1y + k_2z &= r_1 \\ y + k_3z &= r_2 \\ z &= r_3 \end{aligned}$$

Thus, we immediately get a solution for z, and the other variables can be found by back substitution.

This algorithm can be programmed for the computer (see Exercise 5).

Algorithm for Gauss Elimination: We use the following algorithm for applying the elementary transformations to the coefficient matrix *and* to the column of constants.

In this algorithm, a *pivot element* for any row is the one that lies on the main diagonal of the coefficient matrix.

1. Divide a row p by its pivot element. This is called *normalizing*.
2. Multiply the row p just normalized by the first element in another row j. Subtract these products from row j. This will make the first element in row j equal to 0.
3. Repeat step 2 for all remaining rows. When finished, the first column should consist of a 1, with all 0's below it.
4. Repeat steps 1 through 3 until the last row has been normalized.
5. One unknown is now equal to the last element in the column of constants. The others are found by back substitution.

Example: Solve the set of equations

$$\begin{aligned} 2x + y + 2z &= 10 \\ x + 2y + 3z &= 14 \\ 3x + 4y - 3z &= 2 \end{aligned}$$

by Gauss elimination.

Solution: We first represent our system by the matrix

$$\begin{pmatrix} 2 & 1 & 2 & 10 \\ 1 & 2 & 3 & 14 \\ 3 & 4 & -3 & 2 \end{pmatrix}$$

Note that this matrix consists of the coefficient matrix as well as the column of constants. It is called an *augmented matrix*.

For the first pass:

1. We divide the first row by its pivot element, 2.

$$\begin{pmatrix} 1 & 1/2 & 1 & 5 \\ 1 & 2 & 3 & 14 \\ 3 & 4 & -3 & 2 \end{pmatrix}$$

2. We multiply the first row by the first element (1) of the second row, and subtract the resulting elements from the second row.

$$\begin{pmatrix} 1 & 1/2 & 1 & 5 \\ 0 & 3/2 & 2 & 9 \\ 3 & 4 & -3 & 2 \end{pmatrix}$$

3. We multiply the first row by 3 and subtract it from the third row.

$$\begin{pmatrix} 1 & 1/2 & 1 & 5 \\ 0 & 3/2 & 2 & 9 \\ 0 & 5/2 & -6 & -13 \end{pmatrix}$$

Thus we have introduced zeros into the first column.

For the second pass:

1. We normalize the second row by dividing it by $\frac{3}{2}$.

When the pivot element happens to be 0, we get division by zero when trying to normalize. Should this happen, write the rows in a different order.

$$\begin{pmatrix} 1 & 1/2 & 1 & 5 \\ 0 & 1 & 4/3 & 6 \\ 0 & 5/2 & -6 & -13 \end{pmatrix}$$

thus obtaining another 1 on the main diagonal.

2. Multiply the second row by $\frac{5}{2}$ and subtract it from the third row.

$$\begin{pmatrix} 1 & 1/2 & 1 & 5 \\ 0 & 1 & 4/3 & 6 \\ 0 & 0 & -28/3 & -28 \end{pmatrix}$$

For the third pass:

1. Normalize the third row.

$$\begin{pmatrix} 1 & 1/2 & 1 & 5 \\ 0 & 1 & 4/3 & 6 \\ 0 & 0 & 1 & 3 \end{pmatrix}$$

The coefficient matrix is now in triangular form. The third row gives

$$z = 3$$

Back-substituting, we get from the second row,

$$y + \frac{4}{3}z = 6$$

$$y = 6 - 4 = 2$$

and from the first row,

$$x + \frac{y}{2} + z = 5$$

$$x = 5 - 1 - 3 = 1$$

Another Method: Let us return to our original system of equations, in matrix form,

$$\mathbf{AX} = \mathbf{B} \tag{95}$$

Instead of putting the coefficient matrix **A** into triangular form, another method is to use Gauss elimination to transform **A** into the unit matrix **I** (1's on the main diagonal, 0's elsewhere), and in so doing transform the column of unknowns **B** into a new column vector which we call **C**.

$$\mathbf{IX} = \mathbf{C}$$

But by Eq. 94, $\mathbf{IX} = \mathbf{X}$, so

$$\mathbf{X} = \mathbf{C}$$

Thus the column of unknowns **X** is equal to the transformed column of constants **C**, and is our solution.

When we transform the coefficient matrix **A** into the unit matrix **I**, the column vector **B** gets transformed into the solution set.	**97**

We use this method now to find the general solution to a system of two linear equations in two unknowns.

Example: Solve the system of equations

$$a_1x + b_1y = c_1$$

$$a_2x + b_2y = c_2$$

by Gauss elimination.

Solution: Our augmented matrix is

$$\begin{pmatrix} a_1 & b_1 & c_1 \\ a_2 & b_2 & c_2 \end{pmatrix}$$

Let us normalize both rows by dividing the first by a_1 and the second by a_2.

$$\begin{pmatrix} 1 & \dfrac{b_1}{a_1} & \dfrac{c_1}{a_1} \\ 1 & \dfrac{b_2}{a_2} & \dfrac{c_2}{a_2} \end{pmatrix}$$

Subtracting the second row from the first gives, after combining fractions,

$$\begin{pmatrix} 1 & \dfrac{b_1}{a_1} & \dfrac{c_1}{a_1} \\ 0 & \dfrac{a_1b_2 - a_2b_1}{a_1} & \dfrac{a_1c_2 - a_2c_1}{a_1} \end{pmatrix}$$

We normalize again by dividing the second row by its second element.

$$\begin{pmatrix} 1 & \dfrac{b_1}{a_1} & \dfrac{c_1}{a_1} \\ 0 & 1 & \dfrac{a_1c_2 - a_2c_1}{a_1b_2 - a_2b_1} \end{pmatrix}$$

Next we multiply the second row by b_1/a_1 and subtract it from the first row. After combining fractions, we get

$$\begin{pmatrix} 1 & 0 & \dfrac{b_2c_1 - b_1c_2}{a_1b_2 - a_2b_1} \\ 0 & 1 & \dfrac{a_1c_2 - a_2c_1}{a_1b_2 - a_2b_1} \end{pmatrix}$$

which is equivalent to

$$\begin{pmatrix} 1 & 0 \\ 0 & 1 \end{pmatrix}\begin{pmatrix} x \\ y \end{pmatrix} \begin{pmatrix} \dfrac{b_2c_1 - b_1c_2}{a_1b_2 - a_2b_1} \\ \dfrac{a_1c_2 - a_2c_1}{a_1b_2 - a_2b_1} \end{pmatrix}$$

Our original matrix has been transformed into the unit matrix, so our column of constants has become the solution set. Our solution is then

$$x = \frac{c_1b_2 - c_2b_1}{a_1b_2 - a_2b_1} \qquad y = \frac{a_1c_2 - a_2c_1}{a_1b_2 - a_2b_1} \qquad \textbf{63}$$

These, of course, are the same equations we got back in Sec. 8-2 when we solved the same set of equations by the addition–subtraction method.

EXERCISE 6

Gauss Elimination

1. Solve any of the systems of equations in Exercise 2 or 3 by Gauss elimination.

Computer

2. Write a program to solve a system of equations by Gauss elimination, using the algorithm given. Use DATA statements to enter the number of equations, and the coefficients. Test your program by solving any of the systems of equations in this chapter.

9-7. THE INVERSE OF A MATRIX

Inverse of a Square Matrix: Recall from Sec. 9-1 that the unit matrix **I** is a square matrix which has 1's along its main diagonal and zeros elsewhere, such as

$$\mathbf{I} = \begin{pmatrix} 1 & 0 & 0 & 0 \\ 0 & 1 & 0 & 0 \\ 0 & 0 & 1 & 0 \\ 0 & 0 & 0 & 1 \end{pmatrix}$$

We now define the *inverse* of any given matrix **A** as another matrix $\mathbf{A}^{-1}$ such that the product of the matrix and its inverse is equal to the unit matrix.

$$\mathbf{A}\mathbf{A}^{-1} = \mathbf{A}^{-1}\mathbf{A} = \mathbf{I} \qquad \mathbf{93}$$

The product of a matrix and its inverse is the unit matrix.

In *ordinary* algebra, if b is some nonzero quantity, then

$$b\left(\frac{1}{b}\right) = bb^{-1} = 1$$

In *matrix* algebra, the unit matrix **I** has properties similar to the number 1, and the inverse of a matrix has similar properties to the *reciprocal* in ordinary algebra.

Example: The inverse of the matrix

$$\mathbf{A} = \begin{pmatrix} 1 & 0 \\ 2 & 0.5 \end{pmatrix} \quad \text{is} \quad \mathbf{A}^{-1} = \begin{pmatrix} 1 & 0 \\ -4 & 2 \end{pmatrix}$$

because

$$\begin{pmatrix} 1 & 0 \\ 2 & 0.5 \end{pmatrix}\begin{pmatrix} 1 & 0 \\ -4 & 2 \end{pmatrix} = \begin{pmatrix} 1(1) + 0(-4) & 1(0) + 0(2) \\ 2(1) + 0.5(-4) & 2(0) + 0.5(2) \end{pmatrix}$$

$$= \begin{pmatrix} 1 & 0 \\ 0 & 1 \end{pmatrix}$$

and

$$\begin{pmatrix} 1 & 0 \\ -4 & 2 \end{pmatrix}\begin{pmatrix} 1 & 0 \\ 2 & 0.5 \end{pmatrix} = \begin{pmatrix} 1(1) + 0(2) & 1(0) + 0(0.5) \\ -4(1) + 2(2) & -4(0) + 2(0.5) \end{pmatrix}$$

$$= \begin{pmatrix} 1 & 0 \\ 0 & 1 \end{pmatrix}$$

Not every matrix has an inverse. If the inverse exists, the matrix is said to be invertible.

Finding the Inverse of a Matrix: One method to find the inverse of a matrix is to apply Gauss elimination to transform it into a unit matrix, while at the same time performing the same operations on a unit matrix. The unit matrix, after the transformations, becomes the inverse of the original matrix.

Example: Find the inverse of the matrix of the preceding example,

$$\mathbf{A} = \begin{pmatrix} 1 & 0 \\ 2 & 0.5 \end{pmatrix}$$

Solution: To the given matrix **A**, we append a unit matrix,

$$\left(\begin{array}{cc|cc} 1 & 0 & 1 & 0 \\ 2 & 0.5 & 0 & 1 \end{array}\right)$$

obtaining a *double-square* matrix. We then double the first row and subtract it from the second.

$$\left(\begin{array}{cc|cc} 1 & 0 & 1 & 0 \\ 0 & 0.5 & -2 & 1 \end{array}\right)$$

We now double the second row.

$$\left(\begin{array}{cc|cc} 1 & 0 & 1 & 0 \\ 0 & 1 & -4 & 2 \end{array}\right)$$

Our original matrix ***A*** *is now the unit matrix, and the original unit matrix is now the inverse of* ***A***.

$$\mathbf{A}^{-1} = \begin{pmatrix} 1 & 0 \\ -4 & 2 \end{pmatrix}$$

Example: Find the inverse of the matrix

$$\mathbf{A} = \begin{pmatrix} a_1 & b_1 \\ a_2 & b_2 \end{pmatrix}$$

Solution: We append the unit matrix

$$\left(\begin{array}{cc|cc} a_1 & b_1 & 1 & 0 \\ a_2 & b_2 & 0 & 1 \end{array}\right)$$

Then we multiply the first row by b_2 and the second by b_1, and subtract the second from the first.

$$\left(\begin{array}{cc|cc} a_1b_2 - a_2b_1 & 0 & b_2 & -b_1 \\ a_2b_1 & b_1b_2 & 0 & b_1 \end{array}\right)$$

We normalize the first row by dividing it by $a_1b_2 - a_2b_1$.

$$\left(\begin{array}{cc|cc} 1 & 0 & \dfrac{b_2}{a_1b_2 - a_2b_1} & \dfrac{-b_1}{a_1b_2 - a_2b_1} \\ a_2b_1 & b_1b_2 & 0 & b_1 \end{array}\right)$$

Then we multiply the first row by a_2b_1 and subtract it from the second. After combining fractions, we get

$$\left(\begin{array}{cc|cc} 1 & 0 & \dfrac{b_2}{a_1b_2 - a_2b_1} & \dfrac{-b_1}{a_1b_2 - a_2b_1} \\ 0 & b_1b_2 & \dfrac{-a_2b_1b_2}{a_1b_2 - a_2b_1} & \dfrac{a_1b_1b_2}{a_1b_2 - a_2b_1} \end{array}\right)$$

We then divide the second row by b_1b_2.

$$\left(\begin{array}{cc|cc} 1 & 0 & \dfrac{b_2}{a_1b_2 - a_2b_1} & \dfrac{-b_1}{a_1b_2 - a_2b_1} \\ 0 & 1 & \dfrac{-a_2}{a_1b_2 - a_2b_1} & \dfrac{a_1}{a_1b_2 - a_2b_1} \end{array}\right)$$

The inverse of our matrix is then

$$\mathbf{A}^{-1} = \frac{1}{a_1b_2 - a_2b_1}\begin{pmatrix} b_2 & -b_1 \\ -a_2 & a_1 \end{pmatrix}$$

after factoring out $1/(a_1b_2 - a_2b_1)$.

Comparing this inverse with the original matrix, we see that:

1. The elements on the main diagonal are interchanged.
2. The elements on the secondary diagonal have reversed sign.
3. It is divided by the determinant of the original matrix.

These facts provide another way to quickly write the inverse of a 2×2 matrix.

Example: Find the inverse of

$$\mathbf{A} = \begin{pmatrix} 5 & -2 \\ 1 & 3 \end{pmatrix}$$

Solution: We interchange the 5 and the 3, and reverse the signs of the -2 and 1,

$$\begin{pmatrix} 3 & 2 \\ -1 & 5 \end{pmatrix}$$

and divide by the determinant of the original matrix

$$5(3) - (-2)(1) = 17$$

So

$$\mathbf{A}^{-1} = \frac{1}{17}\begin{pmatrix} 3 & 2 \\ -1 & 5 \end{pmatrix}$$

Solving Systems of Equations Using the Inverse: Suppose that we have a system of equations, which we represent in matrix form as

$$\mathbf{AX} = \mathbf{B}$$

where **A** is the coefficient matrix, **X** is the column vector of the unknowns, and **B** is the column of constants. Multiplying both sides by the $\mathbf{A}^{-1}$, the inverse of **A**,

$$\mathbf{A}^{-1}\mathbf{AX} = \mathbf{A}^{-1}\mathbf{B}$$

or

$$\mathbf{IX} = \mathbf{A}^{-1}\mathbf{B}$$

since $\mathbf{A}^{-1}\mathbf{A} = \mathbf{I}$. But, by Eq. 94, $\mathbf{IX} = \mathbf{X}$, so

$$\mathbf{X} = \mathbf{A}^{-1}\mathbf{B} \qquad \textbf{98}$$

The solutions to a set of equations can be obtained by multiplying the column of constants by the inverse of the coefficient matrix.

Example: Solve the system of equations

$$x - 2y = -3$$
$$3x + y = 5$$

by the matrix inversion method.

Solution: The coefficient matrix **A** is

$$\mathbf{A} = \begin{pmatrix} 1 & -2 \\ 3 & 1 \end{pmatrix}$$

We now find the inverse, $\mathbf{A}^{-1}$, of **A**. We append the unit matrix

$$\left(\begin{array}{cc|cc} 1 & -2 & 1 & 0 \\ 3 & 1 & 0 & 1 \end{array}\right)$$

and then multiply the first row by -3 and add it to the second row.

$$\left(\begin{array}{cc|cc} 1 & -2 & 1 & 0 \\ 0 & 7 & -3 & 1 \end{array}\right)$$

Then we multiply the second row by 2/7 and add it to the first row.

$$\left(\begin{array}{cc|cc} 1 & 0 & 1/7 & 2/7 \\ 0 & 7 & -3 & 1 \end{array}\right)$$

Finally, we divide the second row by 7.

$$\left(\begin{array}{cc|cc} 1 & 0 & 1/7 & 2/7 \\ 0 & 1 & -3/7 & 1/7 \end{array}\right)$$

We then get our solution by multiplying the inverse by the column of coefficients.

$$\begin{pmatrix} 1/7 & 2/7 \\ -3/7 & 1/7 \end{pmatrix}\begin{pmatrix} -3 \\ 5 \end{pmatrix} = \begin{pmatrix} -3/7 + 10/7 \\ 9/7 + 5/7 \end{pmatrix} = \begin{pmatrix} 1 \\ 2 \end{pmatrix}$$

so,

$$x = 1 \quad \text{and} \quad y = 2$$

Example: Solve the general system of two equations

$$a_1x + b_1y = c_1$$

$$a_2x + b_2y = c_2$$

by matrix inversion.

Solution: We write the coefficient matrix

$$\mathbf{A} = \begin{pmatrix} a_1 & b_1 \\ a_2 & b_2 \end{pmatrix}$$

The inverse of our coefficient matrix has already been found in a preceding section. It is

$$\mathbf{A}^{-1} = \frac{1}{a_1b_2 - a_2b_1}\begin{pmatrix} b_2 & -b_1 \\ -a_2 & a_1 \end{pmatrix}$$

By Eq. 98 we now get the column of unknowns by multiplying the inverse by the column of coefficients.

$$\mathbf{X} = \mathbf{A}^{-1}\mathbf{B}$$

$$= \frac{1}{a_1b_2 - a_2b_1}\begin{pmatrix} b_2 & -b_1 \\ -a_2 & a_1 \end{pmatrix}\begin{pmatrix} c_1 \\ c_2 \end{pmatrix}$$

$$= \begin{pmatrix} \dfrac{b_2c_1 - b_1c_2}{a_1b_2 - a_2b_1} \\ \dfrac{a_1c_2 - a_2c_1}{a_1b_2 - a_2b_1} \end{pmatrix}$$

Thus we again get the now familiar equations,

$$x = \frac{c_1b_2 - c_2b_1}{a_1b_2 - a_2b_1} \qquad y = \frac{a_1c_2 - a_2c_1}{a_1b_2 - a_2b_1} \tag{63}$$

Example: Solve the system

$$\begin{aligned} 3x - y + z + 4w &= 15 \\ -x + 3y + 5z - w &= -9 \\ 2x + y - z + 3w &= 18 \\ x - 2y - 3z + w &= 7 \end{aligned}$$

Solution: We write the coefficient matrix, with the unit matrix appended.

$$\left(\begin{array}{cccc|cccc} 3 & -1 & 1 & 4 & 1 & 0 & 0 & 0 \\ -1 & 3 & 5 & -1 & 0 & 1 & 0 & 0 \\ 2 & 1 & -1 & 3 & 0 & 0 & 1 & 0 \\ 1 & -2 & -3 & 1 & 0 & 0 & 0 & 1 \end{array}\right)$$

We first use Gauss elimination to put the coefficient matrix into triangular form in exactly the same way as we did in Sec. 9-5. We now normalize one row at a time and then introduce zeros below the pivot element. Then we create zeros *above* the main diagonal by the same method. The various steps in the computation are shown in the following computer-generated table.

Even though this printout shows only 2 significant digits, the work was actually done using the full accuracy of the computer.

Original matrix

$$\left(\begin{array}{cccc|cccc} 3.0 & -1.0 & 1.0 & 4.0 & 1.0 & 0.0 & 0.0 & 0.0 \\ -1.0 & 3.0 & 5.0 & -1.0 & 0.0 & 1.0 & 0.0 & 0.0 \\ 2.0 & 1.0 & -1.0 & 3.0 & 0.0 & 0.0 & 1.0 & 0.0 \\ 1.0 & -2.0 & -3.0 & 1.0 & 0.0 & 0.0 & 0.0 & 1.0 \end{array}\right)$$

Normalize row 1

$$\left(\begin{array}{cccc|cccc} \mathbf{1.0} & -0.3 & 0.3 & 1.3 & 0.3 & 0.0 & 0.0 & 0.0 \\ -1.0 & 3.0 & 5.0 & -1.0 & 0.0 & 1.0 & 0.0 & 0.0 \\ 2.0 & 1.0 & -1.0 & 3.0 & 0.0 & 0.0 & 1.0 & 0.0 \\ 1.0 & -2.0 & -3.0 & 1.0 & 0.0 & 0.0 & 0.0 & 1.0 \end{array}\right)$$

Create zeros below pivot element

$$\left(\begin{array}{cccc|cccc} 1.0 & -0.3 & 0.3 & 1.3 & 0.3 & 0.0 & 0.0 & 0.0 \\ \mathbf{0.0} & 2.7 & 5.3 & 0.3 & 0.3 & 1.0 & 0.0 & 0.0 \\ \mathbf{0.0} & 1.7 & -1.7 & 0.3 & -0.7 & 0.0 & 1.0 & 0.0 \\ \mathbf{0.0} & -1.7 & -3.3 & -0.3 & -0.3 & 0.0 & 0.0 & 1.0 \end{array}\right)$$

Normalize row 2

$$\left(\begin{array}{cccc|cccc} 1.0 & -0.3 & 0.3 & 1.3 & 0.3 & 0.0 & 0.0 & 0.0 \\ 0.0 & \mathbf{1.0} & 2.0 & 0.1 & 0.1 & 0.4 & 0.0 & 0.0 \\ 0.0 & 1.7 & -1.7 & 0.3 & -0.7 & 0.0 & 1.0 & 0.0 \\ 0.0 & -1.7 & -3.3 & -0.3 & -0.3 & 0.0 & 0.0 & 1.0 \end{array}\right)$$

Create zeros below pivot element

$$\left(\begin{array}{cccc|cccc} 1.0 & -0.3 & 0.3 & 1.3 & 0.3 & 0.0 & 0.0 & 0.0 \\ 0.0 & 1.0 & 2.0 & 0.1 & 0.1 & 0.4 & 0.0 & 0.0 \\ 0.0 & \mathbf{0.0} & -5.0 & 0.1 & -0.9 & -0.6 & 1.0 & 0.0 \\ 0.0 & \mathbf{0.0} & 0.0 & -0.1 & -0.1 & 0.6 & 0.0 & 1.0 \end{array}\right)$$

Normalize row 3

$$\left(\begin{array}{cccc|cccc} 1.0 & -0.3 & 0.3 & 1.3 & 0.3 & 0.0 & 0.0 & 0.0 \\ 0.0 & 1.0 & 2.0 & 0.1 & 0.1 & 0.4 & 0.0 & 0.0 \\ 0.0 & 0.0 & \mathbf{1.0} & -0.0 & 0.2 & 0.1 & -0.2 & 0.0 \\ 0.0 & 0.0 & 0.0 & -0.1 & -0.1 & 0.6 & 0.0 & 1.0 \end{array}\right)$$

Create zeros below pivot element

$$\left(\begin{array}{cccc|cccc} 1.0 & -0.3 & 0.3 & 1.3 & 0.3 & 0.0 & 0.0 & 0.0 \\ 0.0 & 1.0 & 2.0 & 0.1 & 0.1 & 0.4 & 0.0 & 0.0 \\ 0.0 & 0.0 & 1.0 & 0.0 & 0.2 & 0.1 & -0.2 & 0.0 \\ 0.0 & 0.0 & \mathbf{0.0} & -0.1 & -0.1 & 0.6 & 0.0 & 1.0 \end{array}\right)$$

Normalize row 4

$$\left(\begin{array}{cccc|cccc} 1.0 & -0.3 & 0.3 & 1.3 & 0.3 & 0.0 & 0.0 & 0.0 \\ 0.0 & 1.0 & 2.0 & 0.1 & 0.1 & 0.4 & 0.0 & 0.0 \\ 0.0 & 0.0 & 1.0 & 0.0 & 0.2 & 0.1 & -0.2 & 0.0 \\ 0.0 & 0.0 & 0.0 & \mathbf{1.0} & 1.0 & -5.0 & 0.0 & -8.0 \end{array}\right)$$

Create zeros above pivot element

$$\left(\begin{array}{cccc|cccc} 1.0 & -0.3 & 0.3 & \mathbf{0.0} & -1.0 & 6.7 & 0.0 & 10.7 \\ 0.0 & 1.0 & 2.0 & \mathbf{0.0} & 0.0 & 1.0 & 0.0 & 1.0 \\ 0.0 & 0.0 & 1.0 & \mathbf{0.0} & 0.2 & 0.0 & -0.2 & -0.2 \\ 0.0 & 0.0 & 0.0 & 1.0 & 1.0 & -5.0 & 0.0 & -8.0 \end{array}\right)$$

$$\left(\begin{array}{cccc|cccc} 1.0 & -0.3 & \mathbf{0.0} & 0.0 & -1.1 & 6.7 & 0.1 & 10.7 \\ 0.0 & 1.0 & \mathbf{0.0} & 0.0 & -0.4 & 1.0 & 0.4 & 1.4 \\ 0.0 & 0.0 & 1.0 & 0.0 & 0.2 & 0.0 & -0.2 & -0.2 \\ 0.0 & 0.0 & 0.0 & 1.0 & 1.0 & -5.0 & 0.0 & -8.0 \end{array}\right)$$

Inversion complete

$$\left(\begin{array}{cccc|cccc} 1.0 & \mathbf{0.0} & 0.0 & 0.0 & -1.2 & 7.0 & 0.2 & 11.2 \\ 0.0 & 1.0 & 0.0 & 0.0 & -0.4 & 1.0 & 0.4 & 1.4 \\ 0.0 & 0.0 & 1.0 & 0.0 & 0.2 & 0.0 & -0.2 & -0.2 \\ 0.0 & 0.0 & 0.0 & 1.0 & 1.0 & -5.0 & 0.0 & -8.0 \end{array}\right)$$

Unit matrix — Inverse matrix

We get our solution by multiplying the inverse matrix by the column of constants.

$$\begin{pmatrix} -1.2 & 7 & 0.2 & 11.2 \\ -0.4 & 1 & 0.4 & 1.4 \\ 0.2 & 0 & -0.2 & -0.2 \\ 1 & -5 & 0 & -8 \end{pmatrix}\begin{pmatrix} 15 \\ -9 \\ 18 \\ 7 \end{pmatrix} = \begin{pmatrix} 1 \\ 2 \\ -2 \\ 4 \end{pmatrix}$$

so

$$x = 1 \qquad y = 2 \qquad z = -2 \qquad \text{and} \qquad w = 4$$

EXERCISE 7

Find the inverse of each matrix.

1. $\begin{pmatrix} 4 & 8 \\ -5 & 0 \end{pmatrix}$

2. $\begin{pmatrix} -2 & 4 \\ 3 & 1 \end{pmatrix}$

3. $\begin{pmatrix} 8 & -3 \\ 1 & 6 \end{pmatrix}$

4. $\begin{pmatrix} 1 & 5 & 3 \\ 2 & 3 & 0 \\ 6 & 8 & 1 \end{pmatrix}$

5. $\begin{pmatrix} -1 & 3 & 7 \\ 4 & -2 & 0 \\ 7 & 2 & -9 \end{pmatrix}$

6. $\begin{pmatrix} 18 & 23 & 71 \\ 82 & 49 & 28 \\ 36 & 19 & 84 \end{pmatrix}$

7. $\begin{pmatrix} 9 & 1 & 0 & 3 \\ 4 & 3 & 5 & 2 \\ 7 & 1 & 0 & 3 \\ 1 & 4 & 2 & 7 \end{pmatrix}$

8. $\begin{pmatrix} -2 & 1 & 5 & -6 \\ 7 & -3 & 2 & -6 \\ -3 & 2 & 6 & 1 \\ 5 & 2 & 7 & 5 \end{pmatrix}$

9. Solve any of the equations of Exercise 2 or 3 by the matrix inversion method.

Computer

10. Write a computer program to find and print the inverse of a matrix. Such a program will have many similarities to the one for Gauss elimination. Test your program by finding the inverse of some of the matrices given in Problems 1 to 8.

CHAPTER TEST

1. Combine:

$$(4 \quad 9 \quad -2 \quad 6) + (2 \quad 4 \quad 9 \quad -1) - (4 \quad 1 \quad -3 \quad 7).$$

2. Evaluate.

$$\begin{vmatrix} 3 & -6 \\ -5 & 4 \end{vmatrix}$$

3. Multiply.

$$(7 \quad -2 \quad 5)\begin{pmatrix} 6 \\ -2 \\ 8 \end{pmatrix}$$

4. Find the inverse.

$$\begin{pmatrix} 5 & -2 & 5 & 1 \\ 2 & 8 & -3 & -1 \\ -4 & 2 & 7 & 5 \\ 9 & 2 & 5 & 3 \end{pmatrix}$$

5. Evaluate.

$$\begin{vmatrix} 6 & 1/2 & -2 \\ 3 & 1/4 & 4 \\ 2 & -1/2 & 3 \end{vmatrix}$$

6. Combine.

$$\begin{pmatrix} 5 & 9 \\ -4 & 7 \end{pmatrix} + \begin{pmatrix} -2 & 6 \\ 3 & 8 \end{pmatrix}$$

7. Evaluate.

$$\begin{pmatrix} -3 & 2 & 0 & 4 \\ -1 & 6 & 5 & -2 \\ 1 & 1 & 0 & 3 \\ -1 & 2 & 0 & 6 \end{pmatrix}$$

8. Multiply.

$$\begin{pmatrix} 8 & 2 & 6 & 1 \\ 9 & 0 & 1 & 4 \\ 1 & 6 & 9 & 3 \\ 6 & 9 & 3 & 5 \end{pmatrix}\begin{pmatrix} 5 \\ 3 \\ 2 \\ 8 \end{pmatrix}$$

9. Solve by determinants.

$$\begin{aligned} x + y + z + w &= -4 \\ x + 2y + 3z + 4w &= 0 \\ x + 3y + 6z + 10w &= 9 \\ x + 4y + 10z + 20w &= 24 \end{aligned}$$

10. Multiply.

$$7\begin{pmatrix} 5 & -3 \\ -4 & 8 \end{pmatrix}$$

11. Solve by Gauss elimination.

$$\begin{aligned} 5x + 3y - 2z &= 5 \\ 3x - 4y + 3z &= 13 \\ x + 6y - 4z &= -8 \end{aligned}$$

12. Find the inverse.

$$\begin{pmatrix} 5 & 1 & 4 \\ -2 & 6 & 9 \\ 4 & -1 & 0 \end{pmatrix}$$

13. Solve by determinants.

$$\begin{aligned} 4x + 3y &= 27 \\ 2x - 5y &= -19 \end{aligned}$$

14. Evaluate.

$$\begin{vmatrix} 8 & 2 & 0 & 1 & 4 \\ 0 & 1 & 4 & 2 & 7 \\ 2 & 6 & 3 & 8 & 0 \\ 1 & 4 & 2 & 6 & 5 \\ 4 & 6 & 8 & 3 & 5 \end{vmatrix}$$

15. Multiply:

$$9(6 \quad 3 \quad -2 \quad 8)$$

16. The four loop currents (mA) in a certain circuit satisfy the equations

$$\begin{aligned} 376I_1 + 374I_2 - 364I_3 - 255I_4 &= 284 \\ 898I_1 - 263I_2 + 229I_3 + 274I_4 &= -847 \\ 364I_1 + 285I_2 - 857I_3 - 226I_4 &= 746 \\ 937I_1 + 645I_2 - 387I_3 + 847I_4 &= -254 \end{aligned}$$

Solve for the four currents by matrix inversion.

17. Find the inverse.

$$\begin{pmatrix} 4 & -4 \\ 3 & 1 \end{pmatrix}$$

18. Multiply.

$$\begin{pmatrix} 8 & 4 \\ 9 & -5 \end{pmatrix}\begin{pmatrix} 3 & 7 \\ -3 & 6 \end{pmatrix}$$

Radicals

We had a first look at radicals in Sec. 1-5, where we learned how to compute roots on the calculator. We had also given some definitions of the parts of a radical and the meaning of principal root. You may want to glance back there now to refresh your memory. It would also be a good idea to review the laws of exponents in Sec. 4-3.

10-1. SIMPLIFICATION OF RADICALS

Relation between Exponents and Radicals: A quantity raised to a *fractional exponent* can also be written as a *radical*.

Examples:

$$x^{1/2} = \sqrt{x} \qquad y^{1/4} = \sqrt[4]{y} \qquad z^{3/4} = \sqrt[4]{z^3} \qquad w^{-1/2} = \frac{1}{\sqrt{w}}$$

In general,

Relation between Exponents and Radicals	$a^{1/n} = \sqrt[n]{a}$	**36**
	$a^{m/n} = \sqrt[n]{a^m} = (\sqrt[n]{a})^m$	**37**

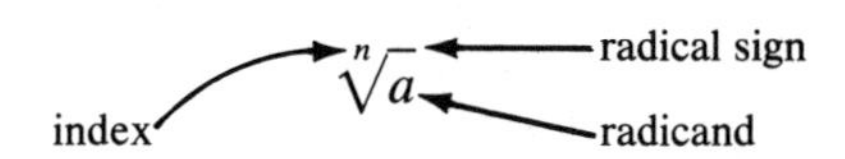

Root of a Product: We have several *rules of radicals*, which are similar to the laws of exponents, and, in fact, are derived from them. The first rule is for products.

Root of a Product	$\sqrt[n]{ab} = \sqrt[n]{a}\sqrt[n]{b}$	**38**

The root of a product equals the product of the roots of the factors.

Example: We may split the radical $\sqrt{9x}$ into two radicals,

$$\sqrt{9x} = \sqrt{9}\sqrt{x} = 3\sqrt{x}$$

Example: Write as a single radical $\sqrt{7}\sqrt{2}\sqrt{x}$.

Solution: By Eq. 38,

$$\sqrt{7}\sqrt{2}\sqrt{x} = \sqrt{7(2)x} = \sqrt{14x}$$

Common Error	There is no similar rule for the root of a sum. $\sqrt[n]{a+b} \neq \sqrt[n]{a} + \sqrt[n]{b}$

Example: $\sqrt{9} + \sqrt{16}$ does *not* equal $\sqrt{25}$.

Common Error	Equation 38 does not hold when a and b are both *negative*.

Example:

$$(\sqrt{-4})^2 = \sqrt{-4}\sqrt{-4} \neq \sqrt{(-4)(-4)}$$
$$\neq \sqrt{16} = +4$$

Instead, convert to *imaginary numbers*, as shown in Sec. 16-1.

Root of a Quotient: Just as the root of a product can be split up into the roots of the factors, the root of a quotient can be expressed as the root of the numerator divided by the root of the denominator.

Root of a Quotient	$\sqrt[n]{\frac{a}{b}} = \frac{\sqrt[n]{a}}{\sqrt[n]{b}}$	**39**

The root of a quotient equals the quotient of the roots of numerator and denominator.

Example: The radical $\sqrt{\frac{w}{25}}$ can be written $\frac{\sqrt{w}}{\sqrt{25}}$ or $\frac{\sqrt{w}}{5}$.

Root of a Power: Another rule that is useful in simplifying radicals is

Root of a Power	$\sqrt[n]{a^m} = (\sqrt[n]{a})^m$	**40**

The root of a number raised to a power equals the power of the root of the number.

Example: Find the value of $\sqrt[4]{(81)^3}$ without a calculator.

Solution: Rather than cubing 81 and then trying to find its fourth root, Eq. 40 allows us to take the fourth root *first* and then cube it.

$$\sqrt[4]{(81)^3} = (\sqrt[4]{81})^3 = (3)^3 = 27$$

Simplest Form for a Radical: A radical is said to be in *simplest form* when:

1. The radicand has been reduced as much as is possible.
2. The index has been made as small as is possible.
3. There are no radicals in the denominator and no fractional radicands.

Removing Factors from the Radicand: Try to factor the radicand so that one or more of the factors is a perfect nth power (where n is the index of the radical). Then use Eq. 38 to split the radical into two or more radicals, some of which can then be reduced.

Example: Simplify $\sqrt{50x^3}$.

Solution: We factor the radicand so that some factors are perfect squares,

$$\sqrt{50x^3} = \sqrt{(25)(2)x^2x}$$

Then by Eq. 38,

$$= \sqrt{25}\sqrt{x^2}\sqrt{2x} = 5x\sqrt{2x}$$

Example: Simplify $\sqrt[3]{24y^5}$.

Solution: We look for factors of the radicand that are perfect cubes.

$$\sqrt[3]{24y^5} = \sqrt[3]{8(3)y^3y^2}$$
$$= \sqrt[3]{8y^3}\,\sqrt[3]{3y^2} = 2y\sqrt[3]{3y^2}$$

When the radicand contains more than one term, try to *factor out* a perfect square.

Example: Simplify

$$\sqrt{4x^2y + 12x^4z}$$

Solution: We factor $4x^2$ from the radicand, and then remove it from under the radical sign.

$$\sqrt{4x^2y + 12x^4z} = \sqrt{4x^2(y + 3x^2z)}$$
$$= 2x\sqrt{y + 3x^2z}$$

Reducing the Index: The index is reduced by first going to exponential form, as in the following example.

Example: Simplify $\sqrt[4]{4x^2y^8}$.

Solution: Going to exponential form,

$$\sqrt[4]{4x^2y^8} = (4x^2y^8)^{1/4}$$

By Eq. 32,

$$= 2^{2/4}x^{2/4}y^{8/4}$$
$$= 2^{1/2}x^{1/2}y^2 = y^2(2x)^{1/2}$$
$$= y^2\sqrt{2x}$$

Rationalizing the Denominator: An expression is considered in simpler form when its denominators contain no radicals. To put it into this form is called *rationalizing* the denominator. We will show how to rationalize the denominator when it is a square root, a cube root, or a root with any index, and where it has more than one term.

If the denominator is a square root, multiply numerator and denominator of the fraction by a quantity that will make the radicand in the denominator a perfect square.

Example:

$$\frac{5}{\sqrt{2}} = \frac{5}{\sqrt{2}} \cdot \frac{\sqrt{2}}{\sqrt{2}} = \frac{5\sqrt{2}}{\sqrt{4}} = \frac{5\sqrt{2}}{2}$$

When the *entire* fraction is under the radical sign, we make the denominator of that fraction a perfect square and remove it from the radical sign.

Example:

$$\sqrt{\frac{3x}{2y}} = \sqrt{\frac{3x(2y)}{2y(2y)}} = \sqrt{\frac{6xy}{4y^2}} = \frac{\sqrt{6xy}}{2y}$$

If the denominator is a *cube* root, we must multiply numerator and denominator by a quantity that will make the quantity under the radical sign a perfect cube.

Example:

$$\frac{7}{\sqrt[3]{4}} = \frac{7}{\sqrt[3]{4}} \cdot \frac{\sqrt[3]{2}}{\sqrt[3]{2}} = \frac{7\sqrt[3]{2}}{\sqrt[3]{8}} = \frac{7\sqrt[3]{2}}{2}$$

The same principle applies regardless of the index. In general, if the index is n, we must make the quantity under the radical sign (in the denominator) a perfect nth power.

Example:

$$\frac{2y}{3\sqrt[5]{x}} = \frac{2y}{3\sqrt[5]{x}} \cdot \frac{\sqrt[5]{x^4}}{\sqrt[5]{x^4}} = \frac{2y\sqrt[5]{x^4}}{3\sqrt[5]{x^5}} = \frac{2y\sqrt[5]{x^4}}{3x}$$

Sometimes the denominator will have more than one term.

Example:

(a)
$$\sqrt{\frac{a}{a^2 + b^2}} = \sqrt{\frac{a}{a^2 + b^2} \cdot \frac{a^2 + b^2}{a^2 + b^2}} = \frac{\sqrt{a(a^2 + b^2)}}{a^2 + b^2}$$

(b)
$$\sqrt{\frac{ab}{a + b}} = \sqrt{\frac{ab \quad (a + b)}{(a + b)\,(a + b)}} = \sqrt{\frac{ab(a + b)}{(a + b)^2}}$$
$$= \frac{\sqrt{ab(a + b)}}{a + b}$$

EXERCISE 1

Exponential and Radical Form

Express in radical form.

1. $a^{1/4}$ **2.** $x^{1/2}$ **3.** $z^{3/4}$

4. $a^{1/2}b^{1/4}$ **5.** $(m - n)^{1/2}$ **6.** $(x^2y)^{-1/2}$

7. $\left(\frac{x}{y}\right)^{-1/3}$ **8.** $a^0b^{-3/4}$

Express in exponential form.

9. $\sqrt{b}$ **10.** $\sqrt[3]{x}$ **11.** $\sqrt{y^2}$

12. $4\sqrt[3]{xy}$ **13.** $\sqrt[n]{a + b}$ **14.** $\sqrt[n]{x^m}$

15. $\sqrt{x^2y^2}$ **16.** $\sqrt[n]{a^nb^{3n}}$

Simplifying Radicals

Write in simplest form.

Do not use your calculator for any of these numerical problems. Leave the answers in radical form.

17. $\sqrt{18}$ **18.** $\sqrt{75}$ **19.** $\sqrt{63}$

20. $\sqrt[3]{16}$ **21.** $\sqrt[3]{-56}$ **22.** $\sqrt[4]{48}$

23. $\sqrt{a^3}$ **24.** $3\sqrt{50x^5}$ **25.** $\sqrt{36x^2y}$

26. $\sqrt[3]{x^2y^5}$ **27.** $x\sqrt[3]{16x^3y}$ **28.** $\sqrt[4]{64m^2n^4}$

29. $3\sqrt[5]{32xy^{11}}$ **30.** $6\sqrt[3]{16x^4}$ **31.** $\sqrt{a^3 - a^3b}$

32. $x\sqrt{x^4 - x^3y^2}$ **33.** $\sqrt{9m^3 + 18n}$ **34.** $\sqrt{2x^3 + x^4y}$

35. $\sqrt{\dfrac{3}{7}}$ **36.** $\sqrt{\dfrac{2}{3}}$ **37.** $\sqrt[3]{\dfrac{1}{4}}$

38. $\sqrt{\dfrac{5}{8}}$ **39.** $\sqrt[3]{\dfrac{2}{9}}$ **40.** $\sqrt[4]{\dfrac{7}{8}}$

41. $\sqrt{\dfrac{1}{2x}}$ **42.** $\sqrt{\dfrac{5m}{7n}}$ **43.** $\sqrt{\dfrac{3a^3}{5b}}$

44. $\sqrt[6]{\dfrac{4x^6}{9}}$ **45.** $\sqrt{x^2 - \left(\dfrac{x}{2}\right)^2}$ **46.** $(x^2 - y^2)\sqrt{\dfrac{x}{x + y}}$

47. A stone is thrown upward with a horizontal velocity of 40 ft/s and an upward velocity of 60 ft/s. At t seconds it will have, by Eq. A16, a horizontal displacement H equal to $40t$ and, by Eq. A17, a vertical displacement V equal to $60t - 16t^2$. The straight-line distance S from the stone to the launch point is found by the Pythagorean theorem. Write an equation for S in terms of t, and simplify.

10-2. OPERATIONS WITH RADICALS

Addition and Subtraction of Radicals: Radicals are called *similar* if they have the same index and the same radicand, such as $5\sqrt[3]{2x}$ and $3\sqrt[3]{2x}$. We add and subtract radicals by *combining similar radicals*.

Example:

$$5\sqrt{y} + 2\sqrt{y} - 4\sqrt{y} = 3\sqrt{y}$$

Radicals that do not appear to be similar at first may turn out to be so after simplification.

Examples:

(a) $$\sqrt{18x} - \sqrt{8x} = 3\sqrt{2x} - 2\sqrt{2x} = \sqrt{2x}$$

(b) $$\sqrt[3]{24y^4} + \sqrt[3]{81x^3y} = 2y\sqrt[3]{3y} + 3x\sqrt[3]{3y}$$

Factoring, $$= (2y + 3x)\sqrt[3]{3y}$$

(c) $$5y\sqrt{\frac{x}{y}} - 2\sqrt{xy} + x\sqrt{\frac{y}{x}}$$

$$= 5y\sqrt{\frac{xy}{y^2}} - 2\sqrt{xy} + x\sqrt{\frac{yx}{x^2}}$$

$$= 5\sqrt{xy} - 2\sqrt{xy} + \sqrt{xy} = 4\sqrt{xy}$$

Multiplication of Radicals: Radicals having the *same index* can be multiplied using Eq. 38.

Examples:

(a) $$\sqrt{x}\,\sqrt{2y} = \sqrt{2xy}$$

(b) $$(5\sqrt[3]{3a})(2\sqrt[3]{4b}) = 10\sqrt[3]{12ab}$$

(c) $$(2\sqrt{m})\,(5\sqrt{n})\,(3\sqrt{mn}) = 30\sqrt{m^2n^2} = 30\,mn$$

Radicals having *different indices* are multiplied by first going to exponential form, then multiplying using Eq. 32 and finally returning to radical form.

Example:

$$\sqrt{a}\,\sqrt[4]{b} = a^{1/2}b^{1/4}$$

$$= a^{2/4}b^{1/4} = (a^2b)^{1/4}$$

Or, in radical form, $$= \sqrt[4]{a^2b}$$

When the radicands are the same, the work is even easier.

Example: Multiply $\sqrt[3]{x}$ by $\sqrt[6]{x}$.

Solution:

$$\sqrt[3]{x}\,\sqrt[6]{x} = x^{1/3}x^{1/6} = x^{2/6}x^{1/6}$$

By Eq. 29, $$= x^{3/6} = x^{1/2}$$

$$= \sqrt{x}$$

Multinomials containing radicals are multiplied in the way we learned in Sec. 4-4.

Example:

$$(3 + \sqrt{x})(2\sqrt{x} - 4\sqrt{y}) = 3(2\sqrt{x}) - 3(4\sqrt{y}) + 2\sqrt{x}\sqrt{x} - 4\sqrt{x}\sqrt{y}$$
$$= 6\sqrt{x} - 12\sqrt{y} + 2x - 4\sqrt{xy}$$

To raise a radical to a *power*, we simply multiply the radical by itself the proper number of times.

Example: Cube the expression

$$2x\sqrt{y}$$

Solution:

$$(2x\sqrt{y})^3 = (2x\sqrt{y})(2x\sqrt{y})(2x\sqrt{y})$$
$$= 2^3x^3(\sqrt{y})^3$$
$$= 8x^3y^{3/2} = 8x^3\sqrt{y^3}$$

Division of Radicals: Radicals having the same indices can be divided using Eq. 39.

Examples:

(a)
$$\frac{\sqrt{4x^2}}{\sqrt{2x}} = \sqrt{\frac{4x^2}{2x}} = \sqrt{2x}$$

(b)
$$\frac{\sqrt{3a^4} + \sqrt{9a^3} - \sqrt{6a^2}}{\sqrt{3a^2}} = \sqrt{\frac{3a^4}{3a^2}} + \sqrt{\frac{9a^3}{3a^2}} - \sqrt{\frac{6a^2}{3a^2}}$$
$$= \sqrt{a^2} + \sqrt{3a} - \sqrt{2} = a + \sqrt{3a} - \sqrt{2}$$

It is a common practice to rationalize the denominator after division.

Example:

$$\frac{\sqrt{10x}}{\sqrt{5y}} = \sqrt{\frac{10x}{5y}} = \sqrt{\frac{2x}{y}}$$

Rationalizing the denominator,

$$\sqrt{\frac{2x}{y}} = \sqrt{\frac{2x}{y} \cdot \frac{y}{y}} = \frac{\sqrt{2xy}}{y}$$

If the indices are different, we go to exponential form, as we did for multiplication. Divide using Eq. 33 and then return to radical form.

Example:

$$\frac{\sqrt[3]{a}}{\sqrt[4]{b}} = \frac{a^{1/3}}{b^{1/4}} = \frac{a^{4/12}}{b^{3/12}} = \left(\frac{a^4}{b^3}\right)^{1/12}$$

Returning to radical form,

$$\left(\frac{a^4}{b^3}\right)^{1/12} = \sqrt[12]{\frac{a^4}{b^3}}$$

$$= \frac{1}{b}\sqrt[12]{a^4b^9}$$

after rationalizing the denominator.

Thus, the conjugate of the binomial $a + b$ is $a - b$. The conjugate of $x - y$ is $x + y$.

When dividing by a binomial containing *square* roots, multiply the divisor and the dividend by the *conjugate* of that binomial (a binomial that differs from the original dividend only in the sign of one term). This operation will remove all square roots from the denominator.

Example: Divide $3 + \sqrt{x}$ by $2 - \sqrt{x}$.

Solution: The conjugate of the divisor $2 - \sqrt{x}$ is $2 + \sqrt{x}$. Multiplying divisor and dividend by $2 + \sqrt{x}$, we get

$$\frac{3 + \sqrt{x}}{2 - \sqrt{x}} = \frac{3 + \sqrt{x}}{2 - \sqrt{x}} \cdot \frac{2 + \sqrt{x}}{2 + \sqrt{x}}$$

$$= \frac{6 + 3\sqrt{x} + 2\sqrt{x} + \sqrt{x}\sqrt{x}}{4 + 2\sqrt{x} - 2\sqrt{x} - \sqrt{x}\sqrt{x}}$$

$$= \frac{6 + 5\sqrt{x} + x}{4 - x}$$

EXERCISE 2

Addition and Subtraction of Radicals

Combine as indicated.

As in Exercise 1, do not use your calculator for any of these numerical problems. Leave the answers in radical form.

1. $2\sqrt{24} - \sqrt{54}$

2. $\sqrt{300} + \sqrt{108} - \sqrt{243}$

3. $\sqrt{24} - \sqrt{96} + \sqrt{54}$

4. $\sqrt{128} - \sqrt{18} + \sqrt{32}$

5. $2\sqrt{50} + \sqrt{72} + 3\sqrt{18}$

6. $\sqrt[3]{384} - \sqrt[3]{162} + \sqrt[3]{750}$

7. $2\sqrt[3]{2} - 3\sqrt[3]{16} + \sqrt[3]{-54}$

8. $3\sqrt[3]{108} + 2\sqrt[3]{32} - \sqrt[3]{256}$

9. $\sqrt[3]{625} - 2\sqrt[3]{135} - \sqrt[3]{320}$

10. $5\sqrt[3]{320} + 2\sqrt[3]{40} - 4\sqrt[3]{135}$

11. $\sqrt[4]{768} - \sqrt[4]{48} - \sqrt[4]{243}$

12. $3\sqrt{\frac{5}{4}} + 2\sqrt{45}$

13. $\sqrt{128x^2y} - \sqrt{98x^2y} + \sqrt{162x^2y}$

14. $3\sqrt{\frac{1}{3}} - 2\sqrt{\frac{3}{4}} + 4\sqrt{3}$

15. $7\sqrt{\frac{27}{50}} - 3\sqrt{\frac{2}{3}}$

16. $4\sqrt{50} - 2\sqrt{72} + \frac{3}{\sqrt{2}}$

17. $\sqrt{a^2x} + \sqrt{b^2x}$

18. $\sqrt{2b^2xy} - \sqrt{2a^2xy}$

19. $\sqrt{x^2y} - \sqrt{4a^2y}$

20. $\sqrt{80a^3} - 3\sqrt{20a^3} - 2\sqrt{45a^3}$

21. $4\sqrt{3a^2x} - 2a\sqrt{48x}$

22. $\sqrt[3]{125x^2} - 2\sqrt[3]{8x^2}$

Multiplication of Radicals

Multiply.

23. $3\sqrt{3}$ by $5\sqrt{3}$

24. $2\sqrt{3}$ by $3\sqrt{8}$

25. $\sqrt{8}$ by $\sqrt{160}$

26. $\sqrt{\frac{5}{8}}$ by $\sqrt{\frac{3}{4}}$

27. $4\sqrt[3]{45}$ by $2\sqrt[3]{3}$

28. $3\sqrt{3}$ by $2\sqrt[3]{2}$

29. $3\sqrt{2}$ by $2\sqrt[3]{3}$

30. $(\sqrt{5} - \sqrt{3})$ by $2\sqrt{3}$

31. $2\sqrt[3]{3}$ by $5\sqrt[4]{4}$

32. $2\sqrt[3]{24}$ by $\sqrt[9]{\frac{8}{27}}$

33. $2x\sqrt{3a}$ by $3\sqrt{y}$

34. $3\sqrt[3]{9a^2}$ by $\sqrt[3]{3abc}$

35. $\sqrt[5]{4xy^2}$ by $\sqrt[5]{8x^2y}$

36. $\sqrt{\frac{a}{b}}$ by $\sqrt{\frac{c}{d}}$

37. $\sqrt{a}$ by $\sqrt[4]{b}$

38. $(3 + 2\sqrt{2})$ by $(3 - 3\sqrt{2})$

39. $\sqrt{xy}$ by $2\sqrt{xz}$ and $\sqrt[3]{x^2y^2}$

40. $\sqrt[3]{a^2b}$ by $\sqrt[3]{2a^2b^2}$ and $\sqrt{3a^3b^2}$

Powers

Square the following expressions.

41. $3\sqrt{y}$

42. $4\sqrt[3]{4x^2}$

43. $3x\sqrt[3]{2x^2}$

44. $5 + 4\sqrt{x}$

45. $3 - 5\sqrt{a}$

46. $\sqrt{a} + 5a\sqrt{b}$

Cube the following radicals.

47. $5\sqrt{2x}$

48. $2x\sqrt{3x}$

49. $5\sqrt[3]{2ax}$

Division of Radicals

Divide.

50. $8 \div 3\sqrt{2}$

51. $6\sqrt{72} \div 12\sqrt{32}$

52. $(4\sqrt[3]{3} - 3\sqrt[3]{2}) \div \sqrt[3]{2}$

53. $5\sqrt[3]{12} \div 10\sqrt{8}$

54. $9\sqrt[3]{18} \div 3\sqrt{6}$

55. $12\sqrt[3]{4a^3} \div 4\sqrt[3]{2a^5}$

56. $12\sqrt{256} \div 18\sqrt{2}$

57. $(4\sqrt[3]{4} + 2\sqrt[3]{3} + 3\sqrt[3]{6}) \div \sqrt[3]{6}$

58. $2 \div \sqrt[3]{6}$

59. $\sqrt{72} \div 2\sqrt[4]{64}$

60. $8 \div 2\sqrt[3]{4}$

61. $8\sqrt[3]{ab} \div 4\sqrt{ac}$

62. $\sqrt[3]{4ab} \div \sqrt[4]{2ab}$

63. $2\sqrt[4]{a^2b^3c^2} \div 4\sqrt[3]{ab^2c^2}$

64. $6\sqrt[3]{x^4y^7z} \div \sqrt[4]{xy}$

65. $(3 + \sqrt{2}) \div (2 - \sqrt{2})$

66. $5 \div \sqrt{3x}$

67. $4\sqrt{x} \div \sqrt{a}$

10-3. RADICAL EQUATIONS

We will do more difficult radical equations in Chapter 11.

An equation in which the unknown is under a radical sign is called a *radical equation*. To solve a radical equation it is necessary to transpose the radical term until it is alone on one side of the equal sign and then to raise both sides to whatever power will eliminate the radical.

Example: Solve for x:

$$\sqrt{x - 5} - 4 = 0$$

Solution: Transposing,

$$\sqrt{x - 5} = 4$$

Squaring both sides,

$$x - 5 = 16$$

$$x = 21$$

Check:

$$\sqrt{21 - 5} - 4 \stackrel{?}{=} 0$$

$$\sqrt{16} - 4 = 0$$

$$4 - 4 = 0 \quad \text{checks}$$

Example: Solve for x:

$$\sqrt{x^2 - 6x} = x - 9$$

Solution: Squaring,

$$x^2 - 6x = x^2 - 18x + 81$$

$$12x = 81$$

$$x = \frac{81}{12} = \frac{27}{4}$$

Check:

Remember that we take only the principal (positve) root.

$$\sqrt{\left(\frac{27}{4}\right)^2 - 6\left(\frac{27}{4}\right)} \stackrel{?}{=} \frac{27}{4} - 9$$

$$\sqrt{\frac{729}{16} - \frac{162}{4}} \stackrel{?}{=} \frac{27}{4} - \frac{36}{4} = -\frac{9}{4}$$

$$\sqrt{\frac{81}{16}} = \frac{9}{4} \neq -\frac{9}{4} \qquad \text{does not check}$$

The given equation has no solution.

Common Error	The squaring process often introduces *extraneous roots.* These are discarded because they do not satisfy the original equation.

If the equation has more than one radical, isolate one at a time and square both sides, as in the following example.

Tip: *It is usually better to isolate and square the* most complicated *radical first.*

Example: Solve for x:

$$\sqrt{x - 32} + \sqrt{x} = 16$$

Solution: Transposing,

$$\sqrt{x - 32} = 16 - \sqrt{x}$$

Squaring,

$$x - 32 = (16)^2 - 32\sqrt{x} + x$$

We transpose again to isolate the radical,

$$32\sqrt{x} = 256 + 32 = 288$$

$$\sqrt{x} = 9$$

Squaring again,

$$x = 81$$

Check:

$$\sqrt{81 - 32} + \sqrt{81} \stackrel{?}{=} 16$$

$$\sqrt{49} + 9 \stackrel{?}{=} 16$$

$$7 + 9 = 16 \qquad \text{checks}$$

Radical equations having indices other than 2 are solved in a similar way.

Example: Solve for x:

$$\sqrt[3]{x - 5} = 2$$

Solution: Cubing both sides,

$$x - 5 = 8$$

$$x = 13$$

Check:

$$\sqrt[3]{13 - 5} \stackrel{?}{=} 2$$

$$\sqrt[3]{8} = 2 \qquad \text{checks}$$

EXERCISE 3

Radical Equations

Solve for x and check.

1. $\sqrt{x} = 6$

2. $\sqrt{x} + 5 = 9$

3. $\sqrt{7x + 8} = 6$

4. $\sqrt{3x - 2} = 5$

5. $\sqrt{x + 1} = \sqrt{2x - 7}$

6. $2 = \sqrt[4]{1 + 3x}$

7. $\sqrt{3x + 1} = 5$

8. $\sqrt[3]{2x} = 4$

9. $\sqrt{x - 3} = \dfrac{4}{\sqrt{x - 3}}$

10. $x + 2 = \sqrt{x^2 + 6}$

11. $\sqrt{x^2 - 7.25} = 8.75 - x$

12. $\sqrt{x - 15.5} = 5.85 - \sqrt{x}$

13. $\dfrac{6}{\sqrt{3 + x}} = \sqrt{x + 3}$

14. $\sqrt{12 + x} = 2 + \sqrt{x}$

15. Solve Eq. A53,

$$Z = \sqrt{R^2 + \left(\omega L - \frac{1}{\omega C}\right)^2}$$

for C.

Computer

16. If you have written a program to solve equations by the *midpoint method*, as suggested in Sec. 6-4, use your program to solve any of the radical equations above. If not, you may want to write that program now, for it will be very useful in later chapters as well.

CHAPTER TEST

Simplify.

1. $\sqrt{52}$

2. $\sqrt{108}$

3. $\sqrt[3]{162}$

4. $\sqrt[4]{9}$

5. $\sqrt[6]{4}$

6. $\sqrt{81a^2x^3y}$

7. $3\sqrt[4]{81x^5}$

8. $\sqrt{ab^2 - b^3}$

9. $\sqrt[3]{(a - b)^5x^4}$

10. $5\sqrt{\dfrac{7x}{12y}}$

11. $\sqrt{\dfrac{a - 2}{a + 2}}$

Perform the indicated operations and simplify. Don't use a calculator.

12. $4b\sqrt{3y} \cdot 5\sqrt{x}$

13. $\sqrt{x} \cdot \sqrt{x^3 - x^4y}$

14. $(\sqrt{x} + \sqrt{y})(\sqrt{x} - \sqrt{y})$

15. $\dfrac{3 - 2\sqrt{3}}{2 - 5\sqrt{2}}$

16. $\dfrac{\sqrt{x} - \sqrt{y}}{\sqrt{x} + \sqrt{y}}$

17. $\dfrac{x + \sqrt{x^2 - y^2}}{x - \sqrt{x^2 - y^2}}$

18. $\sqrt{98x^2y^2} - \sqrt{128x^2y^2}$

19. $4\sqrt[3]{125x^2} + 3\sqrt[3]{8x^2}$

20. $3\sqrt{x^2 - y^2} - 2\sqrt{\dfrac{x + y}{x - y}}$

21. $\sqrt{a}\sqrt[3]{b}$

22. $\sqrt[3]{2abc^2} \cdot \sqrt{abc}$

23. $(3 + 2\sqrt{x})^2$

24. $(4x\sqrt{2x})^3$

25. $3\sqrt{50} - 2\sqrt{32}$

26. $3\sqrt{72} - 4\sqrt{8} + \sqrt{128}$

27. $5\sqrt{\dfrac{9}{8}} - 2\sqrt{\dfrac{25}{18}}$

28. $2\sqrt{2} \div \sqrt[3]{2}$

29. $\sqrt{2ab} \div \sqrt{4ab^2}$

30. $9 \div \sqrt[3]{7x^2}$

31. $3\sqrt{9} \cdot 4\sqrt{8}$

Solve for x and check.

32. $\dfrac{\sqrt{x} - 8}{\sqrt{x} - 6} = \dfrac{\sqrt{x} - 4}{\sqrt{x} + 2}$

33. $\sqrt{x + 6} = 4$

34. $\sqrt{2x - 7} = \sqrt{x - 3}$

35. $\sqrt[5]{2x + 4} = 2$

36. $\sqrt{4x^2 - 3} = 2x - 1$

37. $\sqrt{x} + \sqrt{x - 9.75} = 6.23$

38. $\sqrt[3]{21.5x} = 2.33$

Quadratic Equations

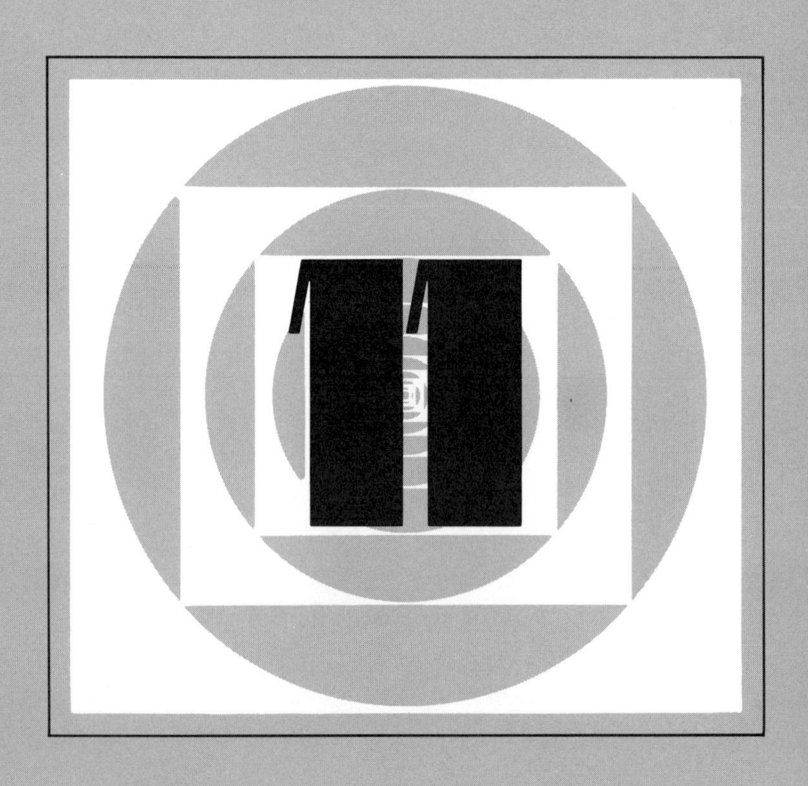

So far we have learned how to solve linear (first-degree) equations, and sets of first-degree equations. But many technical problems require us to solve more complicated equations than that. In this chapter we study second-degree equations (quadratics), and later we cover logarithmic, exponential, and trigonometric equations. That does not include every equation type that exists, but many of the common types.

Quadratic equations arise in many technical problems, such as the simple falling-body problem: "An object is thrown downward with a speed of 15 ft/s. How long will it take to fall 100 ft?"

The displacement of a freely falling body is given by Eq. A17, $s = v_0 t + \frac{1}{2}at^2$, and if we substitute $s = 100$, $v_0 = 15$, and $a = 32$,

$$100 = 15t + 16t^2$$

Now how do we solve that equation? How shall we find the values of t that will make the equation balance? Even this simple problem requires us to solve a quadratic equation.

Let us leave this problem and return to it after we have learned how to solve quadratics. Before we start, you might want to leaf through Chapter 7 if you have gotten rusty on factoring. Look especially at the sections on common factors, trinomials, and the perfect square trinomial.

11-1. TERMINOLOGY

Quadratic Equation: A polynomial equation of second degree is called a *quadratic equation.*

Examples: The equations

$$4x^2 - 5x + 2 = 0 \qquad x^2 = 58$$
$$9x^2 - 5x = 0 \qquad 2x^2 - 7 = 0$$

are quadratic equations.

General Form: A quadratic is in *general form* when it is written in the form

General Form of a Quadratic	$ax^2 + bx + c = 0$	**99**

where a, b, and c are constants.

Example: Write the quadratic equation

$$7 - 4x = \frac{5x^2}{3}$$

in general form.

Solution: Subtracting $5x^2/3$ from both sides and writing the terms in descending order of the exponents, we obtain

$$-\frac{5x^2}{3} - 4x + 7 = 0$$

Quadratics in general form are usually written without fractions and with the first term positive. Multiplying by -3, we get

$$5x^2 + 12x - 21 = 0$$

The equation is now in general form, with $a = 5$, $b = 12$, and $c = -21$.

Number of Roots: A quadratic always has *two solutions*, or roots. The two roots are sometimes equal, or they may be imaginary or complex numbers. We will see later that in practical problems, one of the roots must often be discarded.

Checking: Check the apparent solutions to a quadratic as you would check the solutions to any equation—by substituting back in the original equation.

Example: Are the values $x = 1$ and $x = 2$ the solutions to the quadratic equation $x^2 - 3x + 2 = 0$?

Solution: Substituting $x = 1$,

$$1^2 - 3(1) + 2 \stackrel{?}{=} 0$$

$$1 - 3 + 2 = 0 \qquad \text{checks}$$

Substituting $x = 2$,

$$2^2 - 3(2) + 2 \stackrel{?}{=} 0$$

$$4 - 6 + 2 = 0 \qquad \text{checks}$$

EXERCISE 1

Write the following quadratics in general form, and list the constants a, b, and c.

1. $x(x - 9) = 5$

2. $3x = 2 - 4x^2$

3. $\dfrac{4x}{3} = 7 - x^2$

4. $3x(x + 2) - (x - 2)(x + 3) = 0$

5. $5x^2 - 2x = 6 + 3x - x^2$

Check these proposed solutions in the given equations.

6. $x^2 - x - 6 = 0 \qquad x = -2, x = 3$

7. $2x^2 - 3x - 2 = 0 \qquad x = -\frac{1}{2}, x = 2$

8. $x^2 - 7x + 10 = 0 \qquad x = 2, x = 4$

11-2. SOLVING QUADRATICS BY FACTORING

A quadratic having no x term is called a pure quadratic, *such as $x^2 = 25$. Solve by taking the square root of both sides. Here $x = \pm 5$.*

Incomplete Quadratics: A quadratic is called *incomplete* when the constant term c is zero. To solve an incomplete quadratic, remove the common factor x from each term, and set each factor equal to zero.

Example: Solve

$$x^2 + 5x = 0$$

Solution: Factoring,

$$x(x + 5) = 0$$

We use this idea often in this chapter. If we have the product of two quantities a and b equal to zero, $ab = 0$, this equation will be true if $a = 0$ ($0 \cdot b = 0$) or if $b = 0$ ($a \cdot 0 = 0$), or if both are zero ($0 \cdot 0 = 0$).

Note that this expression will be true if either or both of the two factors equals zero. We therefore set each factor in turn equal to zero.

$$x = 0, \qquad x + 5 = 0$$

$$x = -5$$

The two solutions are thus $x = 0$ and $x = -5$.

Common Error	Do not cancel an x from the terms of an incomplete quadratic. That will cause a root to be lost. If, in the last example, we had said $$x^2 = -5x$$ and dividing by x, getting $$x = -5$$ we would have gotten the correct root $x = -5$ but would have lost the root $x = 0$.

The Complete Quadratic: We now consider a quadratic that has all its terms in place, the complete quadratic

$$ax^2 + bx + c = 0 \tag{99}$$

Finding the Roots: First write the quadratic in general form, Eq. 99. Factor the trinomial (if possible) by the methods of Chapter 7 and set each factor equal to zero.

Example: Solve by factoring:

$$x^2 - x - 6 = 0$$

Solution: Factoring by means of Eq. 45

$$(x - 3)(x + 2) = 0$$

This equation will be satisfied if either or both of the two factors $(x - 3)$ and $(x + 2)$ are zero. We therefore equate each factor to zero.

$$x - 3 = 0, \qquad x + 2 = 0$$

so the roots are

$$x = 3 \qquad \text{and} \qquad x = -2$$

Check: Substituting each root into the original equation:

When $x = 3$:

$$3^2 - 3 - 6 \stackrel{?}{=} 0$$

$$9 - 3 - 6 = 0 \qquad \text{checks}$$

When $x = -2$:

$$(-2)^2 - (-2) - 6 \stackrel{?}{=} 0$$

$$4 + 2 - 6 = 0 \qquad \text{checks}$$

Example: Solve by factoring

$$4x^2 - 10x + 6 = 0$$

Look back at Sec. 7-4 if you have forgotten how to factor this type of trinomial.

Solution: We first divide both sides by 2, getting

$$2x^2 - 5x + 3 = 0$$

Factoring,

$$(x - 1)(2x - 3) = 0$$

So

$$x = 1 \quad \text{and} \quad x = \frac{3}{2}$$

Often an equation must first be simplified before factoring.

Example: Solve for x,

$$x(x - 8) = 2x(x - 1) + 9$$

Solution: Removing parentheses gives

$$x^2 - 8x = 2x^2 - 2x + 9$$

Transposing and collecting terms, we get

$$x^2 + 6x + 9 = 0$$

Factoring,

$$(x + 3)(x + 3) = 0$$

which gives the double root,

$$x = -3, -3$$

Radical Equations: In Chapter 10 we solved simple radical equations. We isolated a radical on one side of the equation and then squared both sides. Here we solve equations in which this squaring operation results in a quadratic equation.

Example: Solve for x,

$$3\sqrt{x - 1} - \frac{4}{\sqrt{x - 1}} = 4$$

Solution: We clear fractions by multiplying through by $\sqrt{x - 1}$.

$$3(x - 1) - 4 = 4\sqrt{x - 1}$$

$$3x - 7 = 4\sqrt{x - 1}$$

Squaring both sides,

$$9x^2 - 42x + 49 = 16(x - 1)$$

Removing parentheses and collecting terms gives

$$9x^2 - 58x + 65 = 0$$

Factoring,

$$(x - 5)(9x - 13) = 0$$

$$x = 5 \quad \text{and} \quad x = \frac{13}{9}$$

Check: When $x = 5$:

$$3\sqrt{5 - 1} - \frac{4}{\sqrt{5 - 1}} \stackrel{?}{=} 4$$

$$3(2) - \frac{4}{2} = 4 \quad \text{checks}$$

When $x = \frac{13}{9}$:

$$\sqrt{\frac{13}{9} - 1} = \sqrt{\frac{4}{9}} = \frac{2}{3}$$

So

$$3 \cdot \frac{2}{3} - \frac{4}{\frac{2}{3}} \stackrel{?}{=} 4$$

$$2 - 6 \neq 4 \quad \text{does not check}$$

EXERCISE 2

Incomplete Quadratics

Solve each incomplete quadratic.

1. $2x = 5x^2$

2. $2x - 40x^2 = 0$

3. $3x(x - 2) = x(x + 3)$

4. $2x^2 - 6 = 66$

5. $5x^2 - 3 = 2x^2 + 24$

6. $7x^2 + 4 = 3x^2 + 40$

7. $(x + 2)^2 = 4x + 5$

8. $5x^2 - 2 = 3x^2 + 6$

9. $8.25x^2 - 2.93x = 0$

10. $284x = 827x^2$

Complete Quadratics

Solve each quadratic by factoring.

Leave your answers in fractional or radical form.

11. $x^2 + 2x - 15 = 0$

12. $x^2 + 6x - 16 = 0$

13. $x^2 - x - 20 = 0$

14. $x^2 + 13x + 42 = 0$

15. $x^2 - x - 2 = 0$

16. $x^2 + 7x + 12 = 0$

17. $x^2 + 3x + 2 = 0$

18. $x^2 - 4x - 21 = 0$

19. $x^2 - 7x - 18 = 0$

20. $x^2 + 6x + 8 = 0$

21. $x^2 + 12x + 32 = 0$

22. $x^2 - 10x - 39 = 0$

23. $x^2 - 12x - 64 = 0$

24. $x^2 + 14x + 33 = 0$

25. $x^2 - 13x + 40 = 0$

26. $x^2 - x - 42 = 0$

27. $2x^2 + 5x - 12 = 0$

28. $3x^2 - x - 2 = 0$

29. $5x^2 + 14x - 3 = 0$

30. $5x^2 + 3x - 2 = 0$

Rearrange each quadratic into general form and solve by factoring.

31. $2 + x - x^2 = 0$

32. $20 + 19x - 6x^2 = 0$

33. $-26x + 5 - 24x^2 = 0$

34. $15x^2 = 8 + 14x$

35. $2x^3 + x^2 - 18 = x(2x^2 + 3)$

36. $\frac{x}{2} + \frac{2}{x} - \frac{5}{2} = 0$

37. $\frac{2 - x}{3} = 2x^2$

38. $(x - 6)(x + 6) = 5x$

39. $x(x - 5) = 36$

40. $(2x - 3)^2 = 2x - x^2$

Arrange the following fractional equations in general form and solve by factoring.

41. $x^2 - \frac{5}{6}x = \frac{1}{6}$

42. $\frac{7}{x + 4} - \frac{1}{4 - x} = \frac{2}{3}$

43. $\frac{2x - 1}{x + 3} = \frac{x + 3}{2x + 1}$

44. $1 - \frac{x}{2} = 5 - \frac{36}{x + 2}$

45. $3 - x^2 = \frac{2x - 76}{3}$

46. $\frac{x}{x - 1} - \frac{x - 1}{x} = \frac{3}{2}$

47. $\frac{2x}{3} - \frac{5}{4x} = \frac{7x}{9} - \frac{21}{4x}$

48. $\frac{2x + 5}{2x - 5} = \frac{7x - 5}{2x}$

49. $x^2 + \frac{x}{2} = \frac{2x^2}{5} - \frac{x}{5} + \frac{13}{10}$

50. $\frac{2x - 1}{x - 1} + \frac{1}{6} = \frac{2x - 3}{x - 2}$

Radical Equations

Solve each radical equation for x and check.

51. $\sqrt{5x^2 - 3x - 41} = 3x - 7$

52. $3\sqrt{x - 1} - \frac{4}{\sqrt{x - 1}} = 4$

53. $\sqrt{x + 5} = \frac{12}{\sqrt{x + 12}}$

54. $\sqrt{7x + 8} - \sqrt{5x - 4} = 2$

55. $2x + \sqrt{4x^2 - 7} = \frac{21}{\sqrt{4x^2 - 7}}$

Computer

56. Use your program for solving equations by the midpoint method (Sec. 6-4) to find the roots of any of the quadratics in this exercise.

11-3. SOLVING QUADRATICS BY COMPLETING THE SQUARE

If a quadratic equation is not factorable, it is possible to manipulate it into factorable form by a procedure called *completing the square*. The form into which we shall put our expression is the *perfect square trinomial*, Eqs. 47 and 48, that we studied in Chapter 7.

If you recall the perfect square trinomial (see Sec. 7-5):

1. The first and last terms were perfect squares.
2. The middle term was twice the product of the square roots of the outer terms.

To complete the square, we manipulate our given expression so that these two conditions are met. This is best shown by an example.

Example: Solve the quadratic $x^2 - 8x + 6 = 0$ by completing the square.

Solution: Subtracting 6 from both sides, we obtain

$$x^2 - 8x = -6$$

We complete the square by adding the square of half the coefficient of the x term to both sides. The coefficient of x is -8. We take half of -8 and square it, getting $(-4)^2$ or 16. Adding 16 to both sides,

$$x^2 - 8x + 16 = -6 + 16 = 10$$

Factoring by Eq. 48,

$$(x - 4)^2 = 10$$

Taking the square root of both sides,

$$x - 4 = \pm\sqrt{10}$$

Finally, adding 4 to both sides,

$$x = 4 \pm \sqrt{10}$$

Common Error	When adding the quantity needed to complete the square to the left-hand side, it is easy to forget to add the same quantity to the right-hand side. $x^2 - 4x \boxed{+ 4} = 5 \boxed{+ 4}$ ← don't forget

If the x^2 term has a coefficient other than 1, divide through by this coefficient before completing the square.

Example: Solve

$$2x^2 + 4x - 3 = 0$$

Solution: Transposing the -3 and dividing by 2,

$$x^2 + 2x = \frac{3}{2}$$

Completing the square,

$$x^2 + 2x + 1 = \frac{3}{2} + 1$$

$$(x + 1)^2 = \frac{5}{2}$$

$$x + 1 = \pm\sqrt{\frac{5}{2}} = \frac{\pm\sqrt{10}}{2}$$

$$x = -1 \pm \frac{1}{2}\sqrt{10}$$

EXERCISE 3

Solve each quadratic by completing the square.

1. $x^2 + 7x - 3 = 0$

2. $x^2 - 3x - 5 = 0$

3. $x^2 - 8x + 2 = 0$

4. $x^2 + 4x - 9 = 0$

5. $4x^2 - 3x - 5 = 0$

6. $2x^2 + 3x - 3 = 0$

7. $3x^2 + 6x - 4 = 0$

8. $5x^2 - 2x - 1 = 0$

9. $4x^2 + 7x - 5 = 0$

10. $8x^2 - 4x - 3 = 0$

11-4. SOLVING QUADRATICS BY FORMULA

Of the several methods we have for solving quadratics, the most useful is the quadratic formula. It will work for any quadratic, regardless of the type of roots, and can easily be programmed for the computer.

Derivation of the Quadratic Formula: We wish to find the roots of the equation

$$ax^2 + bx + c = 0$$

by completing the square. We start by subtracting c from both sides and dividing by a:

$$x^2 + \frac{b}{a}x = -\frac{c}{a}$$

The method of completing the square is really too cumbersome to be a practical tool for solving quadratics. The main reason we learn it is to derive the quadratic formula.

Completing the square,

$$x^2 + \frac{b}{a}x + \left(\frac{b}{2a}\right)^2 = \frac{b^2}{4a^2} - \frac{c}{a}$$

Factoring, by Eq. 47,

$$\left(x + \frac{b}{2a}\right)^2 = \frac{b^2 - 4ac}{4a^2}$$

Taking the square root,

$$y + \frac{b}{2a} = \pm\frac{\sqrt{b^2 - 4ac}}{2a}$$

Transposing, we get

Quadratic Formula	$x = \dfrac{-b \pm \sqrt{b^2 - 4ac}}{2a}$	**100**

as the roots of $ax^2 + bx + c = 0$.

Using the Quadratic Formula: Simply put the given equation into general form (Eq. 99); list a, b, and c; and substitute them into the formula.

Example: Solve $2x^2 - 5x - 3 = 0$ by the quadratic formula.

Solution: The equation is already in general form, with

$$a = 2 \qquad b = -5 \qquad c = -3$$

Substituting into Eq. 100,

$$x = \frac{-(-5) \pm \sqrt{(-5)^2 - 4(2)(-3)}}{2(2)}$$

$$= \frac{5 \pm \sqrt{25 + 24}}{4}$$

$$= \frac{5 \pm \sqrt{49}}{4} = \frac{5 \pm 7}{4} = \frac{5 + 7}{4} \quad \text{and} \quad \frac{5 - 7}{4}$$

$$= \frac{12}{4} \quad \text{and} \quad \frac{-2}{4}$$

$$= 3 \quad \text{and} \quad -\frac{1}{2}$$

Tip	Always rewrite a quadratic in general form before trying to use the quadratic formula.

Example: Solve $5x - 3x^2 + 6 = 0$ by the quadratic formula.

Solution: The constants a, b, and c are *not* 5, -3, and 6. Rewriting the equation in general form,

$$-3x^2 + 5x + 6 = 0$$

where

$$a = -3 \qquad b = 5 \qquad c = 6$$

Substituting into Eq. 100,

$$x = \frac{-5 \pm \sqrt{25 - 4(-3)(6)}}{2(-3)} = \frac{-5 \pm \sqrt{97}}{-6}$$

$$\cong 2.475 \quad \text{and} \quad -0.808$$

We now have the tools to finish the falling-body problem started in the introduction to this chapter.

Example: Solve the equation

$$100 = 15t + 16t^2$$

(where t is in seconds).

Solution: We first go to general form,

$$16t^2 + 15t - 100 = 0$$

Substituting into the quadratic formula,

$$t = \frac{-15 \pm \sqrt{225 - 4(16)(-100)}}{2(16)}$$

$$= \frac{-15 \pm \sqrt{6625}}{32} = \frac{-15 \pm 81.4}{32}$$

$$= 2.075 \text{ s} \quad \text{and} \quad -3.012 \text{ s}$$

In this problem, a negative elapsed time makes no sense, so we discard the negative root. Thus, it will take 2.075 s for an object to fall 100 ft when thrown downward with a speed of 15 ft/s.

Some equations will have to be simplified before they can be put into general form and solved.

Example: Solve for x,

$$\frac{3x - 1}{4x + 7} = \frac{x + 1}{x + 7}$$

Solution: We start by multiplying both sides by the LCD, $(4x + 7)(x + 7)$. We get

$$(3x - 1)(x + 7) = (x + 1)(4x + 7)$$

or

$$3x^2 + 20x - 7 = 4x^2 + 11x + 7$$

Collecting terms gives

$$x^2 - 9x + 14 = 0$$

Applying the quadratic formula,

$$x = \frac{9 \pm \sqrt{81 - 4(1)(14)}}{2} = \frac{9 \pm \sqrt{81 - 56}}{2} x$$

$$= 7 \qquad \text{and} \qquad x = 2$$

Non-Real Roots: Quadratics resulting from practical problems will usually yield real roots if properly set up. Other quadratics, however, could give non-real roots, as shown in the next example.

Example: Solve

$$x^2 + 2x + 5 = 0$$

Solution: From Eq. 100,

$$x = \frac{-2 \pm \sqrt{4 - 4(1)(5)}}{2} = \frac{-2 \pm \sqrt{-16}}{2}$$

See Sec. 16-1 for a complete discussion of imaginary and complex numbers.

The expression $\sqrt{-16}$ can be written $\sqrt{16}\sqrt{-1}$ or $j4$, where $j = \sqrt{-1}$. So

$$x = \frac{-2 \pm j4}{2} = -1 \pm j2$$

Predicting the Nature of the Roots: As seen from the preceding example, when the quantity in Eq. 100 under the radical sign $(b^2 - 4ac)$ is negative, the roots are complex. It should also be evident that if $b^2 - 4ac$ is zero, the roots will be equal. This quantity, called the *discriminant*, can thus be used to predict what kind of roots an equation will give, without having to actually find the roots.

The Discriminant	If $b^2 - 4ac > 0$, the roots are real and unequal. If $b^2 - 4ac = 0$ the roots are real and equal. If $b^2 - 4ac < 0$, the roots are not real.	**101**

Do not lose too much sleep over the discriminant. If you need to know the nature of the roots of a quadratic and cannot remember Eq. 101 all you have to do is calculate the roots themselves. This is not much more work than finding the discriminant.

Example: What sort of roots can you expect from the equation

$$3x^2 - 5x + 7 = 0?$$

Solution: Computing the discriminant, we obtain

$$b^2 - 4ac = (-5)^2 - 4(3)(7) = 25 - 84 = -59$$

A negative discriminant tells us that the roots are not real.

EXERCISE 4

Quadratic Formula

Solve by quadratic formula, to three significant digits.

1. $x^2 - 12x + 28 = 0$

2. $x^2 - 6x + 7 = 0$

3. $x^2 + x - 19 = 0$

4. $x^2 - x - 13 = 0$

5. $3x^2 + 12x - 35 = 0$

6. $29.4x^2 - 48.2x - 17.4 = 0$

7. $36x^2 + 3x - 7 = 0$

8. $28x^2 + 29x + 7 = 0$

9. $49x^2 + 21x - 5 = 0$

10. $16x^2 - 16x + 1 = 0$

11. $3x^2 - 10x + 4 = 0$

12. $x^2 - 34x + 22 = 0$

Arrange in general form and solve by quadratic formula.

13. $3x^2 + 5x = 7$

14. $4x + 5 = x^2 + 2x$

15. $x^2 - 4 = 4x + 7$

16. $x^2 - 6x - 14 = 3$

17. $(4.2x - 5.8)(7.2x - 9.2) = 8.2x + 9.9$

18. $3x^2 - 25x = 5x - 73$

19. $2x^2 + 100 = 32x - 11$

20. $33 - 3x^2 - 10x = 0$

21. $x(2x - 3) = 3x(x + 4) - 2$

22. $2.95(x^2 + 8.27x) = 7.24x(4.82x - 2.47) + 8.73$

23. $6x - 300 = 205 - 3x^2$

24. $(2x - 1)^2 + 6 = 6(2x - 1)$

Nature of the Roots

Determine if the roots of each quadratic are real or complex, and whether they are equal or unequal.

25. $x^2 - 5x - 11 = 0$

26. $x^2 - 6x + 15 = 0$

27. $2x^2 - 4x + 7 = 0$

28. $2x^2 + 3x - 2 = 0$

29. $3x^2 - 3x + 5 = 0$

30. $4x^2 + 4x + 1 = 0$

Computer

31. Write a program that will compute and print the roots of any quadratic once the constants a, b, and c are entered. Have the computer evaluate the discriminant before computing the roots. If the discriminant is negative, cause it to stop the run. This will prevent the program from crashing when the computer tries to take the square root of a negative number.

32. Use your program for solving equations by the midpoint method (Sec. 6-4) to find the roots of any of the quadratics in this exercise.

11-5. APPLICATIONS AND WORD PROBLEMS

Now that we have the tools to solve any quadratic, let us go on to problems from technology that require us to solve these equations.

At this point you might want to take a quick look at Sec. 5-3 and review some of the suggestions for setting up and solving word problems. You should set up these problems just as you did then.

When you solve the resulting quadratic you will get two roots, of course. If one of the roots does not make sense in the physical problem (like a beam having a length of −2000 ft), throw it away. But do not be too hasty. Often a second root will give an unexpected but equally good answer.

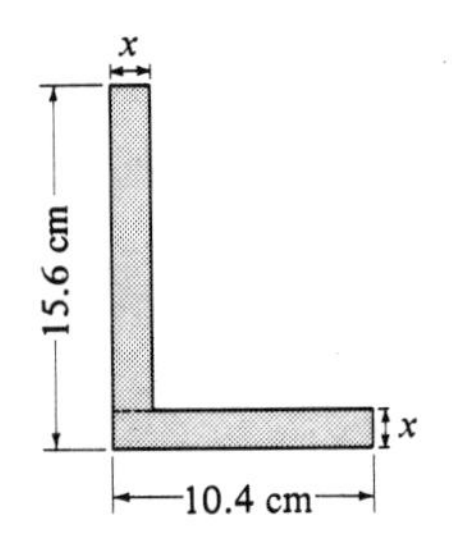

FIGURE 11-1

Example: The angle iron in Fig. 11-1 has a cross-sectional area of 53.4 cm^2. Find the thickness x.

Solution: We divide the area into two rectangles, as shown by the dashed line. One rectangle has an area of $10.4x$ and the other has an area of $(15.6 - x)x$. Since the sum of these areas must be 53.4,

$$10.4x + (15.6 - x)x = 53.4$$

Putting this equation into standard form, we get

$$10.4x + 15.6x - x^2 = 53.4$$
$$x^2 - 26.0x + 53.4 = 0$$

By the quadratic formula,

$$x = \frac{26.0 \pm \sqrt{676 - 4(53.4)}}{2} = \frac{26.0 \pm 21.5}{2}$$

So

$$x = \frac{26.0 - 21.5}{2} = 2.25 \text{ cm}$$

and

$$x = \frac{26.0 + 21.5}{2} = 23.8 \text{ cm}$$

We discard 23.8 cm because it is an impossible solution in this problem.

Check: Does our answer meet the requirements of the original statement? Let us compute the area of the angle iron using our value of 2.25 cm for the thickness. We get

$$\text{area} = 2.25(13.35) + 2.25(10.4)$$
$$= 30.0 + 23.4 = 53.4 \text{ cm}^2$$

which is the required area.

Example: A certain train is to be replaced with a "bullet" train that goes 40 mi/h faster than the old train, and which will make the 850 mi run in 3.0 h less time. Find the speed of each train.

Solution: Let

$$x = \text{rate of old train, mi/h}$$

Then

$$x + 40 = \text{rate of bullet train, mi/h}$$

The time it takes the old train to travel 288 mi at x mi/h is, by Eq. A16,

$$\text{time} = \frac{\text{distance}}{\text{rate}} = \frac{850}{x} \quad \text{h}$$

The time for the bullet train is then $(850/x - 3)$ h. Applying Eq. A16 for the bullet train,

$$\text{rate} \times \text{time} = \text{distance}$$

$$(x + 40)\left(\frac{850}{x} - 3\right) = 850$$

Removing parentheses,

$$850 - 3x + \frac{34{,}000}{x} - 120 = 850$$

Collecting terms and multiplying through by x gives

$$-3x^2 + 34{,}000 - 120x = 0$$

or

$$x^2 + 40x - 11{,}333 = 0$$

Solving for x by the quadratic formula,

$$x = \frac{-40 \pm \sqrt{1600 - 4(-11{,}333)}}{2}$$

If we drop the negative root, we get

$$x = \frac{-40 + 217}{2} = 88.3 \text{ mi/h} = \text{speed of slow train}$$

and

$$x + 40 = 128.3 \text{ mi/h} = \text{speed of bullet train}$$

EXERCISE 5

Number Problems

The "numbers" in these problems are all positive integers.

1. What number added to its reciprocal gives $2\frac{1}{6}$?
2. Find three consecutive numbers such that the sum of their squares will be 434.
3. Find two numbers whose difference is 7 and the difference of whose cubes is 1267.

4. Find two numbers whose sum is 11 and whose product is 30.

5. Find two numbers whose difference is 10 and the sum of whose squares is 250.

6. A number increased by its square is equal to 9 times the next higher number. Find the number.

Geometry Problems

Work these problems to three significant digits.

7. A rectangle is to be 2 m longer than it is wide and to have an area of 24 m^2. Find its dimensions.

8. One leg of a right triangle is 3 cm greater than the other leg, and the hypotenuse is 15 cm. Find the legs of the triangle.

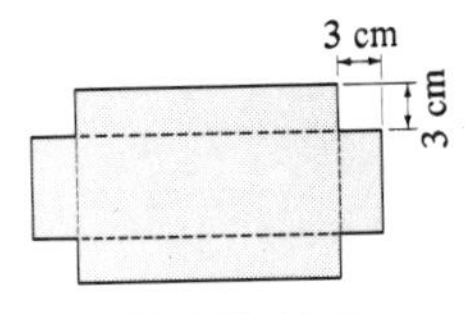

FIGURE 11-2

9. A rectangular sheet of brass is twice as long as it is wide. Squares, 3 cm $\times$ 3 cm, are cut from each corner, and the ends are turned up to form an open box (Fig. 11-2) having a volume of 648 cm^3. What are the dimensions of the original sheet of brass?

10. The length, width, and height of a cubical shipping container are all decreased by 1.0 ft, thereby decreasing the volume of the cube by 37 ft^3. What was the volume of the original container?

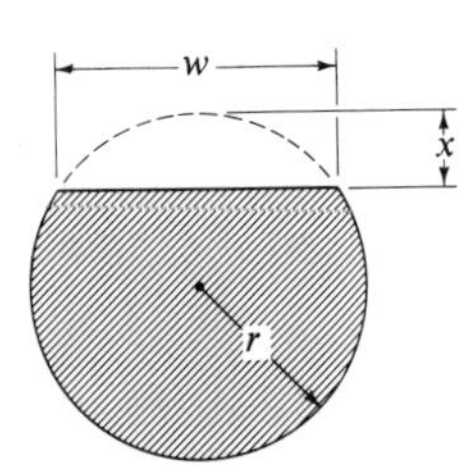

FIGURE 11-3

11. Find the dimensions of a rectangular field that has a perimeter of 724 m and an area of 32,400 m^2.

12. Write a formula for the depth of cut x needed to produce a flat of width w on a cylindrical bar of radius r, as in Fig. 11-3.

13. A rectangular field is surrounded by a fence 160 ft long. The cost of this fence, at 96 cents/ft, was one-tenth as many dollars as there are square feet in the area of the field. What are the dimensions of the field?

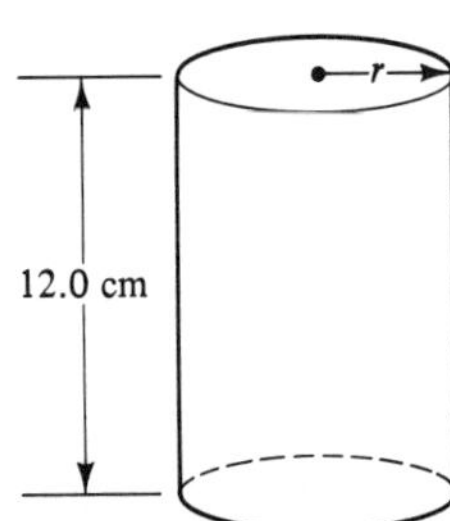

FIGURE 11-4

14. The cylinder in Fig. 11-4 has a surface area of 846 cm^2, including the ends. Find its radius.

15. A cylindrical tank having a diameter of 75.5 in. is placed so that it touches a wall, as in Fig. 11-5. Find the radius of the largest pipe that can fit into the space between the tank, the wall, and the floor.

Uniform Motion

16. A truck travels 350 mi to a delivery point, unloads, and, now empty, returns to the starting point at a speed 8.0 mi/h greater than on the outward trip. What was the speed of the outward trip if the total round-trip driving time was 14.4 h?

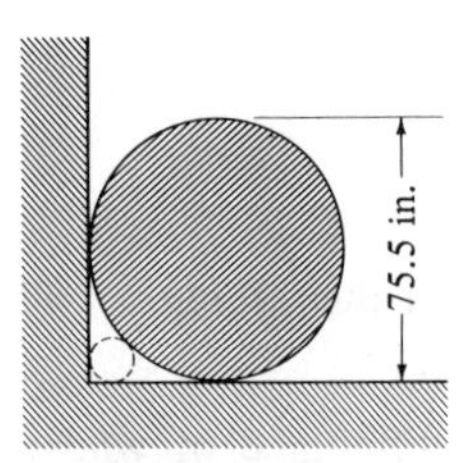

FIGURE 11-5

17. An airplane flies 355 mi to city A. Then, with better winds, it continues on to city B, 448 mi from A, at a speed 15.8 mi/h greater than on the first leg of the trip. The total flying time was 5.2 h. Find the speed at which the plane traveled to city A.

18. An express bus travels a certain 250-mi route in 1.0 h less time than it takes a local bus to travel a 240-mi route. Find the speed of each bus if the speed of the local is 10 mi/h less than the express.

19. A trucker calculates that if he increases his average speed by 20 km/h he could travel his 800-km route in 2.0 h less time than usual. Find his usual speed.

20. A boat sails 30 km at a uniform rate. If the rate had been 1 km/h more, the time of the sailing would have been 1 h less. Find the rate of travel.

Work Problems

21. A certain punch press requires 3 h longer to stamp a box of parts than does a newer-model punch press. After the older press has been punching a box of parts for 5 h, it is joined by the newer machine. Together, they finish the box of parts in 3 additional hours. How long does it take each machine, working alone, to punch a box of parts?

22. Two water pipes together can fill a certain tank in 8.4 h. The smaller pipe alone takes 2.5 h longer than the larger pipe to fill that same tank. How long would it take the larger pipe alone to fill the tank?

23. A laborer built 35 m of stone wall. If he had built 2 m less each day, it would have taken him 2 days longer. How many meters did he build each day, working at his usual rate?

24. A woman worked a certain number of days, receiving for her pay $180. If she had received $1 per day less than she did, she would have had to work 3 days longer to earn the same sum. How many days did she work?

Simply Supported Beam

25. For a simply supported beam of length l having a distributed load of w lb/ft (Fig. 11-6), the bending moment M at any distance x from one end is given by

$$M = \tfrac{1}{2}wlx - \tfrac{1}{2}wx^2$$

Find the locations on the beam where the bending moment is zero.

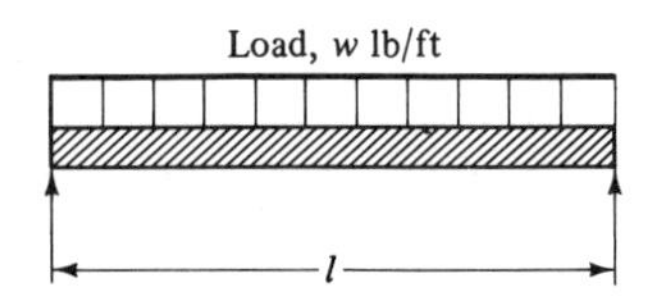

FIGURE 11-6. Simply supported beam with a uniformly distributed load.

26. A simply supported beam, 25 ft long, carries a distributed load of 1550 lb/ft. At what distance from an end of the beam will the bending moment be 112,000 ft lb?

Freely Falling Body

Use Eq. A17 for these falling-body problems, but be careful of the signs. If you take the upward *direction as positive, g will be* negative.

27. An object is thrown upward with a velocity of 145 ft/s. When will it be 85 ft above its initial position?

28. An object is thrown upward with an initial speed of 120 m/s. Find the time for it to return to its starting point. (The acceleration due to gravity, g, in the metric system, is $g = 980\ \text{cm/s}^2$.)

Electrical Problems

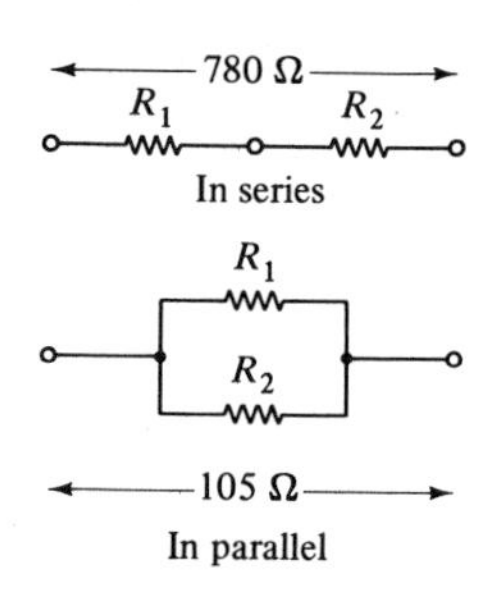

FIGURE 11-7

29. What two resistances (Fig. 11-7) will give a total resistance of 780 Ω when wired in series and 105 Ω when wired in parallel? (See Eqs. A41 and A42.)

30. Find two resistances that will give an equivalent resistance of 9070 Ω in series and 1070 Ω in parallel.

31. In the circuit of Fig. 11-8 the power P dissipated in the load resistor R_1 is

$$P = EI - I^2R$$

If the voltage E is 115 V, and $R = 100\ \Omega$, find the current I needed to produce a power of 29.3 W in the load.

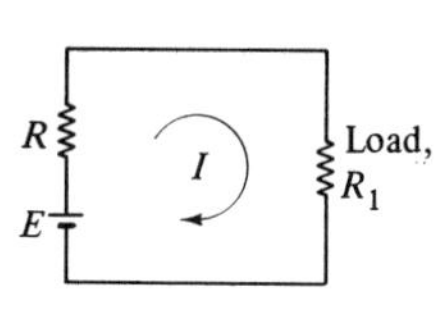

FIGURE 11-8

32. The reactance X of a capacitance C and an inductance L in series (Fig. 11-9) is

$$X = \omega L - \frac{1}{\omega C}$$

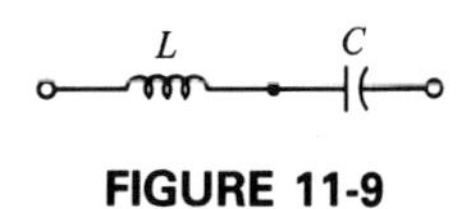

FIGURE 11-9

where ω is the angular frequency in rad/s, and L is in henries, and C is in farads. Find the angular frequency needed to make the reactance equal 1500 Ω, if $L = 0.5$ henry, and $C = 0.2 \times 10^{-6}$ farad (0.2 microfarad).

33. Figure 11-10 shows two currents flowing in a single resistor R. The total current in the resistor will be $I_1 + I_2$, so the power dissipated is

$$P = (I_1 + I_2)^2R$$

If $R = 100\ \Omega$ and $i_2 = 0.2$ A, find the current I_1 needed to produce a power of 9.0 W.

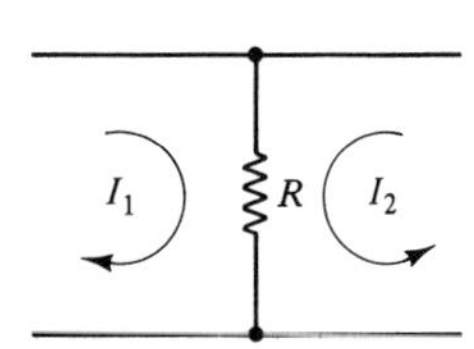

FIGURE 11-10

34. A *square-law device* is one whose output is proportional to the square of the input. A junction field-effect transistor (JFET) (Fig. 11-11) is such a device. The current I that will flow through an n-channel JFET when a voltage V is applied is

$$I = A\left(1 - \frac{V}{B}\right)^2$$

where A is the drain saturation current and B is the gate source pinch-off voltage. Solve this equation for V.

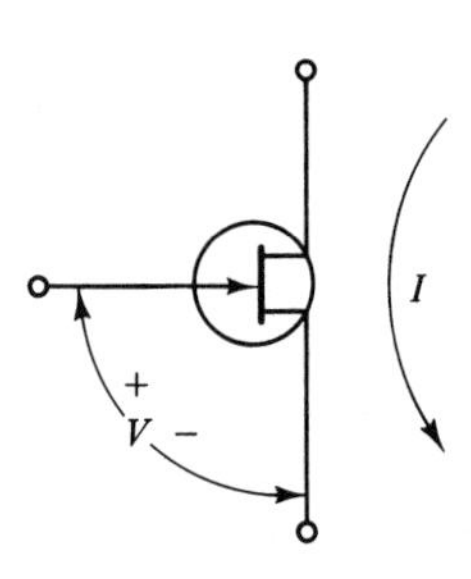

FIGURE 11-11. *N*-channel JFET.

35. A certain JFET has a drain saturation current of 4.8 mA and a gate source pinch-off voltage of −2.5 V. What input voltage is needed to produce a current of 1.5 mA?

11-6. GRAPHING THE QUADRATIC FUNCTION

Quadratic Function: A *quadratic function* is one whose highest-degree term is second degree.

Example: The functions

$$f(x) = 5x^2 - 3x + 2 \qquad f(x) = 9 - 3x^2$$

$$f(x) = x(x + 7) \qquad f(x) = x - 4 - 3x^2$$

are quadratic functions.

The Parabola: When we plot a quadratic function, we get the well-known and extremely useful curve called the *parabola*.

The parabola is one of the four conic sections *(the curves obtained when a cone is intersected by a plane, at various angles), and it has many interesting properties and applications which you will learn when you study analytic geometry.*

Example: Plot the quadratic function $y = x^2 + x - 3$.

Solution: As shown in Sec. 6-4, we compute a table of point pairs:

x	-3	-2	-1	0	1	2
y	3	-1	-3	-3	-1	3

We plot these points (Fig. 11-12) and connect them with a smooth curve, a parabola.

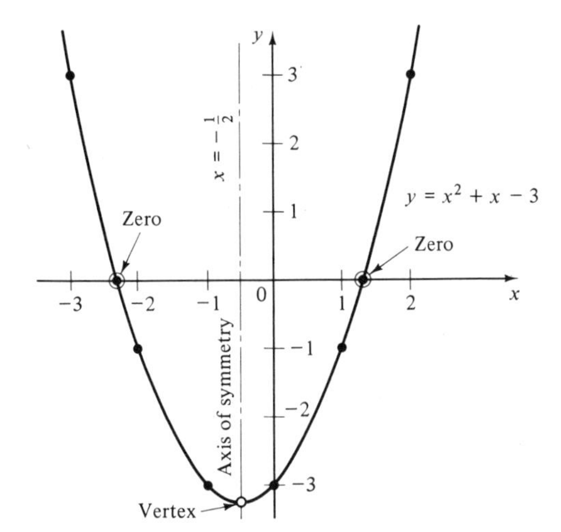

FIGURE 11-12. A parabola.

Note that the curve in this example is symmetrical about the line $x = -\frac{1}{2}$. This line is called the *axis of symmetry*. The point where the parabola crosses the axis of symmetry is called the *vertex*.

EXERCISE 6

Graphing a Quadratic

Plot the following quadratic functions. Find (a) the coordinates of the vertex; (b) the equation of the axis of symmetry; and (c) the approximate values of the zeros.

1. $y = x^2 + 3x - 15$
2. $y = 5x^2 + 14x - 3$
3. $y = 8x^2 - 6x + 1$
4. $y = 2.74x^2 - 3.12x + 5.38$

Computer

5. Use your program for generating a table of point pairs from Chapter 6, Exercise 4, to obtain point pairs for any of the equations in Problems 1 to 4. Plot these points to obtain a graph of the equation.

Parabolic Arch

6. The parabolic arch is often used in construction because of its great strength. The equation of the bridge arch in Fig. 11-13 is $y = 0.0625x^2 - 5x + 100$. Find the distances a, b, c, and d.

7. Plot the parabolic arch of Problem 6, taking values of x every 10 ft.

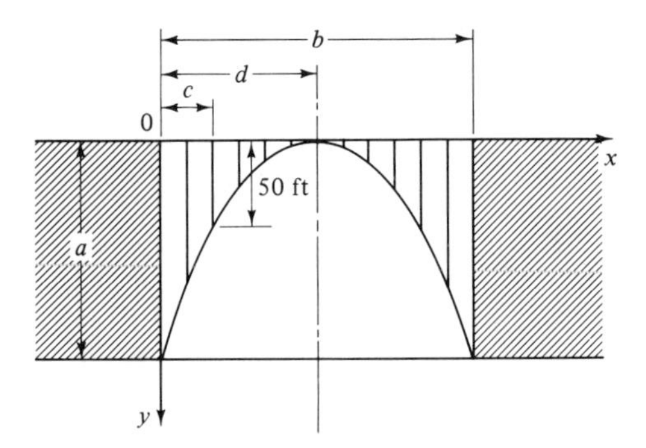

FIGURE 11-13. A parabolic bridge arch. Note that y axis is positive downward.

The parabola has the property that all the rays which are parallel to the axis of symmetry, after being reflected from the parabola, will converge to a single point called the focus. *This property is used in optical and radio telescopes, solar collectors, searchlights, microwave and radar antennas, and other devices.*

Parabolic Reflector

8. The parabolic solar collector in Fig. 11-14 has the equation $x^2 = 125y$. Plot the curve, taking values of x every 10 cm.

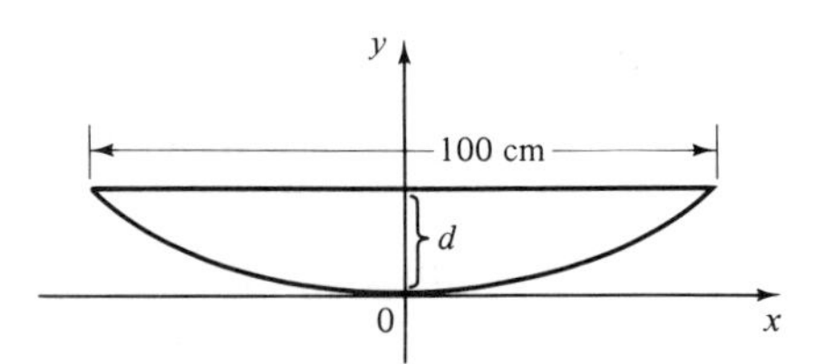

FIGURE 11-14

9. Find the depth d of the collector of Problem 8 at its center.

Vertical Highway Curves

Where there is a change in the slope of a highway surface, such as at the top of a hill or the bottom of a dip, the roadway is often made parabolic in shape to provide a smooth transition between the two different grades.

To construct a vertical highway curve (Fig. 11-15) we make use of the following property of the parabola:

> *The offset C from a tangent to a parabola is proportional to the square of the distance x from the point of tangency:*

or $C = kx^2$, where k is a constant of proportionality, found from a known offset.

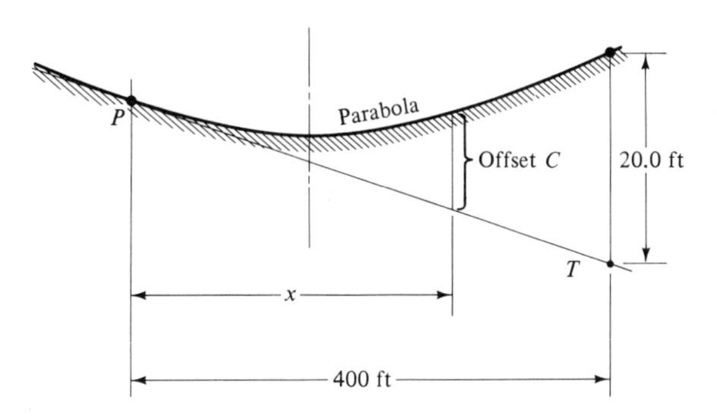

FIGURE 11-15. A dip in a highway.

10. If the offset is 20.0 ft at a point 400 ft from the point of tangency, find the constant of proportionality k.

11. For the highway curve of Problem 10, find the offset at a point 225 ft from the point of tangency.

Simply Supported Beam

12. The bending moment for a simply supported beam carrying a distributed load w (Fig. 11-6) is

$$M = \tfrac{1}{2}wlx - \tfrac{1}{2}wx^2$$

as in Exercise 5, Problem 25. The bending moment curve is therefore a parabola. Plot the curve of M vs. x for a 10-ft-long beam carrying a load of 1000 lb/ft. Take 1-ft intervals along the beam. Graphically locate (a) the points of zero bending moment; (b) the point of maximum bending moment.

Square-Law Devices

13. For the n-channel JFET of Exercise 5, Problem 35, plot the curve of current I vs. applied voltage V. Take the vertical axis for current and the horizontal axis for voltage. Compute I for values of V from -2.5 V to 0 V. Graphically determine the voltage needed to produce a current of 1.0 mA. (The curve obtained is a parabola and is called the *transconductance curve*.)

11-7. EQUATIONS OF QUADRATIC TYPE

Being able to solve equations of quadratic type is not that important (they are rare in practical situations), but being able to recognize form is important. By doing some of the following problems you will be exercising some useful mathematical muscles.

Recognition of Form: The general quadratic, Eq. 99, is

$$a(x)^2 + b(x) + c = 0$$

The form of this equation obviously does not change if we put a different symbol in the parentheses, as long as we put the same one in both places. The equations

$$a(Q)^2 + b(Q) + c = 0$$

and

$$a(\$)^2 + b(\$) + c = 0$$

are still quadratics. The form still does not change if we put more complicated expressions in the parentheses, such as

$$a(x^2)^2 + b(x^2) + c = 0$$

or

$$a(\sqrt{x})^2 + b(\sqrt{x}) + c = 0$$

or

$$a(x^{-1/3})^2 + b(x^{-1/3}) + c = 0$$

The preceding three equations can no longer be called quadratics, however, because they are no longer of second degree. They are called *equations of quadratic type*. They can be easily recognized because the power of x in one term is always twice the power of x in another term.

Solving Equations of Quadratic Type: The solution of these equations is best shown by examples.

Example: Solve

$$x^4 - 5x^2 + 4 = 0$$

Solution: We see that the power of x in one term (4) is twice the power of x in another term (2). Make the substitution; let $w = x^2$. The original equation then becomes the quadratic,

$$w^2 - 5w + 4 = 0$$

Factoring,

$$(w - 1)(w - 4) = 0$$

So

$$w = 1 \quad \text{and} \quad w = 4$$

but

$$w = x^2 = 1 \quad \text{and} \quad w = x^2 = 4$$

So

$$x = \pm 1 \quad \text{and} \quad x = \pm 2$$

Example: Solve the equation

$$2x^{-1/3} + x^{-2/3} + 1 = 0$$

Solution: Inspecting the powers of x, we see that one of them, $-\frac{2}{3}$, is twice the other, $-\frac{1}{3}$. We make the substitution,

$$\text{let } w = x^{-1/3}$$

and our equation becomes

$$2w + w^2 + 1 = 0$$

or

$$w^2 + 2w + 1 = 0$$

which factors into

$$(w + 1)(w + 1) = 0$$

Setting each factor equal to zero yields

$$w + 1 = 0 \quad \text{and} \quad w + 1 = 0$$

So

$$w = -1 = x^{-1/3}$$

Cubing both sides we get

$$x^{-1} = (-1)^3 = -1$$

$$x^{-1} = \frac{1}{x} = -1$$

So

$$x = \frac{1}{-1} = -1$$

EXERCISE 7

Equations of Quadratic Type

Solve for all values of x.

1. $x^6 - 6x^3 + 8 = 0$

2. $x^4 - 10x^2 + 9 = 0$

3. $x^{-1} - 2x^{-1/2} + 1 = 0$

4. $9x^4 - 6x^2 + 1 = 0$

5. $x^{-2/3} - x^{-1/3} = 0$

6. $x^{-6} + 19x^{-3} = 216$

7. $x^{-10/3} + 244x^{-5/3} + 243 = 0$

8. $x^6 - 7x^3 = 8$

9. $x^{2/3} + 3x^{1/3} = 4$

10. $64 + 63x^{-3/2} - x^{-3} = 0$

11. $81x^{-3/4} - 308 - 64x^{3/4} = 0$

12. $27x^6 + 46x^3 = 16$

13. $x^{1/2} - x^{1/4} = 20$

14. $3x^{-2} + 14x^{-1} = 5$

15. $2.35x^{1/2} - 1.82x^{1/4} = 43.8$

16. $1.72x^{-2} + 4.75x^{-1} = 2.24$

Computer

17. Use your program for finding roots of equations from Sec. 6-4 to solve any of the equations in this exercise.

11-8. SIMPLE EQUATIONS OF HIGHER DEGREE

Polynomial of Degree *n*: In Sec. 11-2 we used factoring to find the roots of a quadratic equation. We now expand this idea to include polynomials of higher degree. A polynomial, remember, has exponents that are all positive integers.

Polynomial of Degree n	$a_0x^n + a_1x^{n-1} + \cdots + a_{n-1}x + a_n$	**102**

The Factor Theorem: In Sec. 11-2 the expression

$$x^2 - x - 6 = 0$$

for example, factored into

$$(x - 3)(x + 2) = 0$$

and hence had the roots

$$x = 3 \quad \text{and} \quad x = -2$$

In a similar way, the cubic equation

$$x^3 - 7x - 6 = 0$$

has the factors

$$(x + 1)(x - 3)(x + 2) = 0$$

and hence has the roots

$$x = -1 \quad x = 3 \quad x = -2$$

In general,

Factor Theorem	If a polynomial equation $f(x) = 0$ has a root r, then $(x - r)$ is a factor of the polynomial $f(x)$. Conversely, if $(x - r)$ is a factor of a polynomial $f(x)$, then r is a root of $f(x) = 0$.	**103**

Number of Roots: A *polynomial equation of degree n has n and only n roots*. Two or more of the roots may be equal, however, and some may be nonreal, giving the appearance of fewer than n roots; but there can never be *more* than n roots.

A graph of the function is very helpful in finding the roots.

Solving Polynomials: If the expression cannot be factored directly, we proceed largely by trial and error to obtain one root. Once a root r is obtained (and verified by substituting in the equation), we obtain a *reduced* or *depressed* equation by dividing the original equation by $(x - r)$. The process is then repeated.

Example: One root of the cubic equation $16x^2 - 13x + 3 = 0$ is $x = -1$. Find the other roots.

Solution: For a cubic, we expect a maximum of three roots. By the factor theorem, since $x = -1$ is a root, then $(x + 1)$ must be a factor of the given equation.

$$(x + 1)(\text{other factor}) = 0$$

We obtain the "other factor" by dividing the given equation by $(x + 1)$. Using synthetic division,

Synthetic division is explained in Sec. 4-5. Of course, "ordinary" division (also Sec. 4-5) would work as well.

$$\begin{array}{r|rrrr} -1 & 16 & 0 & -13 & 3 \\ & & -16 & 16 & -3 \\ \hline & 16 & -16 & 3 & 0 \end{array}$$

The "other factor," or reduced equation, is then

$$16x^2 - 16x + 3 = 0, \text{ a quadratic.}$$

By the quadratic formula,

$$x = \frac{16 \pm \sqrt{16^2 - 4(16)(3)}}{2(16)} = \frac{1}{4} \quad \text{and} \quad \frac{3}{4}$$

Our three roots are therefore $x = -1$, $x = \frac{1}{4}$, and $x = \frac{3}{4}$.

This is also known as the method of Picard, *after Charles Émile Picard, 1856–1941.*

Simple Iteration—A Computer Method: We already have the midpoint method for computer solution of equations, and now we learn the method of *simple iteration.*

With this method, we split the given equation into two or more terms, one of which is the unknown itself, and transpose the unknown to the left side of the equation. We then guess the value of x and substitute it into the right side of the equation to obtain another value of x, which is then substituted into the equation, and so on, until our values converge on the correct value of x.

Example: Find one root of the equation

$$16x^3 - 13x + 3 = 0$$

by simple iteration.

Solution: The equation contains an x in both the first and second terms. Either can be chosen to isolate on the left side of the equation. Let us choose the first term. Transposing,

$$16x^3 = 13x - 3$$

Dividing gives

$$x^3 = 0.8125x - 0.1875$$

if we work to four significant digits. Taking the cube root of both sides, we get

$$x = (0.8125x - 0.1875)^{1/3} \qquad (1)$$

Note that we have not *solved* the equation, because there is still an x on the right side of the equation.

For a first approximation, let $x = 1$. Substituting into the right of Eq. (1) gives

$$x = (0.8125 - 0.1875)^{1/3}$$
$$= 0.8550$$

This, of course, is not a root, but only a *second approximation* to a root. We now take the value 0.8550 and substitute it into the right side of equation (1).

$$x = [0.8125(0.8550) - 0.1875]^{1/3}$$
$$= 0.7975$$

This third approximation is substituted into (1) giving

$$x = [0.8125(0.7975) - 0.1875]^{1/3}$$
$$= 0.7722$$

We continue, getting the following table of values.

0.8550
0.7975
0.7722
0.7605
0.7550
0.7524
0.7512
0.7506
0.7503

Notice that we are converging on the value

$$x = 0.750$$

which is one of the roots found in the preceding example. With a different first guess, we can usually make the computation converge on one of the other roots.

If the computation diverges instead of converges, we split the equation in a different way and try again. In the example here, we would isolate the x in the middle term, rather than the one in the first term.

EXERCISE 8

Equations of Higher Degree

Find the remaining roots of each polynomial, given one root.

1. $x^3 + x^2 - 17x + 15 = 0, \quad x = 1$

2. $x^3 + 9x^2 + 2x - 48 = 0, \quad x = 2$

3. $x^3 - 2x^2 - x + 2 = 0, \quad x = 1$

4. $x^3 + 9x^2 + 26x + 24 = 0, \quad x = -2$

5. $2x^3 + 7x^2 - 7x - 12 = 0, \quad x = -1$

6. $3x^3 - 7x^2 + 4 = 0, \quad x = 1$

7. $9x^3 + 21x^2 - 20x - 32 = 0, \quad x = -1$

8. $16x^4 - 32x^3 - 13x^2 + 29x - 6 = 0, \quad x = -1$

Computer

9. Write a program for solving equations by simple iteration. Use it to find the roots of any of the equations in this exercise.

10. Use your program for finding roots by the midpoint method for the equations in this exercise.

11-9. SYSTEMS OF QUADRATIC EQUATIONS

In Chapter 8 we solved systems of *linear* equations. We now use the same techniques, substitution or addition-subtraction, to solve systems where one or both equations are of second degree.

One Equation Linear and One Quadratic: These are best done by substitution.

Example: Solve for x and y:

$$x + y = 5$$
$$2x^2 + y^2 = 17$$

Solution: From the first equation,

$$x = 5 - y$$

Squaring,

$$x^2 = 25 - 10y + y^2$$

Substituting into the second equation

$$2(25 - 10y + y^2) + y^2 = 17$$
$$50 - 20y + 2y^2 + y^2 = 17$$

we get the quadratic

$$3y^2 - 20y + 33 = 0$$

Factoring,

$$(3y - 11)(y - 3) = 0$$

$$y = \frac{11}{3} \quad \text{and} \quad y = 3$$

Substituting into the first equation,

$$x = \frac{4}{3} \quad \text{and} \quad x = 2$$

So the solutions are

$$x = \frac{4}{3}, y = \frac{11}{3} \qquad \text{and} \qquad x = 2, y = 3$$

Example: Solve for x and y.

$$x + y = 2$$
$$xy + 15 = 0$$

Solution: From the first equation,

$$y = 2 - x$$

Substituting into the second equation,

$$x(2 - x) + 15 = 0$$

from which

$$x^2 - 2x - 15 = 0$$

Factoring,

$$(x - 5)(x + 3) = 0$$
$$x = 5 \qquad \text{and} \qquad x = -3$$

Substituting back,

$$y = 2 - 5 = -3$$

and

$$y = 2 - (-3) = 5$$

so the solutions are

$$x = 5, y = -3 \qquad \text{and} \qquad x = -3, y = 5$$

Both Equations Quadratic: These may be solved either by substitution, or addition–subtraction.

Example: Solve for x and y:

$$3x^2 - 4y = 47$$
$$7x^2 + 6y = 33$$

Solution: We multiply the first equation by 3 and the second by 2:

$$9x^2 - 12y = 141$$
$$14x^2 + 12y = 66$$

Adding,

$$23x^2 = 207$$
$$x^2 = 9$$
$$x = \pm 3$$

Substituting back, we get $y = -5$ for both $x = 3$ and $x = -3$. A plot of the given curves (Fig. 11-16) clearly shows the points of intersection $(3, -5)$ and $(-3, -5)$.

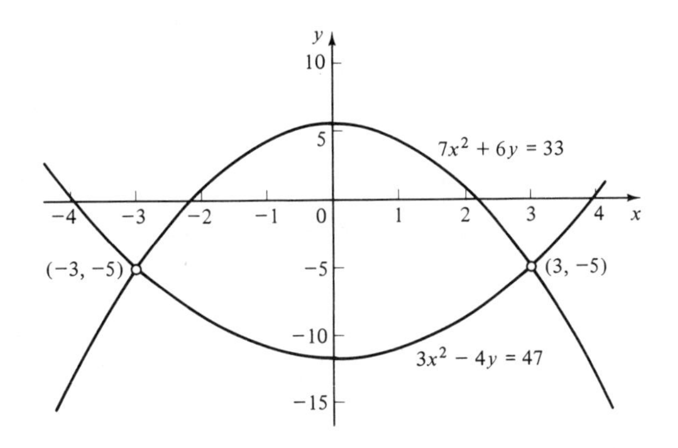

FIGURE 11-16. Intersection of two parabolas. This is a good way to get an approximate solution to any pair of equations that can be readily graphed.

Example: Solve for x and y:

$$3x^2 + 4y^2 = 76$$
$$-11x^2 + 3y^2 = 4$$

Solution: We multiply the first equation by 3 and the second by -4.

$$9x^2 + 12y^2 = 228$$
$$44x^2 - 12y^2 = -16$$

Adding,

$$53x^2 = 212$$
$$x^2 = 4$$
$$x = \pm 2$$

From the first equation, when x is either $+2$ or -2,

$$4y^2 = 76 - 12 = 64$$
$$y = \pm 4$$

So our solutions are

$$x = 2 \quad x = 2 \quad x = -2 \quad x = -2$$
$$y = 4 \quad y = -4 \quad y = 4 \quad y = -4$$

EXERCISE 9

Systems of Quadratic Equations

Solve for x and y.

1. $x - y = 4$, $x^2 - y^2 = 32$

2. $x + y = 5$, $x^2 + y^2 = 13$

3. $x + 2y = 7$, $2x^2 - y^2 = 14$

4. $x - y = 15$, $x = 2y^2$

5. $x + 4y = 14$, $y^2 - 2y + 4x = 11$

6. $x^2 + y = 3$, $x^2 - y = 4$

7. $2x^2 + 3y = 10$, $x^2 - y = 8$

8. $2x^2 - 3y = 15$, $5x^2 + 2y = 12$

9. $x + 3y^2 = 7$, $2x - y^2 = 9$

10. $x^2 - y^2 = 3$, $x^2 + 3y^2 = 7$

11. $x - y = 11$, $xy + 28 = 0$

CHAPTER TEST

Solve each equation

1. $y^2 - 5y - 6 = 0$

2. $2x^2 - 5x = 2$

3. $x^4 - 13x^2 + 36 = 0$

4. $(a + b)x^2 + 2(a + b)x = -a$

5. $w^2 - 5w = 0$

6. $x(x - 2) = 2(-2 - x + x^2)$

7. $\frac{r}{3} = \frac{r}{r + 5}$

8. $6y^2 + y - 2 = 0$

9. $3t^2 - 10 = 13t$

10. $2.73x^2 + 1.47x - 5.72 = 0$

11. $\frac{1}{t^2} + 2 = \frac{3}{t}$

12. $3w^2 + 2w + 11 = 0$

13. $9 - x^2 = 0$

14. $\frac{2y}{3} = \frac{3}{5} + \frac{2}{y}$

15. $2x(x + 2) = x(x + 3) + 5$

16. $\frac{ax^2}{b} + \frac{bx}{c} + \frac{c}{a} = 0$

17. $y + 6 = 5y^{1/2}$

18. $18 + w^2 + 11w - 0$

19. $9y^2 + y = 5$

20. $\frac{5}{w} - \frac{4}{w^2} = \frac{3}{5}$

21. $\frac{z}{2} = \frac{5}{z}$

22. $3x - 6 = \frac{5x + 2}{4x}$

23. $2x^2 + 3x = 2$

24. $x^2 + Rx - R^2 = 0$

25. $27x = 3x^2$

26. $\frac{n}{n + 1} = \frac{2}{n + 3} - 1$

27. $z^{2/3} + 8 = 9z^{1/3}$

28. $6w^2 + 13w + 6 = 0$

29. $5y^2 = 125$

30. $x^2 - 12 = x$

31. $z^2 + 2z - 3 = 0$

32. $\frac{t + 2}{t} = \frac{4}{t}$

33. A person purchased some bags of insulation for \$1000. If she had purchased 5 more bags for the same sum, they would have cost 12 cents less per bag. How many did she buy?

34. The perimeter of a rectangular field is 184 ft and its area 1920 ft^2. Find its dimensions.

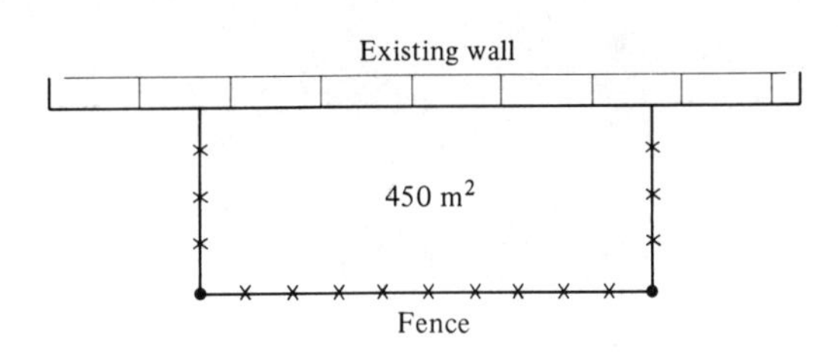

FIGURE 11-17

35. A rectangular yard (Fig. 11-17) is to be enclosed by fence on three sides, and an existing wall is to form the fourth side. The area of the yard is to be 450 m^2, and its length is to be twice its width. Find the dimensions of the yard.

36. Solve for x and y.

$$x - y^2 = 7$$
$$x^2 + 3y^2 = 7$$

37. If one root is $x = -1$, find the other roots of the equation

$$x^4 - 2x^3 - 13x^2 + 14x + 24 = 0$$

38. Plot the parabola $y = 3x^2 + 2x - 6$. Locate the vertex, the axis of symmetry, and any zeros.

39. Solve for x and y.

$$x^2 + y^2 = 55$$
$$x + y = 7$$

Right Triangles and Vectors

With this chapter we begin our study of trigonometry, the branch of mathematics that deals mostly with the solution of triangles. The trigonometric functions are introduced here and used to solve right triangles. Other kinds of triangles are discussed in Chapter 13.

We build on what we learned about triangles in Chapter 1, mainly the Pythagorean theorem and the fact that the sum of the angles of a triangle is 180°. We also make use of coordinate axes, described in Sec. 6-4.

Also introduced in this chapter are vectors, the study of which will be continued in Chapter 13.

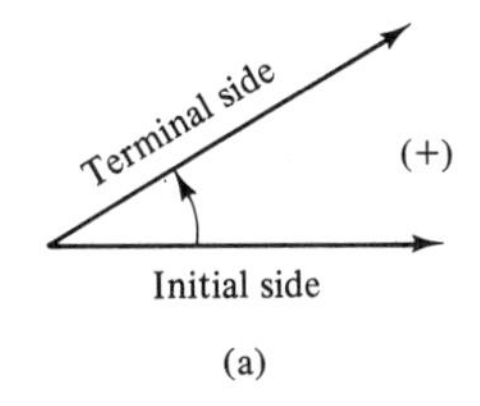

(a)

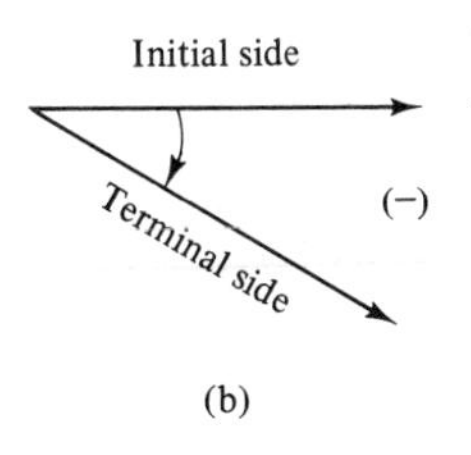

(b)

FIGURE 12-1 Angles formed by rotation.

12-1. ANGLES AND THEIR MEASURES

Angles Formed by Rotation: We can think of an angle as the figure generated by a ray rotating from some initial position (Fig. 12-1) to some terminal position. We usually consider angles formed by rotation in the *counterclockwise* direction as *positive,* and angles formed by *clockwise* rotation as *negative.* A rotation of one *revolution* brings the ray back to its initial position.

Degrees, Minutes, and Seconds: The degree (°) is a unit of angular measure equal to 1/360 of a revolution; thus, 360° = one revolution.

Fractional parts of a degree may be expressed as a common fraction (like $36\frac{1}{2}°$), as a decimal (28.74°) or as *minutes* and *seconds*.

A *minute* (′) is equal to 1/60 of a degree; a *second* (″) is equal to 1/60 of a minute, or 1/3600 of a degree.

Example: Some angles written in degrees, minutes, and seconds are

$$85°18'42'' \qquad 62°12' \qquad 69°55'25.4'' \qquad 75°06'03''$$

Radian Measure: A *central angle* in a circle is one whose vertex is at the center of the circle. An *arc* is a portion of the circle. If an arc is laid off along a circle with a length equal to the radius of the circle, the central angle subtended by this arc is defined as one *radian*.

Thus, an arc having a length of twice the radius will subtend a central angle of 2π radians. Similarly, an arc with a length of 2π times the radius (the circumference) will subtend a central angle of 2π radians. Thus our conversion factor is

$$2\pi \text{ radians} = 1 \text{ revolution} = 360°$$

Conversion of Angles: We convert angles in the same way as we convert any other units. A summary of the conversions needed is:

You may prefer to use π rad = 180°. Some calculators have special keys for angle conversions. Check your manual.

Angle Conversions	$1 \text{ rev} = 360° = 2\pi \text{ rad}$ $1° = 60'$ $1' = 60''$	**117**

Example: Convert 47.6° to radians and revolutions.

Solution: By Eq. 117,

$$47.6 \text{ deg}\left(\frac{2\pi \text{ rad}}{360 \text{ deg}}\right) = 0.831 \text{ rad}$$

Note that we keep the *same number of significant digits* as we had in the original angle. To convert to revolutions,

$$47.6 \text{ deg}\left(\frac{1 \text{ rev}}{360 \text{ deg}}\right) = 0.132 \text{ rev}$$

Example: Convert 1.8473 rad to degrees and revolutions.

Solution: By Eq. 117,

$$1.8473 \text{ rad}\left(\frac{360 \text{ deg}}{2\pi \text{ rad}}\right) = 105.84°$$

and

$$1.8473 \text{ rad}\left(\frac{1 \text{ rev}}{2\pi \text{ rad}}\right) = 0.29401 \text{ rev}$$

Conversions involving degrees, minutes, and seconds require several steps.

Example: Convert 28°17′37″ to decimal degrees and radians.

Solution: We separately convert the minutes and the seconds to degrees, and add them.

$$37 \text{ sec}\left(\frac{1 \text{ deg}}{3600 \text{ sec}}\right) = 0.0103°$$

When the angle is known to the nearest second, work to four decimal places, and vice versa.

$$17 \text{ min}\left(\frac{1 \text{ deg}}{60 \text{ min}}\right) = 0.2833°$$

$$28° = \underline{28.0000°}$$

Adding,

$$28.2936°$$

Now by Eq. 117,

$$28.2936°\left(\frac{2\pi \text{ rad}}{360 \text{ deg}}\right) = 0.493817 \text{ rad}$$

Example: Convert 1.837520 rad to degrees, minutes, and seconds.

Solution: We first convert to decimal degrees.

$$1.837520 \text{ rad}\left(\frac{360°}{2\pi \text{ rad}}\right) = 105.2821°$$

Converting the decimal part (0.2821°) to minutes,

$$0.2821°\left(\frac{60'}{1 \text{ deg}}\right) = 16.93'$$

Converting the decimal part of 16.93′ to seconds,

$$0.93'\left(\frac{60''}{1 \text{ min}}\right) = 56''$$

Another unit of angular measure is the grad. *There are 400 grads in one revolution.*

So

$$1.837520 \text{ rad} = 105°16'56''$$

EXERCISE 1

Angle Conversions

Convert to radians.

1. 27.8°	**2.** 38.7°	**3.** 35°15′
4. 270°27′25″	**5.** 0.55 rev	**6.** 0.276 rev

Convert to revolutions.

7. 4.772 rad	**8.** 2.38 rad	**9.** 68.8°
10. 3.72°	**11.** 77°18′	**12.** 135°27′42″

Convert to degrees (decimal).

13. 2.83 rad	**14.** 4.275 rad	**15.** 0.475 rev
16. 0.236 rev	**17.** 29°27′	**18.** 275°18′35″

Convert to degrees, minutes, and seconds.

19. 4.2754 rad	**20.** 1.773 rad	**21.** 0.44975 rev
22. 0.78426 rev	**23.** 185.972°	**24.** 128.259°

Computer

25. Write a program that will accept an angle in decimal degrees and compute and print the angle in degrees, minutes, and seconds.

12-2. THE TRIGONOMETRIC FUNCTIONS

We do only acute *angles here and discuss larger angles in Chapter 13.*

Angles in Standard Position: An angle is said to be in *standard position* when it is placed on coordinate axes, as in Fig. 12-2, with its vertex at the origin and with its initial side along the positive x axis.

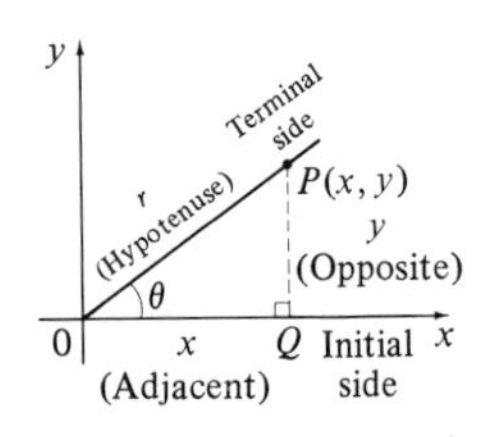

FIGURE 12-2. An angle in standard position.

Trigonometric Ratios: We now select any point P on the terminal side of the angle θ, and we let the coordinates of P be x and y. We form a *right triangle OPQ* by dropping a perpendicular from P to the x axis. The side OP is the *hypotenuse* of the right triangle; we label its length r. Also, side OQ is *adjacent* to angle θ and has a length x, and side PQ is *opposite* to angle θ and has a length y.

If we had chosen to place point P farther out on the terminal side of the angle, the lengths x, y, and r would all be greater than they are now. *But* the *ratio* of any two sides will be the same regardless of where we place P, because all the triangles thus formed are similar to each other. Such a ratio will change only if we change the angle θ. A ratio between any two sides of a right triangle is called a *trigonometric ratio*.

The six trigonometric ratios are defined as follows:

Trigonometric Ratios	sine θ = $\sin\theta = \dfrac{y}{r} = \dfrac{\text{opposite side}}{\text{hypotenuse}}$	**146**
	cosine θ = $\cos\theta = \dfrac{x}{r} = \dfrac{\text{adjacent side}}{\text{hypotenuse}}$	**147**
	tangent θ = $\tan\theta = \dfrac{y}{x} = \dfrac{\text{opposite side}}{\text{adjacent side}}$	**148**
	cotangent θ = $\cot\theta = \dfrac{x}{y} = \dfrac{\text{adjacent side}}{\text{opposite side}}$	**149**
	secant θ = $\sec\theta = \dfrac{r}{x} = \dfrac{\text{hypotenuse}}{\text{adjacent side}}$	**150**
	cosecant θ = $\csc\theta = \dfrac{r}{y} = \dfrac{\text{hypotenuse}}{\text{opposite side}}$	**151**

The Trigonometric Functions: Note that each of the trigonometric ratios is *related to* the angle θ. Take the ratio y/r for example. The values of y/r depend only on the measure of θ, not on the lengths of y or r. To convince yourself of this, draw an angle of, say, 30°, as in Fig. 12-3a. Then form a right triangle by drawing a perpendicular from any point on side *L or from any point on side M* (Fig. 12-3b and c). No matter how you draw the triangle, the side opposite the 30° angle *will always be half the hypotenuse* ($\sin 30° = \frac{1}{2}$). You would get a similar result for any other angle and for any other trigonometric ratio. *The value of the trigonometric ratio depends only on the measure of the angle*. We have a *relation,* such as the relations we studied in Sec. 6-2.

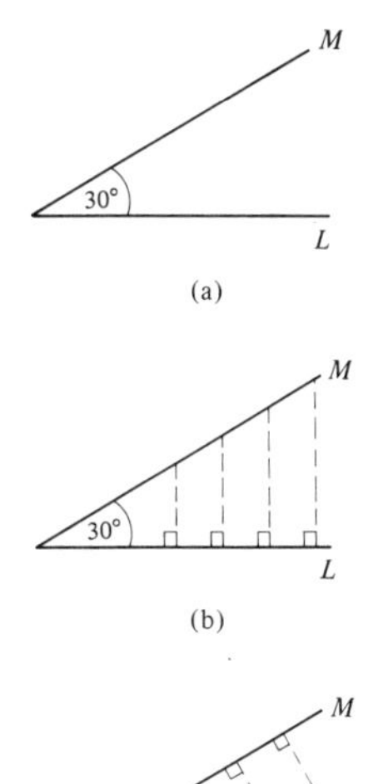

FIGURE 12-3

Furthermore, for any angle θ, there exists *only one* value for each trigonometric ratio. Thus, *each ratio is a function of* θ. We call them the *trigonometric functions* of θ. In functional notation, we could write

$$\sin\theta = f(\theta)$$
$$\cos\theta = g(\theta)$$
$$\tan\theta = h(\theta)$$

and so forth. The angle θ is called the *argument* of the function.

Note that while there is one and only one value of each trigonometric function for a given angle θ, that for any given value of a trigonometric function (say, $\sin\theta = \frac{1}{2}$), there are *indefinitely many* values of θ. We treat this idea more fully in Sec. 13-1.

The trigonometric functions are sometimes called *circular* functions.

Reciprocal Relationships: From the trigonometric functions we see that

$$\sin\theta = \frac{y}{r}$$

and that

$$\csc\theta = \frac{r}{y}$$

Obviously, $\sin\theta$ and $\csc\theta$ are reciprocals.

$$\sin\theta = \frac{1}{\csc\theta}$$

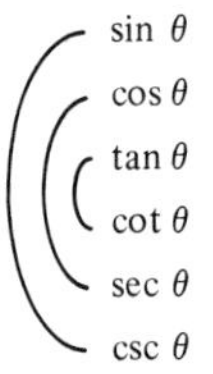

sin θ
cos θ
tan θ
cot θ
sec θ
csc θ

FIGURE 12-4. This diagram shows which functions are reciprocals of each other.

Inspection of the trigonometric functions shows two more sets of reciprocals (see also Fig. 12-4).

Reciprocal Relationships	$\sin\theta = \dfrac{1}{\csc\theta}$	**152a**
	$\cos\theta = \dfrac{1}{\sec\theta}$	**152b**
	$\tan\theta = \dfrac{1}{\cot\theta}$	**152c**

Example: Find $\tan\theta$ if $\cot\theta = 0.638$.

Solution: From the reciprocal relationships,

$$\tan\theta = \frac{1}{\cot\theta} = \frac{1}{0.638} = 1.57$$

This is a good place to use the [1/x] *key on your calculator.*

Finding the Trigonometric Functions of an Angle: The six trigonometric functions of an angle can be written if we know the angle or if we know two of the sides of a right triangle containing the angle.

Example: A right triangle (Fig. 12-5) has sides of 4.37 and 2.83. Write the six trigonometric functions of angle θ.

Solution: We first find the hypotenuse by the Pythagorean theorem,

$$c = \sqrt{(2.83)^2 + (4.37)^2} = \sqrt{27.1}$$
$$= 5.21$$

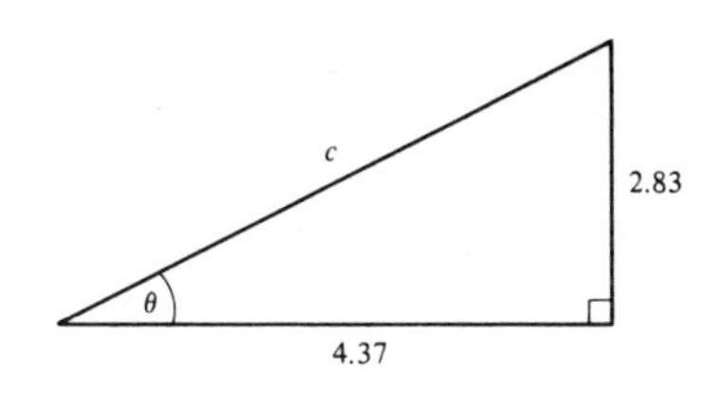

FIGURE 12-5

We also note that the opposite side is 2.83 and the adjacent side is 4.37. Then by Eqs. 146 to 151,

In writing these, do not omit the angle. To write "sin = 0.543" has no meaning.

$$\sin\theta = \frac{\text{opposite side}}{\text{hypotenuse}} = \frac{2.83}{5.21} = 0.543$$

$$\cos\theta = \frac{\text{adjacent side}}{\text{hypotenuse}} = \frac{4.37}{5.21} = 0.839$$

$$\tan\theta = \frac{\text{opposite side}}{\text{adjacent side}} = \frac{2.83}{4.37} = 0.648$$

$$\cot\theta = \frac{\text{adjacent side}}{\text{opposite side}} = \frac{4.37}{2.83} = 1.54$$

$$\sec\theta = \frac{\text{hypotenuse}}{\text{adjacent side}} = \frac{5.21}{4.37} = 1.19$$

$$\csc\theta = \frac{\text{hypotenuse}}{\text{opposite side}} = \frac{5.21}{2.83} = 1.84$$

Example: A point P on the terminal side of an angle B (in standard position) has the coordinates (8.84, 3.15). Write the sine, cosine, and tangent of angle B.

Solution: We are given $x = 8.84$ and $y = 3.15$. We find r from the Pythagorean theorem,

$$r = \sqrt{(8.84)^2 + (3.15)^2} = 9.38$$

Then by Eqs. 146 to 151,

$$\sin B = \frac{y}{r} = \frac{3.15}{9.38} = 0.336$$

$$\cos B = \frac{x}{r} = \frac{8.84}{9.38} = 0.942$$

$$\tan B = \frac{y}{x} = \frac{3.15}{8.84} = 0.356$$

Trigonometric Functions by Calculator: Simply enter the angle and press the proper key, [sin], [cos], or [tan]. Make sure that your calculator is in the "degree" mode if the entered angle is in degrees, or in the "radian" mode if the angle is in radians.

Example: Find the sine of 38.6° by calculator to four significant digits.

Solution: We set the mode switch to "Degrees." Then

The keystrokes are the same for algebraic and reverse Polish calculators.

Press	*Display*
38.6 [sin]	0.6238795967

which we round to 0.6239.

Example: Find tan 1.174 rad by calculator to five significant digits.

Solution: With the mode switch in the "radian" position:

Press	*Display*
1.174 [tan]	2.386509704

which we round to 2.3865.

To find the cotangent, secant, or cosecant, first find the reciprocal function, (tangent, cosine, or sine), and then press the [1/x] key to get the reciprocal.

Example: Find sec 72.8° to four significant digits.

Solution: Set the mode switch to "Degrees." Then

Press	*Display*
72.8 [cos]	0.29570805
[1/x]	3.381713822

which we round to 3.382.

Example: Find the cosecant of 0.748 rad to four significant digits.

Solution: With the mode switch in the "radian" position,

Press	*Display*
.748 [sin] [1/x]	1.470211991

which we round to 1.470.

Common Error	It is easy to forget to set the Degree/Radian mode switch to the proper position. Be sure to check it each time.

Trigonometric Functions on the Computer: Many computer languages such as BASIC have built-in commands for some, but not all, of the trigonometric functions.

Example: The instruction

```
50 PRINT SIN(X)
```

will cause the sine of the number X to be printed, *where* X *is assumed to be in radians.*

If X is in degrees, it must be converted to radians before taking the sine.

Example: The instruction

```
50 PRINT SIN(3.1416*X/180)
```

will print the sine of X degrees.

BASIC usually includes the functions

COS( ) and TAN( )

in addition to SIN( ). As on the calculator, the cotangent, secant, and cosecant are obtained from the reciprocal relations (Eq. 152).

Example: The instruction

```
80 PRINT 1/TAN(X)
```

will print the cotangent of X radians.

Instead of an [arc] *key, some calculators have an* [inv] *key or keys marked* [sin^{-1}], [cos^{-1}], *or* [tan^{-1}].

Finding the Angle When the Trigonometric Function Is Given: Here we reverse the previous operation of finding the trigonometric function. We make use of the [arc] key on the calculator.

Example: If $\sin \theta = 0.7337$, find θ in degrees to three significant digits.

Solution: Switch the calculator into the "degree" mode. Then

Press	*Display*
0.7337 [arc] [sin]	47.19748165

which we round to 47.2°.

Example: If $\tan x = 2.846$, find x in radians to four significant digits.

Solution: With the calculator in the "radian" mode:

Press	*Display*
2.846 [arc] [tan]	1.232901233

which we round to 1.233.

If the cotangent, secant, or cosecant is given, we first take the reciprocal and then use the appropriate reciprocal relationship.

Example: If $\sec A = 1.573$, find A in degrees to four significant digits.

Solution: Put the calculator into the "degree" mode. Then

Press	*Display*
1.573 [1/x]	0.6357279085
[arc] [cos]	50.53 (rounded)

Inverse Trigonometric Functions: The operation of finding the angle when the trigonometric function is given is the *inverse* of finding the function when the angle is given. There is special notation to indicate the inverse trigonometric function. If

$$A = \sin \theta$$

we write

$$\theta = \arcsin A$$

or

$$\theta = \sin^{-1} A$$

which is read "θ is the angle whose sine is A." Similarly, we use the symbols arccos A, $\cos^{-1} A$, arctan A, and so on.

Common Error	Do not confuse the **inverse** with the **reciprocal.** $\frac{1}{\sin \theta} = (\sin \theta)^{-1} \neq \sin^{-1} \theta$

Example: Evalute arctan 1.547 to four significant digits.

Solution: The calculator keystrokes are the same as in the preceding section.

Press	*Display*
(degree mode) 1.547 [arc] [tan]	57.12 (degrees)

In this chapter we limit our work with inverse functions to first-quadrant *angles only and consider any angle in Chapter 13.*

Inverse Trigonometric Functions on the Computer: BASIC has *only one* inverse trigonometric function, the arctangent.

Example: The instruction

```
60 LET A = ATN(B)
```

will set A equal to that angle, *in radians,* that has a tangent of B. Suppose that we know the sine C of some angle A, and want to find A. We know that

$$A = \arcsin C$$

But BASIC has no arcsine function. We can, however, use the arctangent to find the arcsine. We draw a right triangle (Fig. 12-6) with a hypotenuse of 1 and a side C opposite angle A. The sine of A is thus equal to C. By the Pythagorean theorem, the third side is equal to

1
C
A
$\sqrt{1 - C^2}$

FIGURE 12-6

$$\sqrt{1 - C^2}$$

so

$$\tan A = \frac{C}{\sqrt{1 - C^2}}$$

and hence

$$A = \arcsin C = \arctan \frac{C}{\sqrt{1 - C^2}} \qquad \textbf{176}$$

In BASIC,

```
A = ATN(C/SQR(1-C ^ 2))
```

Similarly, if the cosine of some angle is equal to D, then the angle can be found from

Try to derive this.

$$\arccos D = \arctan \frac{\sqrt{1 - D^2}}{D} \qquad \textbf{177}$$

EXERCISE 2

Trigonometric Ratios

Using a protractor, lay off the given angle on coordinate axes. Drop a perpendicular from any point on the terminal side and measure the distances x, y, and r, as in Fig. 12-2. Use these measured distances to write the six trigonometric ratios of the angle. Check your answer with calculator or tables.

1. 30° **2.** 55° **3.** 14° **4.** 37°

Plot the given point on coordinate axes and connect it to the origin. Measure the angle formed with a protractor. Check by taking the tangent of your measured angle. It should equal the ratio y/x.

5. (3, 4) **6.** (5, 3) **7.** (1, 2) **8.** (−3, 4)

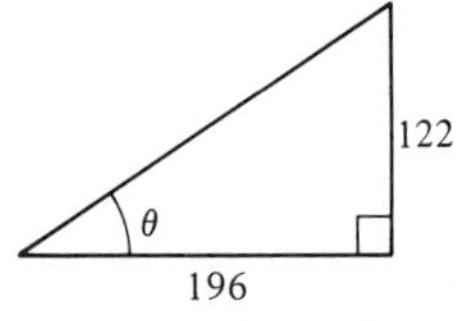

FIGURE 12-7

9. Write the six trigonometric ratios for angle θ in Fig. 12-7.

Each of the given points is on the terminal side of an angle. Compute the distance r from the origin to the point and write the sin, cos, and tan of the angle.

Work to three significant digits on these.

10. (3, 5) **11.** (4, 2) **12.** (5, 3)

13. (2, 3) **14.** (4, 5) **15.** (4.75, 2.68)

Evaluate.

16. sin 47.3° **17.** tan 0.775 rad **18.** sec 29.5°

19. cot 18.2° **20.** csc 87°45′12″ **21.** cos 21.27°

22. cos 86.75° **23.** sec 12°18′ **24.** csc 12.67°

Write the sin, cos, and tan for each angle.

25. 72.8° **26.** 19.2° **27.** 33.1°

28. 0.722 rad **29.** 42.6° **30.** 78°15′23″

Inverse Trigonometric Functions

Find the acute angle (in decimal degrees) whose trigonometric function is,

31. $\sin A = 0.5$ **32.** $\tan D = 1.53$ **33.** $\sin G = 0.528$

34. $\cot K = 1.774$ **35.** $\sin B = 0.483$ **36.** $\cot E = 0.847$

Without using tables or calculator, write the sin, cos, and tan of angle A. Leave your answer in fractional form.

37. $\sin A = \dfrac{3}{5}$ **38.** $\cot A = \dfrac{12}{5}$ **39.** $\cos A = \dfrac{12}{13}$

Evaluate the following, giving your answer in decimal degrees to three significant digits.

40. arcsin 0.635 **41.** arcsec 3.86 **42.** $\tan^{-1} 2.85$

43. $\cot^{-1} 1.17$ **44.** $\cos^{-1} 0.229$ **45.** arccsc 4.26

Computer

46. Tables of trigonometric functions are usually computed using series approximations, such as

$$\sin x = x - \frac{x^3}{3!} + \frac{x^5}{5!} - \frac{x^7}{7!} + \cdots \qquad \textbf{203}$$

where x is the angle *in radians,* and 3! (read 3 factorial) is

$$3! = 3 \cdot 2 \cdot 1 = 6$$

and

$$5! = 5 \cdot 4 \cdot 3 \cdot 2 \cdot 1 = 120$$

and so on. Write a program to compute and print the sines of the angles from 0 to 90°, taking values every 5° and using only the first three terms to the series. Compare your results with those obtained by calculator.

47. Write a program to evaluate the cosine of an angle x using the series expansion

$\cos x = 1 - \frac{x^2}{2!} + \frac{x^4}{4!} - \frac{x^6}{6!} + \cdots$	**204**

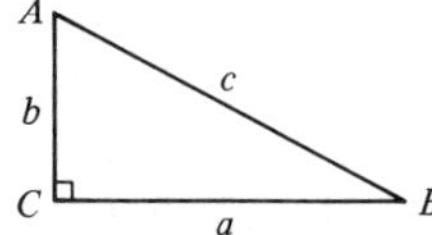

FIGURE 12-8. A right triangle. We will usually *label* a right triangle as shown here. We label the angles with capital letters *A, B,* and *C,* with *C* always the right angle. We label the sides with lowercase letters *a, b,* and *c,* with side *a* opposite angle *A,* side *b* opposite angle *B,* and side *c* (the hypotenuse) opposite angle *C* (the right angle).

12-3. SOLUTION OF RIGHT TRIANGLES

Right Triangles: The right triangle (Fig. 12-8) was introduced in Sec. 1-11 where we also had the formula

Pythagorean Theorem	$c^2 = a^2 + b^2$	**145**

Furthermore, since angle C is always 90°, Eq. 139 becomes

Sum of the Angles	$A + B = 90°$	

To these, we now add

Trigonometric Functions	$\sin \theta = \frac{\text{opposite side}}{\text{hypotenuse}}$	**146**
	$\cos \theta = \frac{\text{adjacent side}}{\text{hypotenuse}}$	**147**
	$\tan \theta = \frac{\text{opposite side}}{\text{adjacent side}}$	**148**

These five equations will be our tools for solving any right triangle.

Solving Right Triangles When One Side and One Angle Are Known: To *solve* a triangle means to find all missing sides and angles (although in most practical problems we need find only one missing side or angle). We can solve any right triangle if we know one side and either another side or one angle.

To solve a right triangle when one side and one angle are known:

1. Make a sketch.

2. Find the missing angle by using Eq. 139.
3. Relate the known side to one of the missing sides by one of the trigonometric ratios. Solve for the missing side.
4. Repeat step 3 to find the second missing side.
5. Check your work with the Pythagorean theorem.

Example: Solve right triangle ABC if $A = 18.6°$ and $c = 135$.

Solution: We make a sketch (Fig. 12-9). Then, by Eq. 139,

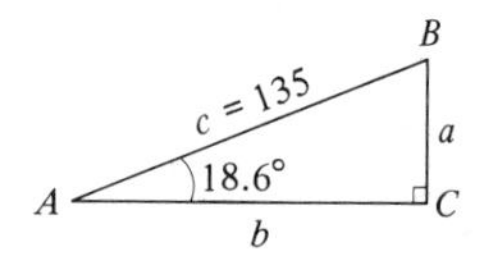

FIGURE 12-9

$$B = 90 - A = 90 - 18.6° = 71.4°$$

Let us now find side a. We must use one of the trigonometric ratios. *But how do we know which one to use*? And further, *which of the two angles, A or B, should we write the trig ratio for*?

It is simple. First, always work with the *given angle,* because if you made a mistake in finding angle B and then used it to find the sides, they would be wrong also. Then, to decide which trigonometric ratio to use, we note that side a is *opposite* to angle A and that the given side is the *hypotenuse*. Thus, our trig function must be one that relates the *opposite side* to the *hypotenuse*. Our obvious choice is Eq. 146.

Realize that either of the two legs can be called opposite *or* adjacent, *depending on which angle we are referring them to.*

$$\sin A = \frac{\text{opposite side}}{\text{hypotenuse}}$$

Substituting the given values, we obtain

$$\sin 18.6° = \frac{a}{135}$$

By calculator, $\sin 18.6° = 0.3190$, so

$$0.3190 = \frac{a}{135}$$

Solving for a,

$$a = 0.3190(135) = 43.1$$

We now find side b. Note that side b is *adjacent* to angle A. We therefore use Eq. 147,

$$\cos 18.6° = \frac{b}{135}$$

So

$$b = 135 \cos 18.6° = 128$$

We have thus found all the missing parts of the triangle. For a check, we see if the three sides will satisfy the Pythagorean theorem.

As an approximate check of any triangle, see if the longest side is opposite the largest angle and if the shortest side is opposite the smallest angle. Also check that the hypotenuse is greater than either leg but less than their sum.

Check:

$$(43.1)^2 + (128)^2 \stackrel{?}{=} (135)^2$$
$$18{,}242 \stackrel{?}{=} 18{,}225$$

Since we are working to three significant digits, this is close enough for a check.

Example: In right triangle ABC, angle $B = 55.2°$ and $a = 207$. Solve the triangle.

Solution: We make a sketch (Fig. 12-10). By Eq. 139,

$$A = 90 - 55.2° = 34.8°$$

FIGURE 12-10

By Eq. 147,

$$\cos 55.2° = \frac{207}{c}$$

$$c = \frac{207}{\cos 55.2°} = \frac{207}{0.5707} = 363$$

Then by Eq. 148,

$$\tan 55.2° = \frac{b}{207}$$

$$b = 207 \tan 55.2° = 207(1.439)$$
$$= 298$$

Checking with the Pythagorean theorem,

$$(363)^2 \stackrel{?}{=} (207)^2 + (298)^2$$

$$131{,}769 \stackrel{?}{=} 131{,}653 \quad \text{checks within three significant digits}$$

Tip	Whenever possible, use the *given information* for each computation, rather than some quantity previously calculated. This way, any errors in the early computation will not be carried along.

Solving Right Triangles When Two Sides Are Known

1. Draw a diagram of the triangle.
2. Write the trigonometric ratio that relates one of the angles to the two given sides. Solve for the angle.
3. Repeat step 2 for the other missing angle.
4. Find the missing side by the Pythagorean theorem.
5. Check that the two angles are complementary, and check the computed side with a trigonometric ratio, as in the following example.

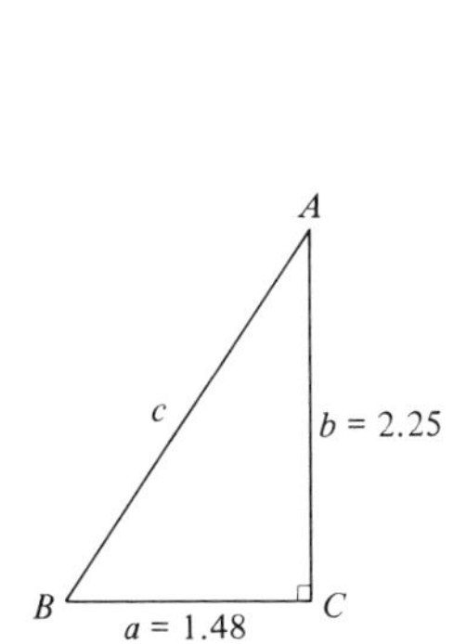

FIGURE 12-11

Example: Solve right triangle ABC if $a = 1.48$ and $b = 2.25$.

Solution: We sketch the triangle (Fig. 12-11). To find angle A, we note that the 1.48 side is *opposite* angle A and that the 2.25 side is *adjacent* to angle A. The trig ratio relating *opposite* and *adjacent* is the *tangent* (Eq. 148).

$$\tan A = \frac{1.48}{2.25} = 0.6578$$

Now using the [arc] or [inv] key on the calculator,

$$A = 33.3°$$

Solving now for angle B,

$$\tan B = \frac{2.25}{1.48} = 1.520$$

$$B = 56.7°$$

We find side c by the Pythagorean theorem,

$$c^2 = (1.48)^2 + (2.25)^2 = 7.253$$

$$c = 2.69$$

Check:

$$33.3° + 56.7° = 90° \qquad \text{checks}$$

$$\sin 33.3° \stackrel{?}{=} \frac{1.48}{2.69}$$

$$0.549 \cong 0.550 \qquad \text{checks within } 0.2\%$$

Cofunctions: The sine of angle A in Fig. 12-8 is

$$\sin A = \frac{a}{c}$$

But a/c is *also* the cosine of the complementary angle B. Therefore,

Similarly,

Cofunctions	$\sin A = \cos B$	**154a**
	$\cos A = \sin B$	**154b**
	$\tan A = \cot B$	**154c**
	$\cot A = \tan B$	**154d**
	$\sec A = \csc B$	**154e**
	$\csc A = \sec B$	**154f**

In general, *a trigonometric function of an acute angle is equal to the corresponding cofunction of the complementary angle.*

Example: If the cosine of the acute angle A in a right triangle ABC is equal to 0.725, find the sine of the other acute angle, B.

Solution: By Eq. 154b,

$$\sin B = \cos A = 0.725$$

Applications: There are, of course, a huge number of applications for the right triangle, a few of which are given in the following exercise. A typical application is that of finding a distance that cannot be measured directly.

Example: From a plane at an altitude of 2750 ft, the pilot observes the angle of depression of a lake to be 18.6°. How far is the lake from a point on the ground directly beneath the plane?

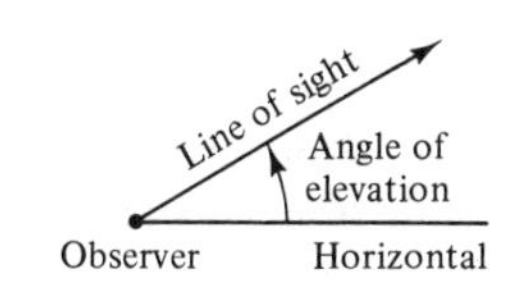

(a) Angle of elevation

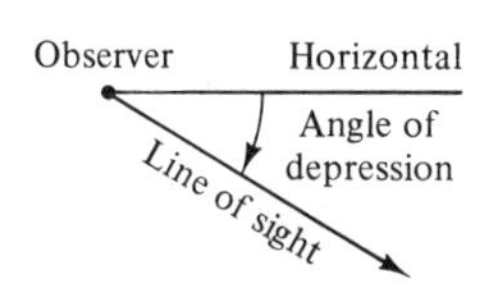

(b) Angle of depression

FIGURE 12-12. Angles of elevation and depression.

Solution: We first note that an *angle of depression,* as well as an *angle of elevation,* is defined as the angle between a line of sight and the *horizontal,* as shown in Fig. 12-12. We then make a sketch for our problem (Fig. 12-13). Then

$$\tan 18.6° = \frac{2750}{x}$$

$$x = \frac{2750}{\tan 18.6°} = 8171 \text{ ft}$$

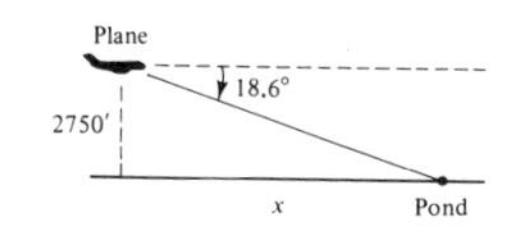

FIGURE 12-13

EXERCISE 3

Right Triangles with One Side and One Angle Given

Sketch each right triangle and find all the missing parts. Assume the triangles to be labeled as in Fig. 12-8. Work to three significant digits.

1. $a = 155 \quad A = 42.9°$

2. $b = 82.6 \quad B = 61.4°$

3. $a = 1.74 \quad B = 31.9°$

4. $b = 7.74 \quad A = 22.5°$

5. $a = 284 \quad A = 1.13$ rad

6. $b = 73.2 \quad B = 0.655$ rad

Right Triangles with Two Sides Given

Sketch each right triangle and find all missing parts. Work to three significant digits and express the angles in decimal degrees.

7. $a = 382 \quad b = 274$

8. $a = 3.88 \quad c = 5.37$

9. $b = 3.97 \quad c = 4.86$

10. $a = 63.9 \quad b = 84.3$

11. $a = 27.4 \quad c = 37.5$

12. $b = 746 \quad c = 957$

Cofunctions

Express as a function of the complementary angle.

13. sin 38°

14. cos 73°

15. tan 19°

16. sec 85.6°

17. cot 63.2°

18. csc 82.7°

19. tan 35°14′ **20.** $\cos \frac{\pi}{8}$ rad **21.** sin 0.875 rad

Measuring Inaccessible Distances

22. From a point on the ground 255 m from the base of a tower, the angle of elevation to the top of the tower is 57.6°. Find the height of the tower.

23. A pilot 4220 m directly above the end of a train observes that the angle of depression of the end of the train is 68.2°. Find the length of the train.

24. From the top of a lighthouse 156 ft above the surface of the water, the angle of depression of a boat is observed to be 28.7°. Find the horizontal distance from the boat to the lighthouse.

25. An observer in an airplane 1520 ft above the surface of the ocean observes that the angle of depression of a ship is 28.8°. Find the horizontal distance from the plane to the ship.

Structures

26. A guy wire from the top of an antenna is anchored 53.5 ft from the base of the antenna and makes an angle of 85.2° with the ground. Find (a) the height h of the antenna; (b) the length L of the wire.

27. Find the angle θ and length AB in the truss shown in Fig. 12-14.

28. A house (Fig. 12-15) is 8.75 m wide and its roof has an angle of inclination of 34.0°. Find the length L of the rafters.

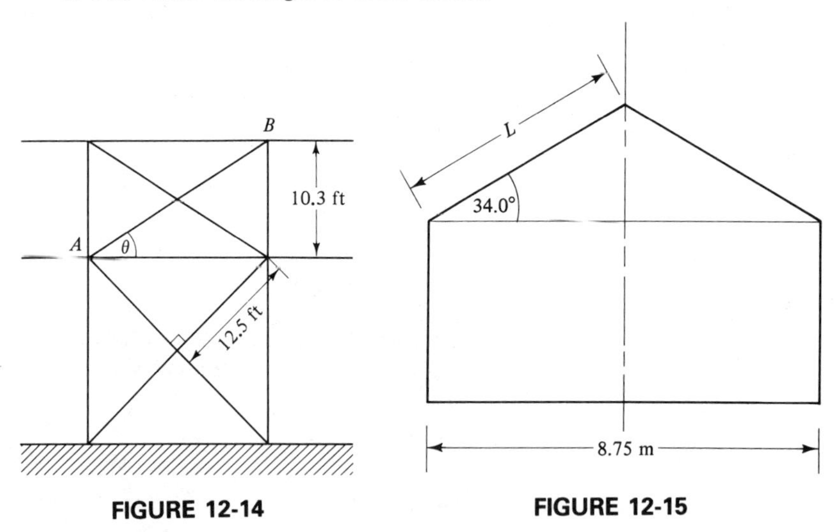

FIGURE 12-14 **FIGURE 12-15**

29. A guy wire 82 ft long is stretched from the ground to the top of a telephone pole 65 ft high. Find the angle between the wire and the pole.

Geometry

30. What is the angle between a diagonal of a cube and a diagonal of a face of the cube?

31. Find the angles between the diagonals of a rectangle whose dimensions are 580 units × 940 units.

32. Find the area of a parallelogram if the lengths of the sides are 255 units and 482 units and if one angle is 83.2°.

33. Find the length of a side of a regular hexagon inscribed in a 125-cm-radius circle.

Shop Trigonometry

34. The bolt circle of Fig. 12-16 is to be made on a jig borer. Find the dimensions x_1, y_1, x_2, and y_2.

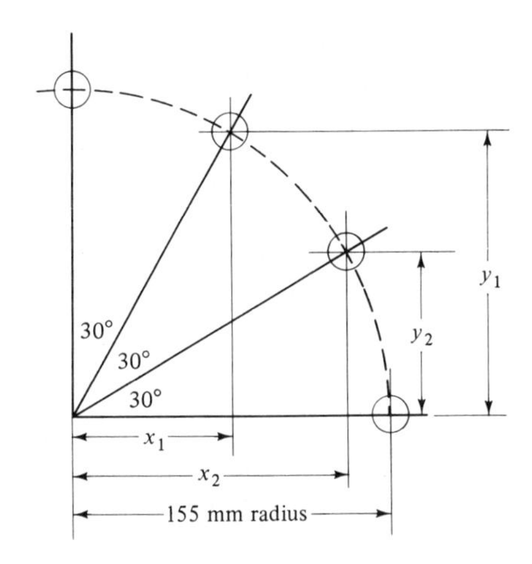

FIGURE 12-16. A bolt circle.

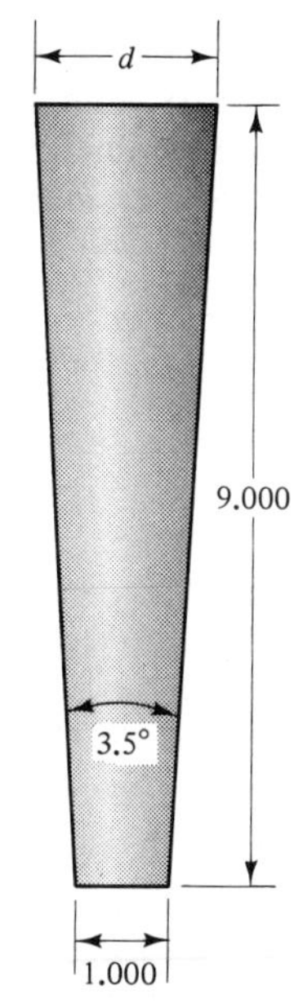

FIGURE 12-17. Tapered shaft.

35. A bolt circle with a radius of 36.000 cm contains 24 holes equally spaced. Find the straight-line distance between the holes so that they may be stepped off with dividers.

36. A 9.000-in.-long bar (Fig. 12-17) is to taper 3.5° and be 1.000 in. in diameter at the narrow end. Find the diameter d of the larger end.

37. To find a circle diameter when the center is inaccessible, a scale can be placed across the curve to be measured, as in Fig. 12-18 and the chord

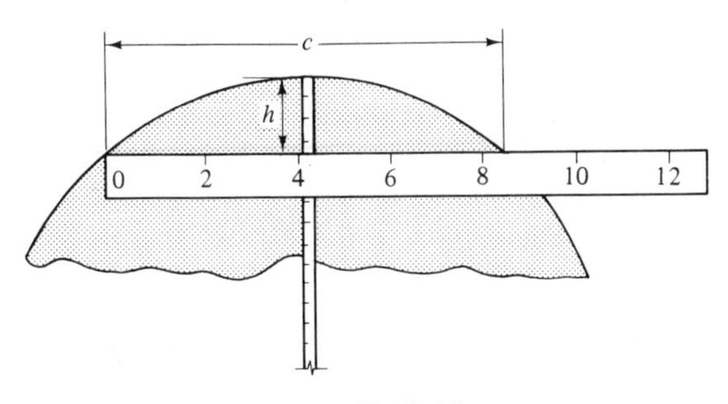

FIGURE 12-18

c and the perpendicular distance h measured. Find the radius of curvature if the chord length is 8.25 cm and h is 1.16 cm.

38. A common way of measuring dovetails is with the aid of round plugs, as in Fig. 12-19. Find the distance x over the 0.500-in.-diameter plugs.

39. A bolt head (Fig. 12-20) measures 0.750 cm across the flats. Find the distance p across the corners and the width r of each flat.

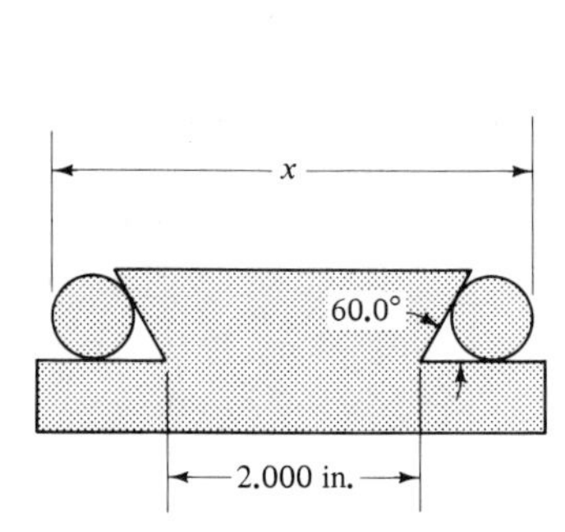

FIGURE 12-19. Measuring a dovetail.

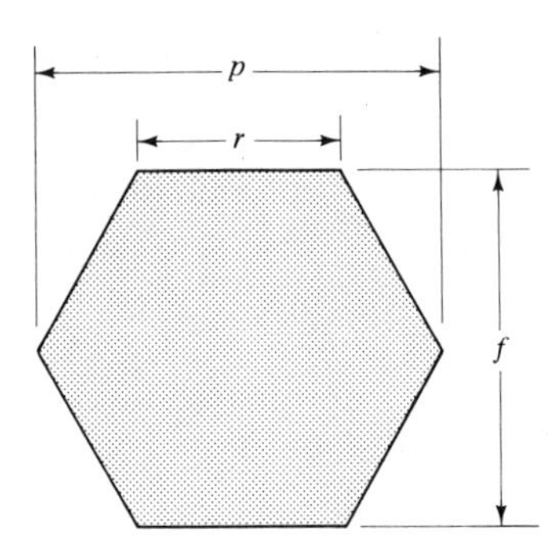

FIGURE 12-20

Computer

40. Write a program to compute the perimeter of a regular polygon of n sides inscribed in a circle having a diameter of one unit. Let n vary from 3 to 20, and for each n print (a) the number of sides, n; (b) the perimeter of the polygon; and (c) the difference between the circumference of the circle and the perimeter of the polygon.

41. Write a program to solve any right triangle. Have the program accept as input a known side and angle, or two known sides, and compute and print all the sides and angles.

12-4. VECTORS

Definition: A *vector quantity* is one that has *direction* as well as magnitude. For example, a velocity cannot be completely described without giving the direction of motion, as well as the speed. Other common vector quantities are force and acceleration.

Quantities that have magnitude but no direction are called *scalar* quantities. These include time, volume, mass, and so on.

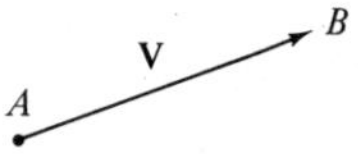

Representation of Vectors: A vector is represented by a line segment (Fig. 12-21) whose length is proportional to the magnitude of the vector, and whose direction is the same as the direction of the vector quantity.

FIGURE 12-21. Representation of a vector.

Vectors are represented by boldface type. Thus, **B** is a vector quantity, whereas B is a scalar quantity. A vector can be designated by a single letter, or by two letters representing the endpoints, with the starting point given first. Thus, the vector in Fig. 12-21 can be labeled

$$\mathbf{V} \quad \text{or} \quad \mathbf{AB}$$

Boldface letters are not practical in handwritten work. Instead, it is customary to place an arrow *over vector quantities.*

or, when handwritten,

$$\vec{V} \quad \text{or} \quad \overrightarrow{AB}$$

The *magnitude* of a vector can be designated with absolute value symbols, or by ordinary (nonboldface) type. Thus, the magnitude of vector **V** in Fig. 12-21 is

$$|\mathbf{V}| \quad \text{or} \quad V \quad \text{or} \quad |\mathbf{AB}| \quad \text{or} \quad AB$$

A vector can also be designated by the *coordinates of its endpoint.*

Glance back at Sec. 9-1, where we defined a vector as a matrix with only one row or one column.

Examples:

(a) The ordered pair (3, 5) is a vector. It designates a line segment drawn from the origin to the point (3, 5).

(b) The *ordered triple* of numbers (4, 6, 3) is a vector in three dimensions. It designates a point drawn from the origin to the point (4, 6, 3).

(c) The set of numbers (7, 3, 6, 1, 5) is a vector in five dimensions. We cannot draw it but it is handled mathematically in the same way as the others.

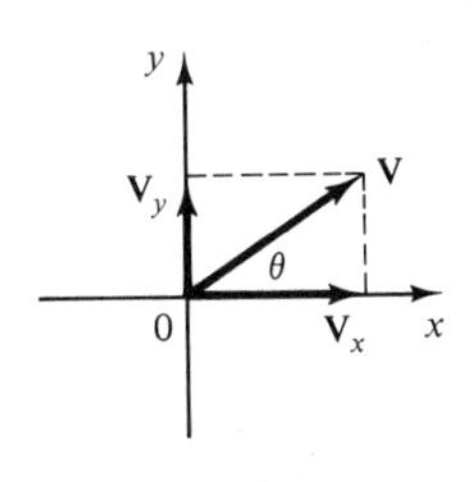

(a)

(b)

FIGURE 12-22. Rectangular components of a vector.

Rectangular Components of a Vector: Any vector can be replaced by two (or more) vectors which, acting together, exactly duplicate the effect of the original vector. They are called the *components* of the vector. The components are usually perpendicular to each other and are then called *rectangular components*. To *resolve* a vector means to replace it by its components.

Example: Figure 12-22a shows a vector **V** resolved into its x component $\mathbf{V}_x$ and its y component $\mathbf{V}_y$.

The component of a vector along an axis is also referred to as the projection *of the vector onto that axis.*

Example: Figure 12-22b shows a block on an inclined plane with its weight **W** resolved into a component **N** normal to the plane and a component **T** tangential (parallel) to the plane.

Resolution of Vectors: We see in Fig. 12-22 that a vector and its rectangular components form a *right triangle*. Thus, we can resolve a vector into its rectangular components with the right-triangle trigonometry of this chapter.

Example: The vector **V** in Fig. 12-22a has a magnitude of 248 units and makes an angle θ of 38.2° with the x axis. Find the x and y components.

Solution: By Eq. 146,

$$\sin 38.2° = \frac{\mathbf{V}_y}{248}$$

$$\mathbf{V}_y = 248 \sin 38.2° = 153$$

Similarly,

$$\mathbf{V}_x = 248 \cos 38.2° = 195$$

Some calculators have a key for converting between rectangular and polar coordinates, which can also be used to find resultants and components of vectors. Check your manual.

Example: A cable exerts a force of 558 N at an angle of 47.2° with the horizontal (Fig. 12-23a). Resolve this force into vertical and horizontal components.

Solution: We draw a vector diagram (Fig. 12-23b). Then by Eq. 146,

$$\sin 47.2° = \frac{\mathbf{V}_y}{558}$$

$$\mathbf{V}_y = 558 \sin 47.2° = 409 \text{ N}$$

Similarly,

$$\mathbf{V}_x = 558 \cos 47.2° = 379 \text{ N}$$

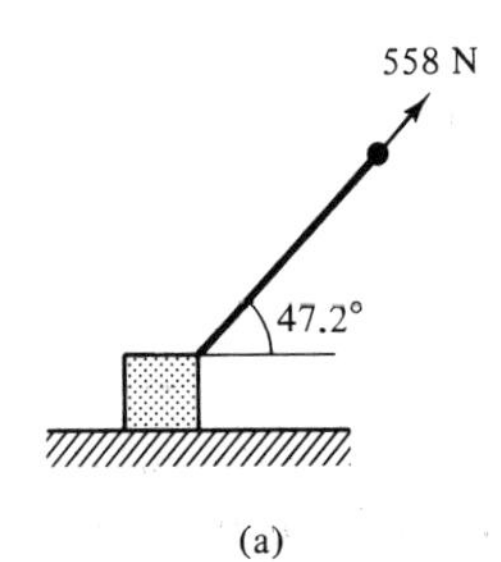

FIGURE 12-23

Resultant: Just as any vector can be *resolved* into components, so can several vectors be *combined* into a single vector called the *resultant*, or *vector sum*. The process of combining vectors into a resultant is called *vector addition*.

When combining or adding two perpendicular vectors, we can use right-triangle trigonometry, just as we did for resolving a vector into rectangular components.

Example: Find the resultant of two perpendicular vectors whose magnitudes are 485 and 627, and the angle it makes with the 627 magnitude vector.

Solution: We draw a vector diagram (Fig. 12-24). Then by the Pythagorean theorem,

$$R = \sqrt{(485)^2 + (627)^2} = 793$$

And by Eq. 148,

$$\tan \theta = \frac{485}{627} = 0.774$$

$$\theta = 37.7°$$

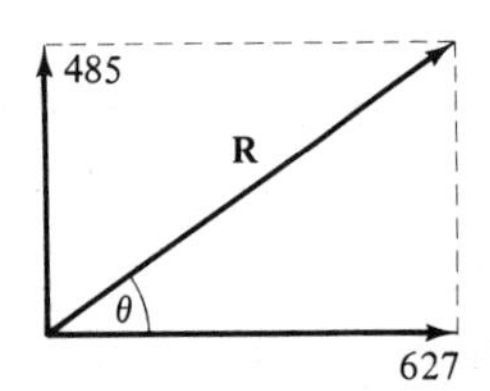

FIGURE 12-24. Resultant of two vectors.

Example: If the components of vector **A** are

$$\mathbf{A}_x = 735 \qquad \text{and} \qquad \mathbf{A}_y = 593$$

find the magnitude of **A** and the angle θ it makes with the x axis.

Solution: By the Pythagorean theorem,

$$\mathbf{A}^2 = \mathbf{A}_x^2 + \mathbf{A}_y^2$$
$$= (735)^2 + (593)^2 = 891{,}900$$
$$\mathbf{A} = 944$$

and

$$\tan\theta = \frac{593}{735}$$
$$\theta = 38.9°$$

In Chapter 13 we use the trigonometry of oblique triangles to find resultants of nonperpendicular vectors. We also find the resultant of many vectors in any of the four quadrants.

Resultants of Nonperpendicular Vectors: To find the sum of two vectors that are *not* at right angles, we first resolve each vector into its rectangular components and then add the components along each axis, resulting in two rectangular vectors. These two vectors can then be combined as in the preceding section.

Example: Find the resultant of the vectors in Fig. 12-25a.

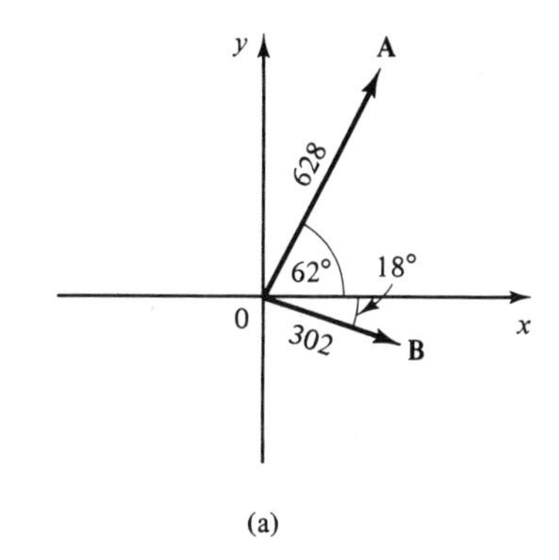

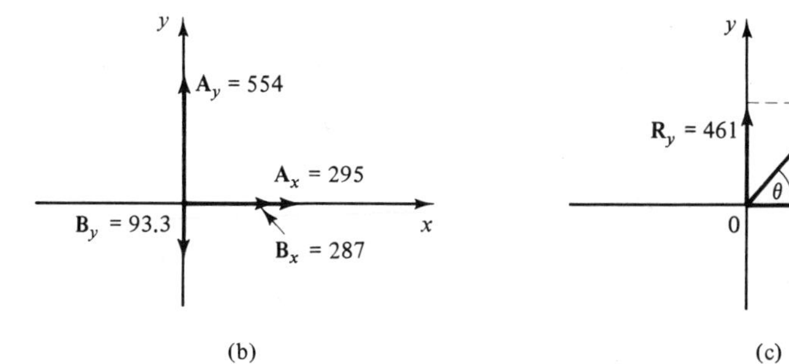

FIGURE 12-25. Resultant of two non-perpendicular vectors.

Solution: We first resolve **A** into x and y components $\mathbf{A}_x$ and $\mathbf{A}_y$.

$$\mathbf{A}_y = 628 \sin 62° = 554$$

and

$$\mathbf{A}_x = 628 \cos 62° = 295$$

Similarly for vector **B**,

$$\mathbf{B}_y = 302 \sin 18° = 93.3$$

and

$$\mathbf{B}_x = 302 \cos 18° = 287$$

These components are shown in Fig. 12-25b. We now combine the vectors along each axis, adding when they are in the same direction and subtracting when in the opposite direction. In the x direction,

$$\mathbf{R}_x = \mathbf{A}_x + \mathbf{B}_x = 295 + 287 = 582$$

and in the y direction,

$$\mathbf{R}_y = \mathbf{A}_y - \mathbf{B}_y = 554 - 93.3 = 461$$

These two vectors and their resultant **R** are shown in Fig. 12-25c. By the Pythagorean theorem,

$$R = \sqrt{(582)^2 + (461)^2} = 742$$

And finally, by Eq. 148,

$$\tan \theta = \frac{461}{582} = 0.792$$

$$\theta = 38.4°$$

When the vectors to be added are *not* on coordinate axes, the work is even easier, as in the following example.

Example: Vectors **A** and **B** have magnitudes of 496 and 628, respectively, and are separated by an angle of 58.4°. Find the magnitude of the resultant **R** and the angle α that it makes with vector **B**.

Solution: We use vector **B** as a reference, and resolve **A** into components $\mathbf{A}_t$ and $\mathbf{A}_n$ which are parallel and perpendicular to **B** (Fig. 12-26).

$$\mathbf{A}_n = 496 \sin 58.4° = 422$$

$$\mathbf{A}_t = 496 \cos 58.4° = 260$$

We then combine $\mathbf{A}_t$ with **B**.

$$\mathbf{A}_t + \mathbf{B} = 260 + 628 = 888$$

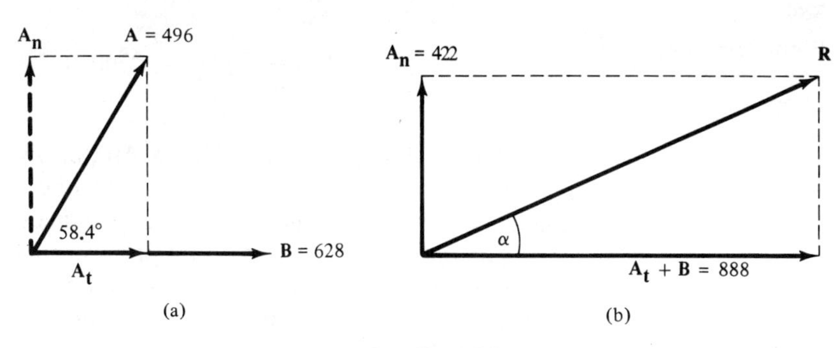

FIGURE 12-26

Then by the Pythagorean theorem,

$$\mathbf{R}^2 = (888)^2 + (260)^2$$

$$\mathbf{R} = 925$$

Finding the angle,

$$\tan \alpha = \frac{422}{888}$$

$$\alpha = 25.4°$$

Applications: Any vector quantities such as force, velocity, and impedance can be combined or resolved by the methods in the preceding section.

Example: A cable running from the top of a telephone pole (Fig. 12-27) creates a horizontal pull of 875 N. A support cable running to the ground is inclined 71.5° from the horizontal. Find the tension in the support cable.

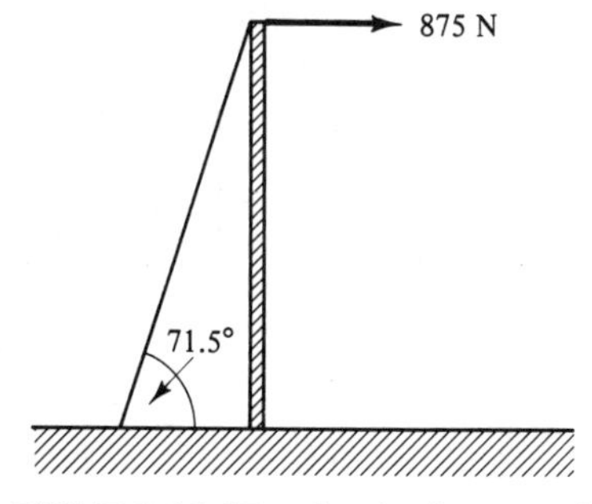

FIGURE 12-27. A telephone pole.

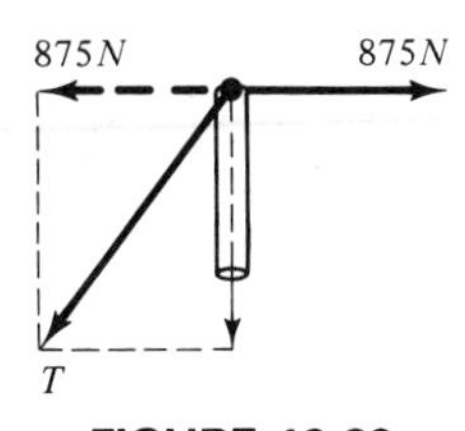

FIGURE 12-28

Solution: We draw the forces acting at the top of the pole in Fig. 12-28. We see that the horizontal component of the tension T in the support cable must equal the horizontal pull of 875 N. So

$$\cos 71.5° = \frac{875}{T}$$

$$T = \frac{875}{\cos 71.5°} = 2760 \text{ N}$$

Impedance, Reactance, and Phase Angle: Vectors find extensive application in electrical technology, and one of the most common applications is in the calculation of impedances.

The *reactance* X is a measure of how much the capacitance and inductance retards the flow of current in an *ac* circuit (Fig. 12-29). It is the difference between the *capacitive reactance* X_C and the *inductive reactance* X_L.

Reactance	$X = X_L - X_C$	**A54**

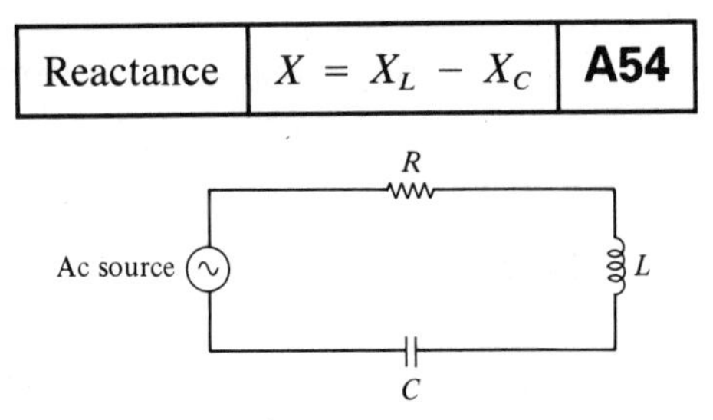

FIGURE 12-29. Series *RLC* circuit.

The *impedance* Z is a measure of how much the flow of current in an *ac* circuit is retarded by all circuit elements. The magnitude of the impedance is related to the total resistance R and reactance X by

Magnitude of Impedance	$\lvert Z \rvert = \sqrt{R^2 + X^2}$	**A55**

The impedance, resistance, and reactance form the three sides of a right triangle, the *vector impedance diagram* of Fig. 12-30. The angle ϕ between Z and R is called the *phase angle*.

Phase Angle	$\phi = \arctan \frac{X}{R}$	**A56**

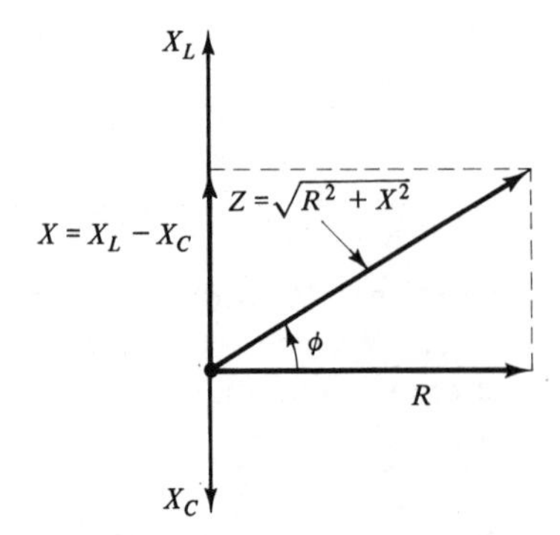

FIGURE 12-30 Vector impedance diagram.

Example: The capacitive reactance of a certain circuit is 2720 Ω, the inductive reactance is 3260 Ω, and the resistance is 1150 Ω. Find the reactance and magnitude of the impedance of the circuit, and the phase angle.

Solution: By Eq. A54,

$$X = 3260 - 2720 = 540\ \Omega$$

By Eq. A55,

$$|Z| = \sqrt{(1150)^2 + (540)^2} = 1270\ \Omega$$

and by Eq. A56,

$$\phi = \arctan \frac{540}{1270} = 23.0°$$

EXERCISE 4

Resolution of Vectors

Given the magnitude of each vector and the angle θ that it makes with the x axis, find the x and y components.

1. magnitude = 4.93 $\theta = 48.3°$
2. magnitude = 835 $\theta = 25°45'$
3. magnitude = 1.884 $\theta = 58.24°$
4. magnitude = 362 $\theta = 138°$

Vector Addition

In the following problems, the magnitudes of perpendicular vectors **A** and **B** are given. Find the resultant and the angle it makes with vector **B**.

5. $A = 483$ $B = 382$
6. $A = 2.85$ $B = 4.82$
7. $A = 7364$ $B = 4837$
8. $A = 46.8$ $B = 38.6$
9. $A = 1.25$ $B = 2.07$

In the following problems, vector **A** has x and y components A_x and A_y. Find the magnitude of **A** and the angle it makes with the x axis.

10. $A_x = 483$ $A_y = 382$
11. $A_x = 6.82$ $A_y = -4.83$
12. $A_x = -58.3$ $A_y = 37.2$
13. $A_x = -2.27$ $A_y = -3.97$

Nonperpendicular Vectors

Given the magnitude and direction of vectors **A** and **B**, find the magnitude R of the resultant and the angle θ it makes with the x axis.

14. $A = 736$ $\theta_A = 14.6°$ $B = 583$ $\theta_B = 66.3°$
15. $A = 4.83$ $\theta_A = 36.7°$ $B = 3.85$ $\theta_B = 115°$
16. $A = 56.3$ $\theta_A = 145°$ $B = 35.8$ $\theta_B = 281°$
17. $A = 746$ $\theta_A = 201°$ $B = 857$ $\theta_B = 298°$

In the following problems, the magnitudes of vectors **A** and **B** and the angle θ between them are given. Find the magnitude of the resultant and the angle it makes with vector **B**.

18. $A = 473 \quad B = 372 \quad \theta = 37.2°$

19. $A = 3.86 \quad B = 6.28 \quad \theta = 63.9°$

20. $A = 55.4 \quad B = 68.3 \quad \theta = 155°$

21. $A = 1.045 \quad B = 2.853 \quad \theta = 169°14'$

Force Vectors

22. What force, neglecting friction, must be exerted to drag a 56.5-N weight up a slope inclined 12.6° from the horizontal?

23. What is the largest weight that a tractor can drag up a slope which is inclined 21.2° from the horizontal if it is able to pull along the incline with a force of 2750 lb? Neglect the force due to friction.

You will need the equations of equilibrium, Eqs. A12 to A14 for some of these problems.

24. A person has just enough strength to pull a 1270-N weight up a certain slope. Neglecting friction, find the angle at which the slope is inclined to the horizontal if the person is able to exert a pull of 550 N.

25. A person wishes to pull a 255-lb weight up an incline to the top of a wall 14.5 ft high. Neglecting friction, what is the length of the shortest incline that can be used if the person's pulling strength is 145 lb?

26. A truck weighing 18.6 tons stands on a hill inclined 15.4° from the horizontal. How large a force must be counteracted by brakes to prevent the truck from rolling downhill?

27. A wagon weighing 8740 N is being pulled at uniform speed up a ramp inclined at 25.5° by a rope parallel to the ramp. If frictional forces are negligible, find the tension in the tow rope.

Velocity Vectors

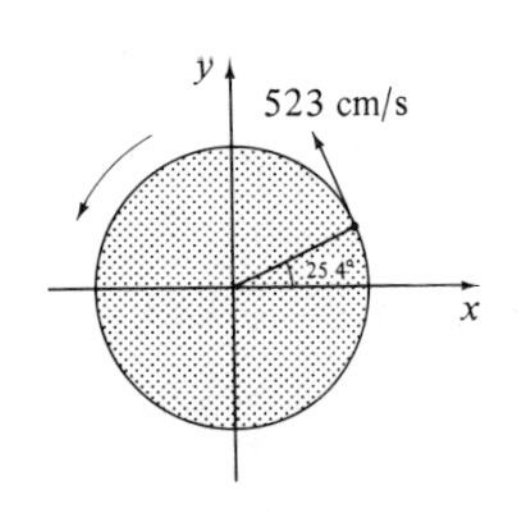

FIGURE 12-31

28. A point on a rotating wheel has a tangential velocity of 523 cm/s. Find the x and y components of the velocity when in the position shown in Fig. 12-31.

29. A certain escalator travels at a rate of 10.6 m/min and its angle of inclination is 32.5°. What is the vertical component of the velocity? How long will it take a passenger to travel 10.0 m vertically?

30. A projectile is launched at an angle of 55.6° to the horizontal with a speed of 7550 m/min. Find the vertical and horizontal components of this velocity.

31. At what speed with respect to the water should a ship head north in order to follow a course N5°15′E if a current is flowing east at the rate of 10.6 mi/h?

32. A river flows at the rate of 4.7 km/h. A rower, who can travel 7.5 km/h in still water, heads directly across the current. Find the rate and the direction of travel of the boat.

Impedance Vectors

33. A circuit has a reactance of 2650 Ω and a phase angle of 44.6°. Find the resistance and the magnitude of the impedance.

34. A circuit has a resistance of 115 Ω and a phase angle of 72 deg. Find the reactance and the magnitude of the impedance.

35. A circuit has an impedance of 975 Ω and a phase angle of 28 deg. Find the resistance and the reactance.

36. A circuit has a reactance of 5.75 Ω and a resistance of 4.22 Ω. Find the magnitude of the impedance and the phase angle.

37. A circuit has a capacitive reactance of 1776 Ω, an inductive reactance of 5140 Ω, and a total impedance of 5560 Ω. Find the resistance and the phase angle.

CHAPTER TEST

Perform the angle conversions, and fill in the missing quantities.

	Decimal degrees	*Degrees–minutes–seconds*	*Radians*	*Revolutions*
1.	38.2			
2.		25°28′45″		
3.			2.745	
4.				0.275

The following points are on the terminal side of an angle θ. Write the six trigonometric functions of the angle to three significant digits, and find the angle in decimal degrees.

5. (3, 7) **6.** (2, 5) **7.** (4, 3) **8.** (2.3, 3.1)

Write the six trigonometric functions of each angle.

9. 72.9° **10.** 35°13′33″ **11.** 1.05 rad

Find acute angle θ in decimal degrees.

12. $\sin \theta = 0.574$ **13.** $\cos \theta = 0.824$ **14.** $\tan \theta = 1.345$

Evaluate each expression. Give your answer as an acute angle in degrees.

15. arctan 2.86 **16.** $\cos^{-1} 0.385$ **17.** arcsec 2.447

Solve right triangle ABC.

18. $a = 746$ and $A = 37.2°$ **19.** $b = 3.72$ and $A = 28.5°$ **20.** $c = 45.9$ and $A = 61.4°$

21. Find the x and y components of a vector that has a magnitude of 885 and makes an angle of 66.3° with the x axis.

22. Find the magnitude of the resultant of two perpendicular vectors that have magnitudes of 54.8 and 39.4 and the angle it makes with the 54.8 vector.

23. A vector has x and y components of 385 and 275. Find the magnitude and direction of that vector.

24. A telephone pole casts a shadow 13.5 m long when the angle of elevation of the sun is 15.4°. Find the height of the pole.

25. From a point 125 ft in front of a church, the angles of elevation of the top and base of its steeple are 22.5° and 19.6°, respectively. Find the height of the steeple.

26. A circuit has a resistance of 125 Ω, an impedance of 256 Ω, an inductive reactance of 312 Ω, and a positive phase angle. Find the capacitive reactance and the phase angle.

Oblique Triangles

In Chapter 12 we defined the trigonometric functions for any angle but used them only for acute angles. Here we study the functions of obtuse angles, negative angles, angles greater than one revolution, and angles with terminal sides on the coordinate axes. We need the trigonometric functions of obtuse angles to solve oblique triangles, and the trigonometric function of angles larger than 180° for vectors and other applications.

An *oblique triangle* is one that does not contain a right angle. In this chapter we derive two new formulas, the *law of sines* and the *law of cosines,* to enable us to quickly solve oblique triangles. We cannot, of course, use the Pythagorean theorem or the six trigonometric functions to solve oblique triangles, although we will use these relationships to derive the law of sines and law of cosines.

We also continue our study of vectors in this chapter. In Chapter 12 we dealt with vectors at right angles to each other; here we consider vectors at any angle.

13-1. TRIGONOMETRIC FUNCTIONS OF ANY ANGLE

Definition of the Trigonometric Functions: We defined the trigonometric functions of any angle in Sec. 12-2 but have so far done problems only with acute angles. We turn now to larger angles.

Figure 13-1a–c shows angles in the second, third, and fourth quadrants, and Fig. 13-1d shows an angle greater than 360°. The trigonometric functions of any of these angles are defined exactly as for an acute angle in quadrant I. From any point P on the terminal side of the angle we drop a perpendicular to the x axis, forming a right triangle with legs x and y and with a hypotenuse r. The six trigonometric ratios are then given by Eqs. 146 to 151, just as before.

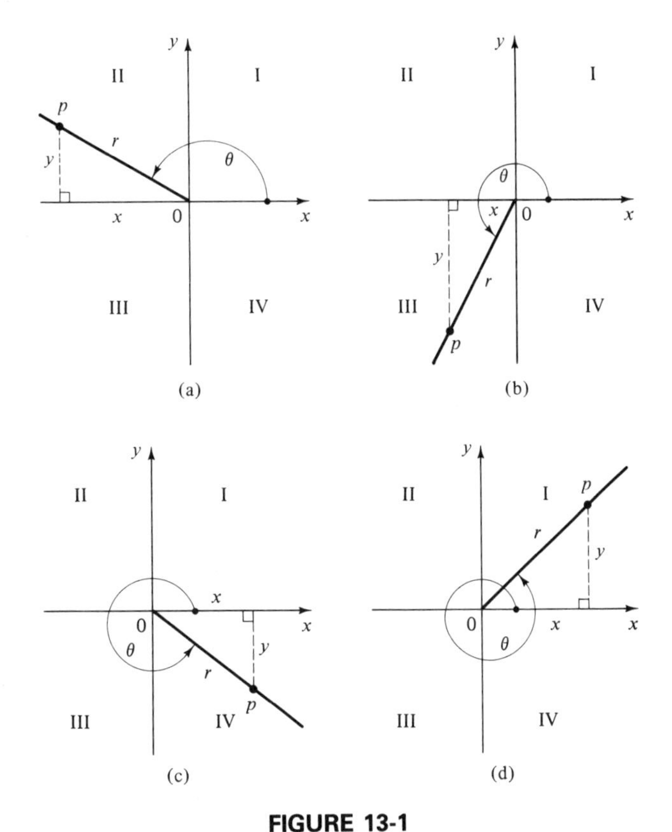

FIGURE 13-1

Example: A point on the terminal side of angle θ has the coordinates $(-3, -5)$. Write the six trigonometric functions of θ to three significant digits.

Solution: We sketch the angle (Fig. 13-2 on page 388) and see that it lies in the third quadrant. We find distance r by the Pythagorean theorem:

$$r^2 = (-3)^2 + (-5)^2 = 9 + 25 = 34$$

$$r = 5.83$$

Then by Eqs. 146 to 151, with $x = -3$, $y = -5$, and $r = 5.83$,

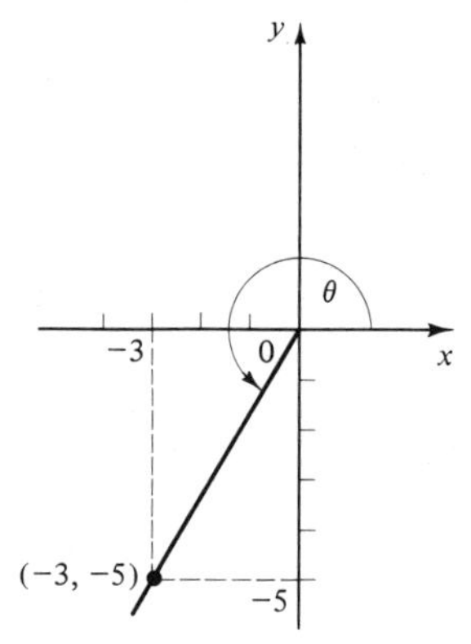

FIGURE 13-2

$$\sin\theta = \frac{y}{r} = \frac{-5}{5.83} = -0.858$$

$$\cos\theta = \frac{x}{r} = \frac{-3}{5.83} = -0.515$$

$$\tan\theta = \frac{y}{x} = \frac{-5}{-3} = 1.67$$

$$\cot\theta = \frac{x}{y} = \frac{-3}{-5} = 0.600$$

$$\sec\theta = \frac{r}{x} = \frac{5.83}{-3} = -1.94$$

$$\csc\theta = \frac{r}{y} = \frac{5.83}{-5} = -1.17$$

Algebraic Signs of the Trigonometric Functions: We saw in Chapter 12 that the trigonometric functions of first quadrant angles were always positive. From the preceding example, it is clear that some of the trigonometric functions of angles in the second, third, and fourth quadrants are negative, because x or y can be negative (r is always positive). Figure 13-3 shows the signs of the trigonometric functions in each quadrant.

Why bother learning the signs when a calculator gives them to us automatically? Because you will need them when using a calculator *to find the inverse of a function, as discussed below.*

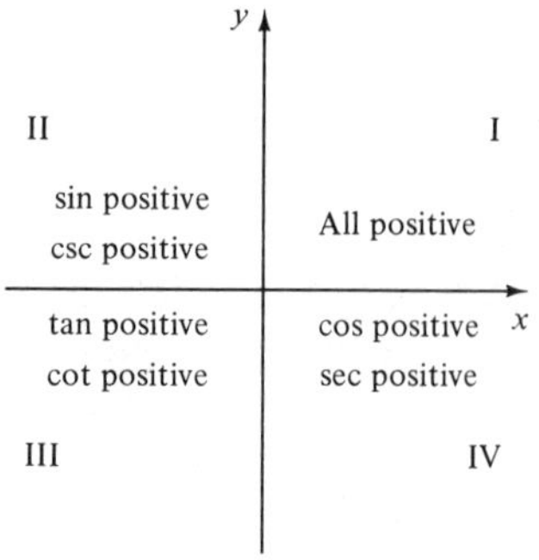

y
II
I
sin positive
csc positive
All positive
tan positive
cot positive
cos positive
sec positive
x
III
IV

FIGURE 13-3. Algebraic signs of the trigonometric functions in all quadrants.

Rather than try to remember which functions are negative in which quadrants, just sketch the angle and note whether x or y is negative. From this you can figure out whether the function you want is positive or negative.

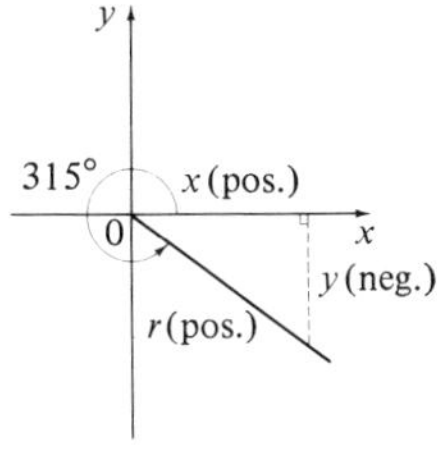

FIGURE 13-4

Example: What is the algebraic sign of csc 315°?

Solution: We make a sketch, such as in Fig. 13-4. It is not necessary to draw the angle accurately, but it must be shown in the proper quadrant, quadrant IV in this case. We note that y is negative and that r is (always) positive. So

$$\csc 315° = \frac{r}{y} = \frac{(+)}{(-)} = \text{negative}$$

Trigonometric Functions of Any Angle by Calculator: Simply enter the angle and press the proper trigonometric key. Be sure that the Degree/Radian switch is in the proper position. The calculator will give the correct algebraic sign.

Example: Find sin 212° to four significant digits.

Solution:

Press	*Display*
Degree mode 212 [sin]	−0.5299 (rounded)

Example: Find tan 5.29 rad to four significant digits.

Solution:

Press	*Display*
Radian mode 5.29 [tan]	−1.534 (rounded)

Use the reciprocal key [1/x] for cot, sec, or csc.

Example: Find sec 124° to four significant digits.

Solution:

Press	*Display*
Degree mode 124 [cos] [1/x]	−1.788 (rounded)

For negative angles, use the change-sign key [CHS] or [+/−] after entering the angle.

Example: Find tan $(-35°)$ to four significant digits.

Solution:

Press	*Display*
Degree mode 35 [+/−]	−35
[tan]	−0.7002 (rounded)

Angles greater than 360° are handled in the same manner as any other angle.

Example: Find cos 412° to four significant digits.

Solution:

Press	*Display*
Degree mode 412 [cos]	0.6157 (rounded)

Example: Find tan 28.46 rad to five significant digits.

Solution:

Press	*Display*
Radian mode 28.46 [tan]	0.18783 (rounded)

Evaluating Trigonometric Expressions:

Example: Evaluate the expression

$$(\sin^2 48° + \cos 62°)^3$$

to four significant digits.

Solution: The notation $\sin^2 48°$ is the same as $(\sin 48°)^2$. Let us carry five digits and round to four in the last step.

$$\begin{aligned}(\sin^2 48° + \cos 62°)^3 &= [(0.74314)^2 + 0.46947]^3 \\ &= (1.0217)^3 = 1.067\end{aligned}$$

Common Error	Do not confuse an exponent which is on the angle with one which is on the entire function. $(\sin \theta)^2 = \sin^2 \theta \neq \sin \theta^2$

Example: Evaluate

$$\sin^3 x + \cos x^4 - (\tan x)^2$$

to three digits, if $x = 1.25$ rad.

Solution: Substituting,

$$\begin{aligned}(\sin 1.25)^3 &+ \cos (1.25)^4 - (\tan 1.25)^2 \\ &= (0.9490)^3 + \cos 2.441 - (3.010)^2 \\ &= 0.8546 + (-0.7647) - 9.058 \\ &= -8.97\end{aligned}$$

Reference Angle: Finding the trigonometric function of any angle by calculator is no problem—the calculator does all the thinking. This is not the case, however, when we are given the function and asked to find the angle, or when we have to find trigonometric functions in a table. For both these operations we will make use of the *reference angle*.

It is also called the working angle.

For an angle in standard position on coordinate axes, the *acute* angle that its terminal side makes with the x axis is called the *reference angle*.

Example: The reference angle m for an angle of 125° is

$$m = 180 - 125 = 55°$$

as in Fig. 13-5.

Example: The reference angle m for an angle of 236° is

$$m = 236° - 180° = 56°$$

as in Fig. 13-6.

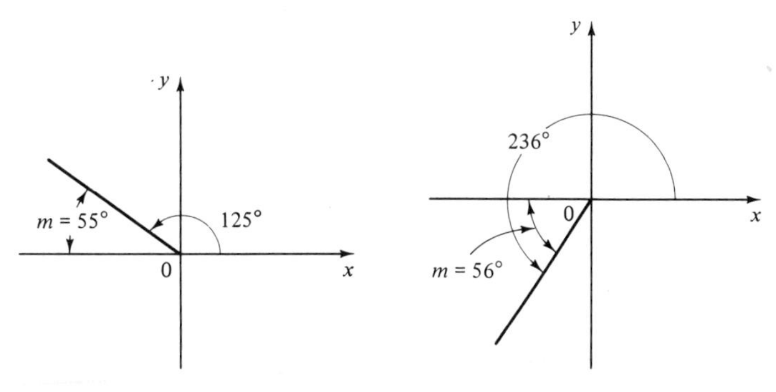

FIGURE 13-5. Reference angle m.

FIGURE 13-6

Example: The reference angle m for an angle of 331.6° is (Fig. 13-7)

$$m = 360 - 331.6 = 28.4°$$

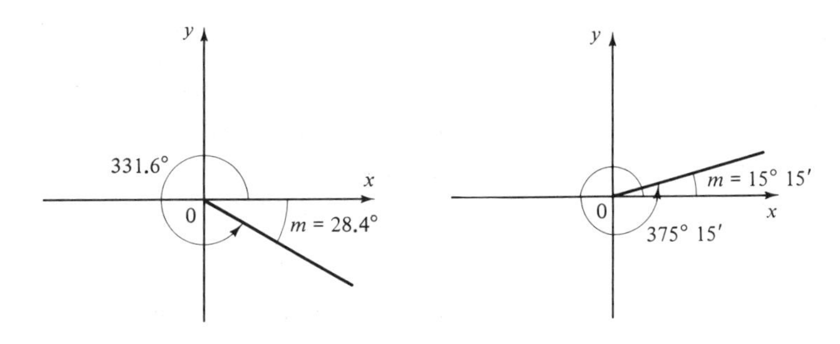

FIGURE 13-7 **FIGURE 13-8**

Example: The reference angle m for an angle of $375°15'$ is

$$m = 375°15' - 360° = 15°15'$$

as in Fig. 13-8.

Common Error	The reference angle is always the angle between the x *axis* and the terminal side, never the y axis.

Finding the Angle When the Function Is Given: We use the same procedure as was given for acute angles in Sec. 12-2. However, a calculator or the table will give us just *one* angle which has the given trigonometric function, whereas we know that there are infinitely many angles that have the same trigonometric function. We must use that angle to find the *reference* angle, which is then used to compute as many larger angles as we wish having the same trigonometric ratio. Usually, we want only the two positive angles less than 360° that have the required trigonometric function.

Example: If $\sin \theta = 0.6293$, find two positive values of θ less than 360°.

Solution: Using the [arc] or [inv] key on the calculator, or from the tables,

$$\theta = 39.0°$$

Check your work by taking the sine. In this case, sin 141.0° = 0.6293. Checks.

The sine is positive in the first and second quadrants. Our first quadrant angle is 39.0°, and our second quadrant angle is, taking 39.0° as the reference angle (Fig. 13-9)

$$\theta = 180 - 39.0 = 141.0°$$

Example: Find the two positive angles less than 360° that have a tangent of -2.25.

Solution: From the calculator, $\theta = -66.0°$, so our reference angle is 66°. The tangent is negative in the second and fourth quadrants.

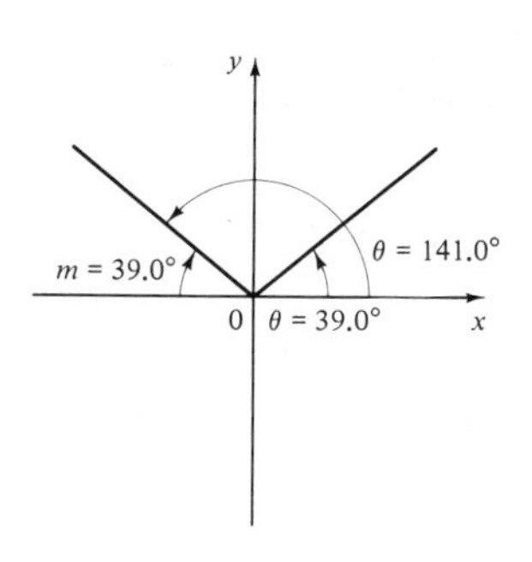

FIGURE 13-9

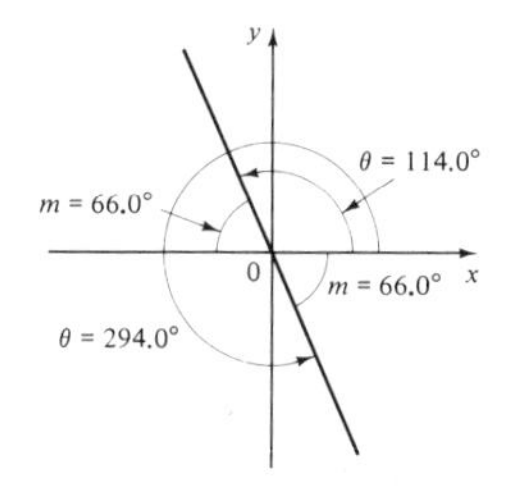

FIGURE 13-10

Our second quadrant angle is

$$180 - 66.0° = 114.0°$$

and our fourth quadrant angle is (Fig. 13-10).

$$360 - 66.0° = 294.0°$$

Example: Find $\cos^{-1} 0.575$.

Solution: From the calculator,

$$\cos^{-1} 0.575 = 54.9°$$

The cosine is positive in the first and fourth quadrants. Our fourth quadrant angle is

$$360° - 54.9° = 305.1°$$

Common Error	Do not confuse the **inverse** with the **reciprocal**. $$\frac{1}{\sin \theta} = (\sin \theta)^{-1} \neq \sin^{-1} \theta$$

When the cotangent, secant, or cosecant is given, we make use of the reciprocal relationships (Eqs. 152) as in the following example.

Example: Find two positive angles less than 360° that have a secant of -4.22.

Solution: If we let the angle be θ, then by Eq. 152b

$$\cos \theta = \frac{1}{\sec \theta} = \frac{1}{-4.22} = -0.237$$

By calculator,

$$\theta = 103.7°$$

The secant is also negative in the third quadrant. Our reference angle is

$$m = 180 - 103.7 = 76.3°$$

so the third quadrant angle is

$$\theta = 180 + 76.3 = 256.3°$$

(see Fig. 13-11).

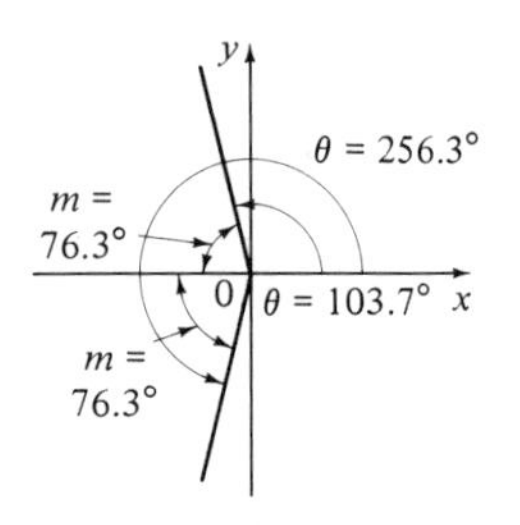

FIGURE 13-11

Example: Evaluate arcsin (−0.528).

Solution: As before, we seek only two positive angles less than 360°. By calculator,

$$\arcsin(-0.528) = -31.9°$$

which is a fourth quadrant angle. As a positive angle, it is

$$\theta = 360° - 31.9° = 328.1°$$

The sine is also negative in the third quadrant. Using 31.9° as our reference angle,

$$\theta = 180° + 31.9° = 211.9°$$

Special Angles: The angles in the following table appear so frequently in problems that it is convenient to be able to write their trigonometric functions from memory.

Angle	Sin	Cos	Tan
0	0	1	0
30	½	$\sqrt{3}/2$ (or 0.8660)	$\sqrt{3}/3$ (or 0.5774)
45	$\sqrt{2}/2$ (or 0.7071)	$\sqrt{2}/2$ (or 0.7071)	1
60	$\sqrt{3}/2$ (or 0.8660)	½	$\sqrt{3}$ (or 1.732)
90	1	0	undefined
180	0	−1	0
270	−1	0	undefined
360	0	1	0

The angles 0°, 90°, 180°, and 360° are called *quadrantal* angles because the terminal side of each of them lies along one of the coordinate axes.

EXERCISE 1

Signs of the Trigonometric Functions

Assume all angles in this exercise to be in standard position.

State in what quadrant or quadrants the terminal side of θ can lie if:

1. $\theta = 123°$

2. $\theta = 272°$

3. $\theta = -47°$

4. $\theta = -216°$

5. $\theta = 415°$

6. $\theta = -415°$

7. $\theta = 845°$

8. $\sin\theta$ is positive

9. $\cos\theta$ is negative

10. $\sec\theta$ is positive

11. $\cos\theta$ is positive and $\sin\theta$ is negative

12. $\sin\theta$ and $\cos\theta$ are both negative

13. $\tan\theta$ is positive and $\csc\theta$ is negative

State whether the following expressions are positive or negative. Do not use your calculator, and try not to refer to your book.

14. $\sin 174°$
15. $\cos 329°$
16. $\tan 227°$
17. $\sec 332°$
18. $\cot 206°$
19. $\csc 125°$
20. $\sin(-47°)$
21. $\tan(-200°)$
22. $\cos 400°$

Give the algebraic signs of the sine, cosine, and tangent of:

23. 110°
24. 206°
25. 335°
26. −48°
27. 500°

Trigonometric Functions

Sketch the angle and write the six trigonometric functions if the terminal side of the angle passes through the point:

28. (3, 5)
29. (−4, 12)
30. (24, −7)
31. (−15, −8)
32. (0, −3)
33. (5, 0)

Write the sine, cosine, and tangent of θ if:

Leave your answers for this set in fractional or radical form.

34. $\sin\theta = -\frac{3}{5}$ and $\tan\theta$ is positive.

35. $\cos\theta = -\frac{4}{5}$ and θ is in the third quadrant.

36. $\csc\theta = -\frac{25}{7}$ and $\cos\theta$ is negative.

37. $\tan\theta = 2$ and θ is not in the first quadrant.

38. $\cot\theta = -\frac{4}{3}$ and $\sin\theta$ is positive.

39. $\sin\theta = \frac{2}{3}$ and θ is not in the first quadrant.

40. $\cos\theta = -\frac{7}{9}$ and $\tan\theta$ is negative.

Write, to four significant digits, the sine, cosine, and tangent of each angle.

41. 101°
42. 216°
43. 331°
44. 125.8°
45. −62.85°
46. −227.4°
47. 486°
48. −527°
49. 114°23′
50. 264°15′45″
51. −166°55′
52. 1.15 rad
53. 2.228 rad
54. 5.397 rad
55. −4.483 rad

Evaluating Trigonometric Expressions

Find the numerical value of each expression to four significant digits.

56. $\sin 35° + \cos 35°$

57. $\sin 125° \tan 225°$

58. $\cos 270° \cos 150° + \sin 270° \sin 150°$

59. $\dfrac{\sin 155°}{1 + \cos 155°}$

60. $\sin^2 75°$

61. $\tan^2 125° - \cos^2 125°$

62. $(\cos 206° + \sin 206°)^2$

63. $\sqrt{\sin 112° - \cos 112°}$

Inverse Trigonometric Functions

Find, to the nearest tenth of a degree, all positive angles less than 360° whose trigonometric function is:

64. $\sin \theta = \frac{1}{2}$

65. $\tan \theta = -1$

66. $\cot \theta = -\sqrt{3}$

67. $\cos \theta = 0.8372$

68. $\csc \theta = -3.85$

69. $\tan \theta = 6.372$

70. $\cos \theta = -\frac{1}{2}$

71. $\cot \theta = 0$

72. $\cos \theta = -1$

Evaluate.

73. $\arcsin (-0.736)$

74. $\text{arcsec } 2.85$

75. $\cos^{-1} 0.827$

76. $\text{arccot } 5.22$

77. $\arctan (-4.48)$

78. $\csc^{-1} 5.02$

Applications

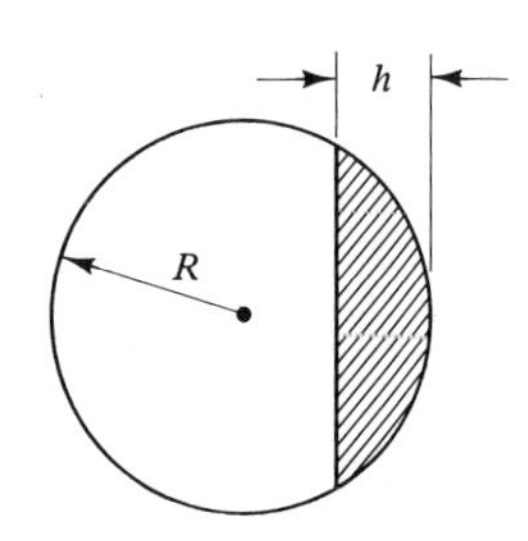

FIGURE 13-12. Segment of a circle.

79. The area of a *segment* of a circle (Fig. 13-12) is

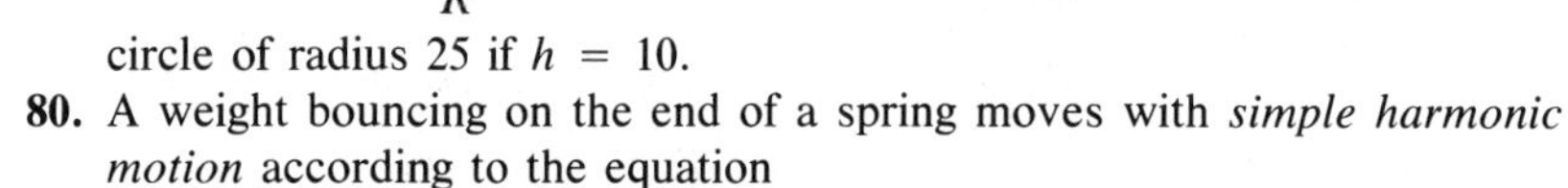

$$\text{area of segment} = R^2 \arccos \frac{R - h}{R} - (R - h)\sqrt{2Rh - h^2}$$

where $\arccos \dfrac{R - h}{R}$ is in radians. Compute the area of a segment in a circle of radius 25 if $h = 10$.

80. A weight bouncing on the end of a spring moves with *simple harmonic motion* according to the equation

$$y = 4 \cos 25t \quad \text{inches}$$

Find the displacement y when $t = 2$ s. (In this equation, the angle $25t$ must be in radians.)

81. The angular distance D (measured at the earth's center) between two points on the earth's surface is found by

$$\cos D = \sin L_1 \sin L_2 + \cos L_1 \cos L_2 \cos (M_1 - M_2)$$

where L_1, L_2 and M_1, M_2 are the respective latitudes and longitudes of the two points. Find the angular distance between Pittsburgh (latitude 40°N, longitude 81°W) and Houston (latitude 30°N, longitude 95°W). See Fig. 13-13 for a definition of latitude and longitude.

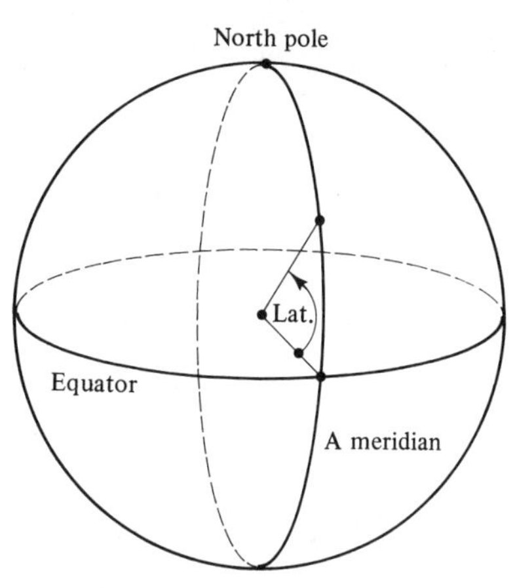

FIGURE 13-13. *Latitude* is the angle (measured at the earth's center) between a point on the earth and the equator. *Longitude* is the angle between the meridian passing through a point on the earth and a reference meridian passing through Greenwich, England.

Computer

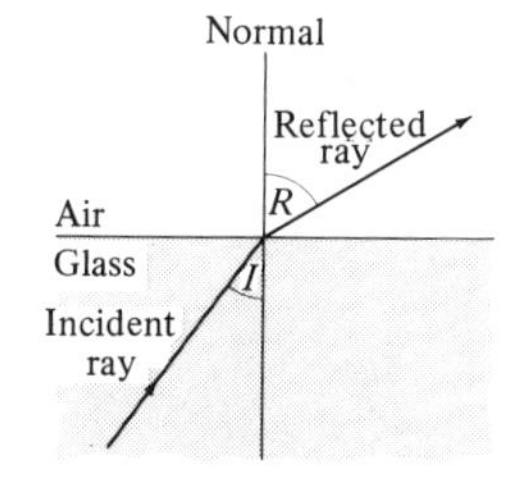

FIGURE 13-14. Refraction of light.

82. Figure 13-14 shows a ray of light passing from glass to air. The angle of refraction R is related to the angle of incidence I by Snell's law,

$$\frac{\sin R}{\sin I} = \text{constant}$$

Using a value for the constant (called the index of refraction) of 1.5, write a program that will compute and print angle R for values of I from 0 to 90°. Have the computer inspect $\sin R$ each time it is computed, and when its value exceeds 1.00, stop the computation and print "TOTAL INTERNAL REFLECTION."

13-2. LAW OF SINES

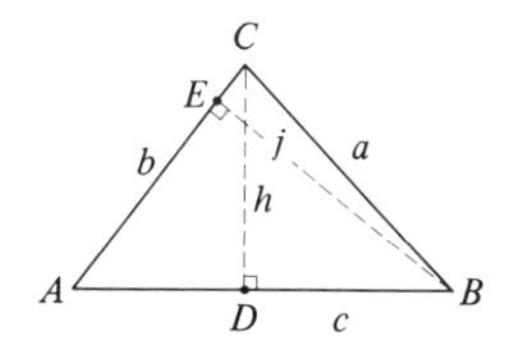

FIGURE 13-15. Derivation of law of sines.

Derivation: We first derive the law of sines for an oblique triangle in which all three angles are acute, such as in Fig. 13-15. We start by breaking the given triangle into two right triangles by drawing altitude h to side AB.

In right triangle ACD,

$$\sin A = \frac{h}{b} \qquad \text{or} \qquad h = b \sin A$$

In right triangle BCD,

$$\sin B = \frac{h}{a} \qquad \text{or} \qquad h = a \sin B$$

So

$$b \sin A = a \sin B$$

Dividing by $\sin A \sin B$,

$$\frac{a}{\sin A} = \frac{b}{\sin B}$$

Similarly, drawing altitude j to side AC, and using triangles BEC and AEB, we get

$$j = a \sin C = c \sin A$$

or

$$\frac{a}{\sin A} = \frac{c}{\sin C}$$

Combining this with the previous result, we obtain

Law of Sines	$\frac{a}{\sin A} = \frac{b}{\sin B} = \frac{c}{\sin C}$	**140**

The sides of a triangle are proportional to the sines of the opposite angles.

When one of the angles of the triangle is obtuse, as in Fig. 13-16, the derivation is nearly the same. We draw an altitude h to the extension of side AB. Then:

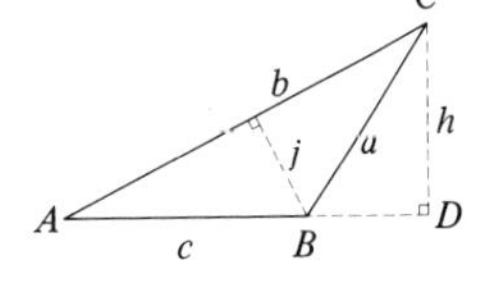

FIGURE 13-16

In right triangle ACD: $\sin A = \frac{h}{b}$, or $h = b \sin A$

In right triangle BCD: $\sin (180 - B) = \frac{h}{a}$

but $\sin (180 - B) = \sin B$, so

$$h = a \sin B$$

The derivation then proceeds exactly as before.

Solving a Triangle with the Law of Sines: To use the law of sines to solve an oblique triangle, we must have *a known side opposite to a known angle,* as well as another side or angle.

Example: Solve triangle ABC where $A = 32.5°$, $B = 49.7°$, and $a = 226$.

Solution: We sketch the triangle (Fig. 13-17). The missing angle is found by Eq. 139.

$$C = 180 - 32.5 - 49.7 = 97.8°$$

Then by the law of sines,

$$\frac{226}{\sin 32.5°} = \frac{b}{\sin 49.7°}$$

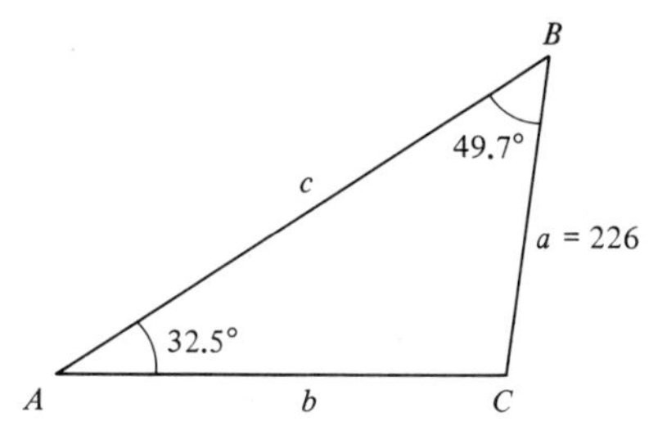

FIGURE 13-17

A diagram drawn more or less to scale can serve as a good check of your work and reveal inconsistencies in the given data. As another rough check, see that the longest side is opposite the largest angle and that the shortest side is opposite the smallest angle.

Solving for b, we get

$$b = \frac{226 \sin 49.7°}{\sin 32.5°} = 321$$

Again using the law of sines,

$$\frac{226}{\sin 32.5°} = \frac{c}{\sin 97.8°}$$

So

$$c = \frac{226 \sin 97.8°}{\sin 32.5°} = 417$$

In the preceding example, *two angles* and *one side* were given. We can also use the law of sines when *one angle* and *two sides* are given, provided that the given angle is opposite one of the given sides.

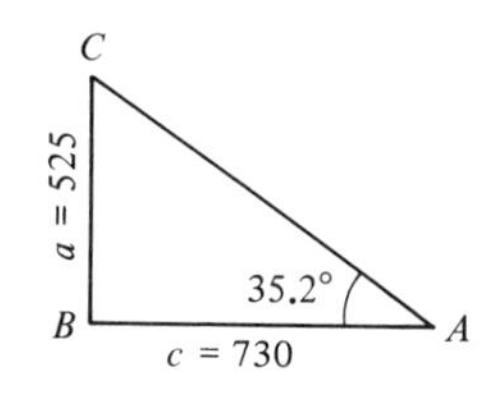

FIGURE 13-18

Example: Solve triangle ABC if $A = 35.2°$, $a = 525$, and $c = 730$.

Solution: We sketch the triangle (Fig. 13-18). We cannot tell from our sketch if angle B is greater or less than 90°, but we *can* tell that angle C is definitely acute, so we find it first.

Tip	When solving a triangle, always *find the acute angle first.*

It is sometimes more convenient, such as in this problem, to use the reciprocals *of the expressions in the law of sines.*

By the law of sines,

$$\frac{\sin C}{730} = \frac{\sin 35.2°}{525}$$

$$\sin C = \frac{730 \sin 35.2°}{525} = 0.8015$$

$$C = 53.3°$$

By Eq. 139,

$$B = 180° - 35.2° - 53.3° = 91.5°$$

Using the law of sines once again, we get

$$\frac{b}{\sin 91.5°} = \frac{525}{\sin 35.2°}$$

$$b = \frac{525 \sin 91.5°}{\sin 35.2°} = 910$$

The Ambiguous Case: In the preceding example (two sides and one angle given) the side that was opposite the given angle (side a) was *larger* than the other given side (side c). If side a were *smaller* than side c, there would have been *two possible solutions*, as in the following example.

Example: Solve triangle ABC where $A = 27.6°$, $a = 112$, and $c = 165$.

Solution: When we draw the triangle (Fig. 13-19) we see that there are two possible positions for side a, both of which satisfy the given data. We will solve both cases. By the law of sines,

$$\frac{\sin C}{165} = \frac{\sin 27.6°}{112}$$

$$\sin C = \frac{165 \sin 27.6°}{112} = 0.6825$$

$$C = 43.0°$$

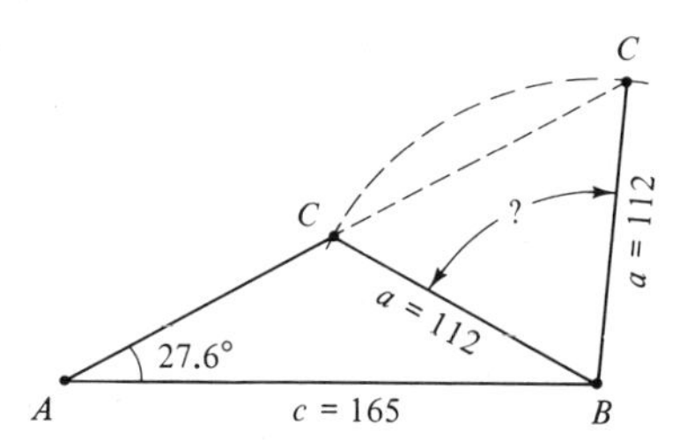

FIGURE 13-19. The ambiguous case.

Recall from Sec. 13-1 that there are two angles less than 360° for which the sine is positive. One of them, θ, is in the first quadrant, and the other, 180° − θ, is in the second quadrant.

This is *one* of the possible values for C. The other is

$$C = 180° - 43.0° = 137.0°$$

We now find the two corresponding values for side b and angle B.

When $C = 43.0°$:

$$B = 180° - 27.6° - 43.0° = 109.4°$$

So

$$\frac{b}{\sin 109.4°} = \frac{112}{\sin 27.6°}$$

from which

$$b = 228$$

When $C = 137.0°$:

$$B = 180° - 27.6° - 137.0° = 15.4°$$

So

$$\frac{b}{\sin 15.4°} = \frac{112}{\sin 27.6°}$$

from which

$$b = 64.2$$

Common Error	In a problem like the preceding one, it is easy to forget the second possible solution ($C = 137.0°$), especially since a calculator will give only the acute angle when taking the arc sine.

EXERCISE 2

Law of Sines

Data for triangle *ABC* is given in the following table. Solve for the missing parts.

	Angles			*Sides*		
	A	*B*	*C*	*a*	*b*	*c*
1.		46°15′		228	304	
2.		1.15 rad			1.59	1.46
3.			0.75 rad		15	21
4.	21°27′	35°10′		276		
5.	1.08 rad		0.82 rad	7.65		
6.	126°	27°			119	
7.		31.6°	44.8°		11.7	
8.	$\frac{\pi}{12}$		$\frac{2\pi}{5}$			375
9.		125°	32°			58
10.	24.14°	38.27°		5562		
11.		55°23′	18°12′		77.85	
12.	0.7762 rad		1.115 rad			1.065
13.	$\frac{\pi}{10}$	$\frac{\pi}{15}$		50.75		
14.			45.55°		1137	1586

Computer

15. Write a program to solve a triangle by the law of sines. Have it *menu-driven* with the following choices.

1. Two angles and an opposite side given (AAS)
2. Two sides and one angle given (SSA)
3. Two angles and the included side given (ASA)

These triangles are shown in Fig. 13-21. For each, have the computer ask for the given sides and angles, and then compute and print the missing sides and angles.

13-3. LAW OF COSINES

Derivation: Consider an oblique triangle *ABC* in Fig. 13-20. As we did for the law of sines, we start by dividing the triangle into two right triangles by drawing an altitude h to side AC.

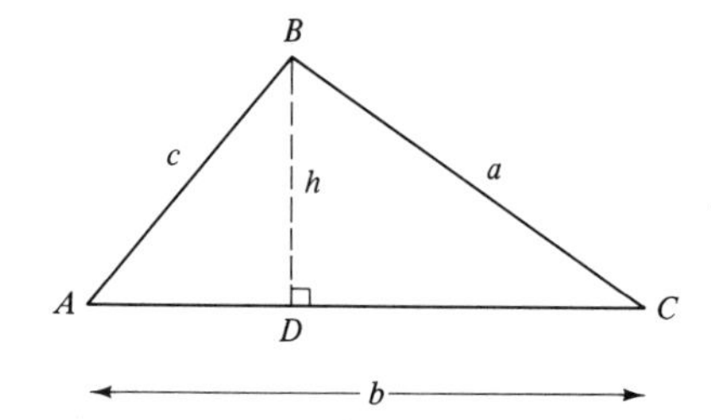

FIGURE 13-20. Derivation of law of cosines.

In right triangle *ABD*,

$$c^2 = h^2 + (AD)^2$$

But $AD = b - CD$. Substituting, we get

$$c^2 = h^2 + (b - CD)^2 \tag{1}$$

Now, in right triangle *BCD*, by Eq. 147

$$\frac{CD}{a} = \cos C$$

or

$$CD = a \cos C$$

Substituting $a \cos C$ for CD in (1) yields

$$c^2 = h^2 + (b - a \cos C)^2$$

Squaring,

$$c^2 = h^2 + b^2 - 2ab \cos C + a^2 \cos^2 C \tag{2}$$

Let us leave this expression for the moment, and write the Pythagorean theorem for the same triangle *BCD*:

$$h^2 = a^2 - (CD)^2$$

Again substituting $a \cos C$ for CD, we obtain

$$\begin{aligned} h^2 &= a^2 - (a \cos C)^2 \\ &= a^2 - a^2 \cos^2 C \end{aligned}$$

Substituting this expression for h^2 back into (2), we get

$$c^2 = a^2 - a^2 \cos^2 C + b^2 - 2ab \cos C + a^2 \cos^2 C$$

So

$$c^2 = a^2 + b^2 - 2ab \cos C$$

We can repeat the derivation, with perpendiculars drawn to side AB and to side BC, and get two more forms of the law of cosines:

Law of Cosines	$a^2 = b^2 + c^2 - 2bc \cos A$ $b^2 = a^2 + c^2 - 2ac \cos B$ $c^2 = a^2 + b^2 - 2ab \cos C$	**141**

The square of any side equals the sum of the squares of the other two sides minus twice the product of the other sides and the cosine of the opposite angle.

When to Use the Law of Sines or Law of Cosines: It is sometimes not clear whether to use the law of sines or the law of cosines to solve a triangle. We use the law of sines when we have a *known side opposite a known angle.* We use the law of cosines only when the law of sines does not work, that is, for all other cases. In Fig. 13-21, the heavy lines indicate the known information and may help in choosing the proper law.

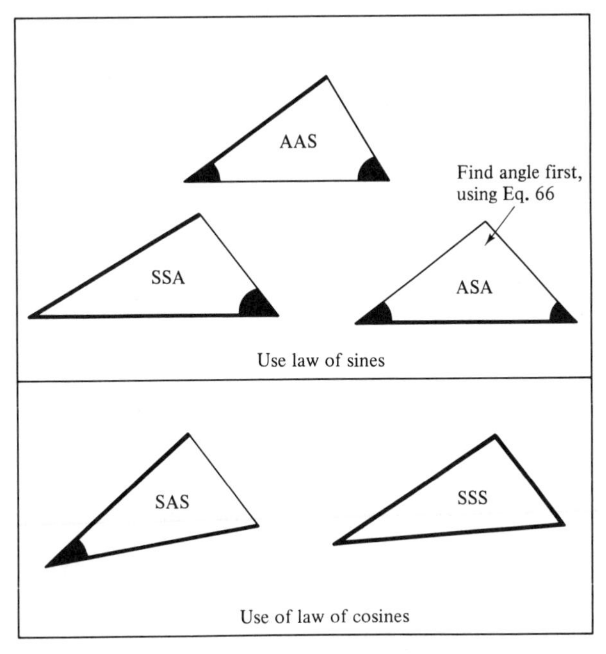

FIGURE 13-21. When to use the law of sines or law of cosines.

Using the Cosine Law When Two Sides and the Included Angle Are Known: We can solve triangles by the law of cosines if we know *two sides and the angle between them,* or if we know *three sides.* We consider the first of these in the following example.

Example: Solve triangle ABC where $a = 184$, $b = 125$, and $C = 27.2°$.

Notice that we cannot initially use the law of sines because we do not have a known side opposite a known angle.

Solution: We make a sketch (Fig. 13-22). Then by the law of cosines,

$$c = \sqrt{(184)^2 + (125)^2 - 2(184)(125)\cos 27.2°}$$
$$= 92.6$$

Now that we have a known side opposite a known angle, we can use the law of sines to find angle A or angle B. Which shall we find first?

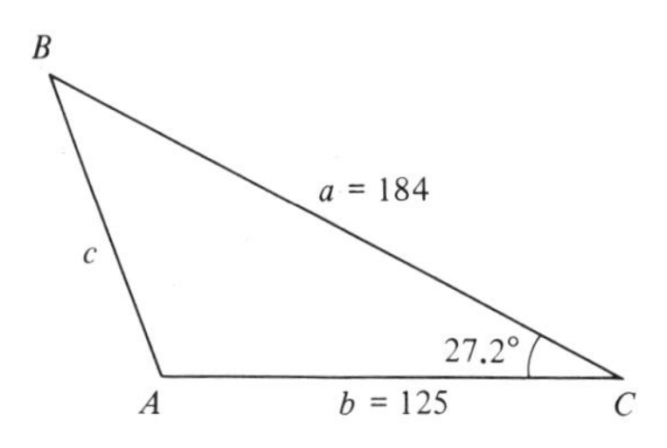

FIGURE 13-22

Use the law of sines to *find the acute angle first* (angle B in this example). If, instead, you solve for the obtuse angle first, you may forget to subtract the angle obtained from the calculator from 180°. Further, if one of the angles is so close to 90° that you cannot tell from your sketch if it is acute or obtuse, find the other angle first and then subtract the two known angles from 180° to obtain the third angle.

By the law of sines,

$$\frac{\sin B}{125} = \frac{\sin 27.2°}{92.6}$$
$$\sin B = \frac{125 \sin 27.2°}{92.6} = 0.617$$
$$B = 38.1°$$

and by Eq. 139,

$$A = 180° - 27.2° - 38.1° = 114.7°$$

Example: Solve triangle ABC where $b = 16.4$, $c = 10.6$, and $A = 128.5°$.

Solution: We make a sketch (Fig. 13-23). Then by the law of cosines,

$$a^2 = (16.4)^2 + (10.6)^2 - 2(16.4)(10.6)\cos 128.5°$$

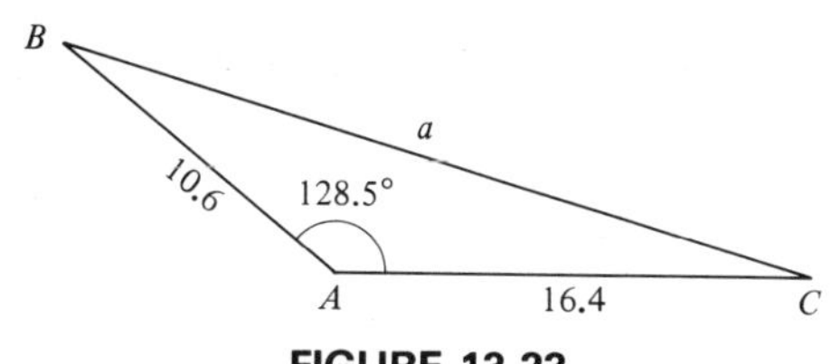

FIGURE 13-23

Common Error	The cosine of an obtuse angle is *negative*. Be sure to use the proper algebraic sign when applying the law of cosines to an obtuse angle.

In our example,

$$\cos 128.5° = -0.6225$$

So

$$a = \sqrt{(16.4)^2 + (10.6)^2 - 2(16.4)(10.6)(-0.6225)} = 24.4$$

By the law of sines,

$$\frac{\sin B}{16.4} = \frac{\sin 128.5°}{24.4}$$

$$\sin B = \frac{16.4 \sin 128.5°}{24.4} = 0.526$$

$$B = 31.7°$$

Then, by Eq. 139,

$$C = 180° - 31.7° - 128.5° = 19.8°$$

Using the Cosine Law When Three Sides Are Known: When three sides of an oblique triangle are known, we can use the law of cosines to solve for one of the angles. A second angle is found using the sine law, and the third angle is found by subtracting the other two from 180°.

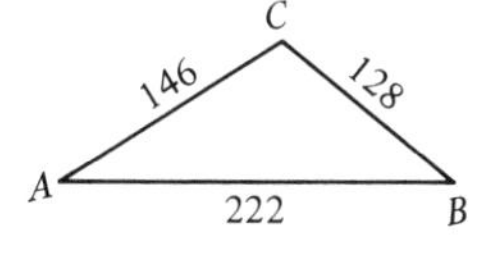

FIGURE 13-24

Example: Solve triangle ABC in Fig. 13-24, where $a = 128$, $b = 146$, and $c = 222$.

Solution: We write the law of cosines for any of the angles, say A.

$$a^2 = b^2 + c^2 - 2bc \cos A$$

$$(128)^2 = (146)^2 + (222)^2 - 2(146)(222) \cos A$$

Solving for $\cos A$, we get

$$\cos A = 0.8364$$

$$A = 33.2°$$

Here again, we find the acute angle B before the obtuse angle C. We could also avoid ambiguity if we had first solved for the angle opposite the longest side (angle C in this case). The law of cosines would tell us if it is acute or obtuse.

Then by the sine law,

$$\frac{\sin B}{146} = \frac{\sin 33.2°}{128}$$

$$\sin B = \frac{146 \sin 33.2°}{128} = 0.6246$$

$$B = 38.7°$$

Finally,

$$C = 180° - 38.7° - 33.2° = 108.1°$$

EXERCISE 3

Law of Cosines

Data for triangle *ABC* are given in the following table. Solve for the missing parts.

	Angles			*Sides*		
	A	*B*	*C*	*a*	*b*	*c*
1.			27.3°	128	152	
2.		51.4°		1.95		1.46
3.	68.3°				18.3	21.7
4.				728	906	663
5.			106°	15.7	11.2	
6.		128°		1.16		1.95
7.	135°				275	214
8.			35°12′	77.3	81.4	
9.				11.3	15.6	12.8
10.		41°44′		199		202
11.	115°18′				46.8	51.3
12.				1.475	1.836	2.017
13.			1.17 rad	9.08	6.75	
14.		2.25 rad		186		179
15.	2.75 rad				1.77	1.99
16.				41.8	57.2	36.7
17.			41.77°	1445	1502	
18.		108.8°		7.286		6.187
19.				97.3	81.4	88.5
20.	36.29°				47.28	51.36

Computer

21. Write a program to solve a triangle by the law of cosines. Have it *menu-driven* with the following choices.

1. Two sides and the included angle given (SAS)
2. Three sides given (SSS)

These triangles are shown in Fig. 13-21. For each, have the computer ask for the given sides and angles, and then compute and print the missing sides and angles.

13-4. APPLICATIONS

As with right triangles, oblique triangles have many applications in technology, as you will see in the exercises for this section. Follow the same procedures for setting up these problems as we used for other word problems, and solve the resulting triangle by the law of sines or the law of cosines, or both.

If an *area* of an oblique triangle is needed, either compute all the sides and use Hero's formula (Eq. 138) or find an altitude with right triangle trigonometry and use Eq. 137.

Example: Find the area of the gusset in Fig. 13-25.

Solution: We first find θ:

$$\theta = 90° + 35.0° = 125.0°$$

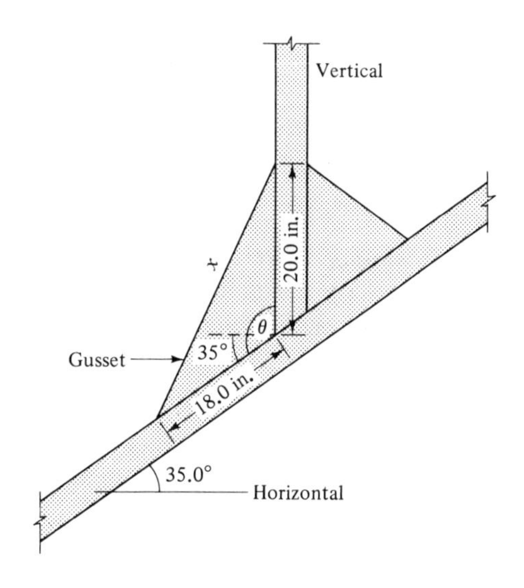

FIGURE 13-25

We now know two sides and the angle between them, so can find side x by the law of cosines:

$$x^2 = (18.0)^2 + (20.0)^2 - 2(18.0)(20.0) \cos 125° = 1137$$
$$x = 33.7 \text{ in.}$$

We find the area of the gusset by Hero's formula (Eq. 138),

$$s = \tfrac{1}{2}(18.0 + 20.0 + 33.7) = 34.9$$

$$\text{area} = \sqrt{34.9(34.9 - 18.0)(34.9 - 20.0)(34.9 - 33.7)}$$
$$= 103 \text{ in}^2.$$

Example: A ship takes a sighting on two buoys. At a certain instant the bearing of buoy A is N 44°14′ W and of buoy B is N 62°10′ E. The distance between the buoys is 3.60 km and bearing of B from A is N 87°52′ E. Find the distance of the ship from each buoy.

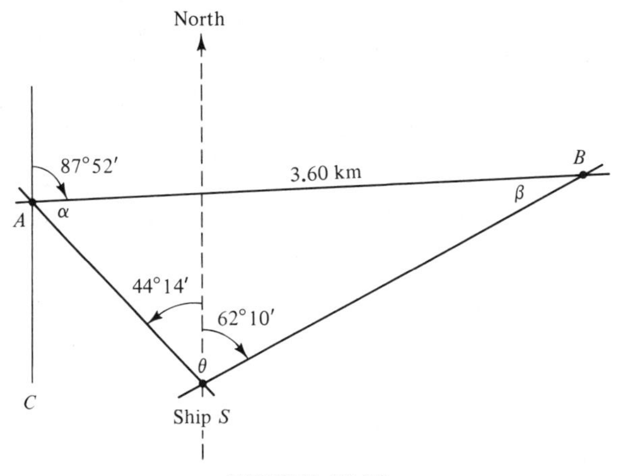

FIGURE 13-26

Solution: We draw a diagram (Fig. 13-26). Notice how the compass directions are laid out, starting from north and turning in the indicated direction. Finding the angles of the triangle ABS,

$$\theta = 44°14' + 62°10' = 106°24' = 106.4°$$

switching to decimal degrees for convenience. And since angle SAC is 44°14′,

$$\alpha = 180° - 87°52' - 44°14' = 47°54' = 47.9°$$

and

$$\beta = 180° - 106.4° - 47.9° = 25.7°$$

From the law of sines,

$$\frac{SA}{\sin 25.7°} = \frac{SB}{\sin 47.9°} = \frac{3.60}{\sin 106.4°}$$

from which

$$SA = \frac{3.60 \sin 25.7°}{\sin 106.4°} = 1.63 \text{ km}$$

and

$$SB = \frac{3.60 \sin 47.9°}{\sin 106.4°} = 2.78 \text{ km}$$

EXERCISE 4

Determining Inaccessible Distances

This set of problems contains no vectors, which are covered in the next section.

1. Two stakes, A and B, are 88.6 m apart. From a third stake C, the angle ACB is 85.4°, and from A, the angle BAC is 74.3°. Find the distance from C to each of the other stakes.
2. From a point on level ground, the angles of elevation of the top and the bottom of an antenna standing on top of a building are 32.6° and 27.8°, respectively. If the building is 125 ft high, how tall is the antenna?
3. A tower stands vertically on sloping ground whose inclination with the horizontal is 11.6°. From a point 42.0 m downhill from the tower (measured along the slope) the angle of elevation of the top of the tower is 18.8°. How tall is the tower?
4. A vertical antenna stands on a slope that makes an angle of 8.7° with the horizontal. From a point directly uphill from the antenna the angle of elevation of its top is 61°. From a point 16 m farther up the slope (measured along the slope) the angle of elevation of its top is 38°. How tall is the antenna?
5. A triangular lot measures 115 m, 187 m, and 215 m along its sides. Find the angles between the sides.
6. Two boats are 45.5 km apart. Both are traveling toward the same point, which is 87.6 km from one of them and 77.8 km from the other. Find the angle at which their paths intersect.

Navigation

Hint: *Draw your diagram after t hours have elapsed.*

7. A ship is moving at 15 km/h in the direction N 15° W. A helicopter with a speed of 22 km/h is due east of the ship. In what direction should the helicopter travel if it is to meet the ship?
8. City A is 215 miles N 12° E from city B. The bearing of city C from B is S 55° E. The bearing of C from A is S 15° E. How far is C from A? From B?
9. A ship is moving in a direction S 24°15′ W at a rate of 8.6 mi/h. If a launch that travels at 15.4 mi/h is due west of the ship, in what direction should it travel in order to meet the ship?
10. A ship is 9.5 km directly east of a port. If the ship sails southeast for 2.5 km, how far will it be from the port?
11. From a plane flying due east, the bearing of a radio station is S 31° E at 1 P.M. and S 11° E at 1:20 P.M. The ground speed of the plane is 625 km/h. Find the distance of the plane from the station at 1 P.M.

Structures

12. A power pole on level ground is supported by two wires that run from the top of the pole to the ground. One wire is 18.5 m long and makes an

angle of 55.6° with the ground, and the other wire is 17.8 m long. Find the angle the second wire makes with the ground.

13. A 71.6-m-high antenna mast (Fig. 13-27) is to be placed on sloping ground with the cables making an angle of 42.5° with the top of the mast. Find the lengths of the two cables.

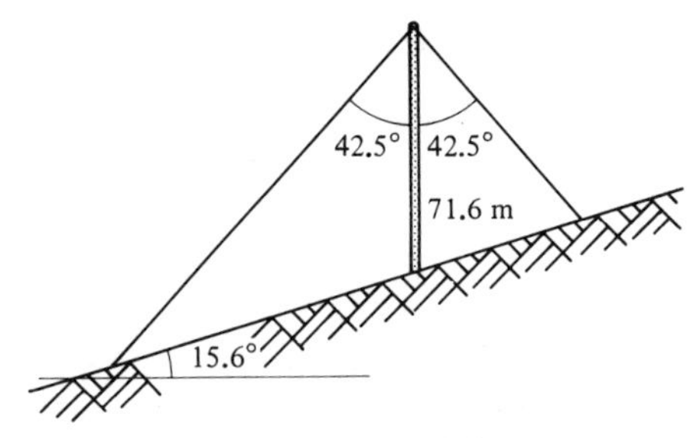

FIGURE 13-27

14. In the roof truss in Fig. 13-28 find the lengths of members *AB*, *BD*, *AC*, and *AD*.

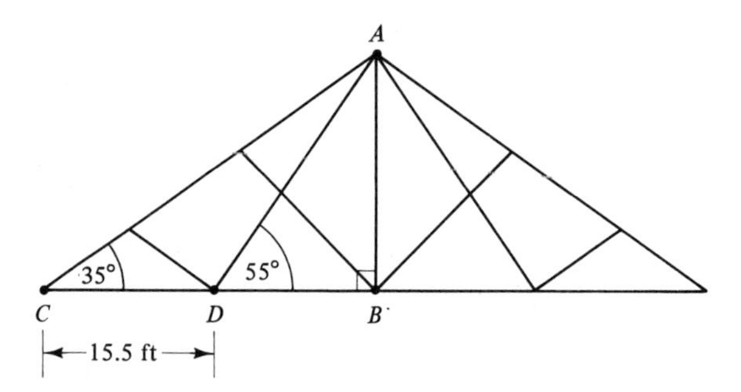

FIGURE 13-28. Roof truss.

15. From a point on level ground between two power poles of the same height, cables are stretched to the top of each pole. One cable is 52.6 ft long, the other is 67.5 ft long, and the angle of intersection between the two cables is 125°. Find the distance between the poles.

16. A pole standing on level ground makes an angle of 85.8° with the horizontal. The pole is supported by a 22-ft prop whose base is 12.5 ft from the base of the pole. Find the angle made by the prop with the horizontal.

Mechanisms

17. In the slider crank mechanism of Fig. 13-29, find the distance x between the wrist pin W and the crank center C, when $\theta = 35.7°$.

18. In the four-bar linkage of Fig. 13-30, find angle θ when angle *BAD* is 41.5°.

19. Two links, *AC* and *BC* (Fig. 13-31) are pivoted at *C*. How far apart are *A* and *B* when angle *ACB* is 66.3°?

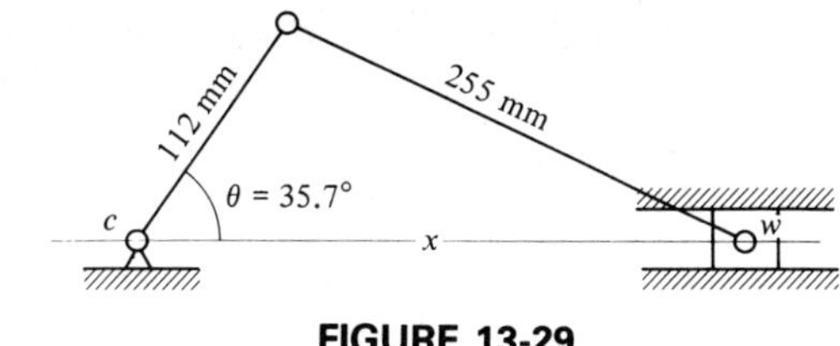

FIGURE 13-29

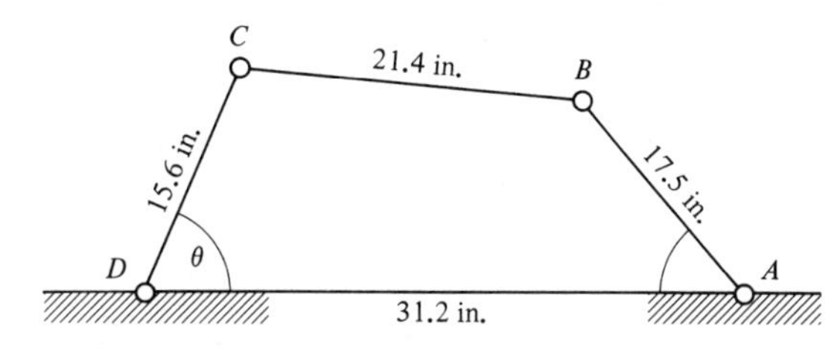

FIGURE 13-30

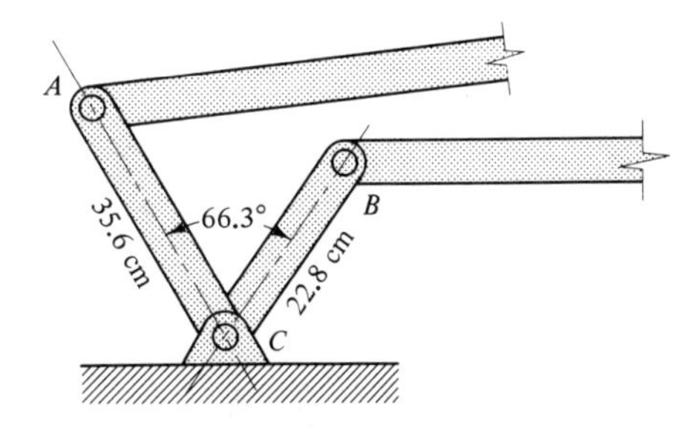

FIGURE 13-31

Geometry

20. Find angles A, B, and C in the quadrilateral in Fig. 13-32.

21. Find side AB in the quadrilateral in Fig. 13-33.

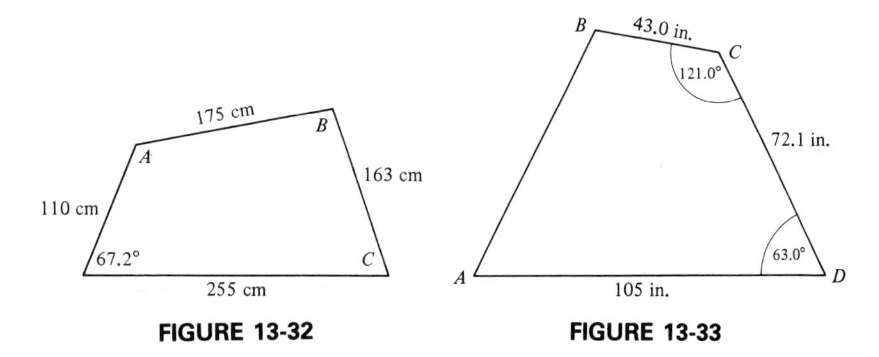

FIGURE 13-32

FIGURE 13-33

22. Two sides of a parallelogram are 22.8 and 37.8 m and one of the diagonals is 42.7 m. Find the angles of the parallelogram.

23. Find the lengths of the sides of a parallelogram if its diagonal, which is 125 mm long, makes angles with the sides of 22.7° and 15.4°.

24. Find the lengths of diagonals of a parallelogram two of whose sides are 3.75 m and 1.26 m; their included angle is 68.4°.

25. In triangle ABC, $A = 62.3°$, $b = 112$, and the median from C to the midpoint of c is 186. Find c.

A median of a triangle is a line joining a vertex to the midpoint of the opposite side.

26. The sides of a triangle are 124, 175, and 208. Find the length of the median drawn to the longest side.

27. The angles of a triangle are as 3:4:5 and the shortest side is 994. Solve the triangle.

28. The sides of a triangle are in the ratio of 2:3:4. Find the cosine of the largest angle.

29. Two solar panels are to be placed as in Fig. 13-34. Find the minimum distance x so that the first panel will not cast a shadow on the second when the angle of elevation of the sun is 18.5°.

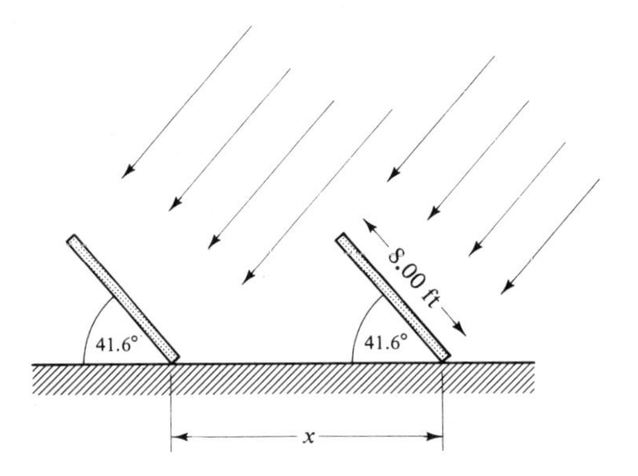

FIGURE 13-34. Solar panels.

30. Find the overhang x so that the window in Fig. 13-35 will be in complete shade when the sun is 60° above the horizontal.

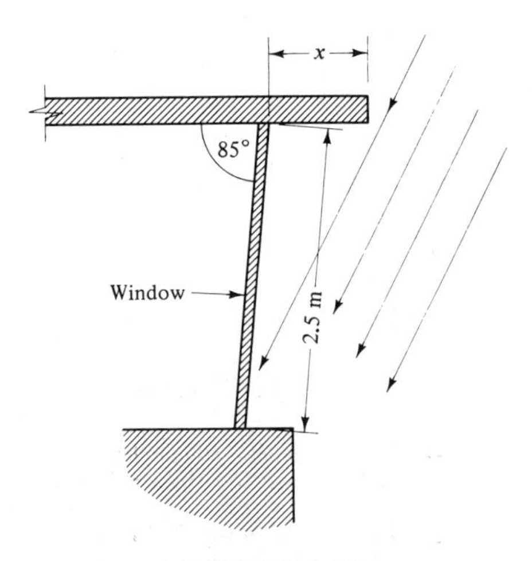

FIGURE 13-35

Computer

31. Combine your triangle-solving programs from exercises 2 and 3 into a single program to solve any triangle. The menu should contain the five options shown in Fig. 13-21. Also compute the area of the triangle by Hero's formula.

32. For the slider crank mechanism of Fig. 13-29, compute and print the position x of the wrist pin for values of θ from 0 to 180°, in 15° steps.

13-5. ADDITION OF VECTORS

In Sec. 12-4 we added two nonperpendicular vectors by first resolving each into components, then adding the components, and finally resolving the components into a single resultant. Now, by using the law of sines and the law of cosines, we can combine two nonperpendicular vectors directly, with much less time and effort. However, when more than two vectors must be added, it is faster to resolve each into its x and y components, combine the x components and the y components, and then find the resultant of those two perpendicular vectors.

In this section we show both methods.

Vector Diagram: We can illustrate the resultant, or vector sum, of two vectors by means of a diagram. Suppose that we wish to add the vectors **A** and **B** (Fig. 13-36a). If we draw the two vectors *tip to tail,* as in Fig. 13-36b, the resultant **R** will be the vector that will complete the triangle when drawn from the tail of the first vector to the tip of the second vector. It does not matter whether vector **A** or vector **B** is drawn first; the same resultant will be obtained either way, as shown

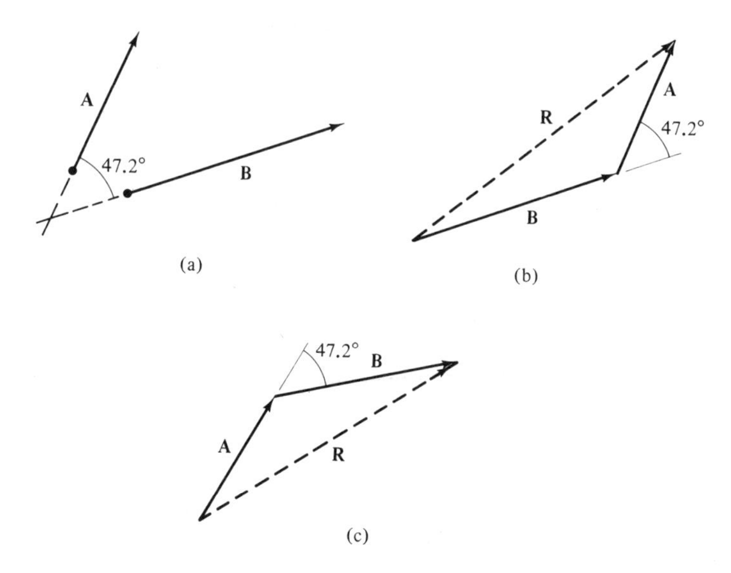

FIGURE 13-36. Addition of vectors.

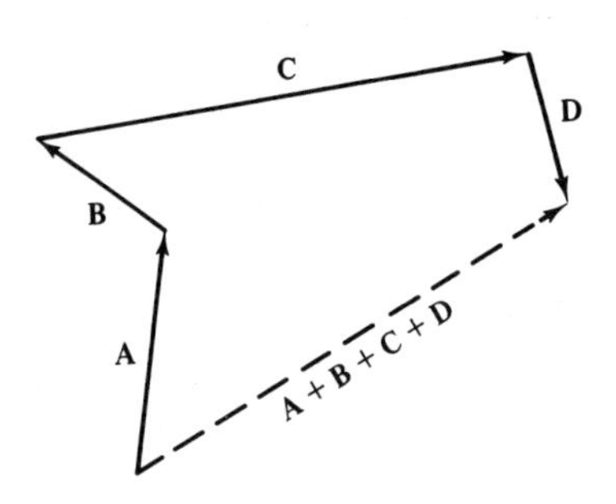

FIGURE 13-37. Vector addition of several vectors.

in Fig. 13-36c. This method can also be used to illustrate the addition of several vectors, as in Fig. 13-37.

The *parallelogram method* will give the same result. To add the same two vectors **A** and **B** as before, we first draw the given vectors *tail to tail* (Fig. 13-38) and complete a parallelogram by drawing lines parallel to the given vectors. The resultant **R** is then the diagonal of the parallelogram drawn from the intersections of the tails of the original vectors.

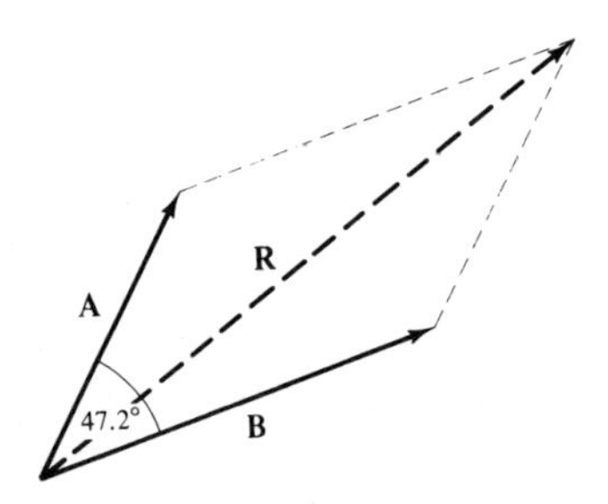

FIGURE 13-38. Parallelogram method.

Finding the Resultant of Two Nonperpendicular Vectors: Whichever method we use for drawing the vectors, *the resultant is one side of an oblique triangle*. To find the length of the resultant and the angle it makes with one of the original vectors, we simply solve the oblique triangle by the methods we learned earlier in this chapter.

Example: Two vectors, **A** and **B**, make an angle of 47.2° with each other (Fig. 13-38). If their magnitudes are $A = 125$ and $B = 146$, find the magnitude of the resultant **R** and the angle **R** makes with vector **B**.

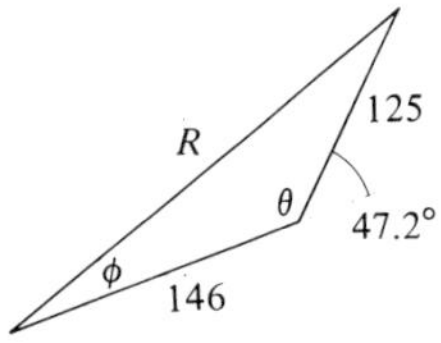

Solution: We make a vector diagram, either tip to tail or by the parallelogram method. Either way, we must solve the oblique triangle in Fig. 13-39 for R and ϕ. Finding θ,

FIGURE 13-39

$$\theta = 180° - 47.2° = 132.8°$$

By the law of cosines,

$$R^2 = (125)^2 + (146)^2 - 2(125)(146)\cos 132.8°$$
$$= 61{,}450$$
$$R = 248$$

Then by the law of sines,

$$\frac{\sin\phi}{125} = \frac{\sin 132.8}{248}$$
$$\sin\phi = \frac{125\sin 132.8}{248} = 0.3698$$
$$\phi = 21.7°$$

Addition of Several Vectors: The law of sines and the law of cosines are good for adding *two* nonperpendicular vectors. However, when *several* vectors are to be added, we usually break each into its x and y components and combine them, as in the following example.

Example: Find the resultant of the vectors shown in Fig. 13-40a.

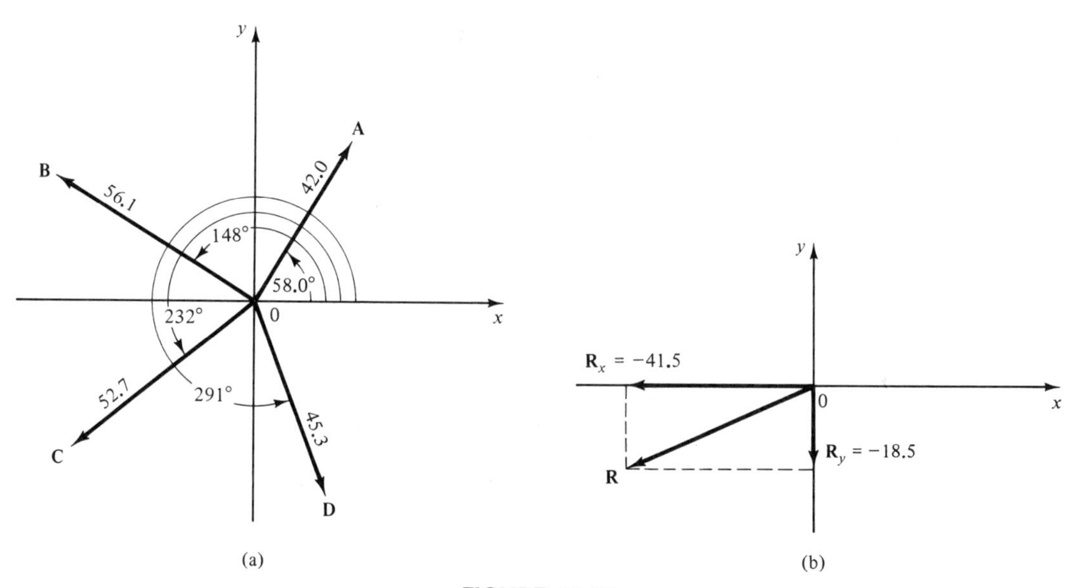

FIGURE 13-40

Solution: The x component of a vector of magnitude V at an angle θ is

$$V\cos\theta$$

and the y component is

$$V\sin\theta$$

We compute and tabulate the x and y components of each original vector, and find the sums of each.

Vector	x *component*	y *component*
A	$42.0 \cos 58.0° = 22.3$	$42.0 \sin 58.0° = 35.6$
B	$56.1 \cos 148° = -47.6$	$56.1 \sin 148° = 29.7$
C	$52.7 \cos 232° = -32.4$	$52.7 \sin 232° = -41.5$
D	$45.3 \cos 291° = \underline{16.2}$	$45.3 \sin 291° = \underline{-42.3}$
	$\mathbf{R}_x = -41.5$	$\mathbf{R}_y = -18.5$

These two vectors are shown in Fig. 13-40b. We find their resultant by the Pythagorean theorem,

$$R^2 = (-41.5)^2 + (-18.5)^2 = 2065$$
$$R = 45.4$$

and the angle θ by

$$\theta = \arctan \frac{\mathbf{R}_y}{\mathbf{R}_x}$$
$$= \arctan \frac{-18.5}{-41.5}$$
$$= 24.0° \quad \text{or} \quad 204°$$

Since our resultant is in the third quadrant, our final result is

$$\mathbf{R} = 45.4 \angle 204°$$

EXERCISE 5

Resultants of Vectors

The magnitudes of vectors **A** and **B** are given in the table, as well as the angle between the vectors. For each, find the magnitude R of the resultant and the angle that the resultant makes with vector **B**.

	Magnitudes		
	A	*B*	*Angle*
1.	244	287	21.8°
2.	1.85	2.06	136°
3.	55.9	42.3	55°28′
4.	1.006	1.745	148°22′
5.	4483	5829	1.75 rad
6.	35.2	23.8	2.55 rad

Force Vectors

7. Two forces of 18.6 N and 21.7 N are applied to a point on a body. The angle between the forces is 44.6°. Find the magnitude of the resultant and the angle it makes with the larger force.

8. Two forces whose magnitudes are 187 lb and 206 lb act on an object. The angle between the forces is 88.4°. Find the magnitude of the resultant force.

9. A force of 125 N pulls due west on a body, and a second force pulls N 28°44′ W. The resultant force is 212 N. Find the second force and the direction of the resultant.

10. Forces of 675 lb and 828 lb act on a body. The smaller force acts due north; the larger force acts N 52°15′ E. Find the direction and magnitude of the resultant.

11. Two forces, of 925 N and 1130 N, act on an object. Their lines of action make an angle of 67.2° with each other. Find the magnitude and direction of their resultant.

12. Two forces, of 136 lb and 251 lb, act on an object with an angle of 53.9° between their lines of action. Find the magnitude of their resultant and its direction.

13. The resultant of two forces of 1120 N and 2210 N is 2870 N. What angle does the resultant make with each of the two forces?

14. Three forces are in equilibrium: 212 N, 325 N, and 408 N. Find the angles between their lines of action.

Velocity Vectors

See Fig. 13-41 for definitions of the terms used in these problems.

15. As an airplane heads west with an air speed of 325 mi/h, a wind with a speed of 35 mi/h causes the plane to travel slightly south of west with a ground speed of 305 mi/h. In what direction is the wind blowing? In what direction does the plane travel?

16. A boat heads S 15° E on a river that flows due west. The boat travels S 11° W with a speed of 25 km/h. Find the speed of the current and the speed of the boat in still water.

17. A pilot wishes to fly in a northeasterly direction. The wind is from the west at 36 km/h and the plane's speed in still air is 388 km/h. Find the heading and the ground speed.

18. The heading of a plane is N 27°44′ E, and its air speed is 255 mi/h. If the wind is blowing from the south with a velocity of 42 mi/h, find the actual direction of travel of the plane and its ground speed.

19. A plane flies with a heading of N 48° W and an air speed of 584 km/h. It is driven from its course by a wind of 58 km/h from S 12° E. Find the ground speed and the drift angle of the plane.

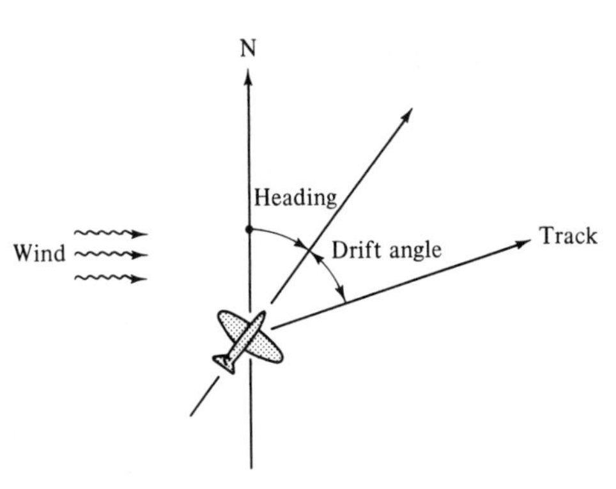

FIGURE 13-41. Flight terminology. The *heading* of an aircraft is the direction in which the craft is pointed. Due to air currents, it usually will not travel in that direction but in an actual path called the *track*. The angle between the heading and the track is the *drift angle*. The *air speed* is the speed relative to the surrounding air, and the *ground speed* is the craft's speed relative to the earth.

Current and Voltage Vectors

20. In Sec. 14-5 we will see that it is possible to represent an alternating current or voltage by a vector whose length is equal to the maximum amplitude of the current or voltage, placed at an angle that we later define as the *phase angle*. Then to add two alternating currents or voltages, we *add the vectors* representing those voltages or currents in the same way that we add force or velocity vectors.

A current $\mathbf{I}_1$ is represented by a vector of magnitude 12.5 at an angle of 15.6°, and a second current $\mathbf{I}_2$ is represented by a vector of magnitude 7.38 at an angle of 132°, as in Fig. 13-42. Find the magnitude and direction of the sum of these currents, represented by the vector **I**.

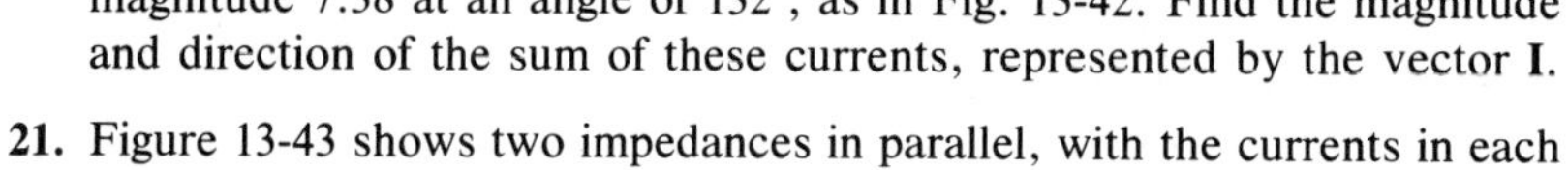

FIGURE 13-42

FIGURE 13-43

21. Figure 13-43 shows two impedances in parallel, with the currents in each represented by

$$\mathbf{I}_1 = 18.4 \text{ at } 51.5°$$

and

$$\mathbf{I}_2 = 11.3 \text{ at } 0°$$

The current **I** will be the sum of $\mathbf{I}_1$ and $\mathbf{I}_2$. Find the magnitude and direction of the vector representing **I**.

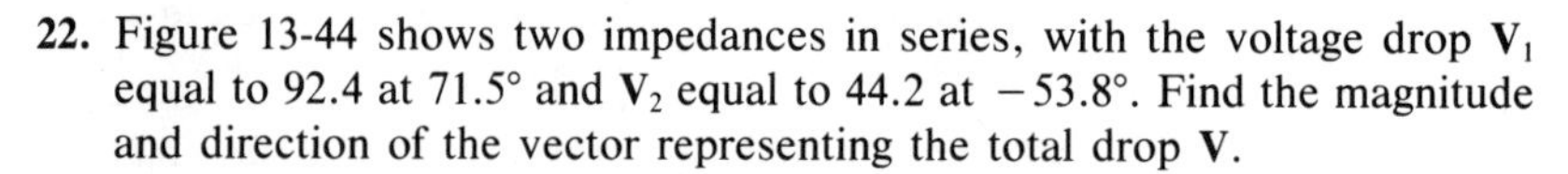

22. Figure 13-44 shows two impedances in series, with the voltage drop $\mathbf{V}_1$ equal to 92.4 at 71.5° and $\mathbf{V}_2$ equal to 44.2 at −53.8°. Find the magnitude and direction of the vector representing the total drop **V**.

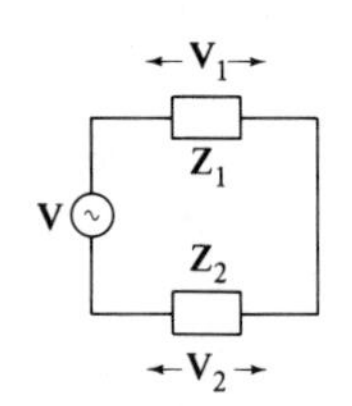

FIGURE 13-44

Addition of Several Vectors

Find the resultant of the following vectors.

23. 273 $\angle 34.0°$, 179 $\angle 143°$, 203 $\angle 225°$, 138 $\angle 314°$

24. 72.5 $\angle 284°$, 28.5 $\angle 331°$, 88.2 $\angle 104°$, 38.9 $\angle 146°$

Computer

25. Write a program that will accept as input the magnitude and direction of any number of vectors. Have the computer resolve each vector into x and y components, combine these components into a single x component and y component, and compute and print the magnitude and direction of the resultant. Try your program on Problems 23 and 24.

26. A certain airplane has an airspeed of 225 mi/h, and the wind is from the northeast at 45 mi/h. Write a program to compute and print the groundspeed and actual direction of travel of the plane, for headings that vary from due north completely around the compass, in steps of 15°.

CHAPTER TEST

Solve oblique triangle ABC, if:

1. $C = 135°$ $a = 44.9$ $b = 39.1$
2. $A = 92.4°$ $a = 129$ $c = 83.6$
3. $B = 38.4°$ $a = 1.84$ $c = 2.06$
4. $B = 22°38'$ $a = 2840$ $b = 1170$
5. $A = 132°$ $b = 38.2$ $c = 51.8$

In what quadrant(s) will the terminal side of θ lie if:

6. $\theta = 227°$ 7. $\theta = -45°$ 8. $\theta = 126°$ 9. $\theta = 3$ rad 10. $\tan\theta$ is negative

Without using book or calculator, state the algebraic sign of:

11. tan 275° 12. sec (−58°) 13. cos 183° 14. $\cos \pi/4$ 15. sin 300°

Write the six trigonometric functions, to 3 digits, for the angle whose terminal side passes through the point

16. (−2, 5) 17. (−3, −4) 18. (5, −1)

Two vectors of magnitudes A and *B are separated by an angle θ. Find the resultant and the angle the resultant makes with vector* **B**.

19. $A = 837$ $B = 527$ $\theta = 58.2°$
20. $A = 2.58$ $B = 4.82$ $\theta = 82.7°$
21. $A = 44.9$ $B = 29.4$ $\theta = 155°$
22. $A = 8374$ $B = 6926$ $\theta = 115.4°$
23. From a ship sailing north at the rate of 18 km/h, the bearing of a lighthouse is N 18°15′ E. Ten minutes later the bearing is N 75°46′ E. How far is the ship from the lighthouse at the time of the second observation?

Write the six trigonometric functions to four significant digits of:

24. 273° 25. 175° 26. 334°36′ 27. 127°22′ 28. 1.99 rad

Evaluate to four significant digits the expression

29. $\sin 35° \cos 35°$ **30.** $\tan^2 68°$ **31.** $(\cos 14° + \sin 14°)^2$

32. What angle does the slope of a hill make with the horizontal if a vertical tower 18.5 m tall, located on the slope of the hill, is found to subtend an angle of 25.5° from a point 35.0 m directly downhill from the foot of the tower?

33. Three forces are in equilibrium. One force of 457 lb acts in the direction N 28° W. The second force acts N 37° E. Find the direction of the third force of magnitude 638 lb.

Find to the nearest tenth of a degree all values of θ less than 360°.

34. $\cos \theta = 0.736$ **35.** $\tan \theta = -1.16$ **36.** $\sin \theta = 0.774$

Evaluate to the nearest tenth of a degree.

37. arcsin 0.737 **38.** $\tan^{-1} 4.37$ **39.** $\cos^{-1} 0.174$

40. A ship wishes to travel in the direction N 38° W. The current is from due east at 4.2 mi/h and the speed of the ship in still water is 18.5 mi/h. Find the direction in which the ship should head, and the speed of the ship in the actual direction of travel.

41. Two forces, of 483 lb and 273 lb, act on a body. The angle between the lines of action of these forces is 48.2°. Find the magnitude of the resultant and the angle it makes with the 483-lb force.

42. Find the vector sum of two voltages, $\mathbf{V}_1 = 134 \angle 24.5°$ and $\mathbf{V}_2 = 204 \angle 85.7°$.

Additional Topics in Trigonometry

This is our final chapter on trigonometry, and into it we put an assortment of topics. We start with *radian measure,* which we then use to calculate *arc lengths* and velocities in *rotating bodies*. Next we study *periodic waveforms*. We graph the *sine and cosine waves* and show how the constants in the equation affect the shape and position of the curves. This is followed by an important application in alternating current. We then introduce *polar coordinates* as an alternative to graphing in the rectangular coordinate system, and show how to convert from one to the other. This is followed by the fundamental *trigonometric identities,* which we use to simpify expressions and to help us solve *trigonometric equations*.

14-1. RADIAN MEASURE AND ARC LENGTH

We have already defined radians. Refresh your memory if necessary by glancing back at Sec. 12-1.

Radian Measure in Terms of π: In our use of radian measure so far, we have expressed radians in decimal form, such as

$$30 \text{ degrees} = 0.5236 \text{ rad} \quad \text{(rounded)}$$

It is also very common to express radian measure in terms of π. We know that 180° equals π radians, so

$$90° = \frac{\pi}{2} \text{ rad}$$

$$45° = \frac{\pi}{4} \text{ rad}$$

$$15° = \frac{\pi}{12} \text{ rad}$$

and so on.

Example: Express 135° in radian measure in terms of π.

Solution:

$$135° \left(\frac{\pi \text{ rad}}{180°}\right) = \frac{3\pi}{4} \text{ rad}$$

Example: Convert $7\pi/9$ rad to degrees.

Solution:

$$\frac{7\pi}{9} \text{ rad} \left(\frac{180°}{\pi \text{ rad}}\right) = 140°$$

Trigonometric Functions of Angles in Terms of π: When finding the trigonometric function of an angle expressed in terms of π on the calculator, it is necessary first to convert the angle to decimal form.

Example: Find cos $(5\pi/12)$ to four significant digits.

Solution: The keystrokes, with the calculator in the "radian" mode, are:

Press	*Display*
5 [×] [π] [÷] 12 [=]	1.308996939
[cos]	0.2588 (rounded)

Example: Evaluate to four significant digits.

$$5 \cos \frac{2\pi}{5} + 4 \sin^2 \frac{3\pi}{7}$$

Solution: From the calculator,

$$\cos \frac{2\pi}{5} = 0.30902 \qquad \text{and} \qquad \sin \frac{3\pi}{7} = 0.97493$$

so

$$5 \cos \frac{2\pi}{5} + 4 \sin^2 \frac{3\pi}{7} = 5(0.3092) + 4(0.97493)^2$$
$$= 5.347$$

Arc Length: An angle in radians is the ratio of the arc it subtends to the radius of the circle in which it is the central angle. Thus, in Fig. 14-1 if s is the length of arc and r is the radius of the circle,

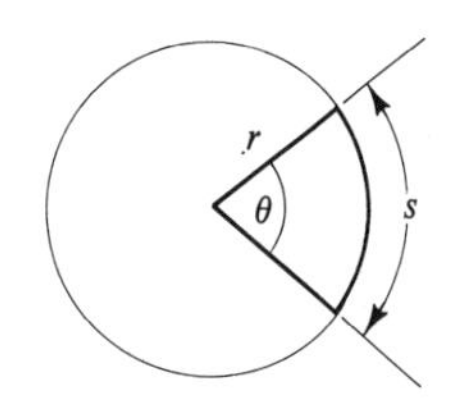

FIGURE 14-1. Relationship between arc length, radius, and central angle.

$\theta = \frac{s}{r}$ (θ must be in radians)	**115**

The central angle in a circle (in radians) is equal to the ratio of the intercepted arc and the radius of the circle.

We can use this formula to find any one of the quantities θ, s, or r when the other two are known.

Example: Find the angle that would intercept an arc of 27.0 in a circle of radius 21.0.

Solution: From Eq. 115,

$$\theta = \frac{s}{r} = \frac{27.0}{21.0} = 1.29 \text{ rad}$$

Example: Find the arc length intercepted by a central angle of 62.5° in a 10.4 cm radius circle.

Solution: Converting the angle to radians, we get

$$62.5°\left(\frac{\pi \text{ rad}}{180°}\right) = 1.09 \text{ rad}$$

By Eq. 115,

$$s = r\theta = 10.4(1.09) = 11.3 \text{ cm}$$

Example: Find the radius of a circle in which an angle of 2.06 rad intercepts an arc of 115 ft.

Solution: By Eq. 115,

$$r = \frac{s}{\theta} = \frac{115}{2.06} = 55.8 \text{ ft}$$

Common Error	Students sometimes confuse the decimal value of π with the degree equivalent of π radians. What is the value of π? 180 or 3.1416 . . .? Remember that the decimal value of π is $$\pi \cong 3.1416$$ and that when converting π to degrees, $$\pi \text{ radians} = 180 \text{ degrees}$$

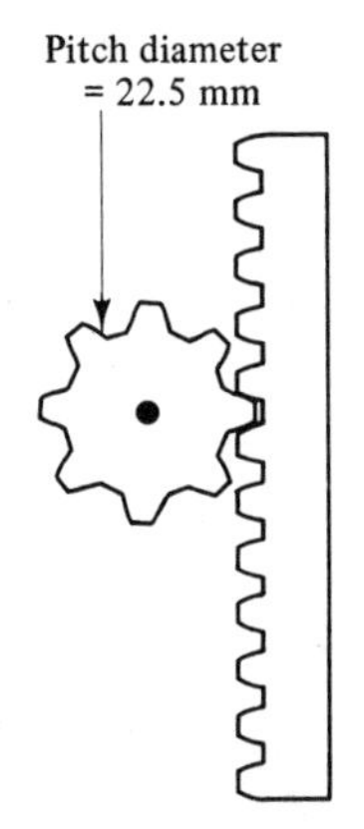

FIGURE 14-2. Rack and pinion.

Example: How far will the rack in Fig. 14-2 move when the pinion rotates 300.0°?

Solution: Converting to radians,

$$\theta = 300.0° \left(\frac{\pi \text{ rad}}{180°}\right) = 5.236 \text{ rad}$$

Then by Eq. 115,

$$s = r\theta = 11.25(5.236) = 58.91 \text{ mm}$$

Area of a Sector: The area of a sector having a radius r, an arc length s, and a central angle θ (in radians) is

Area of Sector	$A = \frac{rs}{2} = \frac{r^2\theta}{2}$	**116**

Example: Find the area of a sector having a radius of 8.25 m and a central angle of 46.8°.

Solution: The central angle must be in radians. By Eq. 117,

$$\theta = 46.8° \left(\frac{2\pi \text{ rad}}{360 \text{ deg}}\right) = 0.817 \text{ rad}$$

Then by Eq. 116,

$$\text{area} = \frac{(8.25)^2(0.817)}{2} = 27.8 \text{ m}^2$$

Common Error	Remember, when using Eqs. 115 and 116 that the angle must be in *radians*.

EXERCISE 1

Radian Measure

Convert each angle to radian measure in terms of π.

1. 60° **2.** 135° **3.** 66°

4. 240° **5.** 126° **6.** 105°

7. 78° **8.** 305° **9.** 400°

10. 150° **11.** 81° **12.** 189°

Convert each angle in radian measure to degrees to three significant digits.

13. $\frac{\pi}{8}$ **14.** $\frac{2\pi}{3}$ **15.** $\frac{9\pi}{11}$

16. $\frac{3\pi}{5}$ **17.** $\frac{\pi}{9}$ **18.** $\frac{4\pi}{5}$

19. $\frac{7\pi}{8}$ **20.** $\frac{5\pi}{9}$ **21.** $\frac{2\pi}{15}$

22. $\frac{6\pi}{7}$ **23.** $\frac{\pi}{12}$ **24.** $\frac{8\pi}{9}$

Evaluate to four significant digits.

25. $\sin \frac{\pi}{3}$ **26.** $\tan \frac{\pi}{8}$ **27.** $\cos \frac{\pi}{4}$

28. $\tan \left(-\frac{2\pi}{3}\right)$ **29.** $\cos \frac{3\pi}{5}$ **30.** $\sin \left(-\frac{7\pi}{8}\right)$

31. $\sec \frac{5\pi}{12}$ **32.** $\csc \frac{4\pi}{3}$ **33.** $\cot \frac{8\pi}{9}$

34. $\tan \frac{9\pi}{11}$ **35.** $\cos \left(-\frac{6\pi}{5}\right)$ **36.** $\sin \frac{4\pi}{9}$

Evaluate to four significant digits.

37. $\sin \frac{\pi}{6} + \cos \frac{\pi}{6}$ **38.** $7 \tan \frac{\pi}{9}$ **39.** $\cos^2 \frac{3\pi}{4}$

40. $\frac{\pi}{6} \sin \frac{\pi}{6}$ **41.** $\sin \frac{\pi}{8} \tan \frac{\pi}{8}$ **42.** $3 \sin \frac{\pi}{9} \cos^2 \frac{\pi}{9}$

Arc Length

In the following table, s is the length of arc subtended by a central angle θ in a circle of radius r. Fill in the missing values.

	r	θ	s
43.	4.83	$\frac{2\pi}{5}$	
44.	11.5	1.36 rad	
45.	284	46°24′	
46.	2.87 m	1.55 rad	
47.	64.8 in.	38.5°	
48.	28.3		32.5
49.	263 mm		582 mm
50.	21.5 ft		18.2 ft
51.	3.87 m		15.8 ft
52.		$\frac{\pi}{12}$	88.1
53.		77.2°	1.11
54.		2.08 rad	3.84 m
55.		12°55′	28.2 ft
56.		$\frac{5\pi}{6}$	125 mm

Assume the earth to be a sphere with a radius of 3960 mi. Actually, the distance from pole to pole is about 27 mi less than the diameter at the equator.

Applications

57. A certain town is at a latitude of 35.2° N. Find the distance in miles from the town to the north pole.

58. The hour hand of a clock is 85.5 mm long. How far does the tip of the hand travel between 1:00 A.M. and 11:00 A.M.?

59. Find the radius of a circular railroad track that will cause a train to change direction by 17.5° in a distance of 180 m.

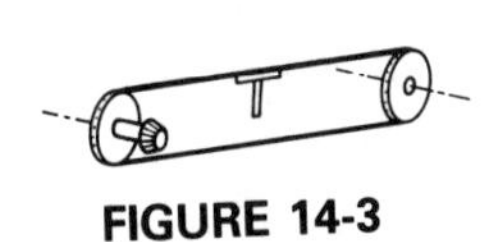

FIGURE 14-3

60. The pulley attached to the tuning knob of a radio (Fig. 14-3) has a radius of 35 mm. How far will the needle move if the knob is turned a quarter of a revolution?

61. Find the length of contact ABC between the belt and pulley in Fig. 14-4.

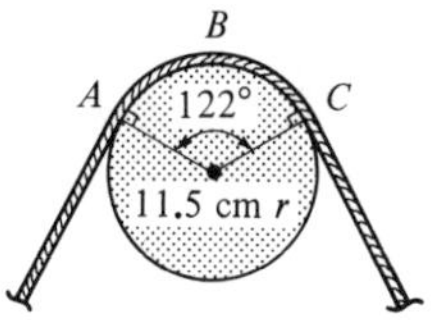

FIGURE 14-4. Belt and pulley.

62. One "track" on a magnetic disk used for computer data storage is located at a radius of 155 mm from the center of the disk. If 10 "bits" of data can be stored in 1 mm of track, how many bits can be stored in the length of track subtending an angle of $\pi/12$ rad?

63. If we assume the earth's orbit around the sun to be circular, with a radius of 93 million miles, how many miles does the earth travel (around the sun) in 125 days?

64. Find the latitude of a city that is 1265 mi from the equator.

65. A 1.25-m-long pendulum swings 5.75° on each side of the vertical. Find the length of arc traveled by the end of the pendulum.

Area of a Sector

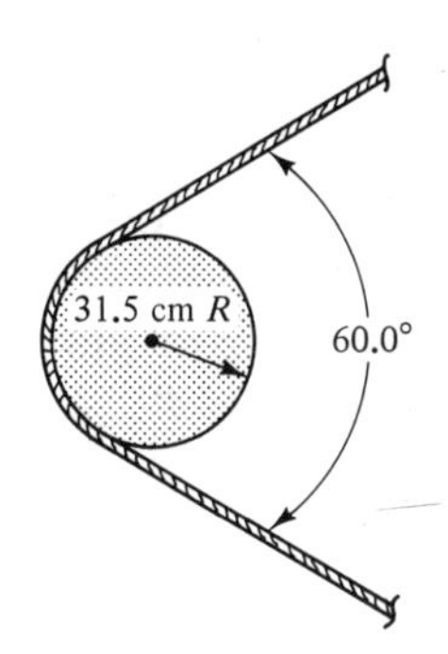

FIGURE 14-5. Brake drum.

66. Find the areas of the sectors bounded by the radii and the given arcs in Problems 43, 48, and 52.

67. A brake band is wrapped around a drum (Fig. 14-5). If the band has a width of 92.0 mm, find the area of contact between the band and the drum.

68. Sheet metal is to be cut from the pattern of Fig. 14-6a and bent to form the frustum of a cone (Fig. 14-6b), with top and bottom open. Find the dimensions r and R and the angle θ in degrees.

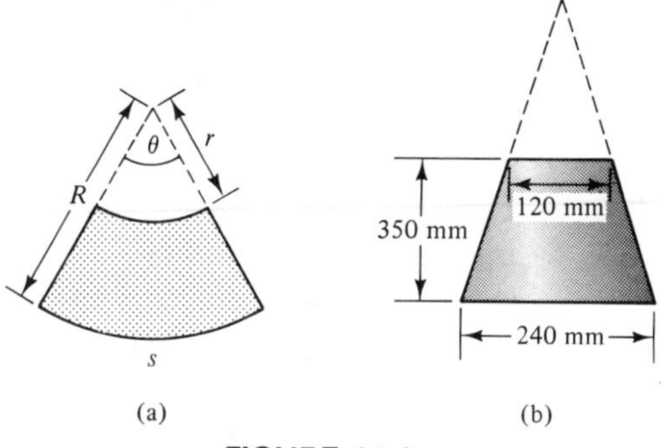

FIGURE 14-6

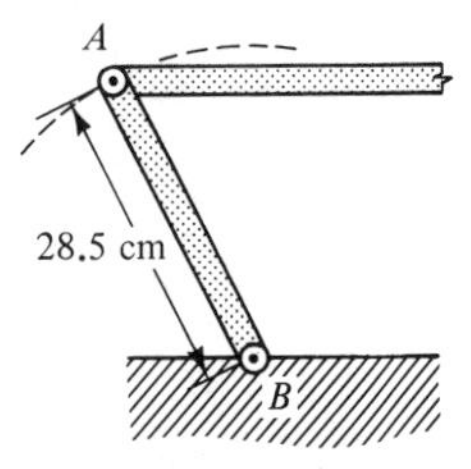

FIGURE 14-7

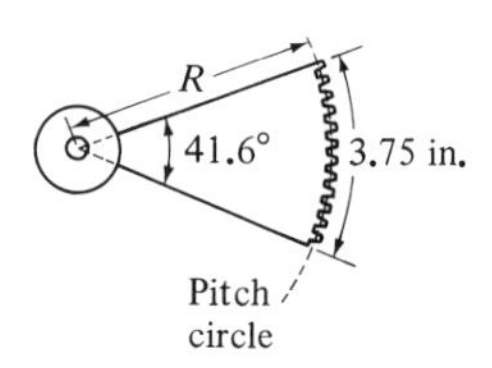

FIGURE 14-8. Sector gear.

69. An isosceles triangle is to be inscribed in a circle of radius 1.000. Find the angles of the triangle if its base subtends an arc of length 1.437.

70. A satellite is in a circular orbit 1.25 mi above the equator of the earth. How many miles must it travel for its longitude to change by 85.0°?

71. City B is due north of city A. City A has a latitude of 14°37′N and city B has a latitude of 47°12′N. Find the distance in kilometers between the cities.

72. The link AB in the mechanism of Fig. 14-7 rotates through an angle of 28.3°. Find the distance traveled by point A.

73. Find the radius R of the sector gear of Fig. 14-8.

74. A circular highway curve has a radius of 325.500 ft and a central angle of 15°25′15″ measured to the centerline of the road. Find the length of the curve.

Computer

75. For the angles from 0 to 10°, with steps every $\frac{1}{2}$ degree, compute and print the angle in radians, the sine of the angle, and the tangent of the angle. What do you notice about these three columns of figures? What is the largest angle for which the sine and tangent are equal to the angle in radians, to three significant digits?

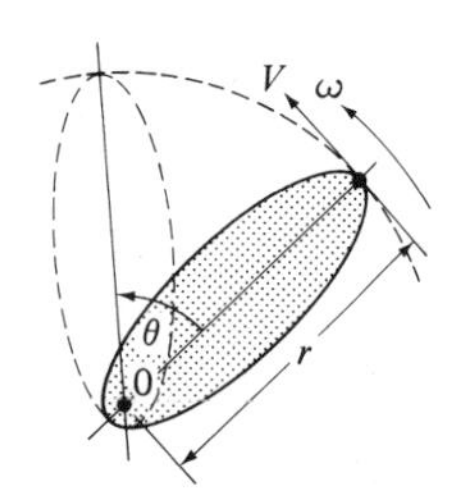

FIGURE 14-9. Rotating body.

14-2. UNIFORM CIRCULAR MOTION

Angular Velocity: Let us consider a rigid body (Fig. 14-9) that is rotating about a point O. The *angular velocity* ω is a measure of the *rate* at which the object rotates. The motion is called *uniform* when the angular velocity is constant. The units of angular velocity are degrees, radians, or revolutions, per unit time.

Angular Displacement: The angle θ through which the body rotates in time t is called the *angular displacement*. It is related to ω and t by

Angular Displacement	$\theta = \omega t$	**A22**

The angular displacement is the product of the angular velocity and the elapsed time.

Example: A wheel is rotating with an angular velocity of 1800 rev/min. Find the angular displacement in 1.50 s.

Solution: We first make the units of time consistent. Converting yields

$$\omega = \frac{1800 \text{ rev}}{\text{min}} \cdot \frac{1 \text{ min}}{60 \text{ s}} = 30 \text{ rev/s}$$

Then by Eq. A22

$$\theta = \omega t = \frac{30 \text{ rev}}{\text{s}}(1.5 \text{ s}) = 45 \text{ rev}$$

Example: Find the angular velocity in revolutions per minute of a pulley that rotates 275° in 0.750 s.

Solution: By Eq. A22

$$\omega = \frac{\theta}{t} = \frac{275°}{0.750 \text{ s}} = 367 \text{ deg/s}$$

Converting to rev/min, we obtain

$$\omega = \frac{367 \text{ deg}}{\text{s}} \cdot \frac{60 \text{ s}}{\text{min}} \cdot \frac{1 \text{ rev}}{360 \text{ deg}} = 61.2 \text{ rev/min}$$

Example: How long will it take a spindle rotating at 3.55 rad/s to make 1000 revolutions?

Solution: Converting revolutions to radians, we get

$$\theta = 1000 \text{ rev} \cdot \frac{2\pi \text{ rad}}{1 \text{ rev}} = 6280 \text{ rad}$$

Then by Eq. A22

$$t = \frac{\theta}{\omega} = \frac{6280 \text{ rad}}{3.55 \text{ rad/s}} = 1770 \text{ s} = 29.5 \text{ min}$$

Linear Speed: For any point on the rotating body, the linear displacement per unit time along the circular path is called the *linear speed.* The linear speed is zero for a point at the center of rotation and is directly proportional to the distance from the point to the center of rotation. If ω is expressed in radians per unit time, the linear speed v is

Linear Speed	$v = \omega r$	**A21**

The linear speed of a point on a rotating body is proportional to the distance from the center of rotation.

Example: A wheel is rotating at 2450 rev/min. Find the linear speed of a point 35.0 cm from the center.

Solution: We first express the angular velocity in terms of radians.

$$\omega = \frac{2450 \text{ rev}}{\text{min}} \cdot \frac{2\pi \text{ rad}}{\text{rev}} = 15{,}400 \text{ rad/min}$$

What became of "radians" in our answer? Shouldn't the final units be rad cm/min? No. Remember that radians is a dimensionless ratio; it is the ratio of two lengths (arc length and radius) whose units cancel.

Then, by Eq. A21,

$$v = \omega r = \frac{15{,}400 \text{ rad}}{\text{min}}(35.0 \text{ cm}) = 539{,}000 \text{ cm/min}$$
$$= 89.8 \text{ m/s}$$

Common Error	Remember when using Eq. A21 that the angular velocity must be expressed in *radians* per unit time.

Just as "radians" appeared to vanish in the last example, here they seem to appear out of nowhere.

Example: A belt having a speed of 885 in./min turns a 12.5-in.-radius pulley. Find the angular velocity of the pulley in rev/min.

Solution: By Eq. A21,

$$\omega = \frac{v}{r} = \frac{885 \text{ in.}}{\text{min}} \div 12.5 \text{ in.} = 70.8 \text{ (rad)/min}$$

Converting yields

$$\omega = \frac{70.8 \text{ rad}}{\text{min}} \cdot \frac{1 \text{ rev}}{2\pi \text{ rad}} = 11.3 \text{ rev/min}$$

EXERCISE 2

Angular Velocity

Fill in the missing values.

	rev/min	*rad/s*	*deg/s*
1.	1850		
2.		5.85	
3.			77.2
4.		$3\pi/5$	
5.			48.1
6.	22,600		

7. A flywheel makes 725 revolutions in a minute. How many degrees does it rotate in 1.00 s?

8. A propeller on a wind generator rotates 60° in 1.00 s. Find the angular velocity of the propeller in revolutions per minute.

9. A gear is rotating at 2550 rev/min. How many seconds will it take to rotate through an angle of 2.00 rad?

Linear Speed

10. A milling machine cutter has a diameter of 75.0 mm and is rotating at 5650 rev/min. What is the linear speed at the edge of the cutter?

11. A sprocket 3.00 in. in diameter is driven by a chain that moves at a speed of 55.5 in./s. Find the angular velocity of the sprocket in rev/min.

12. A capstan on a magnetic tape drive rotates at 3600 rad/min and drives the tape at a speed of 45.0 m/min. Find the diameter of the capstan in millimeters.

13. A blade on a water turbine turns 155° in 1.25 s. Find the linear speed of a point on the tip of the blade 0.750 m from the axis of rotation.

14. A steel bar 6.50 in. in diameter is being ground in a lathe. The surface speed of the bar is 55.0 ft/min. How many revolutions will the bar make in 10.0 s?

15. Assuming the earth to be a sphere 7920 mi in diameter, calculate the linear speed in miles per hour of a point on the equator due to the rotation of the earth about its axis.

16. Assuming the earth's orbit about the sun to be a circle with a radius of 93×10^6 mi, calculate the linear speed of the earth around the sun.

17. A car is traveling at a rate of 65.5 km/h and has tires that have a radius of 31.6 cm. Find the angular velocity of the wheels of the car.

18. A wind generator has a propeller 21.7 ft in diameter and the gearbox between the propeller and the generator has a gear ratio of 1:44 (with the generator shaft rotating faster than the propeller). Find the tip speed of the propeller when the generator is rotating at 1800 rev/min.

19. A winch has a drum 30.0 cm in diameter. The steel cable wrapped around the drum is to be pulled in at a rate of 8.0 ft/s. Ignoring the thickness of the cable, find the angular speed of the drum, in revolutions per minute.

20. Find the linear velocity of the tip of a 5.50-in.-long minute hand of a clock.

21. A satellite has a circular orbit around the earth with a radius of 4250 mi. The satellite makes a complete orbit of the earth in 3 days, 7 h, and 35 min. Find the linear speed of the satellite.

14-3. THE SINE CURVE

Periodic Functions: A curve that *repeats* its shape over and over is called a *periodic* curve or *periodic waveform,* as in Fig. 14-10. The function of which it is a graph is thus called a *periodic function*. Each repeated portion of the curve is called a *cycle*.

Period: The x distance taken for one cycle is called the *period*. The x axis will represent either an *angle* (in degrees or radians), or *time*.

Example: The period of the curve in Fig. 14-11 is 200 ms.

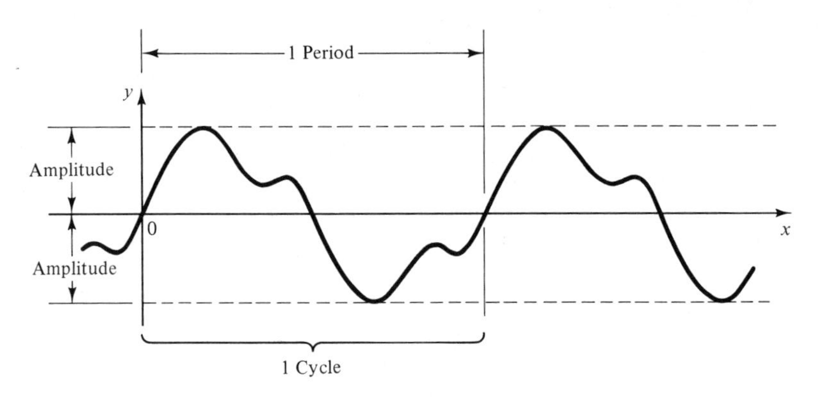

FIGURE 14-10. A periodic waveform.

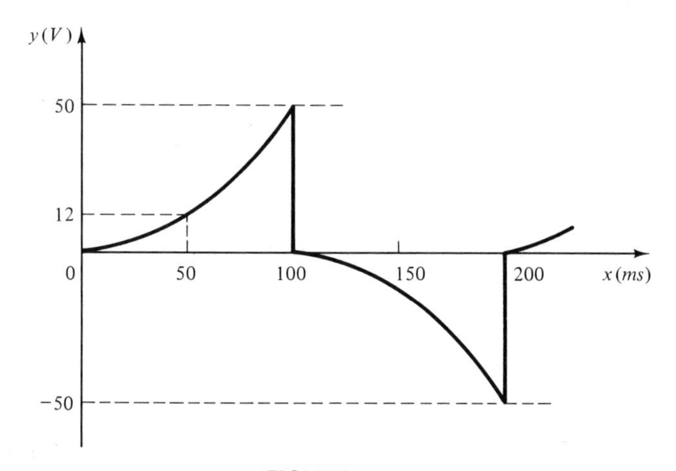

FIGURE 14-11

Amplitude: Half the difference between the greatest and least values of a waveform is called its *amplitude* (Fig. 14-10). The *peak-to-peak* value of a curve is equal to twice the amplitude. The *instantaneous value* of a waveform is its y value at any given x.

Example: For the voltage curve in Fig. 14-11,

$$\text{amplitude} = 50 \text{ V}$$

$$\text{peak-to-peak value} = 100 \text{ V}$$

$$\text{instantaneous value when } x \text{ is } 50 = 12 \text{ V}$$

General Sine Curve: The general equation of the sine curve contains three constants, a, b, and c.

General Sine Curve	$y = a \sin (bx + c)$	**199**

We will show, one at a time, the effect that each constant has on the

The sine curve or sine wave is also called a sinusoid.

sine curve. We start with the simplest case, where $a = 1$, $b = 1$, and $c = 0$.

$$y = \sin x$$

Graph of the Sine Curve $y = \sin x$: For now we will graph the sine curve just as we graphed other functions in Sec. 6-4. We choose values of x and substitute into the equation to get corresponding values of y, and plot the resulting table of point pairs. A little later we will learn a faster method.

The units of angular measure used for the x axis are sometimes degrees and sometimes radians. We will show examples of each.

Example: Plot the sine curve $y = \sin x$, for values of x from 0 to 360°.

Solution: Let us choose 30° intervals and make a table of point pairs.

x	0	30°	60°	90°	120°	150°	180°	210°	240°	270°	300°	330°	360°
y	0	0.5	0.866	1	0.866	0.5	0	−0.5	−0.866	−1	−0.866	−0.5	0

Plotting these points gives the periodic curve shown in Fig. 14-12. The period here is 360°.

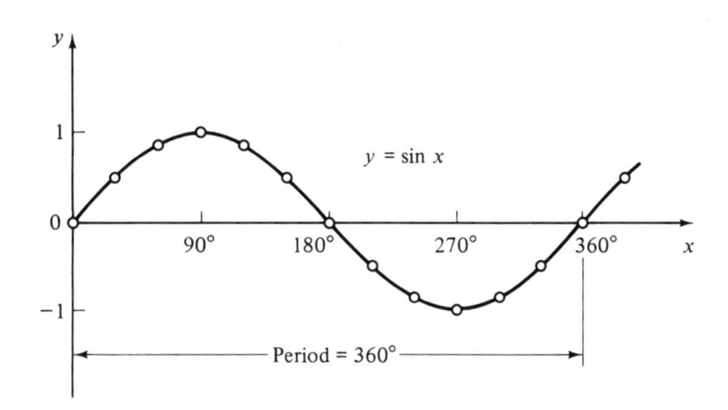

FIGURE 14-12. The sine curve in degrees.

We now do an example in radians.

It is not easy to work in radian measure when we are used to thinking of angles in degrees. You might want to glance back at Sec. 14-1.

Example: Plot the sine curve $y = \sin x$ for values of x from 0 rad to 7 rad.

Solution: We could choose integer values for x, such as 1, 2, 3, . . . and so on, or follow the usual practice of using multiples and fractions of π. This will give us points at the peaks and valleys of the curve, and where the curve crosses the x axis. We take intervals of $\pi/6$ and make the following table of point pairs.

x (rad)	0	$\frac{\pi}{6}$	$\frac{\pi}{3}$	$\frac{\pi}{2}$	$\frac{2\pi}{3}$	$\frac{5\pi}{6}$	π	$\frac{7\pi}{6}$	$\frac{4\pi}{3}$	$\frac{3\pi}{2}$	$\frac{5\pi}{3}$	$\frac{11\pi}{6}$	2π
y	0	0.5	0.866	1	0.866	0.5	0	−0.5	−0.866	−1	−0.866	−0.5	0

Our curve (Fig. 14-13) is the same as before, except for the scale on the x axis. There, we show both integer values of x, and multiples of π.

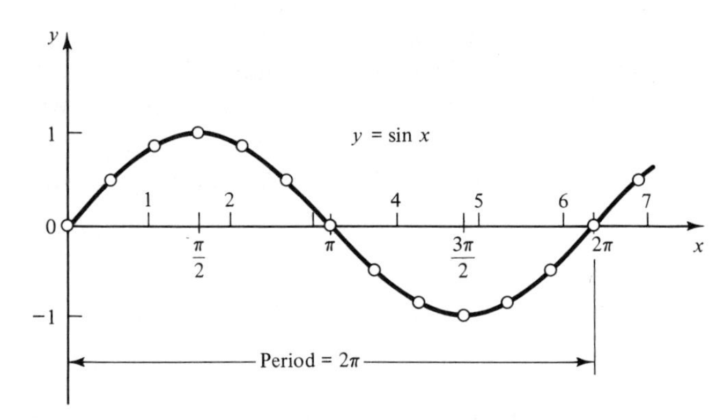

FIGURE 14-13. The sine curve in radians.

We show the same curve (Fig. 14-14) with both radian and degree scales on the x axis. Notice that one cycle of the curve fits into a rectangle whose height is twice the amplitude of the sine curve and whose width is equal to the period. We use this later for rapid sketching of periodic curves.

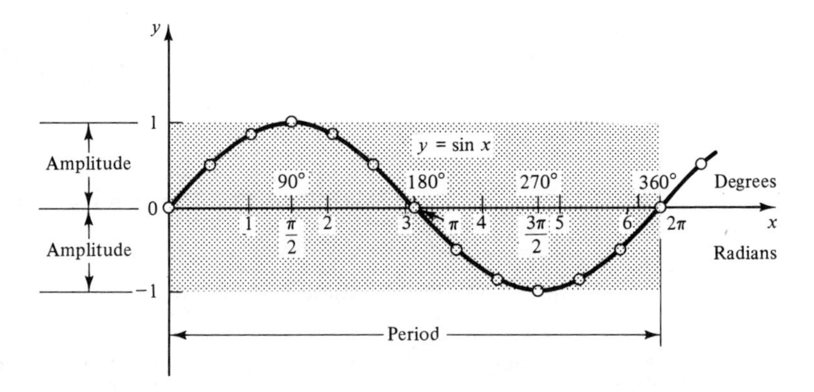

FIGURE 14-14. The sine curve in radians and degrees.

Graph of *a* sin *x*: The constant a in our general sine wave

$$y = \underset{\text{amplitude}}{a} \sin (bx + c)$$

is the *amplitude* of the sine wave. In the preceding examples, the constant a was equal to 1, and we got sine waves with amplitudes of 1. It can, of course, have another value, as in the following example. For now, we still let $b = 1$ and $c = 0$.

Example: Plot the sine curve $y = 3 \sin x$ for $x = 0$ to 2π.

Solution: When we make a table of point pairs, it becomes apparent that all the y values will be three times as large as the ones in the preceding example. These points are plotted in Fig. 14-15.

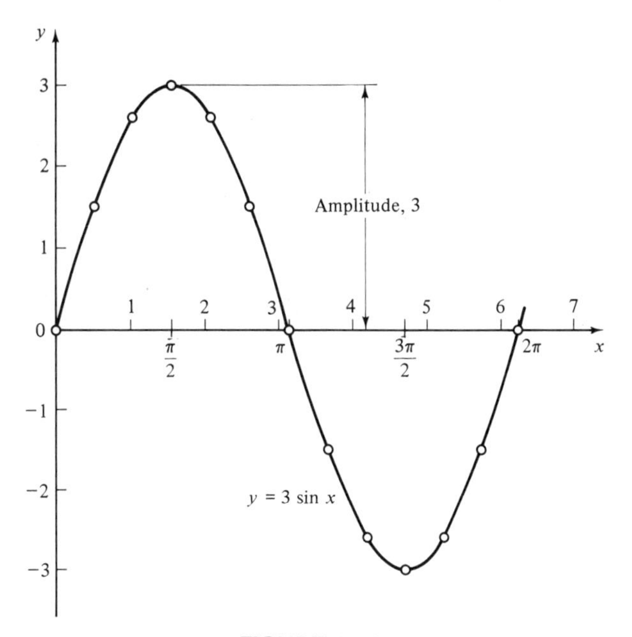

FIGURE 14-15

Graph of *a* sin *bx*: For the sine curve $y = a \sin x$, we saw that the curve repeated in an interval of 360° (or 2π radians). We called this interval the *period*.

Similarly, the curve

$$y = a \sin bx$$

will have a period of $360°/b$, if x is in degrees, or $2\pi/b$ radians, if x is in radians.

Period	$P = \dfrac{360}{b}$ degrees/cycle $= \dfrac{2\pi}{b}$ radians/cycle	**200**

Example: The period for the sine curve $y = \sin 4x$ is

$$P = \frac{360°}{4} = 90°/\text{cycle}$$

or

$$P = \frac{2\pi}{4} = \frac{\pi}{2} \quad \text{radians/cycle}$$

The *frequency* f is the reciprocal of the period.

Frequency	$f = \dfrac{b}{360}$ cycles/degree $= \dfrac{b}{2\pi}$ cycles/radian	**201**

Example: Determine the amplitude, period, and frequency of the sine curve $y = 2 \sin(\pi x/2)$, where x is in radians, and graph one cycle of the curve.

Solution: The amplitude is 2, and the period is

$$P = \frac{2\pi}{\pi/2} = 4 \text{ rad/cycle}$$

or, in degrees,

$$P = \frac{360°}{\pi/2} = 229.2°/\text{cycle}$$

We make a table of point pairs.

x (rad)	0	$\frac{1}{2}$	1	$\frac{3}{2}$	2	$\frac{5}{2}$	3	$\frac{7}{2}$	4
y	0	1.41	2	1.41	0	−1.41	−2	−1.41	0

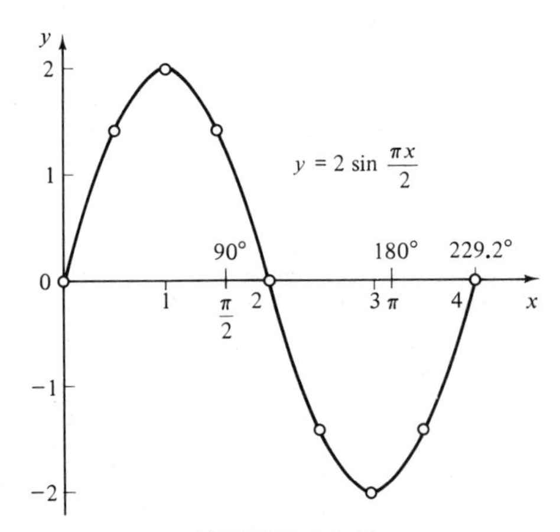

FIGURE 14-16

These points are plotted in Fig. 14-16. The frequency is

$$f = \frac{1}{P} = \frac{1}{4} \text{ cycles/rad}$$

In degrees,

$$f = \frac{1}{229.2°} = 0.00436 \text{ cycle/deg}$$

Graph of $y = a \sin (bx + c)$: The curve $y = a \sin bx$, as we have seen, passes through the origin. However, the curve $y = a \sin (bx + c)$ will be *shifted* in the x direction by an amount called the *phase displacement.*

Example: Graph one cycle of the sine curve $y = \sin (3x - \pi/2 \text{ rad})$.

Solution: As usual, we make a table of point pairs.

x	0	$\frac{\pi}{12}$	$\frac{\pi}{6}$	$\frac{\pi}{4}$	$\frac{\pi}{3}$	$\frac{5\pi}{12}$	$\frac{\pi}{2}$	$\frac{7\pi}{12}$	$\frac{2\pi}{3}$	$\frac{3\pi}{4}$	$\frac{5\pi}{6}$
y	-1	-0.707	0	0.707	1	0.707	0	-0.707	-1	-0.707	0

We see from the graph in Fig. 14-17 that the curve does not pass through the origin, but has been shifted by an amount $\pi/6$. In our equation this can be obtained by dividing the constant c by b,

$$-\frac{\pi}{2} \div 3 = -\frac{\pi}{6}$$

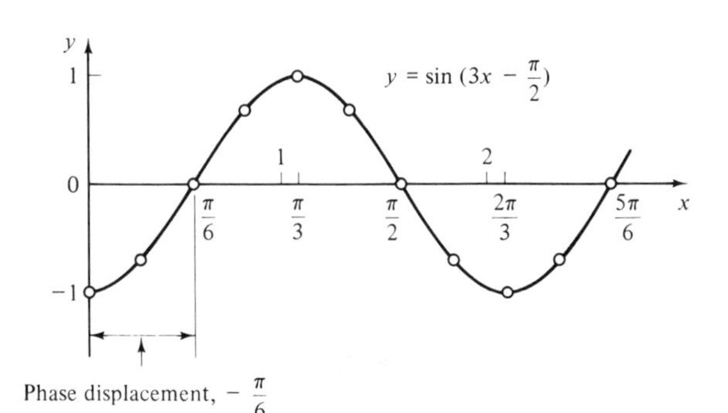

FIGURE 14-17. Sine curve with phase displacement.

When the phase displacement is positive, *the entire curve will be shifted to the* left. *A* negative *phase displacement will give a shift to the right.*

In general,

phase displacement $= \dfrac{c}{b}$	**202**

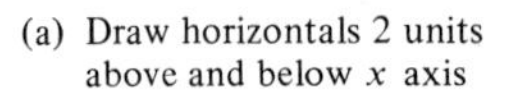

(a) Draw horizontals 2 units above and below x axis

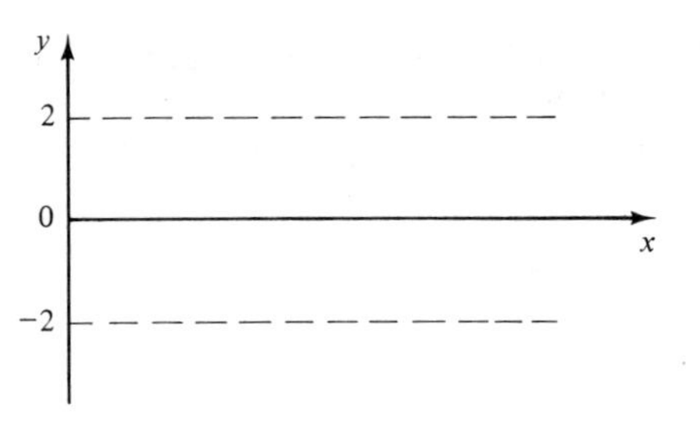

(b) Draw a vertical at a distance P (120°) from the origin

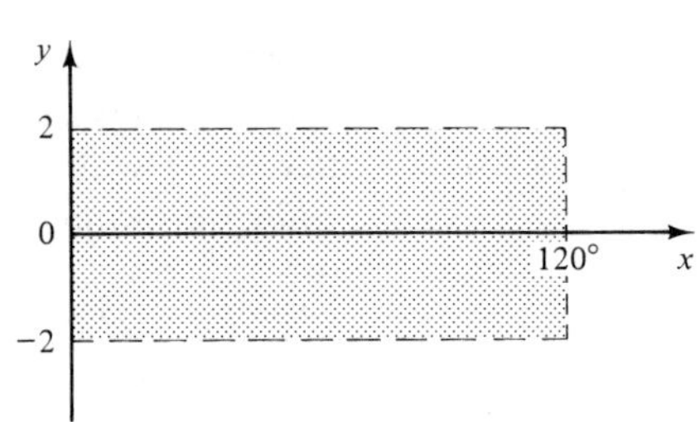

(c) Divide the period into 4 equal parts

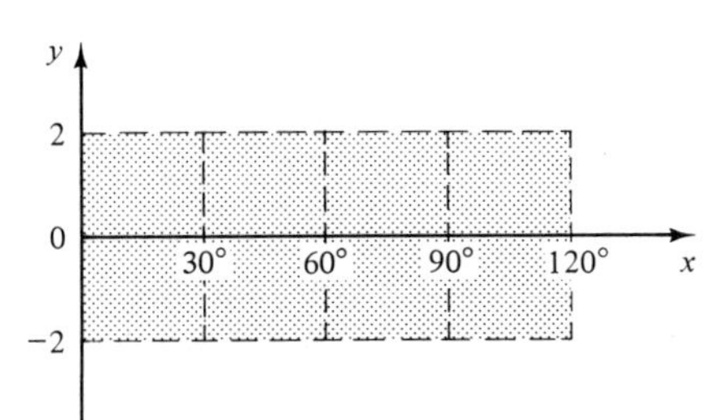

(d) Sketch the curve lightly

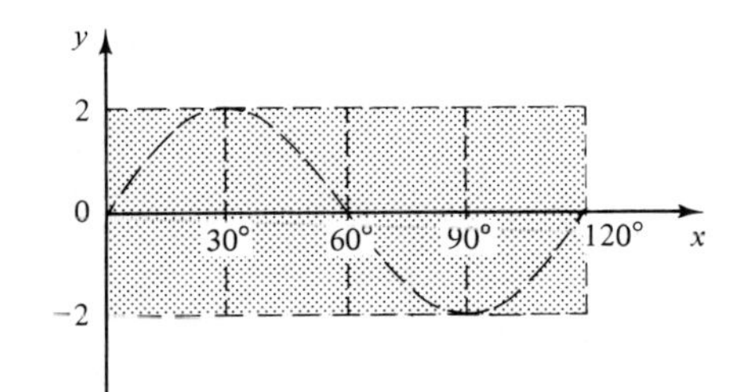

(e) Shift the curve by an amount of the phase displacement (20°), left for a positive phase displacement and right for a negative one

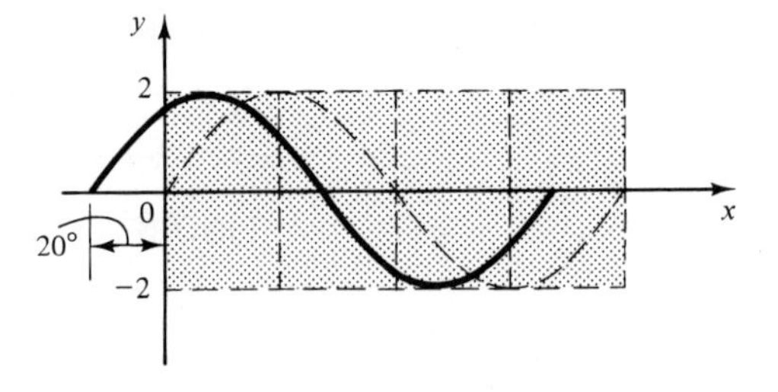

FIGURE 14-18. Quick sketching of the sine curve.

Where no units are indicated for c, assume it to be in radians.

The phase displacement will have the same units, radians or degrees, as the constant c.

Quick Sketching of the Sine Curve: We can plot any curve as in the preceding examples, by computing and plotting a set of point pairs. But a faster way to get a sketch is to draw a rectangle of width P and height $2a$, and then sketch the sine curve (whose *shape* does not vary) within that box. First determine the amplitude, period, and phase displacement. Then:

1. Draw two horizontal lines each at a distance a from the x axis.
2. Draw a vertical line at a distance P from the origin.
3. Subdivide the period P into four equal parts. Label the x axis with these points and draw vertical lines through them.
4. Lightly sketch in the sine curve.
5. Shift the curve by the amount of the phase displacement.

Example: Make a quick sketch of $y = 2 \sin (3x + 60°)$.

Solution: We have $a = 2$, $b = 3$, and $c = 60°$. From Eq. 200, the period is

$$P = \frac{360°}{3} = 120°$$

and by Eq. 202,

$$\text{phase displacement} = \frac{60°}{3} = 20°$$

The steps for sketching the curve are shown in Fig. 14-18.

We now do an example in radians.

Example: Make a quick sketch of the curve

$$y = 1.5 \sin (5x - 2)$$

Solution: From the given equation

$$a = 1.5, \qquad b = 5, \qquad \text{and} \qquad c = -2 \text{ rad}$$

so

$$P = \frac{2\pi}{5} = 1.26 \text{ rad}$$

and

$$\text{phase dispacement} = -\frac{2}{5} = -0.4 \text{ rad}$$

We draw a rectangle (Fig. 14-19a) whose height is $2 \times 1.5 = 3$ units, and whose width is 1.26 rad, and draw three equally spaced verticals

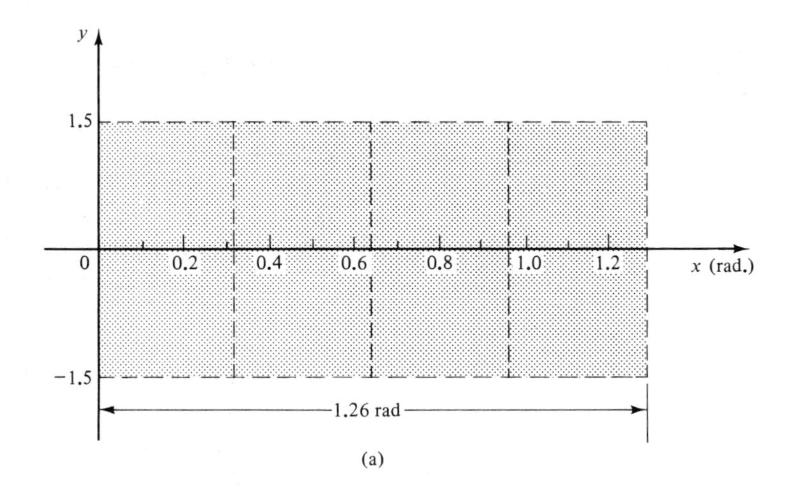

(a)

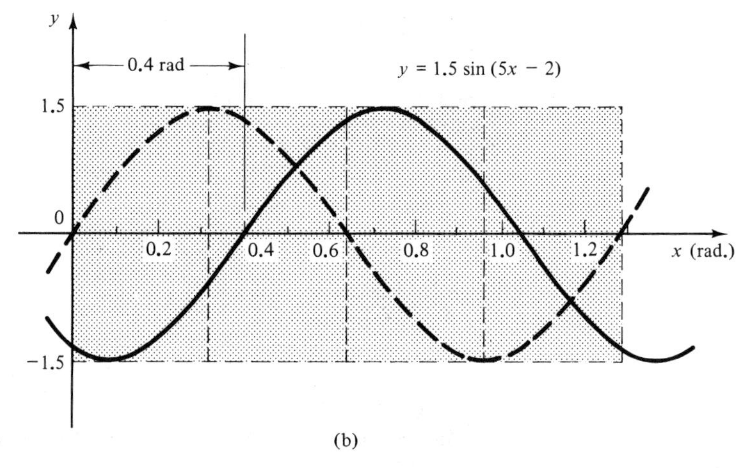

(b)

FIGURE 14-19

at 0.315 rad, 0.63 rad, and 0.945 rad. We sketch a sine wave (shown dashed) into this rectangle, and then shift it 0.4 rad to the right to get the final curve (Fig. 14-19 b).

EXERCISE 3

Periodic Waveforms

Find the period and amplitude for each periodic waveform.

1. Figure 14-20a.

2. Figure 14-20b.

The Sine Curve

Graph each sine wave and find the period, amplitude, and phase displacement. Work either in radians or degrees, and do either a "point-by-point" plot or a "quick" graph, as directed by your instructor.

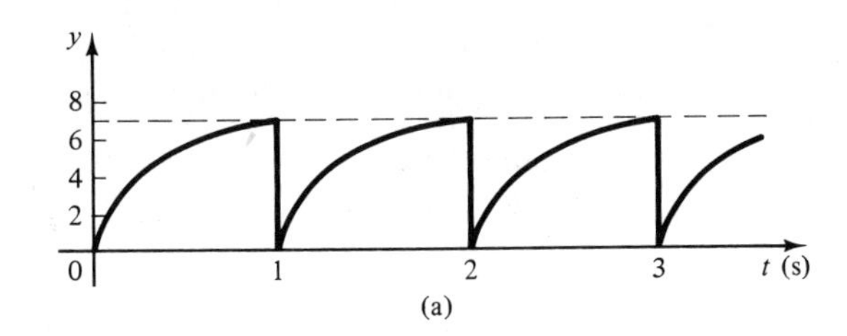

(a)

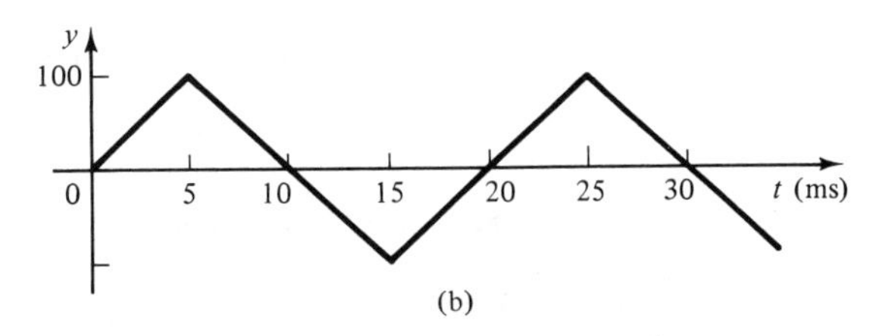

(b)

FIGURE 14-20

3. $y = 2 \sin x$
4. $y = 1.5 \sin x$
5. $y = -3 \sin x$
6. $y = 4.25 \sin x$
7. $y = \sin 2x$
8. $y = \sin \pi x$
9. $y = \sin \dfrac{x}{3}$
10. $y = \sin 3\pi x$
11. $y = 2 \sin 3x$
12. $y = -3 \sin \pi x$
13. $y = 0.5 \sin 2x$
14. $y = 100 \sin \dfrac{3x}{2}$
15. $y = \sin (x - 1)$
16. $y = \sin (x - 90°)$
17. $y = \sin \left(x + \dfrac{\pi}{2}\right)$
18. $y = \sin \left(x + \dfrac{\pi}{8}\right)$
19. $y = 4 \sin (x - 180°)$
20. $y = -2 \sin (x + 90°)$
21. $y = 0.5 \sin \left(\dfrac{x}{2} - 1\right)$
22. $y = 20 \sin (2x - 45°)$
23. Write the equations of the curves in Fig. 14-21.

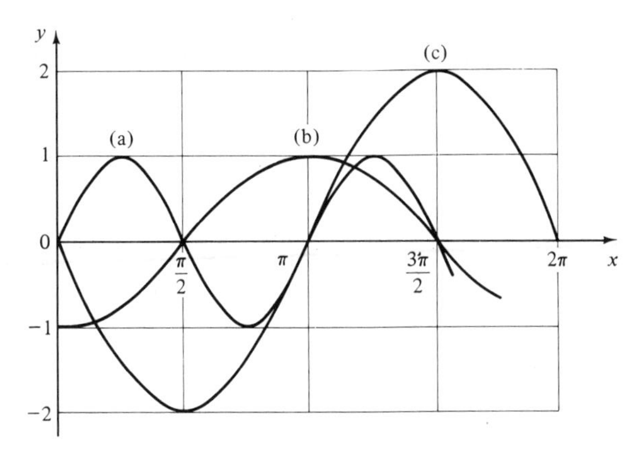

FIGURE 14-21

24. Write the equation of a sine curve that has an amplitude of 3, a period of 4π, and a phase displacement of zero.

25. Write the equation of a sine curve having an amplitude of -2, a period of 6π, and a phase displacement of $\pi/4$ to the left.

Computer

26. Use your program for generating a table of point pairs from Sec. 6-4 on some of the equations in this exercise, and plot the resulting sine waves.

14-4. THE COSINE CURVE

Cosine Curve: Let us graph one cycle of the function $y = \cos x$. Make a table of point pairs.

x	0	$\frac{\pi}{6}$	$\frac{\pi}{3}$	$\frac{\pi}{2}$	$\frac{2\pi}{3}$	$\frac{5\pi}{6}$	π	$\frac{7\pi}{6}$	$\frac{4\pi}{3}$	$\frac{3\pi}{2}$	$\frac{5\pi}{3}$	$\frac{11\pi}{6}$	2π
y	1	0.866	0.5	0	−0.5	−0.866	−1	−0.866	−0.5	0	0.5	0.866	1

Since the sine and cosine curves are identical except for phase shift, we can use either one to describe periodically varying quantities.

The cosine curve is plotted in Fig. 14-22. Note the similarity between it and the sine curve in Fig. 14-12. In fact, they are identical except for a phase displacement of $\pi/2$ (90°). This should not be surprising because from our cofunctions, Eq. 154, we know that

$$\sin \theta = \cos (90° - \theta)$$

so we should expect the curves

$$y = \sin \theta$$

and

$$y = \cos (90° - \theta)$$

to bc identical.

The terms "period," "amplitude," and "frequency" have exactly the same meaning as for the sine curve. The quick plotting method can, of course, be used for the cosine curve also.

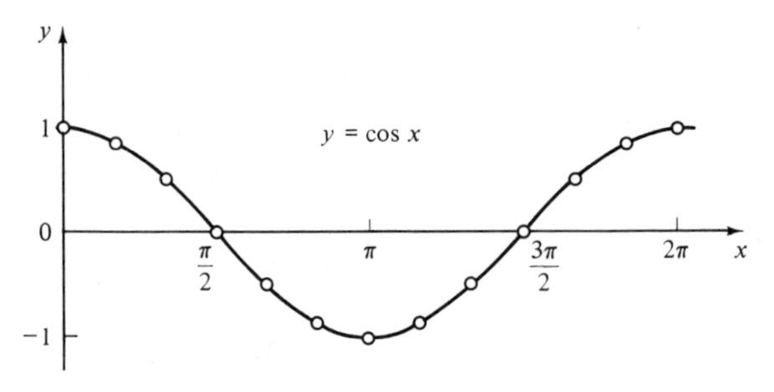

FIGURE 14-22. The cosine curve.

EXERCISE 4

The Cosine Curve

Graph each function, and find the amplitude, period, and phase displacement.

1. $y = 3 \cos x$

2. $y = \cos 2x$

3. $y = 2 \cos 3x$

4. $y = \cos (x + 2)$

5. $y = 3 \cos \left(x - \frac{\pi}{4}\right)$

6. $y = 2 \cos (3x + 1)$

14-5. ALTERNATING CURRENT

The Sine Wave as a Function of Time: The sine curve can be generated in a simple geometric way. Figure 14-23 shows a vector OP rotating with an angular velocity ω. A rotating vector is called a *phasor*. Its angular velocity ω is almost always given in radians per second (rad/s).

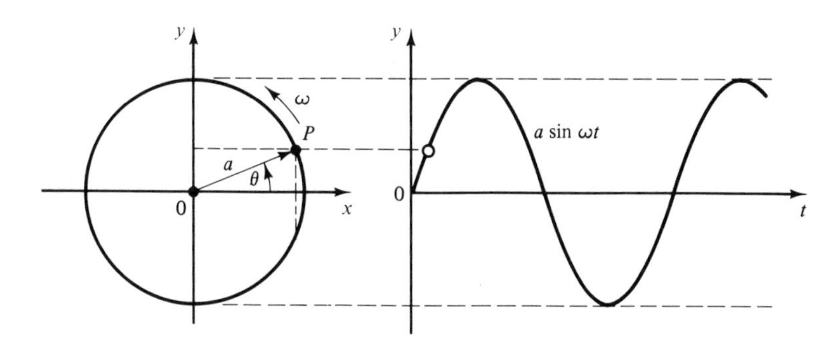

FIGURE 14-23. The sine curve generated by a rotating vector.

If the length of the phasor is a, then its projection on the y axis is

$$y = a \sin \theta$$

But since by Eq. A22 the angle θ at any instant t is equal to ωt,

$$y = a \sin \omega t$$

Further, if the phasor does not start from the x axis but has some phase *displacement* ϕ, we get

$$y = a \sin (\omega t + \phi)$$

Notice in this equation, that *y is a function of time,* rather than of an angle. Thus it is more appropriate for representing alternating current, which we discuss shortly.

Example: Write the equation for a sine wave generated by a phasor of length 8 rotating with an angular velocity of 300 rad/s, and with a phase displacement of 0.

Solution: Substituting, with $a = 8$, $\omega = 300$ rad/s, and $\phi = 0$,

$$y = 8 \sin 300t$$

Period: Recall that the period was defined as the distance along the x axis taken by one cycle of the waveform. When the x axis represented an angle, the period was in radians or degrees. Now that the x axis is time, the period will be in seconds. From Eq. 200,

$$P = \frac{2\pi}{b}$$

But in the equation $y = a \sin \omega t$, $b = \omega$, so

Period of a Sine Wave	$P = \frac{2\pi}{\omega}$	**A60**

where P is in seconds and ω is in rad/s.

Example: Find the period and amplitude of the sine wave

$$y = 6 \sin 10t$$

and make a graph with time as the horizontal axis.

Solution: The period is, from Eq. A60,

$$P = \frac{2\pi}{\omega} = \frac{2(3.142)}{10} = 0.628 \text{ s}$$

Thus, it takes 628 ms for a full cycle, 314 ms for a half cycle, 157 ms for a quarter cycle, and so forth. This sine wave, with amplitude 6, is plotted in Fig. 14-24.

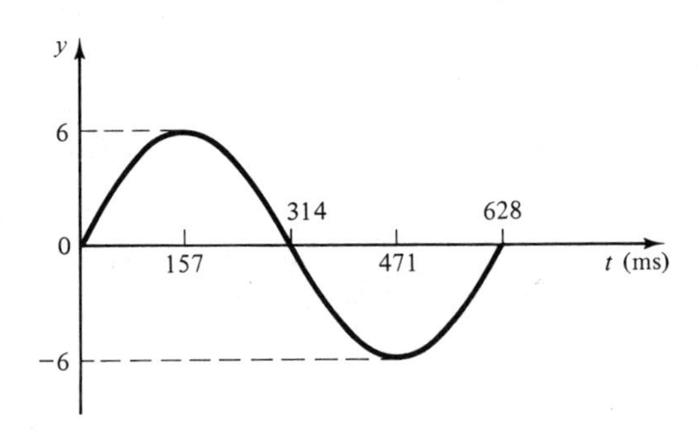

FIGURE 14-24

Frequency: From Eq. 201, we see that the frequency f of a periodic waveform is equal to the reciprocal of the period P. So, for the sine wave,

Frequency of a Sine Wave	$f = \dfrac{1}{P} = \dfrac{\omega}{2\pi}$	**A61**

where P is in seconds and ω is in rad/s.

The unit of frequency is therefore cycles/s, or *hertz* (Hz), where

$$1 \text{ Hz} = 1 \text{ cycle/s}$$

Higher frequencies are often expressed in kilohertz (kHz), where

$$1 \text{ kHz} = 10^3 \text{ Hz}$$

or in megahertz (MHz), where

$$1 \text{ MHz} = 10^6 \text{ Hz}$$

Example: The frequency of the sine wave of the preceding example is

$$f = \frac{1}{P} = \frac{1}{0.628} = 1.59 \text{ Hz}$$

When the period P is not wanted, the angular velocity can be obtained directly from Eq. A61 by noting that $\omega = 2\pi f$.

Example: Find the angular velocity of a 1000-Hz source.

Solution: From Eq. A61

$$\omega = 2\pi f = 2(3.14)(1000) = 6280 \text{ rad/s}$$

Alternating Current: When a loop of wire rotates in a magnetic field, any portion of the wire cuts the field while traveling first in one direction and then in the other direction. Since the polarity of the voltage induced in the wire depends on the direction in which the field is cut, the induced current will travel first in one direction and then in the other. It *alternates*.

The same is true with an armature more complex than a simple loop. We get an *alternating current*. Just as the rotating point P in Fig. 14-23 generated a sine wave, the current induced in the rotating armature will be sinusoidal.

The sinusoidal voltage or current can be described by the equation

$$y = a \sin(\omega t + \phi)$$

or, in electrical terms,

Alternating Current	$i = I_m \sin(\omega t + \phi)$	**A59**

where i is the current at time t, I_m the maximum current (the amplitude), and ϕ the phase angle. And if we let V_m stand for the amplitude of the voltage wave, the instantaneous voltage v becomes

Alternating Voltage	$v = V_m \sin(\omega t + \phi)$	**A58**

Equations A60 and A61 still apply here.

$$\text{period} = \frac{2\pi}{\omega}\ \text{s}$$

$$\text{frequency} = \frac{\omega}{2\pi}\ \text{Hz}$$

Example: Utilities in the United States supply alternating current at a frequency of 60 Hz. Find the angular velocity and the period P.

Solution: By Eq. A60,

$$P = \frac{1}{f} = \frac{1}{60} = 0.0167\ \text{s}$$

and by Eq. A61,

$$\omega = \frac{2\pi}{P} = \frac{2\pi}{0.0167} = 377\ \text{rad/s}$$

This is a good number to remember.

Example: A certain alternating current has an amplitude of 1.5 A and a frequency of 60 hertz (cycles/s). Taking the phase angle as zero, write the equation for the current as a function of time, find the period, and find the current at $t = 0.01$ s.

Solution: From the preceding example,

$$P = 0.0167\text{s} \qquad \text{and} \qquad \omega = 377\ \text{rad/s}$$

So the equation is

$$i = 1.5 \sin(377t)$$

When $t = 0.01$ s,

$$i = 1.5 \sin(377)(0.01) = -0.882\ \text{A}$$

Phase Displacement: When writing the equation of a single alternating voltage or current we are usually free to choose the origin anywhere along the time axis, and so can place it to make the phase displacement equal to zero. However, when there are *two* curves on the same graph that are *out of phase,* we usually locate the origin so that the phase displacement of one curve is zero.

Example: Write the equations for the voltage and current waves in Fig. 14-25.

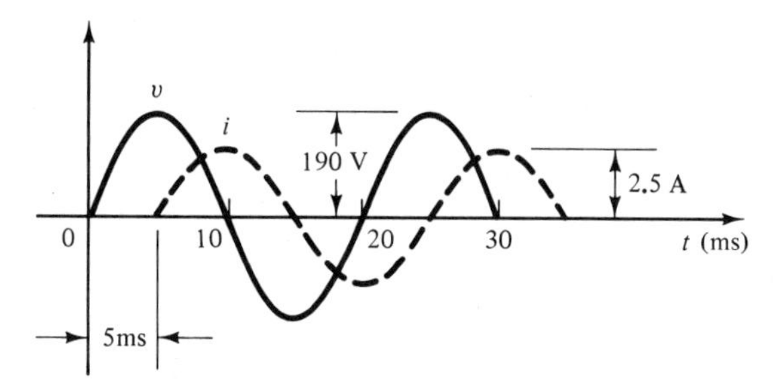

FIGURE 14-25

Solution: For the voltage wave,

$$V_m = 190, \quad \phi = 0, \quad \text{and} \quad P = 0.02 \text{ s}$$

By Eq. A60,

$$\omega = \frac{2\pi}{0.02} = 100\pi = 314 \text{ rad/s}$$

and by Eq. A58,

$$v = 190 \sin 314t \quad V$$

For the current wave, $I_m = 2.5$ A.
Since the curve is shifted to the right by 5 ms,

$$\phi = 0.005 \text{ s}(100\pi \text{ rad/s}) = \frac{\pi}{2} \text{ rad} = 90°$$

The angular frequency is the same as before, so from Eq. A59,

$$i = 2.5 \sin (314t - 90°) \quad \text{A}$$

It is not unusual to see a phase angle given in degrees *in the same equation with the angular frequency in* radians *per second.*

or, with ϕ in radians,

$$i = 2.5 \sin \left(314t - \frac{\pi}{2}\right) \quad \text{A}$$

EXERCISE 5

Find the period and angular velocity of a repeating waveform that has a frequency of:

1. 68 Hz **2.** 10 Hz **3.** 5000 Hz

Find the frequency (in Hz) and angular velocity of a repeating waveform whose period is:

4. 1 s **5.** $\frac{1}{8}$ s **6.** 95 ms

7. If a periodic waveform has a frequency of 60 Hz, how many seconds will it take to complete 200 cycles?

8. Find the frequency in Hz for a wave that completes 150 cycles in 10 s.

Find the period and frequency of a sinusoid that has an angular velocity of:

9. 455 rad/s **10.** 2.58 rad/s **11.** 500 rad/s

Find the period, amplitude, and phase angle for each sine wave.

12. Figure 14-26a.

13. Figure 14-26b.

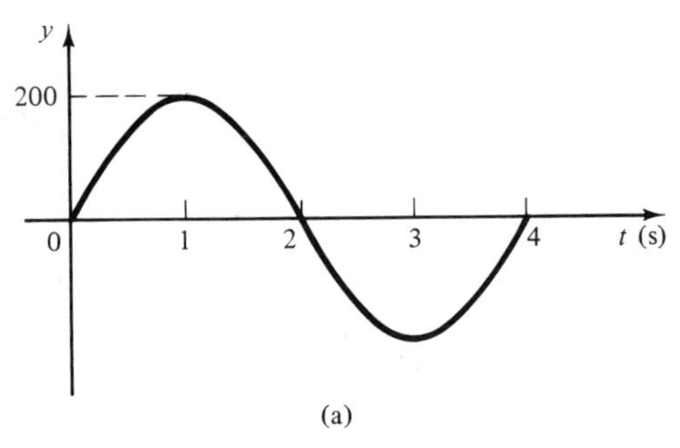

(a)

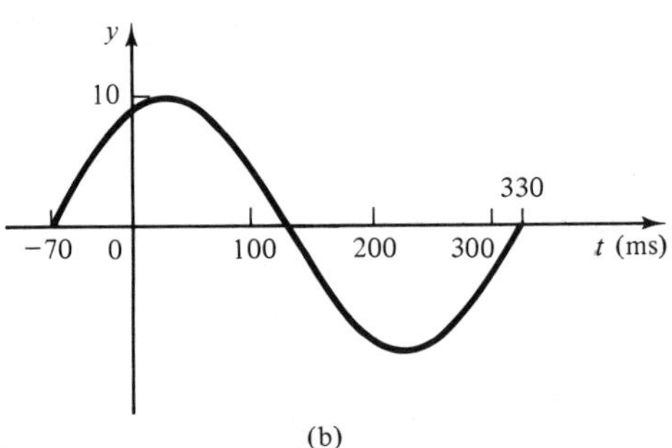

(b)

FIGURE 14-26

14. Write an equation for a sine wave generated by a phasor of length 5 rotating with an angular velocity of 750 rad/s, with a phase angle of 0°.

15. Repeat Problem 14, with a phase angle of 15°.

Plot each sine wave.

16. $y = \sin t$ **17.** $y = 3 \sin 377t$

18. $y = 54 \sin (83t - 20°)$ **19.** $y = 375 \sin \left(55t + \frac{\pi}{4}\right)$

20. An alternating current has the equation

$$i = 25 \sin (635t - 18°) \quad \text{A}$$

Find the maximum current, the period, the frequency, the phase angle, and the instantaneous current at $t = 0.01$ s.

21. Write the equation of an alternating voltage that has a peak value of 155 V, a frequency of 60 Hz, and a phase angle of 22°.

14-6. POLAR COORDINATES

Most of our graphing will continue to be in rectangular coordinates, but in some cases polar coordinates are more convenient.

The Polar Coordinate System: The *polar coordinate system* (Fig. 14-27) consists of a *polar axis,* passing through point O, which is called the *pole*. The location of a point P is given by its distance r from the origin, called the *radius vector,* and by the angle θ, called the *polar angle* (sometimes called the vectorial angle or reference angle). The polar angle is called positive when measured counterclockwise from the polar axis, and negative when measured clockwise.

The *polar coordinates* of a point P are thus r and θ, usually written in the form $P(r, \theta)$.

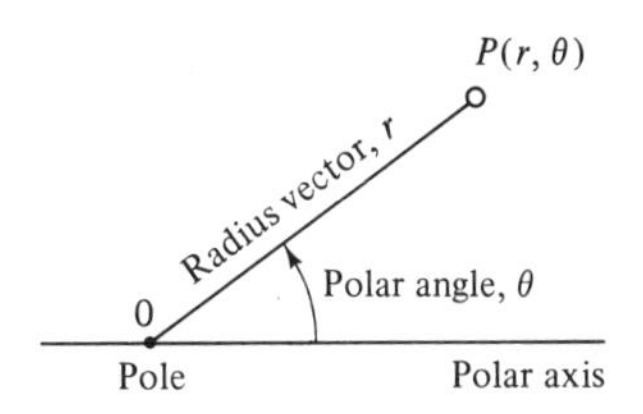

FIGURE 14-27. Polar coordinates.

Plotting Points in Polar Coordinates: For plotting in polar coordinates it is convenient, although not essential, to have *polar coordinate paper* (Fig. 14-28). This paper has concentric circles, equally spaced, and an angular scale in degrees or radians.

Example: The points $P(3, 45°)$, $Q(-2, 320°)$, $R(4, 7\pi/6)$, and $S(2.5, -5\pi/3)$ are plotted in Fig. 14-28 on page 450.

Transforming between Polar and Rectangular Coordinates: The relations between rectangular coordinates and polar coordinates are easily seen when we put both systems on a single diagram (Fig. 14-29). Using the trigonometric functions and the Pythagorean theorem, we get

Rectangular	$x = r \cos \theta$	**158**
	$y = r \sin \theta$	**159**
Polar	$r = \sqrt{x^2 + y^2}$	**160**
	$\theta = \arctan \dfrac{y}{x}$	**161**

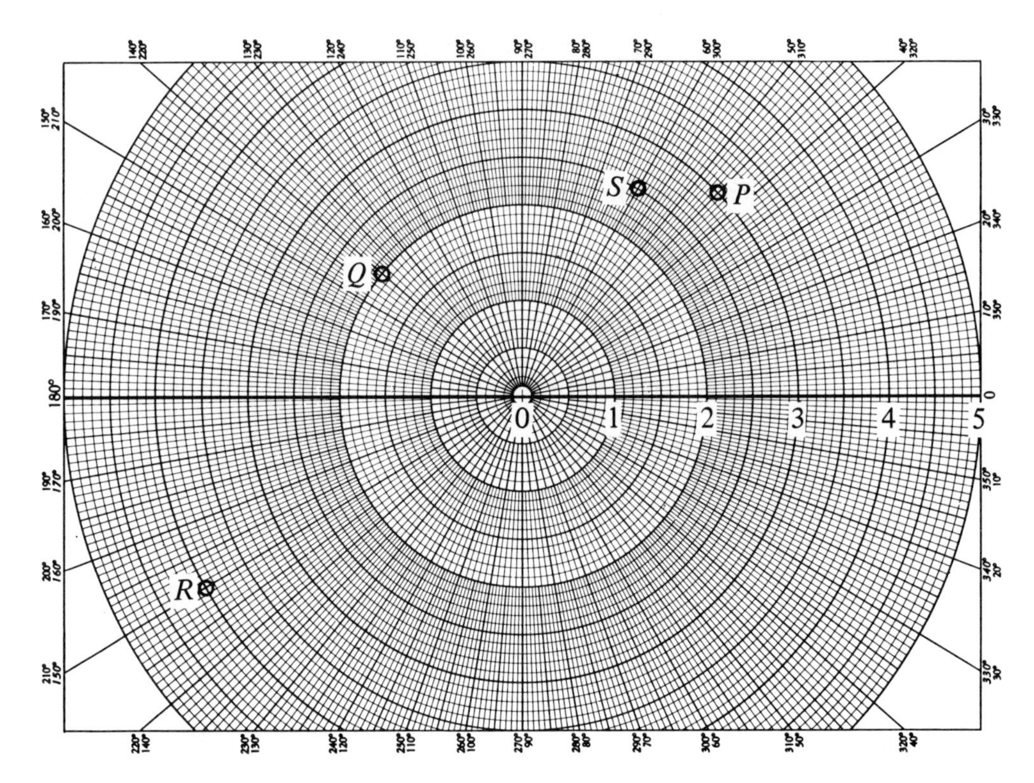

FIGURE 14-28. Polar coordinate graph paper.

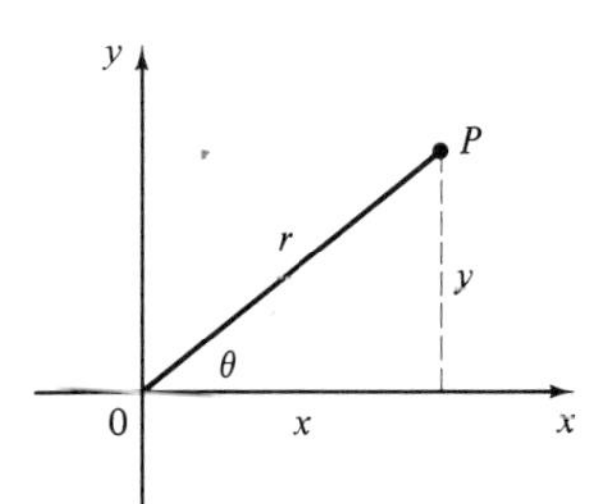

FIGURE 14-29. Rectangular and polar coordinates of a point.

Example: What are the polar coordinates of $P(3, 4)$?

Some calculators have special keys for converting between polar and rectangular coordinates. Check your manual.

Solution: By Eq. 160,

$$r = \sqrt{9 + 16} = 5$$

and by Eq. 161,

$$\theta = \arctan \frac{4}{3} = 53.1°$$

So the polar coordinates of P are $(5, 53.1°)$.

Example: The polar coordinates of a point are (8, 125°). What are the rectangular coordinates?

Solution: By Eq. 158,

$$x = 8 \cos 125° = -4.59$$

and by Eq. 159,

$$y = 8 \sin 125° = 6.55$$

So the rectangular coordinates are (−4.59, 6.55).

EXERCISE 6

Polar Coordinates

Plot the points in polar coordinates.

1. (4, 35°) **2.** (3, 120°) **3.** (2.5, 215°)

4. (3.8, 345°) **5.** $\left(2.7, \frac{\pi}{6}\right)$ **6.** $\left(3.9, \frac{7\pi}{8}\right)$

7. $\left(-3, \frac{\pi}{2}\right)$ **8.** $\left(4.2, \frac{2\pi}{5}\right)$ **9.** (3.6, −20°)

10. (−2.5, −35°) **11.** $\left(-1.8, -\frac{\pi}{6}\right)$ **12.** $\left(3.7, -\frac{3\pi}{5}\right)$

Write the polar coordinates of the points.

13. (3, 6) **14.** (1, 4) **15.** (−4, 3)

16. (2.7, −1.8) **17.** (−4.8, −5.9) **18.** (207, 186)

19. (−312, −509) **20.** (1.08, −2.15)

Write the rectangular coordinates of the points.

21. (5, 47°) **22.** (6.3, 227°) **23.** (445, 312°)

24. $\left(3.6, \frac{\pi}{5}\right)$ **25.** $\left(-4, \frac{3\pi}{4}\right)$ **26.** $\left(18.3, \frac{2\pi}{3}\right)$

27. (15, −35°) **28.** (−12, −48°) **29.** $\left(-9.8, -\frac{\pi}{5}\right)$

14-7. FUNDAMENTAL IDENTITIES

In this section we cover the basic relationships between the six trigonometric ratios. We then go on to use these identities to simplify expressions and to prove more complicated identities. Then in Sec. 14-8 we use the fundamental identities to help solve trigonometric equations.

Reciprocal Relations: We have already encountered the reciprocal relations in Sec. 12-2, and we repeat them here.

Reciprocal Relations	$\sin\theta = \dfrac{1}{\csc\theta}$, or $\csc\theta = \dfrac{1}{\sin\theta}$, or $\sin\theta\csc\theta = 1$	**152a**
	$\cos\theta = \dfrac{1}{\sec\theta}$, or $\sec\theta = \dfrac{1}{\cos\theta}$, or $\cos\theta\sec\theta = 1$	**152b**
	$\tan\theta = \dfrac{1}{\cot\theta}$, or $\cot\theta = \dfrac{1}{\tan\theta}$, or $\tan\theta\cot\theta = 1$	**152c**

Example: Simplify

$$\frac{\cos\theta}{\sec^2\theta}$$

Solution: Using Eq. 152b,

$$\frac{\cos\theta}{\sec^2\theta} = \cos\theta\,(\cos^2\theta) = \cos^3\theta$$

Quotient Relations: Figure 14-30 shows an angle θ in standard position, as when we first defined the trigonometric functions in Sec. 12-2. We see that

$$\sin\theta = \frac{y}{r} \quad \text{and} \quad \cos\theta = \frac{x}{r}$$

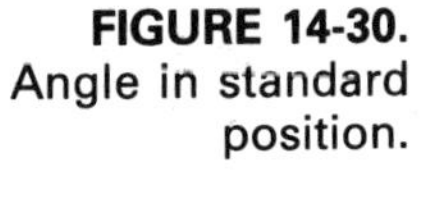

FIGURE 14-30. Angle in standard position.

Dividing,

$$\frac{\sin\theta}{\cos\theta} = \frac{\frac{y}{r}}{\frac{x}{r}} = \frac{y}{x}$$

But $y/x = \tan\theta$, so

Quotient Relations	$\tan\theta = \dfrac{\sin\theta}{\cos\theta}$	**162**
	$\cot\theta = \dfrac{\cos\theta}{\sin\theta}$	**163**

Taking the reciprocal,

Example: Rewrite the expression

$$\frac{\cot\theta}{\csc\theta} - \frac{\tan\theta}{\sec\theta}$$

so that it contains only the sine and cosine.

Solution:

$$\frac{\cot\theta}{\csc\theta}-\frac{\tan\theta}{\sec\theta}=\frac{\cos\theta}{\sin\theta}\cdot\frac{\sin\theta}{1}-\frac{\sin\theta}{\cos\theta}\cdot\frac{\cos\theta}{1}$$

$$=\cos\theta-\sin\theta$$

Pythagorean Relations: We can get three more relations by applying the Pythagorean theorem to the triangle in Fig. 14-30.

$$x^2+y^2=r^2$$

Dividing through by r^2, we get

$$\frac{x^2}{r^2}+\frac{y^2}{r^2}=1$$

or

$$\left(\frac{x}{r}\right)^2+\left(\frac{y}{r}\right)^2=1$$

but $x/r = \cos\theta$ and $y/r = \sin\theta$, so

Recall that $\sin^2\theta$ is a short way of writing $(\sin\theta)^2$.

$\sin^2\theta+\cos^2\theta=1$	**164**

Returning to the equation $x^2+y^2=r^2$, we now divide through by x^2,

$$1+\frac{y^2}{x^2}=\frac{r^2}{x^2}$$

But $y/x = \tan\theta$ and $r/x = \sec\theta$, so

$1+\tan^2\theta=\sec^2\theta$	**165**

Now dividing $x^2+y^2=r^2$ by y^2,

$$\frac{x^2}{y^2}+1=\frac{r^2}{y^2}$$

But $x/y = \cot\theta$ and $r/y = \csc\theta$, so

$1+\cot^2\theta=\csc^2\theta$	**166**

Example:

$$\sin^2\theta-\csc^2\theta-\tan^2\theta+\cot^2\theta+\cos^2\theta+\sec^2\theta$$
$$=(\sin^2\theta+\cos^2\theta)-(\tan^2\theta-\sec^2\theta)+(\cot^2\theta-\csc^2\theta)$$
$$=1-(-1)+(-1)=1$$

Simplification of Trigonometric Expressions: One use of the trigonometric identities is the simplification of expressions, as in the preceding example. We now give a few more examples.

Example: Simplify

$$(\cot^2\theta + 1)(\sec^2\theta - 1)$$

Solution: By Eqs. 165 and 166,

$$(\cot^2\theta + 1)(\sec^2\theta - 1) = \csc^2\theta\tan^2\theta$$

and by Eqs. 152a and 162,

$$= \frac{1}{\sin^2\theta}\cdot\frac{\sin^2\theta}{\cos^2\theta}$$

$$= \frac{1}{\cos^2\theta}$$

and by Eq. 152b,

$$= \sec^2\theta$$

Example: Simplify

$$\frac{\cos\theta}{1-\sin\theta} + \frac{\sin\theta - 1}{\cos\theta}$$

Solution: Combining the two fractions,

$$\frac{\cos\theta}{1-\sin\theta} + \frac{\sin\theta - 1}{\cos\theta} = \frac{\cos^2\theta + (\sin\theta - 1)(1 - \sin\theta)}{(1-\sin\theta)\cos\theta}$$

$$= \frac{\cos^2\theta + \sin\theta - \sin^2\theta - 1 + \sin\theta}{(1-\sin\theta)\cos\theta}$$

From Eq. 164,

$$= \frac{-\sin^2\theta - \sin^2\theta + 2\sin\theta}{(1-\sin\theta)\cos\theta}$$

$$= \frac{-2\sin^2\theta + 2\sin\theta}{(1-\sin\theta)\cos\theta}$$

Factoring the numerator,

$$= \frac{2\sin\theta\,(-\sin\theta + 1)}{(1-\sin\theta)\cos\theta}$$

$$= \frac{2\sin\theta}{\cos\theta}$$

and by Eq. 162,

$$= 2\tan\theta$$

Trigonometric Identities: In Sec. 5-1 we said that an *identity* is an equation that is true for any values of the unknowns. A *trigonometric* identity is one that contains trigonometric terms, such as the eight fundamental identities presented so far in this chapter. We now use these eight identities to *verify* or *prove* whether a given identity is true. We do this by trying to transform one side of the identity (usually the more complicated side) until it is identical to the other side.

Example: Prove the relation

$$\sec\theta\csc\theta = \tan\theta + \cot\theta$$

Proving identities is not easy. It takes a good knowledge of the fundamental identities, lots of practice, and often several false starts. Do not be discouraged if you fail to get them right away.

Solution: We try to transform the right side so that it is identical to the left. By Eqs. 162 and 163,

$$\sec\theta\csc\theta = \tan\theta + \cot\theta$$

$$= \frac{\sin\theta}{\cos\theta} + \frac{\cos\theta}{\sin\theta}$$

Adding the fractions,

$$= \frac{\sin^2\theta + \cos^2\theta}{\sin\theta\cos\theta}$$

By Eq. 164,

$$= \frac{1}{\sin\theta\cos\theta}$$

$$= \frac{1}{\sin\theta}\cdot\frac{1}{\cos\theta}$$

And by Eq. 152a and 152b,

$$= \csc\theta\sec\theta$$

Tip	Each side of an identity must be worked *separately;* we cannot transpose, or multiply both sides by the same thing, the way we do with equations. The reason for this is that we do not yet know if the two sides are in fact equal; that is what we are trying to prove.

Example: Prove the identity,

$$\csc x + \tan x = \frac{\cot x + \sin x}{\cos x}$$

Solution:

$$\csc x + \tan x = \frac{\cot x}{\cos x} + \frac{\sin x}{\cos x}$$

$$= \frac{\dfrac{\cos x}{\sin x}}{\cos x} + \tan x$$

$$= \frac{1}{\sin x} + \tan x$$

$$= \csc x + \tan x$$

EXERCISE 7

Simplifying Expressions

Change to an expression containing only sin and cos.

1. $\tan x - \sec x$

2. $\cot x + \csc x$

3. $\tan \theta \csc \theta$

4. $\sec \theta - \tan \theta \sin \theta$

5. $\dfrac{\tan \theta}{\csc \theta} + \dfrac{\sin \theta}{\tan \theta}$

6. $\cot x + \tan x$

Simplify.

7. $1 - \sec^2 x$

8. $\dfrac{\csc \theta}{\sin \theta}$

9. $\dfrac{\cos \theta}{\cot \theta}$

10. $\sin \theta \csc \theta$

11. $\tan \theta \csc \theta$

12. $\dfrac{\sin \theta}{\csc \theta}$

13. $\sec x \sin x$

14. $\sec x \sin x \cot x$

15. $\csc \theta \tan \theta - \tan \theta \sin \theta$

16. $\dfrac{\cos x}{\cot x \sin x}$

17. $\cot \theta \tan^2 \theta \cos \theta$

18. $\dfrac{\tan x (\csc^2 x - 1)}{\sin x + \cot x \cos x}$

19. $\dfrac{\sin \theta}{\cos \theta \tan \theta}$

20. $\dfrac{\sin^2 x + \cos^2 x}{1 - \cos^2 x}$

21. $\dfrac{1}{\sec^2 x} + \dfrac{1}{\csc^2 x}$

22. $\sin \theta (\csc \theta + \cot \theta)$

23. $\csc x - \cot x \cos x$

24. $1 + \dfrac{\tan^2 \theta}{1 + \sec \theta}$

25. $\dfrac{\sec x - \csc x}{1 - \cot x}$

26. $\dfrac{1}{1 + \sin x} + \dfrac{1}{1 - \sin x}$

27. $\sec^2 x(1 - \cos^2 x)$

28. $\tan x + \dfrac{\cos x}{\sin x + 1}$

29. $\cos \theta \sec \theta - \dfrac{\sec \theta}{\cos \theta}$

30. $\cot^2 x \sin^2 x + \tan^2 x \cos^2 x$

Identities

Prove.

31. $\tan x \cos x = \sin x$

32. $\tan x = \dfrac{\sec x}{\csc x}$

All identities in this chapter can be proven.

33. $\dfrac{\sin x}{\csc x} + \dfrac{\cos x}{\sec x} = 1$

34. $\sin \theta = \dfrac{1}{\cot \theta \sec \theta}$

35. $(\cos^2 \theta + \sin^2 \theta)^2 = 1$

36. $\tan x = \dfrac{\tan x - 1}{1 - \cot x}$

37. $\dfrac{\csc \theta}{\sec \theta} = \cot \theta$

38. $\cot^2 x = \dfrac{\cos x}{\tan x \sin x}$

39. $\cos x + 1 = \dfrac{\sin^2 x}{1 - \cos x}$

40. $\csc x - \sin x = \cot x \cos x$

41. $\cot^2 x - \cos^2 x = \cos^2 x \cot^2 x$

42. $\csc x = \cos x \cot x + \sin x$

43. $1 = (\csc x - \cot x)(\csc x + \cot x)$

44. $\tan x = \dfrac{\tan x + \sin x}{1 + \cos x}$

45. $\dfrac{\tan x + 1}{1 - \tan x} = \dfrac{\sin x + \cos x}{\cos x - \sin x}$

46. $\cot x = \cot x \sec^2 x - \tan x$

47. $\dfrac{\sin \theta + 1}{1 - \sin \theta} = (\tan \theta + \sec \theta)^2$

48. $\dfrac{1 + \sin \theta}{1 - \sin \theta} = \dfrac{1 + \csc \theta}{\csc \theta - 1}$

49. $(\sec \theta - \tan \theta)(\tan \theta + \sec \theta) = 1$

50. $\dfrac{1 + \cot \theta}{\csc \theta} = \dfrac{\tan \theta + 1}{\sec \theta}$

14-8. TRIGONOMETRIC EQUATIONS

Solving Trigonometric Equations: One use for the trigonometric identities studied in this chapter is in the solution of trigonometric equations. A trigonometric equation will usually have infinitely many solutions. For example, the equation $\sin \theta = \frac{1}{2}$ has solutions of $\theta =$ 30° and 150°, but also 390°, 510°, and so on, if we allow angles greater than 360°. In this section we limit our solution to nonnegative values of θ less than 360°.

It is difficult to give general rules for solving the great variety of trigonometric equations possible, but a study of the following examples should be helpful.

Equations Containing a Single Trigonometric Function and a Single Angle: We start with the simplest types, those containing only one trigonometric function (all sine, or all cosine, for example) and having only one angle. First isolate the trigonometric function on one side of the equation and then find the angles.

Example: Solve the equation

$$2 \cos 2\theta - 1 = 0$$

Solution: Transposing, and dividing,

$$\cos 2\theta = \tfrac{1}{2}$$

$$2\theta = 60°, 300°, 420°, 660°, \text{etc.}$$

$$\theta = 30°, 150°, 210°, \text{and } 330°$$

if we limit our solution to angles less than 360°.

If one of the functions is *squared,* you may have an equation in *quadratic form,* which can be solved by the methods of Chapter 11.

Example: Solve the equation

$$2 \sin^2 \theta - \sin \theta = 0$$

Solution: This is an *incomplete quadratic* in sin θ (Sec. 11-2). Factoring,

$$\sin \theta \, (2 \sin \theta - 1) = 0$$

Setting each factor equal to zero,

$\sin \theta = 0$	$2 \sin \theta - 1 = 0$
$\theta = 0°, 180°$	$\sin \theta = \frac{1}{2}$
	$\theta = 30°, 150°$

If the equation is the form of a quadratic that cannot be factored, use the quadratic formula (Eq. 100).

Example: Solve the equation

$$\cos^2 \theta = 3 + 5 \cos \theta$$

Solution: Transposing,

$$\cos^2 \theta - 5 \cos \theta - 3 = 0$$

This will not factor, so, using the quadratic formula,

$$\cos \theta = \frac{5 \pm \sqrt{25 - 4(-3)}}{2}$$

So

$\cos \theta = 5.54$	$\cos \theta = -0.541$
(not possible)	$\theta = 123°, 237°$

Equations with One Angle but More Than One Function: If the equation contains two or more trigonometric functions of the same angle, first transpose all the terms to one side and try to factor that side into factors, *each containing only a single function,* and proceed as above.

Example: Solve

$$\sin \theta \sec \theta - 2 \sin \theta = 0$$

Solution: Factoring,

$$\sin \theta \, (\sec \theta - 2) = 0$$

$\sin \theta = 0$	$\sec \theta - 2 = 0$
$\theta = 0°, 180°$	$\sec \theta = 2$
	$\cos \theta = \frac{1}{2}$
	$\theta = 60°, 300°$

If the expression is *not factorable,* use the fundamental identities to *express everything in terms of a single trigonometric function* and proceed as above.

Example: Solve

$$\sin^2 \theta + \cos \theta = 1$$

Solution: By Eq. 164, $\sin^2 \theta = 1 - \cos^2 \theta$. Substituting,

$$1 - \cos^2 \theta + \cos \theta = 1$$
$$\cos \theta - \cos^2 \theta = 0$$

Factoring,

$$\cos \theta (1 - \cos \theta) = 0$$

$\cos \theta = 0$	$1 - \cos \theta = 0$
$\theta = 90°, 270°$	$\cos \theta = 1$
	$\theta = 0°, 360°$

EXERCISE 8

Trigonometric Equations

Solve each equation for all positive values of the angle less than 360°.

1. $4 \sin^2 \theta = 3$

2. $2 \sin 3\theta = \frac{1}{2}$

3. $2 \sin (\theta + 15°) = 1$

4. $\csc^2 \theta = 4$

5. $2 \cos^2 \theta = 1 + 2 \sin^2 \theta$

6. $2 \sec \theta = -3 - \cos \theta$

7. $4 \sin^4 \theta = 1$

8. $2 \csc \theta - \cot \theta = \tan \theta$

9. $1 + \tan \theta = \sec^2 \theta$

10. $1 + \cot^2 \theta = \sec^2 \theta$

11. $3 \cot \theta = \tan \theta$

12. $\sin^2 \theta = 1 - 6 \sin \theta$

13. $3 \sin \theta - 1 = 2 \sin^2 \theta$

14. $\sin \theta = \cos \theta$

15. $4 \cos^2 \theta + 4 \cos \theta = -1$

16. $\cos \theta \sin 2\theta = 0$

17. $3 \tan \theta = 4 \sin^2 \theta \tan \theta$

18. $3 \sin \theta \tan \theta + 2 \tan \theta = 0$

19. $\sec \theta = -\csc \theta$

20. $\sin \theta = 2 \cos \frac{\theta}{2}$

21. $\sin \theta = 2 \sin \theta \cos \theta$

22. $1 + \sin \theta = \sin \theta \cos \theta + \cos \theta$

Computer

23. Use your program for solving equations by the midpoint method (Sec. 6-4) to find the roots of any of the trigonometric equations in this section.

CHAPTER TEST

Graph one cycle of each curve and find the period, amplitude, and phase displacement.

1. $y = 3 \sin 2x$

2. $y = 5 \cos 3x$

3. $y = 1.5 \sin \left(3x + \frac{\pi}{2}\right)$

4. $y = -5 \cos \left(x - \frac{\pi}{6}\right)$

5. $y = 2.5 \sin \left(4x + \frac{2\pi}{9}\right)$

6. $y = 4 \tan x$

Prove.

7. $\sec^2 \theta + \tan^2 \theta = \sec^4 \theta - \tan^4 \theta$

8. $\dfrac{1 + \csc \theta}{\cot \theta} = \dfrac{\cot \theta}{\csc \theta - 1}$

9. $\tan^2 \theta \sin^2 \theta = \tan^2 \theta - \sin^2 \theta$

10. $\cot \theta + \csc \theta = \dfrac{1}{\csc \theta - \cot \theta}$

11. $\tan \theta + \cot \theta = \cot \theta \sec^2 \theta$

12. $\sin^4 \theta + 2 \sin^2 \theta \cos^2 \theta + \cos^4 \theta = 1$

13. $\dfrac{\sin^3 \theta + \cos^3 \theta}{\sin \theta + \cos \theta} = 1 - \sin \theta \cos \theta$

14. $\cos \theta \cot \theta = \dfrac{\cos \theta + \cot \theta}{\tan \theta + \sec \theta}$

15. Write the equation of a sine curve that has an amplitude of 5, a period of 3π, and a phase displacement of $\pi/6$.

Convert to degrees.

16. $\dfrac{3\pi}{7}$ **17.** $\dfrac{9\pi}{4}$ **18.** $\dfrac{\pi}{9}$ **19.** $\dfrac{2\pi}{5}$ **20.** $\dfrac{11\pi}{12}$

21. Find the central angle in radians that would intercept an arc of 5.83 m in a circle of radius 7.29 m.

22. Find the angular velocity in rad/s of a wheel that rotates 33.5 revolutions in 1.45 min.

23. A wind generator has blades 3.50 m long. Find the tip speed when the blades are rotating at 35 rev/min.

Simplify.

24. $\dfrac{\sin \theta \sec \theta}{\tan \theta}$

25. $\sec^2 \theta - \sin^2 \theta \sec^2 \theta$

26. $\dfrac{\sec \theta}{\sec \theta - 1} + \dfrac{\sec \theta}{\sec \theta + 1}$

27. $\dfrac{\sin^3 \theta + \cos^3 \theta}{\sin \theta + \cos \theta}$

28. $\cos \theta \tan \theta \csc \theta$

29. $(\sec \theta - 1)(\sec \theta + 1)$

30. $\dfrac{\cot \theta}{1 + \cot^2 \theta} \cdot \dfrac{\tan^2 \theta + 1}{\tan \theta}$

31. $\dfrac{\cos \theta}{1 - \tan \theta} + \dfrac{\sin \theta}{1 - \cot \theta}$

32. $(1 - \sin^2 \theta) \sec^2 \theta$

33. $\tan^2 \theta (1 + \cot^2 \theta)$

Plot the points in polar coordinates.

34. $(3.4, 125°)$

35. $(-2.5, 228°)$

36. $\left(1.7, -\frac{\pi}{6}\right)$

Solve for all positive values of θ less than 360°.

37. $1 + 2 \sin^2 \theta = 3 \sin \theta$

38. $3 + 5 \cos \theta = 2 \cos^2 \theta$

39. $\cos \theta - 2 \cos^3 \theta = 0$

40. $\tan \theta = 2 - \cot \theta$

41. $5 \tan \theta = 6 + \tan^2 \theta$

42. $\sin^2 \theta = 1 + 6 \sin \theta$

43. $\tan^2 2\theta = 1 + \sec 2\theta$

44. $16 \cos^4 \dfrac{\theta}{2} = 9$

45. $\tan^2 \theta + 1 = \sec \theta$

46. $\sin \theta = 1 - 3 \cos \theta$

Write the polar coordinates of the points.

47. $(7, 3)$

48. $(-5.3, 3.8)$

49. $(-24, -52)$

Write the rectangular coordinates of the points.

50. $(3.8, 48°)$

51. $\left(-65, \dfrac{\pi}{9}\right)$

52. $(3.8, -44°)$

Exponential and Logarithmic Functions

You may have heard the story about the inventor of chess, who asked that the reward for his invention be a grain of wheat on the first square of the chessboard, two grains on the second square, four on the third, and so on, each square getting double the amount as on the previous square. It turns out that he would receive more wheat than has been grown by man during all history; an illustration of the remarkable properties of the exponential function, which we study in this chapter.

We also study logarithms and show how they are related to the exponential function, and do a few logarithmic computations. We then solve exponential and logarithmic equations and use them in some interesting applications, and do some graphing on logarithmic and semilogarithmic paper.

15-1. THE EXPONENTIAL FUNCTION

Definition: An *exponential function* is one in which the independent variable appears in the exponent. The quantity that is raised to the power is called the *base*.

Realize that these are different from the power function, $y = ax^n$. Here the unknown is in the exponent.

Examples:

$$y = 5^x \qquad y = 3a^x \qquad y = 7^{x-3} \qquad y = 5e^{-2x} \qquad y = (2x)^{x+1}$$

are exponential functions if a and e represent constants.

Graph of the Exponential Function: We may make a graph of an exponential function in the same way as we graphed other functions in Sec. 6-4. Choose values of the independent variable x throughout the region of interest, and for each x compute the corresponding value of the dependent variable y. Plot the resulting table of ordered pairs and connect with a smooth curve.

Example: Graph the exponential function $y = 2^x$ from $x = -2$ to $x = +4$.

Solution: We compute values of y and obtain the following table of point pairs:

x	-2	-1	0	1	2	3	4
y	$\frac{1}{4}$	$\frac{1}{2}$	1	2	4	8	16

These point pairs are plotted in Fig. 15-1.

Note that as x increases, y increases ever more greatly, and that the curve gets steeper and steeper. This is a characteristic of *exponential growth,* which we cover more fully below. Also note that the y intercept is 1 and that there is no x intercept, the x axis being an asymptote.

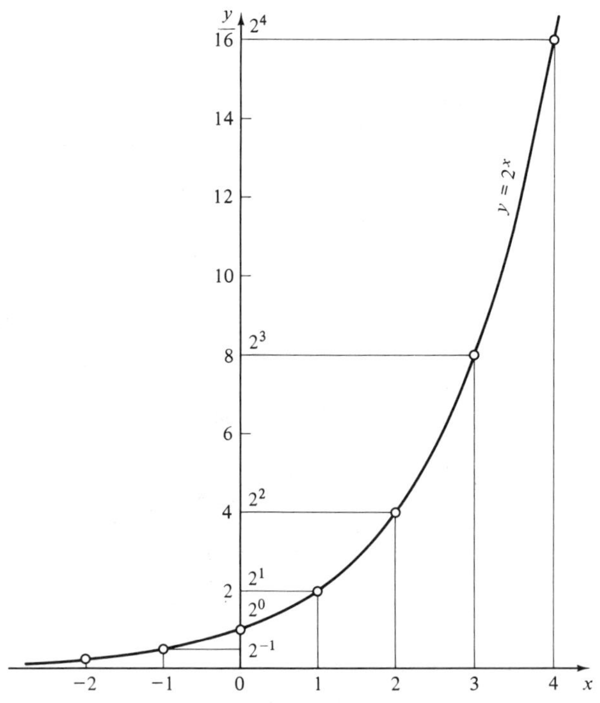

FIGURE 15-1. Graph of the exponential function, $y = 2^x$.

Finally, note that each successive y value is two times the previous one. We say that y *increases geometrically*.

Compound Interest: In this and the next few sections, we will see where some exponential equations come from.

When an amount of money a (called the *principal*) is invested at an interest rate n (expressed as a decimal), it will earn an amount of interest an at the end of the first interest period. If this interest is then added to the principal and both continue to earn interest, we have what is called *compound interest*, the type of interest commonly given on bank accounts. We can arrive at a formula for compound interest by computing the interest for several periods.

Period	*Interest earned at end of period*	*Balance y at end of period*
1	an	$a + an = a(1 + n)$
2	$a(1 + n)n$	$a(1 + n) + an(1 + n) = a(1 + n)^2$
3	$a(1 + n)^2n$	$a(1 + n)^2 + an(1 + n)^2 = a(1 + n)^3$
4	$a(1 + n)^3n$	$a(1 + n)^3 + an(1 + n)^3 = a(1 + n)^4$
.		.
.		.
.		.

We see a pattern emerging. In each case the balance is equal to the initial amount times $(1 + n)$ raised to the period. For example,

period ④ → balance $a(1 + n)^{④}$

This will always be so, and gives us the general formula

Compound Interest	$y = a(1 + n)^t$	**A9**

where t is the number of interest periods.

Example: To what amount will an initial deposit of \$500 accumulate if invested at a compound interest rate of $6\frac{1}{2}\%$ per year for 8 years?

Solution: From Eq. A9,

$$y = 500(1 + 0.065)^8$$
$$= \$827.50$$

Exponential Growth: The formula for compound interest, Eq. A9, is valuable not only in itself, but also because it leads us to an equation for the type of growth (and decline) that occurs in nature. In nature, growth and decline do not occur in steps, once a year, or once a minute, but occur *continuously*.

Let us modify the compound interest formula. Instead of computing interest *once* per period, let us compute it *m times* per period. The exponent in Eq. A9 thus becomes mt. The interest rate n, however, has to be reduced to $1/m$th of its old value, or to n/m.

Compound Interest Computed m Times per Period	$y = a\left(1 + \frac{n}{m}\right)^{mt}$	**A10**

Example: Repeat the computation of the preceding example if the interest is computed monthly rather than annually.

Solution: From Eq. A10, with $m = 12$,

$$y = 500\left(1 + \frac{0.065}{12}\right)^{12(8)} = 500(1.00542)^{96}$$

$$= \$839.83$$

or \$12.33 more than before.

We could, of course, use Eq. A10 to compute the interest if it were compounded every week, or every day, or every second. But what about *continuous* compounding, or continuous growth? What would happen to Eq. A10 if m got very, very large, in fact, infinite? Let us make m increase to large values, but first, to simplify the work, we make the substitution

$$\text{let } k = \frac{m}{n}$$

Equation A10 then becomes

$$y = a\left(1 + \frac{1}{k}\right)^{knt}$$

which, by Eq. 31, can be written

$$y = a\left[\left(1 + \frac{1}{k}\right)^{k}\right]^{nt}$$

Then as m grows large, so will k. Let us try some values on the calculator, and evaluate only the quantity in the square brackets, to four decimal places.

k	$\left(1 + \frac{1}{k}\right)^{k}$
1	2
10	2.5937 ...
100	2.7048 ...
1,000	2.7169 ...
10,000	2.7181 ...
100,000	2.7183 ...
1,000,000	2.7183 ...

The number e is called the limit *of the expression $(1 + 1/k)^k$ as k approaches infinity. The idea of a limit is of central importance in the study of calculus.*

We get the surprising result that as m (and k) continue to grow infinitely large, the value of $(1 + 1/k)^k$ does *not* grow without limit, but approaches the value 2.7183. . . . This important number is given the special symbol e. Thus, when m gets infinitely large, the quantity $(1 + 1/k)^k$ becomes e, and the formula for continuous growth becomes

Exponential Growth	$y = ae^{nt}$	**190**

To raise e to a power on a calculator, simply enter the power and press the $\boxed{e^x}$ key.

To raise e to a power x in BASIC, we use the EXP( ) function. Thus the statement

```
80  PRINT EXP(4.6)
```

will cause the value of $e^{4.6}$ (99.4843) to be printed.

Example: Repeat the computation of the preceding example if the interest were compounded *continuously* rather than monthly.

Banks that offer "continuous compounding" of your account do not have a clerk constantly computing your interest. They use Eq. 190.

Solution: From Eq. 190,

$$y = 500e^{0.065(8)} = 500e^{0.52}$$
$$= \$841.01$$

or \$1.18 more than before.

Common Error	When using Eq. 190 the time units of n and t *must agree.*

Example: A quantity grows exponentially at the rate of 2% per minute. By how many times will it have increased after 4 h?

Solution: The units of n and t do *not* agree (yet).

$$n = 2\% \text{ per } minute \qquad \text{and} \qquad t = 4 \text{ } hours$$

So we convert one of them to agree with the other.

$$t = 4 \text{ h} = 240 \text{ min}$$

By Eq. 190, we find the ratio of y to a,

$$\frac{y}{a} = e^{0.02(240)} = e^{4.8} = 122$$

So the final amount is 122 times as great as the initial amount.

Exponential Decay: When a quantity *decreases* so that the amount of decrease is proportional to the present amount, we have what is called *exponential decay*. Take, for example, a piece of steel just removed from a furnace. The amount of heat leaving the steel depends upon its temperature; the hotter it is, the faster heat will leave, and the faster it will cool. So as it cools, its temperature drops, *causing* *it to cool more slowly*. The result is the typical exponential decay curve shown in Fig. 15-2. Finally, the temperature is said to *asymptotically approach* the room temperature.

This is called Newton's law of cooling.

The equation for exponential decay is the same as for exponential growth, except that the exponent is negative.

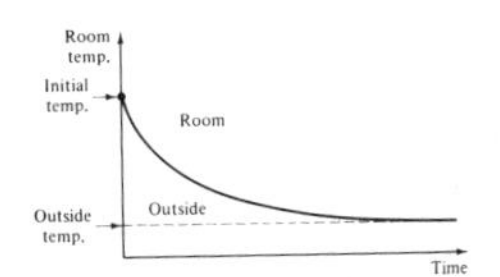

FIGURE 15-2. Exponential decrease in temperature.

Exponential Decay	$y = ae^{-nt}$	**192**

Example: A room initially 80°F above the outside temperature cools exponentially at the rate of 25% per hour. Find the temperature of the room (above the outside temperature) at the end of 135 min.

Solution: We first make the units consistent; 135 min = 2.25 h. Then by Eq. 192,

$$y = 80e^{-0.25(2.25)} = 45.6°\text{F}$$

Exponential Growth to an Upper Limit: Suppose that the bucket in Fig. 15-3 initially contains a gallons of water. Then assume that y_1, the amount remaining in the bucket, decreases exponentially. By Eq. 192,

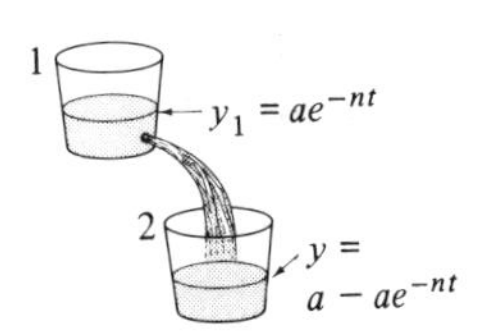

FIGURE 15-3. The amount in bucket 2 grows exponentially to an upper limit a.

$$y_1 = ae^{-nt}$$

The amount of water in bucket 2 thus increases from zero to a final amount a. The amount y in bucket 2 at any instant is equal to the amount which has left bucket 1, or

$$y = a - y_1$$
$$= a - ae^{-nt}$$

or

Exponential Growth to an Upper Limit	$y = a(1 - e^{-nt})$	**193**

Example: The voltage across a certain capacitor grows exponentially from 0 V to a maximum of 75 V, at a rate of 7% per second. Write the equation for the voltage at any instant, and find the voltage at 2.0 s.

Solution: From Eq. 193, with $a = 75$ and $n = 0.07$,

$$v = 75(1 - e^{-0.07t})$$

where v represents the instantaneous voltage. When $t = 2.0$ s,

$$v = 75(1 - e^{-0.14})$$
$$= 9.80 \text{ V}$$

Time Constant: Our equations for exponential growth and decay all contained a rate of growth n. Often it is more convenient to work with the *reciprocal* of n, rather than n itself. This reciprocal is called the time *constant, T*.

Time Constant	$T = \frac{1}{n}$	**194**

Example: If a quantity decays exponentially at a rate of 25% per second, the time constant is

$$T = \frac{1}{0.25}$$
$$= 4 \text{ s}$$

Notice that the time constant has the units of *time*.

Universal Growth and Decay Curves: If we replace n by $1/T$ in Eq. 190, we get

$$y = ae^{-t/T}$$

and then divide both sides by a,

$$\frac{y}{a} = e^{-t/T}$$

We plot this curve (Fig. 15-4) with the dimensionless ratio t/T for our horizontal axis and the dimensionless ratio y/a for the vertical axis. Thus, the horizontal axis is *the number of time constants* that have elapsed, and the vertical axis is *the ratio of the final amount to the initial amount*. It is called the *universal decay curve*.

We can use the universal decay curve to obtain graphical solutions to exponential decay problems.

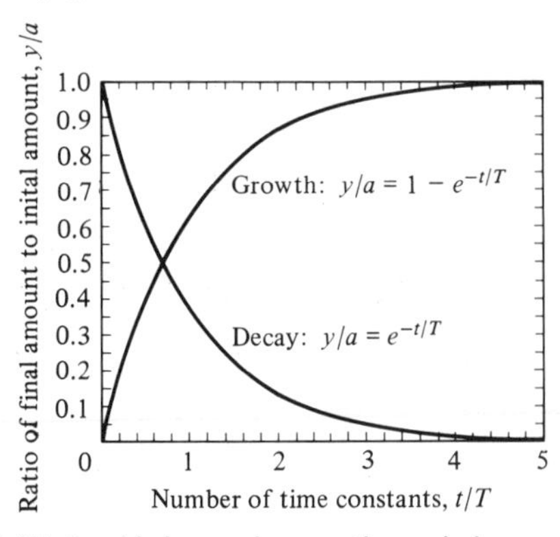

FIGURE 15-4. Universal growth and decay curves.

Example: A current decays from an initial value of 300 mA, at an exponential rate of 20% per second. Find the current after 7 s.

Solution: We have $n = 0.20$, so the time constant is, from Eq. 194,

$$T = \frac{1}{0.20} = 5 \text{ s}$$

Our given time $t = 7$ s is then equivalent to

$$\frac{t}{T} = \frac{7}{5} = 1.4 \text{ time constants}$$

From the universal decay curve, at $t/T = 1.4$, we read

$$\frac{y}{a} = 0.25$$

Since $a = 300$ mA,

$$y = 300(0.25) = 75 \text{ mA}$$

Figure 15-4 also shows the *universal growth curve*

$$\frac{y}{a} = 1 - e^{-t/T}$$

which is used in the same way as the universal decay curve.

EXERCISE 1

Exponential Function

FIGURE 15-5. A rope or chain hangs in the shape of a catenary.

Try it! Mount your finished graph vertically and hang a fine chain from the endpoints. Does its shape (after you have adjusted its length) match that of your graph?

Graph each exponential function over the given domains of x.

1. $y = 0.2(3.2)^x \quad (x = -4 \text{ to } +4)$

2. $y = 5(1 - e^{-x}) \quad (x = 0 \text{ to } 10)$

3. $y = 3(1.5)^{-2x} \quad (x = 0 \text{ to } 5)$

4. $y = 4e^{x/2} \quad (x = 0 \text{ to } 4)$

5. When a rope or chain hangs between two points (Fig. 15-5) the curve it describes, called a *catenary*, is given by

$$y = \frac{a(e^{x/a} + e^{-x/a})}{2}$$

Using a value of $a = 2.5$, plot y for $x = -5$ to $x = 5$.

Compound Interest

6. Find the amount to which \$500 will accumulate in 6 years at a compound interest rate of 6% per year, compounded annually.

The amount a is called the present worth *of the future amount y. This formula, and the three that follow, are standard formulas used in business and finance.*

7. If our compound interest formula, Eq. A9 is solved for a, we get

$$a = \frac{y}{(1 + n)^t}$$

where a is the amount that must be deposited now at interest rate n to produce an amount y in t years. How much money would have to be invested at $7\frac{1}{2}\%$ per year to accumulate to \$10,000 in 10 years?

8. What annual compound interest rate (compounded annually) is needed to enable an investment of \$5000 to accumulate to \$10,000 in 12 years?

9. Using Eq. A10, find the amount to which $1 will accumulate in 20 years at a compound interest rate of 10% per year (a) compounded annually; (b) compounded monthly; (c) compounded daily; and using Eq. 190, (d) compounded continuously.

10. If an amount of money R is deposited *every year* for t years at a compound interest rate n, it will accumulate to an amount y, where

A series of annual payments such as these is called an annuity.

$$y = R\left[\frac{(1+n)^t - 1}{n}\right]$$

If a worker pays $2000 per year into a retirement plan yielding 6% annual interest, what will be the value of this annuity after 25 years?

11. If the equation for an annuity in Problem 10 is solved for R, we get

This is called a sinking fund.

$$R = \frac{ny}{(1+n)^t - 1}$$

where R is the annual deposit required to produce a total amount y at the end of t years. How much would the worker in Problem 10 have to deposit each year (at 6% per year) to have a total of $100,000 after 25 years?

12. The yearly payment R that can be obtained for t years from a present investment a is

This is called capital recovery.

$$R = a\left[\frac{n}{(1+n)^t - 1} + n\right]$$

If a person has $85,000 in a retirement fund, how much can be withdrawn each year so that the fund will be exhausted in 20 years if the amount remaining in the fund is earning $6\frac{1}{4}$% per year?

Exponential Growth

13. A quantity grows exponentially at the rate of 5% per year for 7 years. Find the final amount if the initial amount is 200 units.

14. A yeast manufacturer finds that the yeast will grow exponentially at a rate of 15% per hour. How many pounds of yeast must they start with to obtain 500 lb at the end of 8 h?

15. If the world population were 4 billion in 1975 and grew at an annual rate of 1.7%, what would be the population in 1990?

16. If the rate of inflation is 12% per year, how much could a camera that now costs $225 be expected to cost 5 years from now?

17. The U.S. energy consumption in 1979 is estimated to be 37 million barrels (bbl) per day oil equivalent. If the rate of consumption increases at an annual rate of 7%, what will be the energy need in 1990?

18. A pharmaceutical company is growing an organism to be used in a vaccine that grows at a rate of 2.5% per hour. How many units of this organism must they have initially to have 1000 units after 5 days?

Exponential Growth to an Upper Limit

19. When an object is placed in surroundings that are at a higher temperature c, it will increase in temperature exponentially according to the equation

$$y = a(1 - e^{-nt}) \tag{193}$$

and approach a asymptotically (Fig. 15-6). A steel casting initially at 0°C is placed in a furnace at 800°C. If it increases in temperature at the rate of 5% per minute, find its temperature after 20 min.

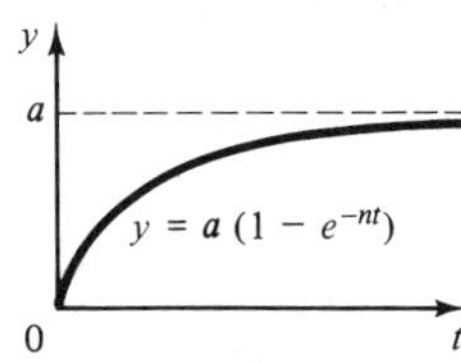

FIGURE 15-6. Exponential growth to an upper limit.

20. The equation given in Problem 19 approximately describes the hardening of concrete, where y is the compressive strength (lb/in.^{2}) t days after pouring. Using values of $a = 4000$ and $n = 0.0696$ in the equation, find the compressive strength after 14 days.

21. When the switch in Fig. 15-7 is closed, the current i will grow exponentially according to the equation

$$i = \frac{E(1 - e^{-Rt/L})}{R} \quad \text{amperes}$$

where L is the inductance in henries and R is the resistance in ohms. Find the current at 0.075 s if $R = 6.25\ \Omega$, $L = 186$ henries, and $E = 250$ V.

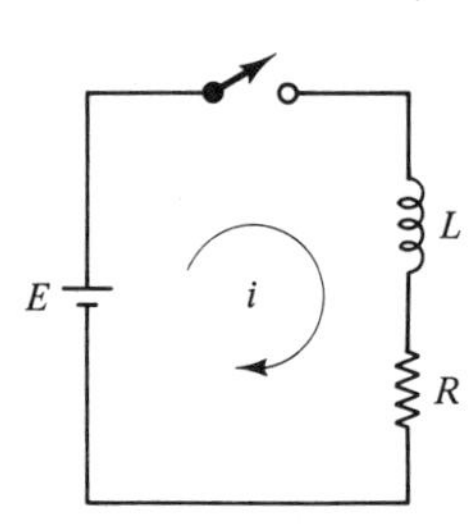

FIGURE 15-7

Exponential Decay

22. A flywheel is rotating at a speed of 1800 rev/min. When the power is disconnected, the speed decreases exponentially at the rate of 32% per minute. Find the speed after 10 min.

23. A steel forging is 1500°F above room temperature. If it cools exponentially at the rate of 2% per minute, how much will its temperature drop in 1 h?

24. As light passes through glass or water, its intensity decreases exponentially according to the equation

$$I = I_0 e^{-kx}$$

where I is the intensity at a depth x and I_0 is the intensity before entering the glass or water. If, for a certain filter glass, $k = 0.5$/cm (which means that each centimeter of filter thickness removes half the light reaching it), find the fraction of the original intensity that will pass through a filter glass 2.0 cm thick.

25. The atmospheric pressure p decreases exponentially with the height h (miles) above the earth according to the function $p = 29.92e^{-h/5}$ inches of mercury. Find the pressure at a height of 30,000 ft.

26. A certain radioactive material loses its radioactivity at the rate of $2\frac{1}{2}$% per year. What fraction of its initial radioactivity will remain after 10 years?

27. When a capacitor C, charged to a voltage v_0, is discharged through a resistor R (Fig. 15-8) the current i will decay exponentially according to the equation

$$i = \frac{v_0}{R}e^{-t/RC} \quad \text{amperes}$$

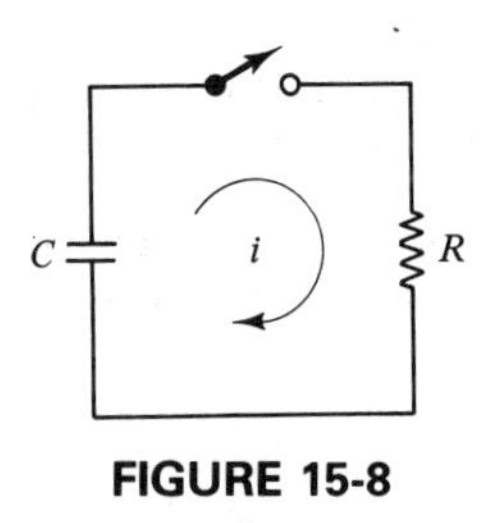

FIGURE 15-8

Find the current after 45 ms (milliseconds) in a circuit where $v_0 = 220$ V, $C = 130$ microfarads, and $R = 2700\ \Omega$.

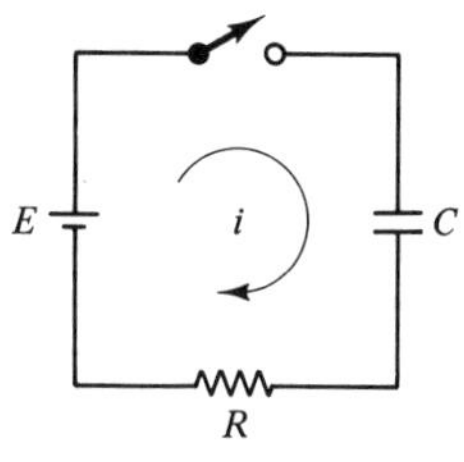

FIGURE 15-9

28. When a fully discharged capacitor C (Fig. 15-9) is connected across a battery, the current i flowing into the capacitor will decay exponentially according to the equation

$$i = \frac{E}{R}e^{-t/RC} \quad \text{amperes}$$

If $E = 115$ V, $R = 350\ \Omega$, and $C = 0.00075$ farad, find the current after 75 ms.

29. In a certain fabric mill, cloth is removed from a dye bath and is then observed to dry exponentially at the rate of 24% per hour. What percent of the original moisture will still be present after 5 h?

Computer

30. Program the computer to print a two-column table, the first column giving the value of k and the second column the value of $(1 + 1/k)^k$. Take a starting value for k equal to 1 and increase it each time by a factor of 10. Stop the run when k is less than 0.00001. Does the value of $(1 + 1/k)^k$ seem to be converging on some value? If not, why?

31. We can get an approximate value for e from the infinite series

$$e = 2 + \frac{1}{2!} + \frac{1}{3!} + \frac{1}{4!} + \cdots \qquad \textbf{195}$$

where 4! (read 4 *factorial*) is $4(3)(2)(1) = 24$. Write a program to compute e using the first five terms of the series.

32. The infinite series often used to calculate e^x is

$$e^x = 1 + x + \frac{x^2}{2!} + \frac{x^3}{3!} + \frac{x^4}{4!} + \cdots \qquad \textbf{196}$$

Write a program to compute e^x by using the first 15 terms of this series, and use it to find e^5.

33. Use your program for finding roots by simple iteration or by the midpoint method to solve any of the equations in Problems 1 to 4.

15-2. LOGARITHMS

In Sec. 15-1 we studied the exponential function

$$y = b^x$$

and given b and x, were able to find y. We are also able, given y and x, to find b. For example, if y were 24 and x were, say, 3.5,

$$24 = b^{3.5}$$

$$b = 24^{1/3.5} = 2.48$$

Suppose, now, that we have y and b and want to find x. Say that $y = 47$ when $b = 3.5$:

$$47 = 3.5^x$$

We will finish this problem in Sec. 15-6.

Try it. You will find that none of the math we have learned so far will enable us to find x. To solve this and similar problems, we need *logarithms,* which are explained in this section.

Definition: The logarithm of some positive number y *is the exponent* to which a base b must be raised to obtain y.

Example: Since $10^2 = 100$, we say that "2 is the logarithm of 100, to the base 10" because 2 is the exponent to which the base 10 must be raised to obtain 100. The logarithm is written

$$\log_{10} 100 = 2$$

We have the same relationship between the same quantities as we had before but a different way of expressing it. We have a similar situation in ordinary language. Take "Mary is John's sister.*" If we had only the word* sister *(and not* brother*) to describe the relationship between children, we would have to say something like "John is the person whom Mary is the sister of." That is as awkward as "the exponent to which the base must be raised to obtain the number." So we invent new words, such as* brother *and* logarithm.

Example: Given the exponential function $y = b^x$, we say that "x is the logarithm of y, to the base b" because x is the exponent to which the base b must be raised to obtain y.

So, in general,

Definition of a Logarithm	If Then	$y = b^x \quad (y > 0)$ $x = \log_b y$	**178**

Converting between Logarithmic and Exponential Form: In working with logarithmic and exponential expressions, we often have to switch between exponential and logarithmic form. We do this by means of the definition of a logarithm, Eq. 178. While converting from one form to the other, remember that (1) the base in exponential form is also the base in logarithmic form, and (2) "the *log* is the *exponent.*"

Example: Change the exponential expression $3^2 = 9$ to logarithmic form.

Solution: The base (3) in the exponential expression remains the base in the logarithmic expression

$$3^2 = 9$$
$$\downarrow$$
$$\log_3 (\ \ ) = (\ \ )$$

and the log is the exponent

$$3^2 = 9$$
$$\log_3 (\ \ ) = 2$$

So our expression is

$$\log_3 9 = 2$$

We could also write the same expression in radical form, $\sqrt{9} = 3$.

Example: Change $\log_e x = 3$ to exponential form.

Solution:

$$\log_e x = 3$$

$$e^3 = x$$

Example: Change $10^3 = 1000$ to logarithmic form.

Solution:

$$\log_{10} 1000 = 3$$

The Logarithmic Function: We wish to graph the logarithmic function $x = \log_b y$, but first let us follow the usual convention of calling the dependent variable y and the independent variable x. Our function then becomes

Logarithmic Function	$y = \log_b x$	**197**

Example: Plot the logarithmic function $y = \log_2 x$ for $x = -1$ to 16.

Solution: We have no means of calculating logs to the base 2, so we switch to exponential form, $x = 2^y$. Now we can select values of y and calculate x (the reverse of our usual plotting procedure, where we select x and calculate y). Make a table of point pairs.

y	-2	-1	0	1	2	3	4
x	$\frac{1}{4}$	$\frac{1}{2}$	1	2	4	8	16

These points are plotted in Fig. 15-10 on page 476.

It becomes apparent that no value of y, positive or negative or zero, is going to yield a negative or zero value for x, so we cannot find a y to give $x = -1$. Thus, the domain of x for $\log_2 x$ is the positive numbers only. We will see that this is true for logs to *any base*. Stated another way: *we cannot take the logarithm of a negative number.*

We will do harder ones in Sec. 15-5.

Solving Simple Logarithmic Equations: Many logarithmic equations can be solved merely by rewriting them in exponential form.

Example: Solve for x:

$$x = \log_5 25$$

Solution: Changing to exponential form,

$$5^x = 25 = 5^2$$

$$x = 2$$

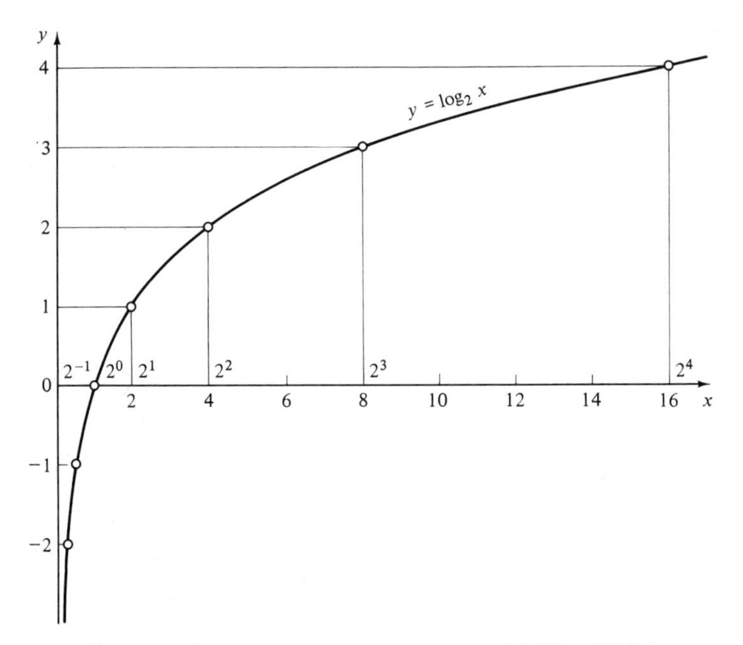

FIGURE 15-10. Compare this graph with Fig. 15-1 for $y = 2^x$. What similarities do you see?

Example: Solve for x:

$$\log_x 4 = \frac{1}{2}$$

Solution: Going to exponential form,

$$x^{1/2} = 4$$

Squaring both sides,

$$x = 16$$

Example: Solve for x:

$$\log_{25} x = -\frac{3}{2}$$

Solution: Writing the equation in exponential form,

$$25^{-3/2} = x$$

By Eq. 35,

$$x = \frac{1}{25^{3/2}} = \frac{1}{5^3} = \frac{1}{125}$$

Example: Solve for x,

$$x = \log_{81} 27$$

Solution: Changing to exponential form,

$$81^x = 27$$

But 81 and 27 are both powers of 3.

$$(3^4)^x = 3^3$$
$$3^{4x} = 3^3$$
$$4x = 3$$
$$x = 3/4$$

EXERCISE 2

Converting between Logarithmic and Exponential Form

Convert to logarithmic form.

1. $3^4 = 81$ **2.** $5^3 = 125$ **3.** $4^6 = 4096$

4. $7^3 = 343$ **5.** $x^5 = 995$ **6.** $a^3 = 6.83$

7. $y^4 = 73.8$ **8.** $x^{2.5} = 7.3$ **9.** $7^x = 83$

10. $6^a = 337$ **11.** $10^x = 4000$ **12.** $e^y = 94.8$

13. $10^x = b$ **14.** $e^y = z$ **15.** $x^a = b$

Convert to exponential form.

16. $\log_{10} 100 = 2$ **17.** $\log_2 16 = 4$ **18.** $\log_5 125 = 3$

19. $\log_4 1024 = 5$ **20.** $\log_3 x = 57$ **21.** $\log_x 54 = 285$

22. $\log_{10} y = 736$ **23.** $\log_e x = 1.378$ **24.** $\log_5 12 = w$

25. $\log_a b = c$ **26.** $\log_x 847 = 3.72$ **27.** $\log_{10} x = y$

28. $\log_x m = n$ **29.** $\log_e 74.8 = x$ **30.** $\log_{10} w = 55$

Logarithmic Function

Graph each logarithmic function from $x = \frac{1}{2}$ to $x = 4$.

31. $y = \log_x x$ **32.** $y = \frac{1}{2} \log_2 x$ **33.** $y = \log_{1/2} x$

34. $y = \log_e x$ **35.** $y = \log_{10} x$ **36.** $y = \frac{1}{4} \log_2 3x$

Simple Logarithmic Equations

Solve for x.

37. $x = \log_3 9$ **38.** $x = \log_2 8$ **39.** $x = \log_8 2$

40. $x = \log_9 27$ **41.** $x = \log_{27} 9$ **42.** $x = \log_4 8$

43. $x = \log_8 4$ **44.** $x = \log_{27} 81$ **45.** $\log_x 8 = 3$

46. $\log_3 x = 4$ **47.** $\log_x 27 = 3$ **48.** $\log_x 16 = 4$

49. $\log_5 x = 2$ **50.** $\log_x 4 = \frac{1}{2}$ **51.** $\log_{36} x = \frac{1}{2}$

52. $\log_2 x = 3$ **53.** $x = \log_{25} 125$ **54.** $x = \log_5 25$

55. $x = \log_{125} 5$ **56.** $x = \log_3 (\frac{1}{3})$ **57.** $x = \log_9 (\frac{1}{27})$

58. $\log_{27} x = \frac{2}{3}$ **59.** $\log_x \frac{1}{25} = -\frac{2}{3}$ **60.** $\log_x \frac{2}{3} = -\frac{1}{3}$

15-3. PROPERTIES OF LOGARITHMS

Products: We wish to write an expression for the log of the product of two numbers, say, M and N.

$$\log_b MN = ?$$

Let us substitute b^c for M, and b^d for N. Then

$$MN = b^c b^d$$

But by Eq. 29,

$$MN = b^{c+d}$$

Writing this expression in logarithmic form,

$$\log_b MN = c + d$$

But, since $b^c = M$, $c = \log_b M$. Similarly, $d = \log_b N$. Substituting, we get

Log of a Product	$\log_b MN = \log_b M + \log_b N$	**179**

In words, *the log of a product equals the sum of the logs of the factors.*

Example: Express log $7x$ as the sum of two logarithms.

Solution: By Eq. 179,

$$\log 7x = \log 7 + \log x$$

Example: Express as a single logarithm:

$$\log 3 + \log x + \log y$$

Solution: By Eq. 179,

$$\log 3 + \log x + \log y = \log 3xy$$

Common Error	The log of a sum is *not* equal to the sum of the logs. $\log (M + N) \neq \log M + \log N$
Common Error	The product of two logs is *not* equal to the sum of the logs. $(\log M)(\log N) \neq \log M + \log N$

Quotients: Let us now consider the quotient M divided by N. Using the same substitution as above,

$$\frac{M}{N} = \frac{b^c}{b^d}$$

By Eq. 30,

$$= b^{c-d}$$

Going to logarithmic form,

$$\log_b \frac{M}{N} = c - d$$

Finally, substituting back,

Log of a Quotient	$\log_b \frac{M}{N} = \log_b M - \log_b N$	**180**

The log of a quotient equals the log of the dividend minus the log of the divisor.

Example:

$$\log \frac{3}{4} = \log 3 - \log 4$$

Example: Express

$$\log \frac{ab}{xy}$$

as the sum or difference of two or more logarithms.

Solution: By Eq. 180,

$$\log \frac{ab}{xy} = \log ab - \log xy$$

and by Eq. 179,

$$\log ab - \log xy = \log a + \log b - (\log x + \log y)$$
$$= \log a + \log b - \log x - \log y$$

Example: The expression $\log 10x^2 - \log 5x$ can be expressed as a single logarithm,

$$\log \frac{10x^2}{5x} \quad \text{or} \quad \log 2x$$

Common Error	Similar errors are made with quotients as with products. $\log (M - N) \neq \log M - \log N$ $\frac{\log M}{\log N} \neq \log M - \log N$

Powers: Consider now the quantity M raised to a power p. Substituting b^c for M, as before,

$$M^p = (b^c)^p$$

and by Eq. 31,

$$= b^{cp}$$

In logarithmic form,

$$\log_b M^p = cp$$

Substituting back, we get

Log of a Power	$\log_b M^p = p \log_b M$	**181**

The log of a number raised to a power equals the power times the log of the number.

Example: The expression

$$7 \log x$$

can be expressed as a single logarithm with a coefficient of 1, as

$$\log x^7$$

Example:

$$\log 3.85^{1.4} = 1.4 \log 3.85$$

Example: Express as a single logarithm,

$$2 \log 5 + 3 \log 2 - 4 \log 1$$

Solution: By Eq. 181,

$$2 \log 5 + 3 \log 2 - 4 \log 1 = \log 5^2 + \log 2^3 - \log 1^4$$

and by Eqs. 179 and 180,

$$= \log \frac{25(8)}{1} = \log 200$$

Example: Express as a single logarithm with a coefficient of 1.

$$2 \log \frac{2}{3} + 4 \log \frac{a}{b} - 3 \log \frac{x}{y}$$

Solution:

$$2 \log \frac{2}{3} + 4 \log \frac{a}{b} - 3 \log \frac{x}{y} = \log \frac{2^2}{3^2} + \log \frac{a^4}{b^4} - \log \frac{x^3}{y^3}$$

$$= \log \frac{4a^4y^3}{9b^4x^3}$$

Example: Rewrite without logarithms, the equation

$$\log (a^2 - b^2) + 2 \log a = \log (a - b) + 3 \log b$$

Solution: We first try to combine all the logarithms into a single logarithm. Transposing gives

$$\log(a^2 - b^2) - \log(a - b) + 2\log a - 3\log b = 0$$

By the laws of logarithms,

$$\log\frac{a^2 - b^2}{a - b} + \log\frac{a^2}{b^3} = 0$$

or

$$\log\frac{a^2(a - b)(a + b)}{b^3(a - b)} = 0$$

Simplifying,

$$\log\frac{a^2(a + b)}{b^3} = 0$$

We then eliminate the logarithm altogether by going from logarithmic to exponential form,

$$\frac{a^2(a + b)}{b^3} = 10^0 = 1$$

or

$$a^2(a + b) = b^3$$

Roots: By a derivation identical to the one for powers, we get

Roots	$\log_b \sqrt[q]{M} = \frac{1}{q}\log_b M$	**182**

The log of the root of a number equals the log of the number divided by the index of the root.

Examples:

(a) $$\log_2 \sqrt[5]{8} = \frac{1}{5}\log_2 8 = \frac{1}{5}(3) = \frac{3}{5}$$

(b) $$\frac{\log x}{2} + \frac{\log z}{4} = \log\sqrt{x} + \log\sqrt[4]{z}$$

Log of 1: Let us take the log to the base b of 1, and call it x,

$$x = \log_b 1$$

Switching to exponential form,

$$b^x = 1$$

so

$$x = 0$$

Therefore,

Log of 1	$\log_b 1 = 0$	**183**

The log of 1 is zero.

Log of the Base: We now take the $\log_b$ of its own base b. Let us call this quantity x.

$$\log_b b = x$$

Going to exponential form,

$$b^x = b$$

So x must equal 1.

Log of the Base	$\log_b b = 1$	**184**

The log (to the base b) of b is equal to 1.

Log of the Base Raised to a Power: Consider the expression

$$\log_b b^n$$

We set this expression equal to x, and as before, go to exponential form,

$$\log_b b^n = x$$
$$b^x = b^n$$
$$x = n$$

So

Log of Base Raised to a Power	$\log_b b^n = n$	**185**

The log (to the base b) of b raised to a power is equal to the power.

Examples:

(a) $$\log_2 2^{4.83} = 4.83$$

(b) $$\log_e e^{2x} = 2x$$

(c) $$\log_{10} 0.0001 = \log_{10} 10^{-4} = -4$$

EXERCISE 3

Properties of Logarithms

Write as the sum or difference of two or more logarithms.

1. $\log \frac{2}{3}$ **2.** $\log 4x$ **3.** $\log ab$

4. $\log \dfrac{x}{2}$ **5.** $\log xyz$ **6.** $\log 2ax$

7. $\log \frac{3x}{4}$ **8.** $\log \frac{5}{xy}$ **9.** $\log \frac{1}{2x}$

10. $\log \frac{2x}{3y}$ **11.** $\log \frac{abc}{d}$ **12.** $\log \frac{x}{2ab}$

Express as a single logarithm with a coefficient of 1. Assume that the logarithms in each problem have the same base.

13. $\log 3 + \log 4$ **14.** $\log 7 - \log 5$

15. $\log 2 + \log 3 - \log 4$ **16.** $3 \log 2$

17. $4 \log 2 + 3 \log 3 - 2 \log 4$ **18.** $\log x + \log y + \log z$

19. $3 \log a - 2 \log b + 4 \log c$ **20.** $\frac{\log x}{3} + \frac{\log y}{2}$

21. $\log \frac{x}{a} + 2 \log \frac{y}{b} + 3 \log \frac{z}{c}$

22. $2 \log x + \log (a + b) - \frac{1}{2} \log (a + bx)$

23. $\frac{1}{2} \log (x + 2) + \log (x - 2)$

Rewrite each equation so that it contains no logarithms.

24. $\log x + 3 \log y = 0$ **25.** $\log_2 x + 2 \log_2 y = x$

26. $2 \log (x - 1) = 5 \log (y + 2)$ **27.** $\log_2 (a^2 + b^2) + 1 = 2 \log_2 a$

28. $\log (x + y) - \log 2 = \log x + \log (x^2 - y^2)$

29. $\log (p^2 - q^2) - \log (p + q) = 2$

15-4. COMMON LOGARITHMS AND LOGARITHMIC COMPUTATION

Also called Briggs' logarithms, *after Henry Briggs (1561–1630).*

Systems of Logarithms: In this chapter we have written logarithms with many different bases; 2, $\frac{1}{2}$, 5, 25, and so on. In practice, however, only two bases are used, the base 10 and the base *e*. Logs to the base *e*, called *natural* logs, are discussed in Sec. 15-5. In this section we study logs to the base 10, called *common* logarithms. Any logarithm in which the base is not written is assumed to be a common logarithm.

$$\log 3.85 \equiv \log_{10} 3.85$$

Common Logarithms by Calculator: Enter the number you wish to take the log of (the *argument*) and press the [log] key.

Example: Find log 74.82 to five significant digits.

Solution: Enter 74.82. Press [log]. Read 1.8740 (rounded).

Example: Find log 0.0375 to four significant digits.

Solution: Enter .0375. Press [log]. Read −1.426.

Example: Find log (−6.73).

Solution: Enter 6.73. Press [+/−]. Press [log]. Read *error indication.* Recall from Sec. 15-2 that we cannot take the logarithm of a negative number.

Antilogarithms by Calculator: Suppose that we have the logarithm of a number, and want to find the number itself: say, log x = 1.852. Switching to exponential form,

$$x = 10^{1.852}$$

Some calculators use [INV] [LOG] *to find antilogarithms.*

This quantity can be found using the [10^x] key on the calculator (or the [y^x] key if you have no [10^x] key).

Enter 1.852. Press [10^x]. Read 71.12 (to four digits).

or

Enter 10. Press [y^x]. Enter 1.852. Press [=]. Read 71.12.

Example: Find x to four digits if log x = −1.738.

Solution: Enter 1.738. Press [+/−]. Press [10^x]. Read .01828.

This may not seem like hot stuff in the age of calculators and computers, but it revolutionized computation in its time. The French mathematician Laplace wrote: "The method of logarithms, by reducing to a few days the labor of many months, doubles, as it were, the life of the astronomer, besides freeing him from the errors and disgust inseparable from long calculation."

Multiplying by Means of Logarithms: We now combine our ability to find common logarithms with the laws of logarithms of Sec. 15-3 to do some logarithmic computation. We start with multiplication.

Example: Multiply (2.947)(9.362) by logarithms.

Solution: We take the log of the product

$$\begin{aligned}\log(2.947 \times 9.362) &= \log 2.947 + \log 9.362 \\ &= 0.46938 + 0.97137 = 1.44075\end{aligned}$$

Taking the antilog,

$$(2.947)(9.362) = 10^{1.44075} = 27.59$$

Division: The procedure is similar to that for multiplication.

Example: Divide using logarithms:

$$28.47 \div 1.883$$

Solution: By Eq. 180,

$$\begin{aligned}\log\frac{28.47}{1.883} &= \log 28.47 - \log 1.883 \\ &= 1.4544 - 0.2749 = 1.1795\end{aligned}$$

Taking the antilog,

$$28.47 \div 1.883 = 10^{1.1795} = 15.12$$

Powers: Use Eq. 181 to reduce a problem of raising to a power into one of multiplication.

Example: Find $35.82^{1.4}$ by logarithms.

Solution: By Eq. 181,

It is customary to use common *logarithms for logarithmic computation, but natural logarithms would work as well.*

$$\log 35.82^{1.4} = 1.4 \log 35.82$$
$$= 1.4(1.5541) = 2.1758$$

Taking the antilog,

$$35.82^{1.4} = 10^{2.1758} = 149.9$$

Roots:

Example: Find $\sqrt[5]{3824}$ by logarithms.

Solution: By Eq. 182,

$$\log \sqrt[5]{3824} = \frac{1}{5} \log 3824 = \frac{3.5825}{5}$$
$$= 0.7165$$

Taking the antilog,

$$\sqrt[5]{3824} = 10^{0.7165} = 5.206$$

EXERCISE 4

Common Logarithms

Find the common logarithm of each number to four decimal places.

1. 5.93	**2.** 9.26	**3.** 48.3
4. 385	**5.** 836	**6.** 1.03
7. 27.4	**8.** 0.0573	**9.** 0.00737
10. 9738	**11.** 34,970	**12.** 5.28×10^6

Find x to three significant digits if the common logarithm of x is equal to:

13. 1.584	**14.** 2.957	**15.** 5.273
16. 0.886	**17.** −2.857	**18.** 0.366
19. −2.227	**20.** 4.973	**21.** 3.979
22. 1.295	**23.** −3.972	**24.** −0.8835

Logarithmic Computation

Evaluate using logarithms.

25. 5.937×92.47

26. 6923×0.003846

27. $3.97 \times 8.25 \times 9.82$

28. $88.25 \div 42.94$

29. $\sqrt[9]{8563}$

30. $4.836^{3.97}$

31. $83.62^{0.572}$

32. $\sqrt[3]{587}$

33. $\sqrt[7]{8364}$

34. $\sqrt[4]{62.4}$

15-5. NATURAL LOGARITHMS

Also called Napierian logarithms, *after John Napier (1550–1617), the discoverer of logarithms.*

The number e (2.718 approximately) has already been discussed in Sec. 15-1. Logarithms using e as a base are called *natural* logarithms. They are written $\ln x$. Thus, $\ln x$ is understood to mean $\log_e x$.

Natural Logarithms by Calculator: To find the natural logarithm of a (positive) number by calculator, enter the number and press the [ln] key.

Example: Find ln 19.3 to four significant digits.

Solution: Enter 19.3. Press [ln]. Read 2.960 (rounded).

Antilogarithms by Calculator: If the natural logarithm of a number is known, we can find the number itself by using the [e^x] key.

Some calculators do not have an [e^x] *key. Instead, use* [inv] [ln].

Example: Find x to four digits if $\ln x = 3.825$.

Solution: Rewriting the expression $\ln x = 3.825$ in exponential form,

$$x = e^{3.825}$$

On the calculator: Enter 3.825. Press [e^x]. Read 45.83 (rounded).

Change of Base: We can convert between natural logarithms and common logarithms (or between logarithms to any base) by the following procedure. Suppose $\log N$ is the common logarithm of some number N. We set it equal to x,

$$\log N = x$$

Going to exponential form,

$$10^x = N$$

Now we take the natural log of both sides,

$$\ln 10^x = \ln N$$

By Eq. 181,

$$x \ln 10 = \ln N$$

$$x = \frac{\ln N}{\ln 10}$$

But $x = \log N$, so

Change of Base	$\log N = \dfrac{\ln N}{\ln 10}$ where $\ln 10 \cong 2.3026$.	**186**

The common logarithm of a number is equal to the natural log of that number divided by the natural log of 10.

Example: Find $\log N$ if $\ln N = 5.824$.

Solution: By Eq. 186

$$\log N = \frac{5.824}{2.3026} = 2.529$$

Example: What is $\ln N$ if $\log N = 3.825$?

Solution: By Eq. 186

$$\ln N = \ln 10 (\log N)$$
$$= 2.3026(3.825) = 8.807$$

Logarithms on the Computer: To find natural logarithms in BASIC, we use the LOG( ) function.

Example: The BASIC instruction

```
50 PRINT LOG(X)
```

will cause the *natural* logarithm of X to be printed.

If the *common* log is wanted, we use Eq. 186.

Example: The BASIC instruction

```
50 PRINT LOG(X)/LOG(10)
```

will cause the common logarithm of X to be printed.

Common Error	The BASIC function LOG can easily be mistaken for the common logarithm.

To find the *antilog* of some number N, we must find e^N when using natural logs, and 10^N when using common logs.

Example: The BASIC instructions

```
20 PRINT EXP(5)
30 PRINT 10^5
```

will cause the natural antilog of 5 to be printed at line 20, and the common antilog at line 30.

EXERCISE 5

Natural Logarithms

Find the natural logarithm of each number to four decimal places.

1. 48.3	**2.** 846	**3.** 2365
4. 1.005	**5.** 0.845	**6.** 4.97
7. 0.00836	**8.** 45,900	**9.** 0.0000462
10. 82,900	**11.** 3.84×10^4	**12.** 8.24×10^{-3}

Find the number whose natural logarithm is:

13. 2.846	**14.** 4.263	**15.** 0.879
16. −2.846	**17.** −0.365	**18.** 5.937
19. 0.936	**20.** −4.97	**21.** 15.84
22. −9.47	**23.** −18.36	**24.** 21.83

Change of Base

Find the common logarithm of the number whose natural logarithm is:

25. 8.36	**26.** −3.846	**27.** 3.775
28. 15.36	**29.** 5.26	**30.** −0.638

Find the natural logarithm of the number whose common logarithm is:

31. 84.9	**32.** 2.476	**33.** −3.82
34. 73.9	**35.** 2.37	**36.** −2.63

Computer

37. The series approximation for the natural logarithm of a positive number x is

$$\ln x = 2a + \frac{2a^3}{3} + \frac{2a^5}{5} + \frac{2a^7}{7} + \cdots \tag{198}$$

where

$$a = \frac{x-1}{x+1}$$

Write a program to compute $\ln x$ using the first 10 terms of the series.

38. The series expansion for b^x is

$b^x = 1 + x \ln b + \frac{(x \ln b)^2}{2!} + \frac{(x \ln b)^3}{3!} + \cdots$ $(b > 0)$	**189**

Write a program to compute b^x using the first x terms of the series.

15-6. SOLVING EXPONENTIAL EQUATIONS WITH LOGARITHMS

Solving Exponential Equations: We return now to the problem we started in Sec. 15-2 but could not finish, that of finding the *exponent* when the other quantities in an exponential equation were known. We tried to solve the equation

$$47 = 3.5^x$$

The key to solving exponential equations is to *take the logarithm of both sides*. Doing so, we get

$$\log 47 = \log 3.5^x$$

By Eq. 181,

$$\log 47 = x \log 3.5$$

$$x = \frac{\log 47}{\log 3.5} = \frac{1.672}{0.5541} = 3.07$$

In this example we took *common* logarithms, but *natural* logarithms would have worked just as well.

Example: Solve for x:

$$3.25^{x+2} = 1.44^{3x-1}$$

Solution: We take the log of both sides. This time choosing to use natural logs, we get,

$$(x + 2) \ln 3.25 = (3x - 1) \ln 1.44$$

By calculator,

$$(x + 2)(1.179) = (3x - 1)(0.3646)$$

or

$$3.233(x + 2) = 3x - 1$$

Removing parentheses and collecting terms,

$$0.233x = -7.467$$

$$x = -32.0$$

Exponential Equations with Base *e*: If the exponential equation contains the base e, taking *natural* logs rather than common logarithms will simplify the work.

Example: Solve for x:

$$157 = 112e^{3x+2}$$

Solution: Dividing by 112,

$$1.402 = e^{3x+2}$$

Taking natural logs,

$$\ln 1.402 = \ln e^{3x+2}$$

By Eq. 181,

$$\ln 1.402 = (3x + 2) \ln e$$

But, by Eq. 184, $\ln e = 1$, so

$$\ln 1.402 = 3x + 2$$

$$x = \frac{\ln 1.402 - 2}{3} = -0.554$$

Example: Solve for x to three significant digits:

$$3e^{x} + 2e^{-x} = 4(e^{x} - e^{-x})$$

Solution: Removing parentheses gives

$$3e^{x} + 2e^{-x} = 4e^{x} - 4e^{-x}$$

Transposing and combining like terms,

$$e^{x} - 6e^{-x} = 0$$

We can simplify further by dividing through by e^{-x}.

$$e^{2x} - 6 = 0$$

$$e^{2x} = 6$$

Taking natural logs,

$$2x \ln e = \ln 6 = 1.792$$

$$x = 0.896$$

Our exponential equation might be of *quadratic type,* and can then be solved by substitution as we did in Sec. 11-1.

Example: Solve for x to three significant digits:

$$3e^{2x} - 2e^{x} - 5 = 0$$

Solution: If we substitute

$$w = e^{x}$$

our equation becomes the quadratic

$$3w^2 - 2w - 5 = 0$$

By the quadratic formula,

$$w = \frac{2 \pm \sqrt{4 - 4(3)(-5)}}{6} = -1 \quad \text{and} \quad \frac{5}{3}$$

Substituting back,

$$e^x = -1 \qquad \text{and} \qquad e^x = \frac{5}{3}$$

So, $x = \ln(-1)$ which we discard and

$$x = \ln\frac{5}{3} = 0.511$$

Exponential Equations with Base 10: Similarly, if the equation contains 10 as a base, the work will be simplified by taking *common* logarithms.

Example: Solve

$$10^{5x} = 2(10^{2x})$$

Solution: Taking common logs of both sides,

$$\log 10^{5x} = \log [2(10^{2x})]$$

By Eqs. 179 and 181,

$$5x \log 10 = \log 2 + 2x \log 10$$

By Eq. 184, $\log 10 = 1$, so

$$5x = \log 2 + 2x$$

$$3x = \log 2$$

$$x = \frac{\log 2}{3} = 0.100$$

Doubling Time: Being able to solve an exponential equation for the exponent allows us to return to the formulas for exponential growth and decay (Sec. 15-1) and derive two interesting quantities: doubling time and half-life.

If a quantity grows exponentially according to the function

$$y = ae^{nt} \tag{190}$$

it will eventually double (y would be twice a). Setting $y = 2a$, we get

$$2a = ae^{nt}$$

or

$$2 = e^{nt}$$

Taking natural logs,

$$\ln 2 = \ln e^{nt} = nt \ln e = nt$$

or

Doubling Time	$t = \dfrac{\ln 2}{n}$	**191**

Since ln 2 = 0.699, *if we let P be the rate of growth expressed as a percent* (not *as a decimal.* $P = 100n$), *we get a convenient* rule of thumb:

$$\text{doubling time} \cong \frac{70}{P}$$

Example: In how many years will a quantity double if it grows at the rate of 5% per year?

Solution: From Eq. 191,

$$\text{doubling time} = \frac{\ln 2}{0.05} = \frac{0.693}{0.05} = 13.9 \text{ years}$$

Half-Life: When a material decays exponentially according to the function

$$y = ae^{-nt} \tag{192}$$

the time it takes for the material to be half gone is called the *half-life*. If we let $y = a/2$ in Eq. 192,

$$\frac{1}{2} = e^{-nt} = \frac{1}{e^{nt}}$$

$$2 = e^{nt}$$

Taking natural logarithms,

$$\ln 2 = \ln e^{nt} = nt$$

Notice that the equation for half-life is the same as for doubling time. The rule of thumb is, of course, also the same.

Half-Life	$t = \dfrac{\ln 2}{n}$	**191**

Example: Find the half-life of a radioactive material that decays exponentially at the rate of 2% per year.

Solution: By Eq. 191,

$$\text{half-life} = \frac{\ln 2}{0.02} = 34.7 \text{ years}$$

EXERCISE 6

Exponential Equations

Solve for x to three significant digits.

1. $2^x = 7$

2. $(7.26)^x = 86.8$

3. $(1.15)^{x+2} = 12.5$

4. $(2.75)^x = (0.725)^{x^2}$

5. $(15.4)^{\sqrt{x}} = 72.8$

6. $e^{5x} = 125$

7. $5.62e^{3x} = 188$

8. $1.05e^{4x+1} = 5.96$

9. $e^{2x-1} = 3e^{x+3}$

10. $14.8e^{3x^2} = 144$

11. $5^{2x} = 7^{3x-2}$

12. $3^{x^2} = 175^{x-1}$

13. $10^{3x} = 3(10^x)$

14. $e^x + e^{-x} = 2(e^x - e^{-x})$

Hint: *Problems 18 to 20 are* equations of quadratic type.

15. $2^{3x+1} = 3^{2x+1}$

16. $10^{x-5} = 10^{x-4} - 5$

17. $7e^{1.5x} = 2e^{2.4x}$

18. $e^{4x} - 2e^{2x} - 3 = 0$

19. $e^{x} + e^{-x} = 2$

20. $e^{6x} - e^{3x} - 2 = 0$

Applications

21. The efficiency of an internal combustion engine (Otto cycle) is

$$E = 1 - \frac{1}{r^{0.4}} \times 100$$

where r is the compression ratio. What compression ratio will give an efficiency of 56%?

22. The current i in a certain circuit is given by

$$i = 6.25e^{-125t} \quad \text{amperes}$$

where t is the time in seconds. At what time will the current be 1.00 A?

23. The temperature above its surroundings of an iron casting initially at 2000°F will be

$$T = 2000e^{-0.062t}$$

after t seconds. Find the time for the casting to be at a temperature of 500°F above its surroundings.

24. The flow rate for water passing over a certain weir (Fig. 15-11) is given by

$$Q = 12.5H^{1.73} \quad \text{ft}^3/\text{h}$$

where H is the water depth above the weir in feet. What will be the water depth for a flow of 25.0 ft^3/h?

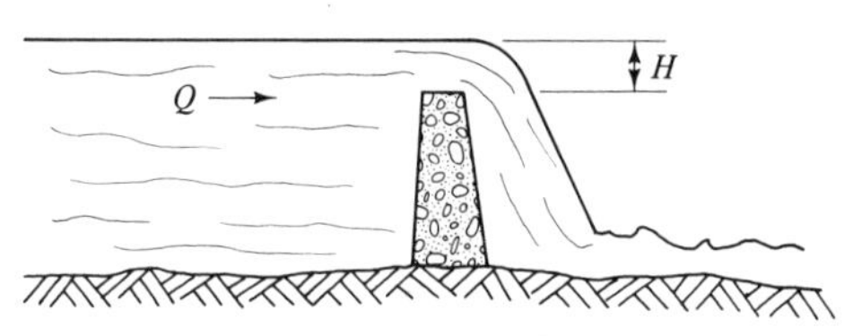

FIGURE 15-11. Flow over a weir.

25. A certain long pendulum, released from a height y_0, will be at a height

$$y = y_0e^{-0.75t}$$

at t seconds. If the pendulum is released at a height of 15 cm, at what time will the height be 5 cm?

26. A population growing at a rate of 2% per year from an initial population of 9000 will grow in t years to an amount

$$P = 9000e^{0.02t}$$

How many years will it take for the population to triple?

27. The barometric pressure in inches of mercury at a height of h feet above sea level is

$$p = 30e^{-kh}$$

where $k = 3.83 \times 10^{-5}$. At what height will the pressure be 10 inches of mercury?

28. The approximate density of seawater at a depth of h miles is

$$d = 64.0e^{0.00676h} \qquad \text{lb/ft}^3$$

Find the depth at which the density will be 64.5 lb/ft^3.

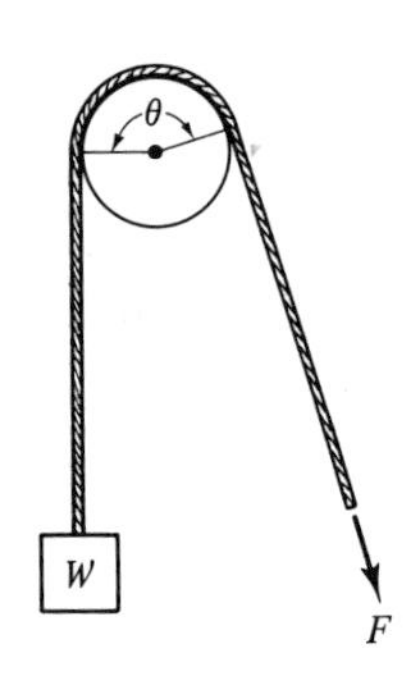

FIGURE 15-12

29. A rope passing over a rough cylindrical beam (Fig. 15-12) supports a weight W. The force F needed to hold the weight is

$$F = We^{-\mu\theta}$$

where μ is the coefficient of friction and θ is the angle of wrap in radians. If $\mu = 0.15$, what angle of wrap is needed for a force of 100 lb to hold a weight of 200 lb?

30. Using the formula for compound interest, Eq. A9, calculate the number of years it will take for a sum of money to triple when invested at a rate of 12% per year.

31. Using the formula for present worth, Exercise 1, Problem 7, in how many years will \$50,000 accumulate to \$70,000 at 15% interest?

32. Using the annuity formula, Exercise 1, Problem 10, find the number of years it will take a worker to accumulate \$100,000 with an annual payment of \$3000 if the interest rate is 8%.

33. Using the capital recovery formula from Exercise 1, Problem 12, calculate the number of years a person can withdraw \$10,000/yr from a retirement fund containing \$60,000 if the rate of interest is $6\frac{3}{4}$%.

34. Find the half-life of a material that decays exponentially at the rate of 3.5% per year.

35. How long will it take the U.S. annual oil consumption to double if it is increasing exponentially at a rate of 7% per year?

36. How long will it take the world population to double at an exponential growth rate of 1.64% per year?

37. What is the maximum annual growth in energy consumption permissible if the consumption is not to double in the next 20 years?

Computer

38. An iron ball with a mass of 1 kg is cut into two equal pieces. It is cut again, the second cut producing four equal pieces, the third cut making eight pieces, and so on. Write a program to compute the number of cuts needed to make each piece smaller than one atom (mass $= 9.4 \times 10^{-26}$ kg). Assume that no material is removed during cutting.

39. Assuming that the present annual world oil consumption is 17×10^9 barrels/yr, that this rate of consumption is increasing at a rate of 5% per year, and that the total world oil reserves are 1700×10^9 barrels, compute and print the following table.

Year	*Annual consumption*	*Oil remaining*
0	17	1700
1	17.85	1682.15
⋮	⋮	⋮

Have the computation stop when the oil reserves are gone.

40. Use your program for finding roots of equations by simple iteration or the midpoint method to solve any of Problems 1 to 20.

15-7. LOGARITHMIC EQUATIONS

A *logarithmic equation* is one that contains a logarithm *of the variable.*

Example: $\log 2x - 5 = 0$ is a logarithmic equation, while $x \log 7 - 3 = 0$ would not usually be called a logarithmic equation.

If only one term in our equation contains a log, we isolate that term on one side of the equation and then go to exponential form.

Example: Solve for x to three significant digits:

$$3 \log (x^2 + 2) - 6 = 0$$

Solution: Transposing and dividing by 3 gives

$$\log (x^2 + 2) = 2$$

Going to exponential form,

$$x^2 + 2 = 10^2 = 100$$

$$x^2 = 98$$

$$x = \pm 9.90$$

If every term contains a log, we use the laws of logarithms to combine them into a single log on each side of the equation, and then take the antilog of both sides.

Example: Solve for x to three significant digits.

$$3 \log x - 2 \log x = 2 \log 5$$

Solution: Using the laws of logarithms gives

$$\log x^3 - \log x^2 = \log 5^2$$

$$\log \frac{x^3}{x^2} = \log x = \log 25$$

Taking the antilog of both sides,

$$x = 25$$

If one or more terms does not contain a log, combine all the terms that do contain a log on one side of the equation and go to exponential form.

Example: Solve for x:

$$\log (5x + 2) - 1 = \log (2x - 1)$$

Solution: Transposing,

$$\log (5x + 2) - \log (2x - 1) = 1$$

By Eq. 180,

$$\log \frac{5x + 2}{2x - 1} = 1$$

Taking the antilog,

$$\frac{5x + 2}{2x - 1} = 10^1 = 10$$

$$5x + 2 = 20x - 10$$

$$12 = 15x$$

$$x = \frac{4}{5}$$

EXERCISE 7

Logarithmic Equations

Solve for x to three significant digits. Check your answers.

1. $\log (2x + 5) = 2$

2. $2 \log (x + 1) = 3$

3. $\log (2x + x^2) = 2$

4. $\log x - 2 \log x = \log 64$

5. $\log 6 + \log (x - 2) = \log (3x - 2)$

6. $\log (x + 2) - \log 36 = \log x$

7. $\log (5x + 2) - \log (x + 6) = \log 4$

8. $\log x + \log 4x = 2$

9. $\ln x + \ln (x + 2) = 1$

10. $\log 8x^2 - \log 4x = 2.54$

11. $2 \log x - \log (1 - x) = 1$

12. $3 \log x - 1 = 3 \log (x - 1)$

13. $\log (x^2 - 4) - 1 = \log (x + 2)$

14. $2 \log x - 1 = \log (20 - 2x)$

15. $\log (x^2 - 1) - 2 = \log (x + 1)$

16. $\ln 2x - \ln 4 + \ln (x - 2) = 1$

17. $\log (4x - 3) + \log 5 = 6$

Applications

This equation is derived from Eq. A9, the compound interest formula. The equations in the next two problems are obtained from the equations for an annuity and for capital recovery from Exercise 1. Can you see how they were derived?

18. An amount of money a invested at a compound interest rate of n per year will take t years to accumulate to an amount y, where t is

$$t = \frac{\log y - \log a}{\log (1 + n)} \quad \text{years}$$

How many years will it take for an investment to triple in value when deposited at 8% per year?

19. If an amount R is deposited once every year at a compound interest rate n, the number of years it will take to accumulate to an amount y is

$$t = \frac{\log\left[\dfrac{ny}{R} + 1\right]}{\log(1 + n)} \quad \text{years}$$

How many years will it take an annual payment of $1500 to accumulate to $13,800 at 9% per year?

20. If an amount a is invested at a compound interest rate n, it will be possible to withdraw a sum R at the end of every year for t years until the deposit is exhausted. The number of years t is given by

$$t = \frac{\log\left[\dfrac{an}{R - an} + 1\right]}{\log(1 + n)}$$

If $200,000 is invested at 12% interest, for how many years can an annual withdrawal of $30,000 be made before the money is used up?

21. The time it takes for an object to cool from a temperature T_0 to temperature T is

$$\text{time} = k \log \frac{T}{T_0} \quad \text{min}$$

A casting initially at 1800°F cools in air to 1000°F and then is immersed in water, where it cools further to temperature T_1. If the constant k for the first stage of cooling is -5.1 and -1.4 for the second stage, the total time will be

$$\text{total time} = -5.1 \log \frac{1000}{1800} - 1.4 \log \frac{T_1}{1000} \quad \text{min}$$

Find T_1 if the total time is 2.00 min.

22. The *magnitude* M of a star of intensity I is

$$M = 2.5 \log \frac{I_1}{I} + 1$$

where I_1 is the intensity of a first-magnitude star. What is the magnitude of a star whose intensity is one-tenth that of a first-magnitude star?

23. The difference in elevation h (ft) between two locations having barometer readings of B_1 and B_2 inches of mercury is given by the logarithmic equation

$$h = 60{,}470 \log \left(\frac{B_2}{B_1}\right)$$

where B_1 is the pressure at the *upper* station. Find the difference in elevation between two stations having barometer readings of 29.14 in. and 26.22 in.

24. What will be the barometer reading 815.0 ft above a station having a reading of 28.22 in.?

25. If the power input to a network or device is P_1 and the power output is

P_2, the amount of decibels gained or lost in the device is given by the logarithmic equation

Decibels Gained or Lost	$G = 10 \log_{10} \dfrac{P_2}{P_1}$ dB	**A63**

A certain amplifier gives a power output of 1000 W for an input of 50 W. Find the dB gain.

26. A transmission line has a loss of 3.25 dB. Find the power transmitted for an input of 2750 kW.

27. What power input is needed to produce a 250-W output with an amplifier having a 50-dB gain?

28. The output of a certain device is half the input. This is a loss of how many dB?

29. The heat loss q per foot of cylindrical pipe insulation (Fig. 15-13) having an inside radius r_1 and outside radius r_2 is given by the logarithmic equation

$$q = \frac{2\pi k(t_1 - t_2)}{\ln (r_2/r_1)} \qquad \text{Btu/h}$$

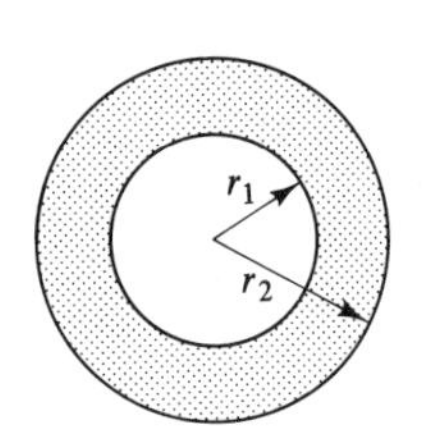

FIGURE 15-13. An insulated pipe.

where t_1 and t_2 are the inside and outside temperatures (°F) and k is the conductivity of the insulation. Find q for a 4-in.-thick insulation having a conductivity of 0.036 wrapped around a 9-in.-diameter pipe at 550°F if the surroundings are at 90°F.

30. If a resource is being used up at a rate that increases exponentially, the time it takes to exhaust the resource (called the exponential expiration time, EET) is

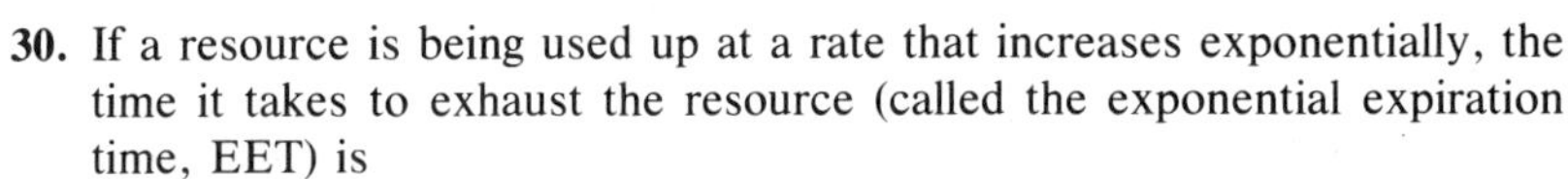

$$\text{EET} = \frac{1}{n} \ln \left(\frac{nR}{r} + 1 \right)$$

where n is the rate of increase in consumption, R the total amount of the resource, and r the initial rate of consumption. If we assume that the United States has oil reserves of 207×10^9 barrels and that our present rate of consumption is 6×10^9 barrels/yr, how long will it take to exhaust these reserves if our consumption increases by 7% per year?

31. pH: The pH value of a solution having a concentration C of hydrogen ions is

$\text{pH} = -\log_{10} C$	**A27**

Find the concentration of hydrogen ions in a solution having a pH of 4.65.

32. A pH of 7 is considered *neutral,* while a lower pH is *acid* and a higher pH is *alkaline*. What is the hydrogen ion concentration at a pH of 7?

33. The acid rain during a certain storm had a pH of 3.5. Find the hydrogen ion concentration. How does it compare with that for a pH of 7?

34. Show that (a) when the pH doubles, the hydrogen ion concentration is *squared,* and (b) when the pH increases by a factor of n, the hydrogen ion concentration increases *to the* nth *power.*

Computer

35. Use your program for solving equations by simple iteration or by the midpoint method to solve any one of Problems 1 to 17.

15-8. GRAPHS ON LOGARITHMIC AND SEMILOGARITHMIC PAPER

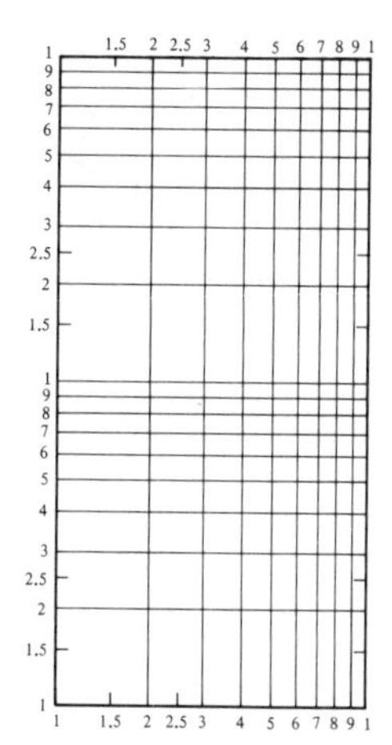

FIGURE 15-14. Logarithmic graph paper.

Logarithmic and Semilogarithmic Paper: Our graphing so far has all been done on ordinary graph paper, on which the lines are equally spaced. For some purposes, though, it is better to use *logarithmic* paper (Fig. 15-14), also called *log-log* paper, or *semilogarithmetic* paper (Fig. 15-15), also called *semilog* paper. Looking at the logarithmic scales of these graphs, we note the following:

1. The lines are not equally spaced. The distance in inches from, say, 1 to 2, is equal to the distance from 2 to 4, which, in turn, is equal to the distance from 4 to 8.
2. Each tenfold increase in the scale, say from 1 to 10 or from 10 to 100, is called a *cycle.* Each cycle requires the same distance in inches along the scale.
3. The log scales do not include zero.

When to Use Logarithmic or Semilog Paper: We use these special papers for the following reasons:

1. When the range of the variables is too large for ordinary paper.
2. When we want to graph a power function or an exponential function, because each of these will plot as a straight line on the appropriate paper, as shown in Fig. 15-16.
3. When we want to find an equation that will approximately represent a set of empirical data.

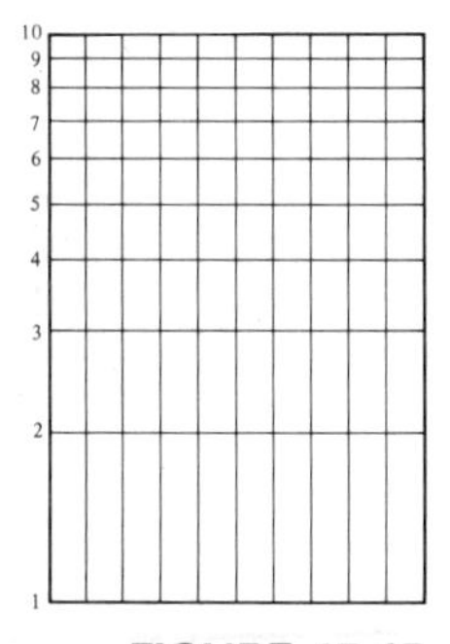

FIGURE 15-15. Semilogarithmic graph paper.

Graphing the Power Function: A *power function* is one whose equation is of the form

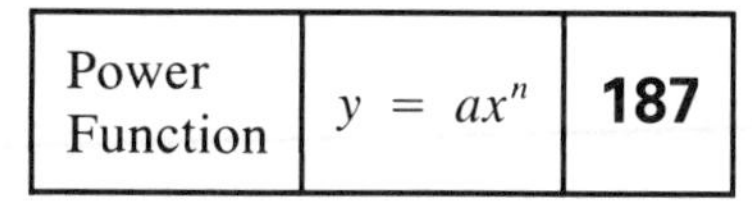

where a and n are nonzero constants. This equation is nonlinear (except when $n = 1$), and the shape of its graph depends upon whether n is positive or negative, and whether n is greater than or less than 1. Figure 15-16 shows the shapes that this curve can have for various ranges of n. If we take the logarithm of both sides of Eq. 187, we get

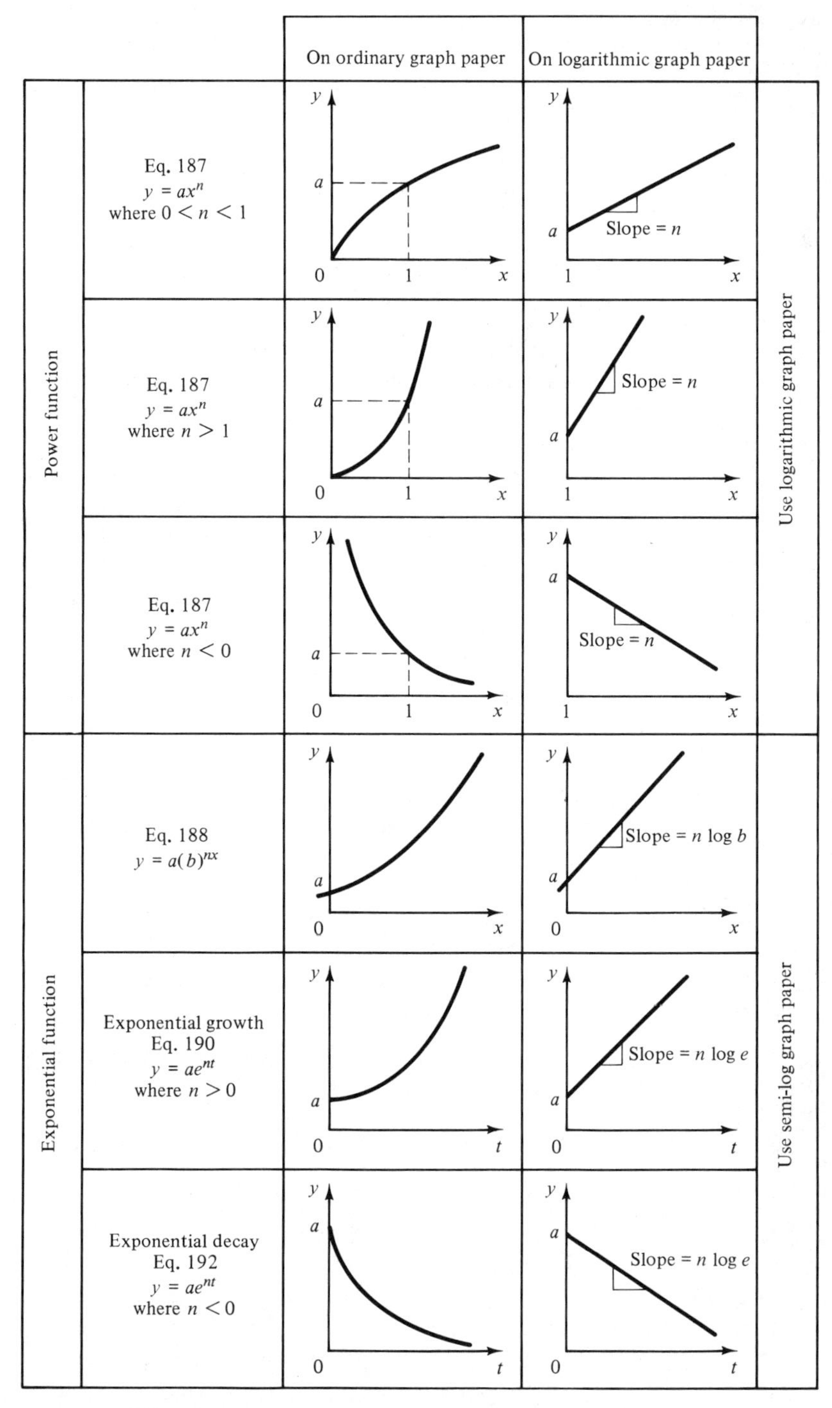

FIGURE 15-16. The power function and the exponential function, graphed on ordinary paper and on logarithmic paper.

$$\log y = \log (ax^n)$$
$$= \log a + n \log x$$

If we now make the substitution

$$Y = \log y$$

and

$$X = \log x$$

our equation becomes

$$Y = nX + \log a$$

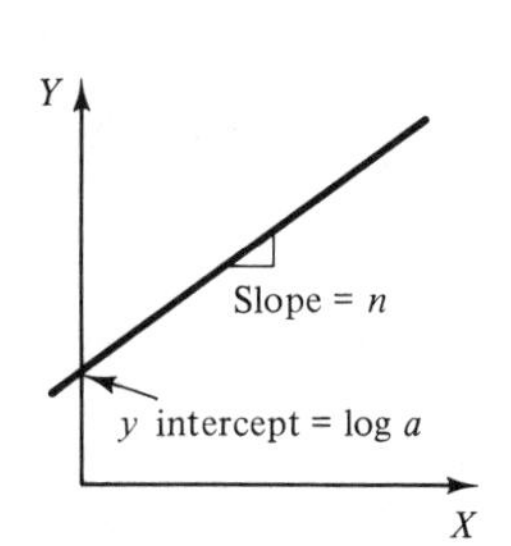

FIGURE 15-17. Graph of $y = ax^n$, where $X = \log x$ and $Y = \log y$.

This equation *is* linear, and graphs as a straight line with a slope of n and a y intercept of $\log a$ (Fig. 15-17). However, we do not have to make the substitutions shown above *if we use logarithmic paper,* where the scales are proportional to the logarithms of the variables x and y. We simply have to plot the original equation on log-log paper, and it will be a straight line with a slope n which cuts the y axis at a.

Example: Plot the equation $y = 2.5x^{1.4}$, for values of x from 1 to 10. Choose graph paper so that the equation plots as a straight line.

Solution: We make a table of point pairs. Since the graph will be a straight line we need only two points, with a third as a check. Here, we will plot 4 points to show that they do lie on a straight line. We choose values of x and for each compute the value of y.

x	1	4	7	10
y	2.5	17.4	38.1	62.8

How to select paper: We choose log-log rather than semilog paper, because we know that the power function plots as a straight line on this paper (see Fig. 15-17). We choose the number of cycles for each scale by looking at the range of values for x and y. Thus, on the x axis we need one cycle. On the y axis, we must go from 2.5 to 62.8. With two cycles we can span a range of 1 to 100. Thus we need log-log paper, one cycle by two cycles.

We mark the scales and plot the points as shown in Fig. 15-18, and get a straight line as expected. We note that the value of y at $x = 0$ is equal to 2.5, and is the same as the coefficient of $x^{1.5}$ in the given equation.

We can get the slope of the straight line by measuring the rise and run with a scale, and dividing rise by run. Or we can use the values from the graph, remembering to take the logarithm (either common or natural will give the same result) of those values. Thus,

$$\text{slope} = \frac{\text{rise}}{\text{run}} = \frac{\ln 62.8 - \ln 2.5}{\ln 10 - \ln 1} = \frac{3.22}{2.30} = 1.40$$

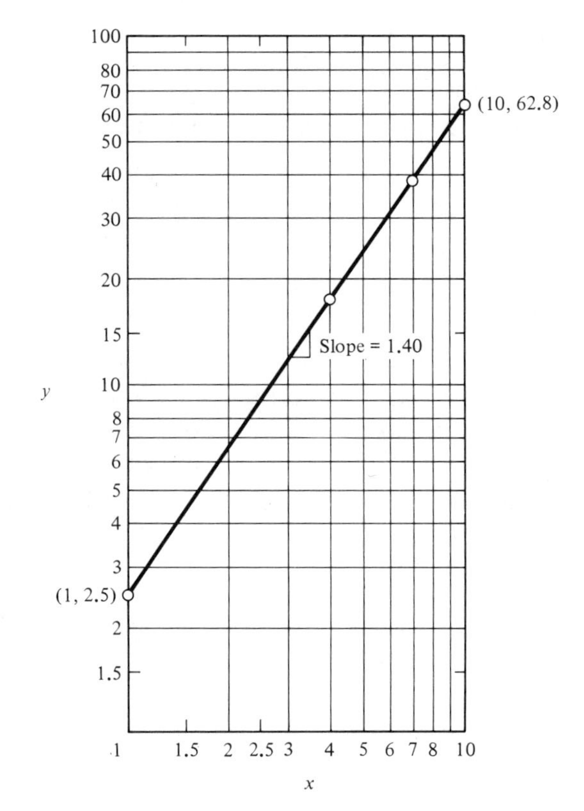

FIGURE 15-18. Graph of $y = 2.5\,x^{1.4}$.

The slope of the line is thus equal to the power of x, as expected from Fig. 15-17. We will use these ideas later when we try to write an equation to fit a set of data.

Common Error	Be sure to take the logs of the values on the x and y axes, when computing the slope.

Graphing The Exponential Function: Consider the exponential function given by Eq. 188

$$y = a(b)^{nx}$$

If we take the logarithm of both sides, we get

$$\begin{aligned} \log y &= \log a + \log b^{nx} \\ &= \log a + nx \log b \end{aligned}$$

If we replace $\log y$ with Y, we get the linear equation

$$Y = (n \log b)x + \log a$$

Thus if we graph the given equation on semilog paper with the logarithmic scale along the y axis, we get a straight line with a slope of $n \log b$ which cuts the y axis at a, Fig. 15-16. Also shown are the special cases where the base b is equal to e (2.718 . . .). Here, the independent variable is shown as t, for exponential growth and decay are usually functions of time.

Example: Plot the exponential function

$$y = 100\, e^{-0.2x}$$

for values of x from 0 to 10.

Solution: We make a table of point pairs

x	0	2	4	6	8	10
y	100	67.0	44.9	30.1	20.2	13.5

We choose semilog paper for graphing the exponential function, and use the linear scale for x. The range of y is from 13.5 to 100, thus we need *one cycle* of the logarithmic scale. The graph is shown in Fig. 15-19.

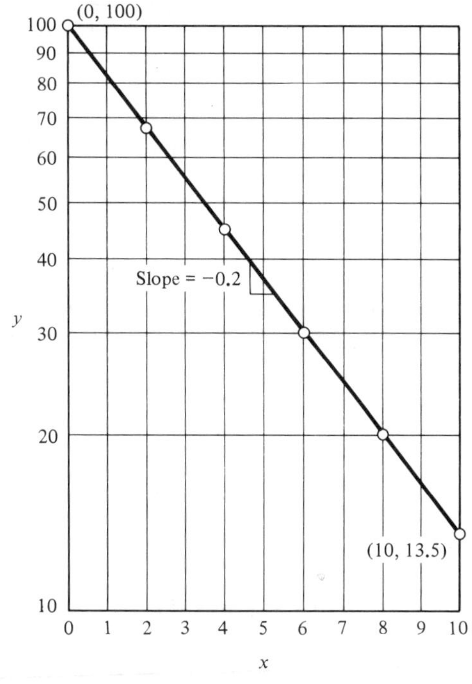

FIGURE 15-19. Graph of $y = 100\, e^{-0.2x}$.

Note that the line obtained has a y intercept of 100. Also, the slope is equal to $n \log e$, or, if we use natural logs, is equal to n.

When computing the slope on semilog paper, we take the logarithms of the values on the log scale, but not on the linear scale.

$$\text{slope} = n = \frac{\ln 13.5 - \ln 100}{10} = -0.2$$

This is the coefficient of x in our given equation.

Empirical Functions: We would plot a set of empirical data on logarithmic or semilog paper when:

1. The range of values is too large for ordinary paper.
2. We suspect that the relation between our variables may be a power function or an exponential function, and we want to find that function.

The process of finding an approximate equation to fit a set of data points is called curve fitting. *Here we are able to do only some very simple cases.*

We show the second case by means of an example.

Example: A test of a certain electronic device shows it to have an output current i versus input voltage v as in Table 15-1. Plot the given empirical data and try to find an approximate formula for y in terms of x.

Table 15-1

v (V)	1	2	3	4	5
i (A)	5.61	15.4	27.7	42.1	58.1

Solution: We first make a graph on linear graph paper (Fig. 15-20) and get a curve that is concave upward. Comparing its shape with the curves in Fig. 15-16, we suspect that the equation of the curve (if we can find one at all) may be either a power function or an exponential function.

Next, we make a plot on semilog paper (Fig. 15-21) and do *not* get a straight line, but a plot on log-log paper (Fig. 15-22) *is* linear. We thus assume that our equation has the form

$$i = av^n$$

In our earlier plot of the power function, we saw that the coefficient a was equal to the value of the function when x was 1, and that the exponent n was equal to the slope of the straight line. In our present example,

$$a = 5.61$$

and the slope of the line is

$$n = \frac{\ln 58.1 - \ln 5.61}{\ln 5 - \ln 1} = 1.45$$

Our equation is then

$$i = 5.61v^{1.45}$$

Finally, we test this formula by computing values of i and comparing them with the original data.

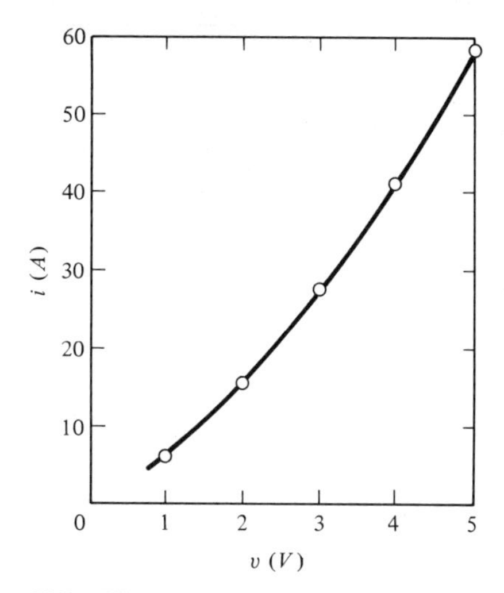

FIGURE 15-20. Plot of Table 15-1 on linear graph paper.

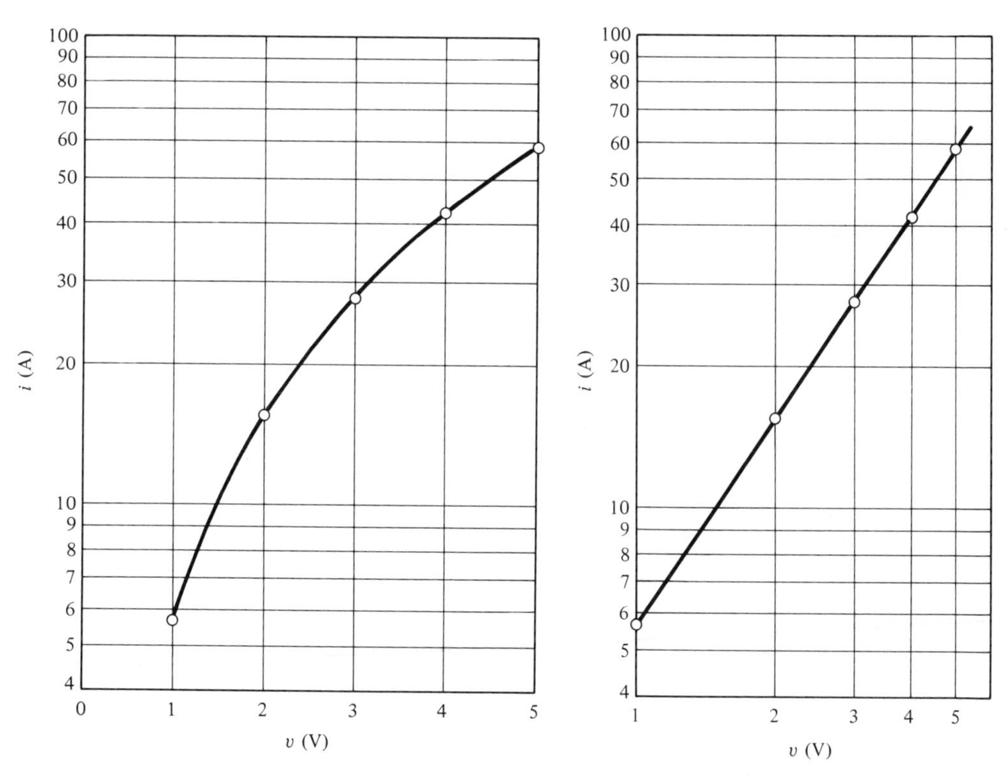

FIGURE 15-21. Plot of Table 15-1 on semilog paper.

FIGURE 15-22. Plot of Table 15-1 on log-log paper.

v	1	2	3	4	5
Original i	5.61	15.4	27.7	42.1	58.1
Calculated i	5.61	15.3	27.6	41.9	57.9

We get values very close to the original.

EXERCISE 8

Graphing the Power Function

Graph each power function on log-log paper, for $x = 1$ to 10.

1. $y = 2x^3$ **2.** $y = 3x^4$ **3.** $y = x^5$

4. $y = x^{3/2}$ **5.** $y = 5\sqrt{x}$ **6.** $y = 2\sqrt[3]{x}$

7. $y = 1/x$ **8.** $y = 3/x^2$ **9.** $y = 2x^{2/3}$

Graph each set of data on log-log paper, determine the coefficients graphically, and write an approximate equation to fit the given data.

10.

x	2	4	6	8	10
y	3.48	12.1	25.2	42.2	63.1

11.

x	2	4	6	8	10
y	240	31.3	9.26	3.91	2.00

Graphing the Exponential Function

Graph each exponential function on semilog paper.

12. $y = 3^x$ **13.** $y = 5^x$ **14.** $y = e^x$

15. $y = e^{-x}$ **16.** $y = 4^{x/2}$ **17.** $y = 3^{x/4}$

18. $y = 3e^{2x/3}$ **19.** $y = 5e^{-x}$ **20.** $y = e^{x/2}$

Graph each set of data on semilog paper, determine the coefficients graphically, and write an approximate equation to fit the given data.

21.

x	0	1	2	3	4	5
y	1.00	2.50	6.25	15.6	39.0	97.7

22.

x	1	2	3	4
y	48.5	29.4	17.6	10.8

Graphing Empirical Functions

Graph each set of data on log-log or semilog paper, determine the coefficients graphically, and write an approximate equation to fit the given data.

23. Current in a tungsten lamp for various voltages.

v (V)	2	8	25	50	100	150	200
i (mA)	24.6	56.9	113	172	261	330	387

24. Difference in temperature T between a cooling body and its surroundings at various times t.

t (s)	0	3.51	10.9	19.0	29.0	39.0	54.0	71.0
T (°F)	20.0	19.0	17.0	15.0	13.0	11.0	8.50	6.90

25. Pressure of 1 lb of saturated steam at various volumes.

v (ft^3)	26.4	19.1	14.0	10.5	8.00
p (lb/in^2)	14.7	20.8	28.8	39.3	52.5

26. Maximum height y reached by a long pendulum t seconds after being set in motion.

t (s)	0	1	2	3	4	5	6
y (in.)	10	4.97	2.47	1.22	0.61	0.30	0.14

CHAPTER TEST

Convert to logarithmic form.

1. $x^{5.2} = 352$　　**2.** $4.8^x = 58$　　**3.** $24^{1.4} = x$

Convert to exponential form.

4. $\log_3 56 = x$　　**5.** $\log_x 5.2 = 124$　　**6.** $\log_2 x = 48.2$

Solve for x.

7. $\log_{81} 27 = x$　　**8.** $\log_3 x = 4$　　**9.** $\log_x 32 = -\frac{5}{7}$

Write as the sum or difference of two or more logarithms.

10. $\log xy$　　**11.** $\log \dfrac{3x}{z}$　　**12.** $\log \dfrac{ab}{cd}$

Express as a single logarithm.

13. $\log 5 + \log 2$　　**14.** $2 \log x - 3 \log y$　　**15.** $\frac{1}{2} \log p - \frac{1}{4} \log q$

Find the common logarithm of each number to four decimal places.

16. 6.83　　**17.** 364　　**18.** 0.00638

Find x if $\log x$ *is equal to:*

19. 2.846　　**20.** 1.884　　**21.** −0.473

Evaluate using logarithms.

22. 5.836×88.24 **23.** $\dfrac{8375}{2846}$ **24.** $(4.823)^{1.5}$

Find the natural logarithm of each number to four decimal places.

25. 84.72 **26.** 2846 **27.** 0.00873

Find the number whose natural logarithm is:

28. 5.273 **29.** 1.473 **30.** -4.837

Find $\log x$ *if:*

31. $\ln x = 4.837$ **32.** $\ln x = 8.352$

Find $\ln x$ *if:*

33. $\log x = 5.837$ **34.** $\log x = 7.264$

Solve for x.

35. $(4.88)^x = 152$ **36.** $e^{2x+1} = 72.4$

37. $3 \log x - 3 = 2 \log x^2$ **38.** $\log (3x - 6) = 3$

39. A sum of $1500 is deposited at a compound interest rate of $6\frac{1}{2}\%$ compounded quarterly. How much will it accumulate to in 5 years?

40. A quantity grows exponentially at a rate of 2% per day for 10 weeks. Find the final amount if the initial amount is 500 units.

41. A flywheel decreases in speed exponentially at the rate of 5% per minute. Find the speed after 20 min if the initial speed is 2250 rev/min.

42. Find the half-life of a radioactive substance that decays exponentially at the rate of $3\frac{1}{2}\%$ per year.

43. Find the doubling time for a population growing at the rate of 3% per year.

Complex Numbers

So far we have dealt entirely with real numbers; in this chapter we consider imaginary and complex numbers. We learn what they are, how to express them in several different forms, and how to perform the basic operations with them.

Why bother with a new type of number when, as we have seen, the real numbers can do so much? We learn complex numbers mainly for the purpose of manipulating *vectors,* which are so important in many branches of technology. Complex numbers (despite their unfortunate name) really simplify computations with vectors, as we see especially with the alternating current calculations at the end of this chapter.

16-1. COMPLEX NUMBERS IN RECTANGULAR FORM

Complex Numbers: A *complex number* is any number, real or imaginary, written in the form

$$a + jb$$

where a and b are real numbers, and $j = \sqrt{-1}$, the *imaginary unit.*

The letter i is often used for the imaginary unit. In technical work, however, we save i for electric current.

Example: The numbers

$$4 + j2 \qquad -7 + j8 \qquad 5.92 - j2.93 \qquad 83 \qquad \text{and} \qquad j27$$

are complex numbers.

Real and Imaginary Numbers: When $b = 0$ in a complex number $a + jb$, we have a *real number*. When $a = 0$, the number is called a *pure imaginary number*.

Examples:

(a) The complex number 48 is also a real number.

(b) The complex number $j9$ is a pure imaginary number.

The two parts of a complex number are called the *real part* and the *imaginary part*.

This is called the rectangular form *of a complex number.*

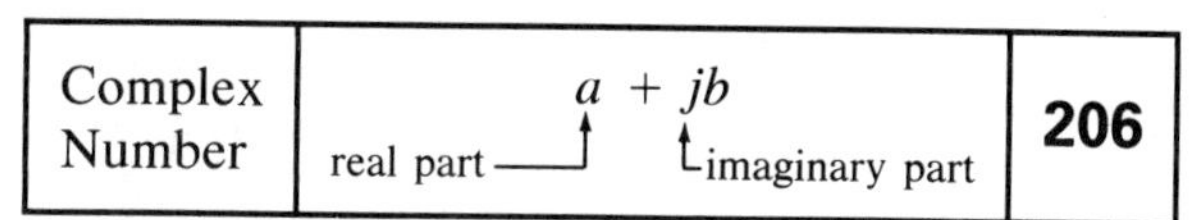

Complex Number	$a + jb$ (real part: a; imaginary part: b)	**206**

Addition and Subtraction of Complex Numbers: To combine complex numbers, separately combine the real parts, and then the imaginary parts, and express the result in the form $a + jb$.

Examples:

(a) $j3 + j5 = j8$

(b) $j2 + (6 - j5) = 6 - j3$

(c) $(2 - j5) + (-4 + j3) = (2 - 4) + j(-5 + 3) = -2 - j2$

(d) $(-6 + j2) - (4 - j) = (-6 - 4) + j[2 - (-1)] = -10 + j3$

Powers of *j*: We often have to evaluate powers of the imaginary unit, especially the square of j. Since

$$j = \sqrt{-1}$$
$$j^2 = \sqrt{-1}\sqrt{-1} = -1$$

Higher powers are easily found.

$$j^3 = j^2 j = (-1)j = -j$$
$$j^4 = (j^2)^2 = (-1)^2 = 1$$
$$j^5 = j^4 j = (1)j = j$$
$$j^6 = j^4 j^2 = (1)j^2 = -1$$

We see that the values are starting to repeat. $j^5 = j$, $j^6 = j^2$, and so on. The first four values keep repeating.

Powers of j	$j = \sqrt{-1}$, $j^2 = -1$, $j^3 = -j$, $j^4 = 1$, $j^5 = j$, etc.	**205**

Example: Evaluate j^{17}.

Solution: Using the laws of exponents, we express j^{17} in terms of one of the first four powers of j.

$$j^{17} = j^{16}j = (j^4)^4 j = (1)^4 j = j$$

Multiplication of Imaginary Numbers: Multiply as with ordinary numbers, but use Eq. 205 to simplify any powers of j.

Examples:

(a) $5 \times j3 = j15$

(b) $j2 \times j4 = j^2 8 = (-1)8 = -8$

(c) $3 \times j4 \times j5 \times j = j^3 60 = (-j)60 = -j60$

(d) $(j3)^2 = j^2 3^2 = (-1)9 = -9$

Multiplication of Complex Numbers: Multiply complex numbers as you would any algebraic expressions, replace j^2 by -1, and put the expression into the form $a + jb$.

Examples:

(a) $3(5 + j2) = 15 + j6$

(b) $(j3)(2 - j4) = j6 - j^2 12 = j6 - (-1)12 = 12 + j6$

(c)
$$\begin{aligned}(3 - j2)(-4 + j5) &= 3(-4) + 3(j5) + (-j2)(-4) + (-j2)(j5)\\ &= -12 + j15 + j8 - j^2 10\\ &= -12 + j15 + j8 - (-1)10\\ &= -2 + j23\end{aligned}$$

(d)
$$\begin{aligned}(3 - j5)^2 &= (3 - j5)(3 - j5)\\ &= 9 - j15 - j15 + j^2 25 = 9 - j30 - 25\\ &= -16 - j30\end{aligned}$$

Tip	Always convert radicals to imaginary numbers *before* performing other operations, or contradictions may result.

Example: Multiply $\sqrt{-4}$ by $\sqrt{-4}$.

Solution: Converting to imaginary numbers, we obtain

$$\sqrt{-4}\sqrt{-4} = (j2)(j2) = j^2 4$$

We will see in the next section that multiplication and division are easier in polar *form.*

Since $j^2 = -1$,

$$j^2 4 = -4$$

It is *incorrect* to write

$$\sqrt{-4}\sqrt{-4} = \sqrt{(-4)(-4)} = \sqrt{16} = +4$$

The Conjugate of a Complex Number: The *conjugate* of a complex number is obtained by changing the sign of its imaginary part.

Examples:

(a) The conjugate of $2 + j3$ is $2 - j3$.

(b) The conjugate of $-5 - j4$ is $-5 + j4$.

(c) The conjugate of $a + jb$ is $a - jb$.

Multiplying any complex number by its conjugate will eliminate the j term.

This has the same form as the difference of two squares (Eq. 41).

Example:

$$(2 + j3)(2 - j3) = 4 - j6 + j6 - j^2 9 = 4 - (-1)(9) = 13$$

Division of Complex Numbers: Division involving single terms, real or imaginary, is shown by example.

Examples:

(a) $$j8 \div 2 = j4$$

(b) $$j6 \div j3 = 2$$

(c) $$(4 - j6) \div 2 = 2 - j3$$

Example: Divide 6 by $j3$.

Solution:

$$\frac{6}{j3} = \frac{6}{j3} \times \frac{j3}{j3} = \frac{j18}{j^2 9} = \frac{j18}{-9} = -j2$$

To divide by a complex number, multiply dividend and divisor by the conjugate of the divisor. This will make the divisor a real number, as in the following example.

This is very similar to rationalizing the denominator *of a fraction containing radicals (Sec. 10-1).*

Example: Divide $3 - j4$ by $2 + j$.

Solution: Writing the division as a fraction,

$$\frac{3 - j4}{2 + j}$$

Multiplying by the conjugate, $2 - j$,

$$\frac{3 - j4}{2 + j} = \frac{3 - j4}{2 + j} \cdot \frac{2 - j}{2 - j}$$

$$= \frac{6 - j3 - j8 + j^2 4}{4 + j2 - j2 - j^2}$$

$$= \frac{2 - j11}{5}$$

$$= \frac{2}{5} - j\frac{11}{5}$$

Quadratics with Complex Roots: When we solved quadratic equations in Chapter 11, the roots were real numbers. Some quadratics, however, will yield complex roots, as in the following example.

Example: Solve for x to three significant digits,

$$2x^2 - 5x + 9 = 0$$

Solution: By the quadratic formula (Eq. 100),

$$x = \frac{5 \pm \sqrt{25 - 4(2)(9)}}{4} = \frac{5 \pm \sqrt{-47}}{4}$$

$$= \frac{5 \pm j6.86}{4} = 1.25 \pm j1.72$$

EXERCISE 1

Complex Numbers

Write as imaginary numbers.

1. $\sqrt{-9}$ **2.** $\sqrt{-81}$ **3.** $\sqrt{-\frac{4}{16}}$ **4.** $\sqrt{-\frac{1}{5}}$

Write as a complex number in rectangular form.

5. $4 + \sqrt{-4}$ **6.** $\sqrt{-25} + 3$ **7.** $\sqrt{-\frac{9}{4}} - 5$ **8.** $\sqrt{-\frac{1}{3}} + 2$

Evaluate the powers of j.

9. j^{11} **10.** j^5 **11.** j^{10} **12.** j^{21} **13.** j^{14}

Combine and simplify.

14. $\sqrt{-9} + \sqrt{-4}$

15. $\sqrt{-4a^2} - a\sqrt{-25}$

16. $(3 - j2) + (-4 + j3)$

17. $(-1 - j2) - (j + 6)$

18. $(a - j3) + (a + j5)$

19. $(p + jq) + (q + jp)$

20. $\left(\frac{1}{2} + \frac{j}{3}\right) + \left(\frac{1}{4} - \frac{j}{6}\right)$

21. $(-84 + j91) - (28 + j72)$

22. $(2.28 - j1.46) + (1.75 + j2.66)$

Multiply and simplify.

23. $\sqrt{-4}\sqrt{-3}$

24. $(3 + j2)(5 - j3)$

25. $(1 + \sqrt{-1})(2 - \sqrt{-1})$

26. $j7 \cdot j9$

27. $(2 - j3)(5 + j)$

28. $(4 + j2)(4 - j2)$

29. $(j5)^2$

30. $(3 - j2)^2$

31. $j6(5 - j2)$

Write the conjugates.

32. $2 - j3$

33. $-5 - j7$

34. $p + jq$

35. $-j5 + 6$

36. $-jm + n$

37. $5 - j8$

Divide and simplify.

38. $j8 \div 4$

39. $9 \div j3$

40. $(4 + j2) \div 2$

41. $8 \div (4 - j)$

42. $(-2 + j3) \div (1 - j)$

43. $(5 - j6) \div (-3 + j2)$

44. $(j7 + 2) \div (j3 - 5)$

45. $(-9 + j3) \div (2 - j4)$

Quadratics with Complex Roots

Solve for x to three significant digits.

46. $3x^2 - 5x + 7 = 0$

47. $2x^2 + 3x + 5 = 0$

48. $x^2 - 2x + 6 = 0$

49. $4x^2 + x + 8 = 0$

Computer

50. Modify your program for solving quadratic equations (Chapter 11, Exercise 4) so that the run will not end when the discriminant is negative, but will instead compute and print complex roots in the form $a + jb$.

16-2. GRAPHING COMPLEX NUMBERS

The Complex Plane: A complex number in rectangular form is made up of two parts: a real part a and an imaginary part jb. To graph a complex number we can use a rectangular coordinate system (Fig. 16-1) in which the horizontal axis is the *real* axis and the vertical axis

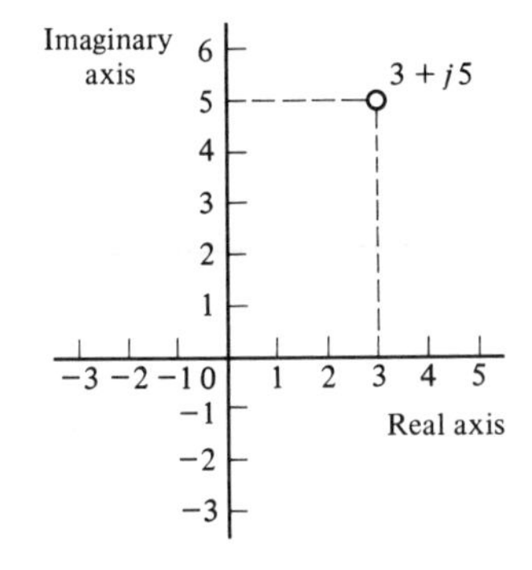

FIGURE 16-1. The complex plane.

is the *imaginary* axis. Such a coordinate system is called the *complex plane*.

Graphing a Complex Number: To plot a complex number $a + jb$ in the complex plane, simply locate a point with a horizontal coordinate of a and a vertical coordinate of b.

Such a plot is called an Argand diagram, *named for Jean Robert Argand (1768–1822).*

Example: The complex numbers $2 + j3$, $-1 + j2$, $-3 - j2$, and $1 - j3$, are plotted in Fig. 16-2.

Imaginary
3 2 + j3
−1 + j2 2
1
−3 −2 −1 0 1 2 3 Real
−1
−2
−3 −j2
−3 1 − j3

FIGURE 16-2

EXERCISE 2

Graph each complex number.

1. $2 + j5$	**2.** $-1 - j3$	**3.** $3 - j2$
4. $2 - j$	**5.** $j5$	**6.** $2.25 - j3.62$
7. $-2.7 - j3.4$	**8.** $5.02 + j$	**9.** $1.46 - j2.45$

16-3. TRIGONOMETRIC AND POLAR FORMS OF A COMPLEX NUMBER

In some books, the terms "polar form" and "trigonometric form" are used interchangeably. Here, we distinguish between them.

Polar Form: In Sec. 14-6 we saw that a point in a plane could be located by *polar coordinates,* as well as by rectangular coordinates, and we learned how to convert from one set of coordinates to the other. We will do the same thing now with complex numbers.

In Fig. 16-3, we plot a complex number $a + jb$. We then let the length of the vector be r, at an angle of θ with the horizontal axis. Now in addition to expressing the complex number in terms of a and b, we can express it in terms of r and θ:

Imaginary
a + jb = r∠θ
b
r
θ
0 a Real

FIGURE 16-3. Polar form of a complex number shown on a complex plane.

Polar Form	$a + jb = r\angle\theta$	**216**

By Eqs. 146 and 147,

$a = r \cos \theta$	**212**

and

$b = r \sin \theta$	**213**

The radius r is called the *absolute value*. It can be found from the Pythagorean theorem:

$r = \sqrt{a^2 + b^2}$	**214**

The angle θ is called the *argument* of the complex number and is found by Eq. 148,

$\theta = \arctan \dfrac{b}{a}$	**215**

Example: Write the complex number $2 + j3$ in polar form.

Solution: The absolute value is

$$r = \sqrt{2^2 + 3^2} \cong 3.61$$

And the argument is found from

If your calculator has keys for converting between rectangular and polar coordinates, you can use these keys to convert complex numbers between rectangular and polar forms.

$$\tan \theta = \frac{3}{2}$$

$$\theta \cong 56.3°$$

So

$$2 + j3 = 3.61\underline{/56.3°}$$

Trigonometric Form: If we substitute the values of a and b from Eqs. 212 and 213 into $a + jb$, we get

$$a + jb = (r \cos \theta) + j(r \sin \theta)$$

Factoring,

The angle θ is sometimes written $\theta = \arg z$, which means "θ is the argument of the complex number z." Also, the expression in parentheses in Eq. 211 is sometimes abbreviated as cis θ: so cis $\theta = \cos \theta + j \sin \theta$.

$a + jb = r(\cos \theta + j \sin \theta)$ rectangular form — trigonometric form	**211**

Example: Write the complex number $2 + j3$ in trigonometric form.

Solution: We already have r and θ from the preceding example,

$$r = 3.61 \qquad \text{and} \qquad \theta = 56.3°$$

So by Eq. 211,

$$2 + j3 = 3.61(\cos 56.3° + j \sin 56.3°)$$

Example: Write the complex number $6(\cos 30° + j \sin 30°)$ in polar and rectangular form.

Solution: By inspection,

$$r = 6 \qquad \text{and} \qquad \theta = 30°$$

So

$$a = r \cos \theta = 6 \cos 30° = 5.20$$

and

$$b = r \sin \theta = 6 \sin 30° = 3.00$$

So our complex number, in rectangular form, is $5.20 + j3.00$, and in polar form, $6\underline{/30°}$.

Arithmetic Operations in Trigonometric and Polar Form: It is common practice to switch back and forth between rectangular and polar forms during a computation, using that form in which a particular operation is easiest. We saw that addition and subtraction are fast and easy in rectangular form, and we now show that multiplication, division, and powers are best done in polar form.

Products: We will use trigonometric form to work out the formula for multiplication and then express the result in the simpler polar form.

Let us multiply $r(\cos\theta + j\sin\theta)$ by $r'(\cos\theta' + j\sin\theta')$.

$$[r(\cos\theta + j\sin\theta)][r'(\cos\theta' + j\sin\theta')]$$
$$= rr'(\cos\theta\cos\theta' + j\cos\theta\sin\theta' + j\sin\theta\cos\theta' + j^2\sin\theta\sin\theta')$$
$$= rr'[(\cos\theta\cos\theta' - \sin\theta\sin\theta') + j(\sin\theta\cos\theta' + \cos\theta\sin\theta')]$$

But, by Eq. 168,

$$\cos\theta\cos\theta' - \sin\theta\sin\theta' = \cos(\theta + \theta')$$

and by Eq. 167,

$$\sin\theta\cos\theta' + \cos\theta\sin\theta' = \sin(\theta + \theta')$$

So

$$r(\cos\theta + j\sin\theta)\cdot r'(\cos\theta' + j\sin\theta')$$
$$= rr'[\cos(\theta + \theta') + j\sin(\theta + \theta')]$$

Switching now to polar form, we get

Products	$r\angle\theta \cdot r'\angle\theta' = rr'\angle\theta + \theta'$	**217**

The absolute value of the product of two complex numbers is the product of their absolute values; and the argument is the sum of the individual arguments.

Example: Multiply $5\angle 30°$ by $3\angle 20°$.

Solution: The absolute value of the product will be $5(3) = 15$ and the argument of the product will be $30° + 20° = 50°$, so

$$5\angle 30° \cdot 3\angle 20° = 15\angle 50°$$

You might try deriving this yourself.

Quotients: The rule for division of complex numbers in trigonometric form is similar to that for multiplication:

Quotients	$\dfrac{r\angle\theta}{r'\angle\theta'} = \dfrac{r}{r'}\angle\theta - \theta'$	**218**

The absolute value of the quotient of two complex numbers is the quotient of their absolute values; and the argument is the difference (numerator minus denominator) of their arguments.

Example:

$$\frac{6\angle 70^\circ}{2\angle 50^\circ} = 3\angle 20^\circ$$

Powers: To raise a complex number to a power, we merely have to multiply it by itself the proper number of times, using Eq. 217.

$$(r\angle\theta)^2 = r\angle\theta \cdot r\angle\theta = r \cdot r\angle\theta + \theta = r^2\angle 2\theta$$

$$(r\angle\theta)^3 = r\angle\theta \cdot r^2\angle 2\theta = r \cdot r^2\angle\theta + 2\theta = r^3\angle 3\theta$$

Do you see a pattern developing? In general, we get

After Abraham DeMoivre (1667–1754).

DeMoivre's Theorem	$(r\angle\theta)^n = r^n\angle n\theta$	**219**

When a complex number is raised to the n*th power, the new absolute value is equal to the original absolute value raised to the* n*th power, and the new argument is n times the original argument.*

Example:

$$(2\angle 10^\circ)^5 = 2^5\angle 5(10^\circ) = 32\angle 50^\circ$$

EXERCISE 3

Trigonometric and Polar Forms

Write each complex number in polar and trigonometric form.

1. $5 + j4$ **2.** $-3 - j7$ **3.** $4 - j3$

4. $8 + j4$ **5.** $-5 - j2$ **6.** $-4 + j7$

7. $-9 - j5$ **8.** $7 - j3$ **9.** $-4 - j7$

Write in rectangular and in polar form.

10. $4(\cos 25^\circ + j \sin 25^\circ)$ **11.** $3(\cos 57^\circ + j \sin 57^\circ)$

12. $2(\cos 110^\circ + j \sin 110^\circ)$ **13.** $9(\cos 150^\circ + j \sin 150^\circ)$

14. $7(\cos 12^\circ + j \sin 12^\circ)$ **15.** $5.46(\cos 47.3^\circ + j \sin 47.3^\circ)$

Write in rectangular and trigonometric form.

16. $5\angle 28^\circ$ **17.** $9\angle 59^\circ$ **18.** $-4\angle 63^\circ$

19. $7\angle -53^\circ$ **20.** $-6\angle -125^\circ$

Multiplication and Division

Multiply.

21. $3(\cos 12^\circ + j \sin 12^\circ)$ by $5(\cos 28^\circ + j \sin 28^\circ)$

22. $7(\cos 48^\circ + j \sin 48^\circ)$ by $4(\cos 72^\circ + j \sin 72^\circ)$

23. $5.82(\cos 44.8^\circ + j \sin 44.8^\circ)$ by $2.77(\cos 10.1^\circ + j \sin 10.1^\circ)$

24. $5\angle 30^\circ$ by $2\angle 10^\circ$

25. $8\angle 45^\circ$ by $7\angle 15^\circ$

26. $2.86\angle 38.2^\circ$ by $1.55\angle 21.1^\circ$

Divide.

27. $8(\cos 46^\circ + j \sin 46^\circ)$ by $4(\cos 21^\circ + j \sin 21^\circ)$

28. $49(\cos 27^\circ + j \sin 27^\circ)$ by $7(\cos 15^\circ + j \sin 15^\circ)$

29. $58.3(\cos 77.4^\circ + j \sin 77.4^\circ)$ by $12.4(\cos 27.2^\circ + j \sin 27.2^\circ)$

30. $24\angle 50^\circ$ by $12\angle 30^\circ$

31. $50\angle 72^\circ$ by $5\angle 12^\circ$

32. $71.4\angle 56.4^\circ$ by $27.7\angle 15.2^\circ$

Evaluate.

33. $[2(\cos 15^\circ + j \sin 15^\circ)]^3$

34. $[9(\cos 10^\circ + j \sin 10^\circ)]^2$

35. $(7\angle 20^\circ)^2$

36. $(1.55\angle 15^\circ)^3$

16-4. VECTOR OPERATIONS USING COMPLEX NUMBERS

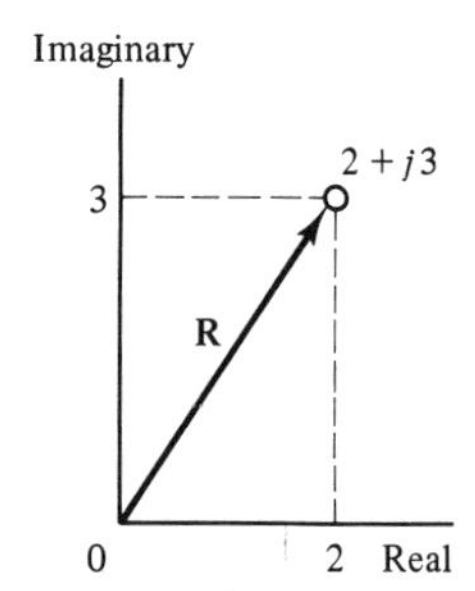

FIGURE 16-4. A vector represented by a complex number.

Vectors Represented by Complex Numbers: One of the major uses of complex numbers is that they can represent vectors, and, as we will soon see, enable us to manipulate them in ways that are easier than we learned in Sec. 13-5 (oblique triangles).

Take the complex number $2 + j3$, for example, which is plotted in Fig. 16-4. If we connect that point with a line to the origin, we can think of the complex number $2 + j3$ as *representing a vector* **R** *having a horizontal component of 2 units and a vertical component of 3 units.* The complex number used to represent a vector can, of course, be expressed in any of the forms of a complex number.

Vector Addition and Subtraction: Let us place a second vector on our diagram, represented by the complex number $3 - j$ (Fig. 16-5). We can add them graphically by the parallelogram method (Sec. 13-5) and we get a resultant $5 + j2$. But let us add the two original complex numbers, $2 + j3$ and $3 - j$.

$$(2 + j3) + (3 - j) = 5 + j2$$

the same as we obtained graphically. In other words, *the resultant of two vectors is equal to the sum of the complex numbers representing those vectors.*

FIGURE 16-5. Vector addition with complex numbers.

Example: Subtract the vectors $25\angle 48^\circ - 18\angle 175^\circ$.

Solution: Vector addition and subtraction is best done in rectangular form. Converting the first vector,

$$a_1 = 25 \cos 48° = 16.7$$
$$b_1 = 25 \sin 48° = 18.6$$

so

$$25\angle 48° = 16.7 + j18.6$$

Similarly for the second vector,

$$a_2 = 18 \cos 175° = -17.9$$
$$b_2 = 18 \sin 175° = 1.57$$

so

$$18\angle 175° = -17.9 + j1.57$$

Combining,

$$(16.7 + j18.6) - (-17.9 + j1.57)$$
$$= (16.7 + 17.9) + j(18.6 - 1.57)$$
$$= 34.6 + j17.0$$

These vectors are shown in Fig. 16-6.

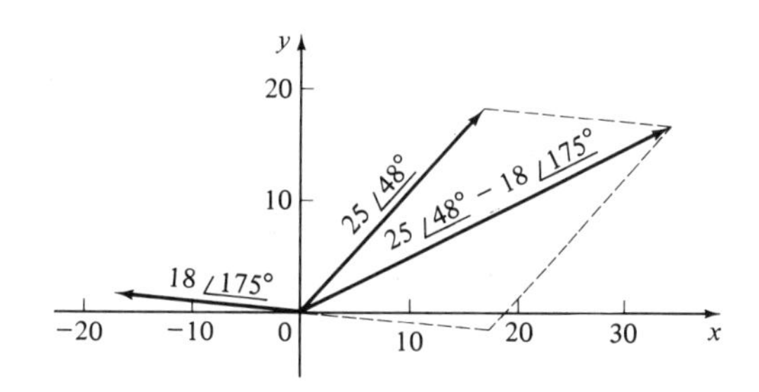

FIGURE 16-6

Multiplication and Division of Vectors: Multiplication or division is best done if the vector is expressed as a complex number in polar form.

Example: Multiply the vectors $2\angle 25°$ and $3\angle 15°$.

Solution: By Eq. 217, the product vector will have a magnitude of $3(2) = 6$ and an angle of $25° + 15° = 40°$. The product is thus

$$6\angle 40°$$

Example: Divide the vector

$$8\angle 44° \text{ by } 4\angle 12°.$$

Solution: By Eq. 218,

$$(8\angle 44°) \div (4\angle 12°) = \frac{8}{4}\angle 44° - 12° = 2\angle 32°$$

The j Operator: Let us multiply a complex number $r\underline{/\theta}$ by j. First we must express j in polar form. By Eq. 211,

$$j = 0 + j1 = 1(\cos 90° + j \sin 90°) = 1\underline{/90°}$$

After multiplying, the new absolute value will be

$$r \cdot 1 = r$$

and the argument will be

$$\theta + 90°$$

So

$$r\underline{/\theta} \cdot j = r\underline{/\theta + 90°}$$

Thus, the only effect of multiplying our original complex number by j was to change the angle by 90°. If our complex number represents a vector, *we may think of j as an operator that causes the vector to rotate through one-quarter revolution.*

EXERCISE 4

Vectors

Express each vector in rectangular and polar form.

	Magnitude	Angle
1.	7	49°
2.	28	136°
3.	193	73.5°

	Magnitude	Angle
4.	34.2	1.1 rad
5.	39	2.5 rad
6.	59.4	−58°

Combine the vectors.

7. $(3 + j2) + (5 - j4)$

8. $(7 - j3) - (4 + j)$

9. $58\underline{/72°} + 21\underline{/14°}$

10. $8.6\underline{/58°} - 4.2\underline{/160°}$

11. $9(\cos 42° + j \sin 42°) + 2(\cos 8° + j \sin 8°)$

12. $8(\cos 15° + j \sin 15°) - 5(\cos 9° - j \sin 9°)$

Multiply the vectors.

13. $(7 + j3)(2 - j5)$

14. $(2.5\underline{/18°})(3.7\underline{/48°})$

15. $2(\cos 25° - j \sin 25°) \cdot 6(\cos 7° - j \sin 7°)$

Divide the vectors.

16. $(25 - j2) \div (3 + j4)$

17. $(7.7\underline{/47°}) \div (2.5\underline{/15°})$

18. $5(\cos 72° + j \sin 72°) \div 3(\cos 31° + j \sin 31°)$

16-5. ALTERNATING-CURRENT CALCULATIONS

Rotating Vectors in the Complex Plane: We have already represented a vector **R** in the complex plane in rectangular form

$$a + jb$$

and in polar form

$$R\angle\theta$$

We can represent a *phasor* (a *rotating* vector, as in Sec. 14-5) by replacing θ with ωt, where ω is the angular velocity and t is time. Our phasor is thus

$$R\angle\omega t$$

Example: A vector of magnitude 11 is rotating with an angular velocity of 5 rad/s. An expression for this phasor in polar form, is

$$11\angle 5t$$

Complex Representation of the Sine or Cosine Wave: In Sec. 14-5 we noted that the sine wave $a \sin \omega t$ was the projection of a phasor on the vertical axis, and that the cosine wave $a \cos \omega t$ was the projection of the phasor on the horizontal axis. If our phasor is now in the complex plane, so that the horizontal axis represents real numbers and the vertical axis imaginary numbers, we can say that the cosine wave

$$V \cos \omega t = \text{real part of } V\angle\omega t$$

and the sine wave

$$V \sin \omega t = \text{imaginary part of } V\angle\omega t$$

Thus, the sine or cosine wave can be represented, not by an entire complex number, but by the imaginary part or the real part of a complex number.

Example: The function $y = 3.5 \cos 7t$ can be represented by a complex number in polar form as

$$y = \text{real part of } 3.5\angle 7t$$

Example: The function $y = 35 \sin (\omega t - 2)$, as a complex number in polar form, is

$$y = \text{imaginary part of } 35\angle \omega t - 2$$

Voltage and Current in Complex Form: Since alternating currents and voltages are sinusoidal quantities, they can be represented by complex numbers in the same way as we represented the sine and cosine waves in the preceding section.

Example: The voltage 75 cos ($\omega t + \pi/6$) can be represented as

$$\text{the real part of } 75\underline{/\omega t + \pi/6}$$

In general, the voltage $v = V_m \cos(\omega t + \phi)$ can be represented by

$$\text{the real part of } V_m\underline{/\omega t + \phi}$$

This phasor is shown in Fig. 16-7a, with the phasor $V_m\underline{/\omega t + \phi}$ shown at a phase displacement ϕ from some reference phasor **R**.

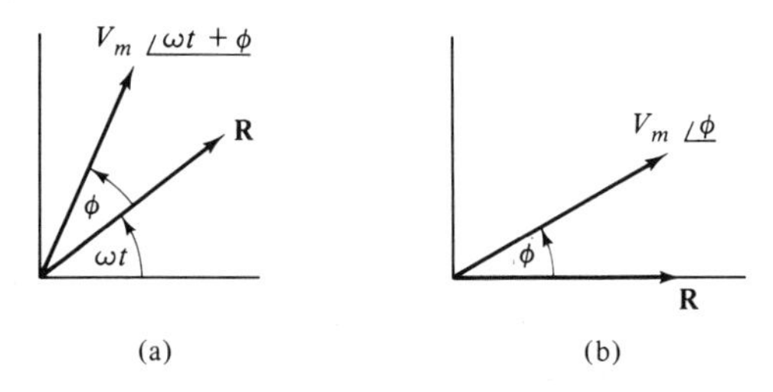

FIGURE 16-7

Realize that a voltage represented by the imaginary part of a complex number is not at all imaginary. *It will turn a motor or give you just as bad a shock as a voltage expressed as the real part of a complex number.*

The orientation of these two vectors is of no importance; they rotate together at an angular velocity ω, keeping a spacing of ϕ between them. To simplify matters it is common practice to take the reference phasor along the horizontal axis, as in Fig. 16-7b. We do this by letting $t = 0$. Our voltage, then, is represented by

$$\text{the real part of } V_m\underline{/\phi}$$

Similarly, the voltage $V_m \sin(\omega t + \phi)$ would be represented by

$$\text{the imaginary part of } V_m\underline{/\phi}$$

Further, we now drop the expressions "real part of" and "imaginary part of," it being understood that we are making use of only one part of a complex number. Thus, the complex or phasor forms of voltage and current are:

Complex Voltage and Current	the voltage $v = V_m \sin(\omega t + \theta_1)$ is represented by $\mathbf{V} = V_m\underline{/\theta_1}$	**A58**
	the current $i = I_m \sin(\omega t + \theta_2)$ is represented by $\mathbf{I} = I_m\underline{/\theta_2}$	**A59**

Note that the complex expressions for voltage and current do not contain t. We say that the voltage and current have been converted from the *time domain* to the *phasor domain*.

In practice, it is customary to divide V_m and I_m by $\sqrt{2}$ in order to get the rms *or* effective values *of voltage and current. We will not introduce that complication here, but work instead with the* peak *values V_m and I_m.*

Example: Convert the sinusoidal voltage $v = 55 \sin(\omega t + 35°)$ to complex form.

Solution: By Eq. A58,

$$\mathbf{V} = 55\underline{/35°}$$

Example: Convert the complex current

$$\mathbf{I} = 2.4\underline{/68°}$$

to sinusoidal form.

Solution: By Eq. A59,

$$i = 2.4 \sin(\omega t + 68°)$$

Complex Impedance: In Sec. 12-4 we drew a vector impedance diagram in which the impedance was the resultant of two perpendicular vectors: the resistance R along the horizontal axis and the reactance X along the vertical axis. If we now use the complex plane, and draw the resistance along the *real* axis and the reactance along the *imaginary* axis, we can represent impedance by a complex number.

Complex Impedance	$\mathbf{Z} = R + jX = Z\underline{/\phi}$	**A57**

where

$$R = \text{circuit resistance}$$

$$X = \text{circuit reactance} = X_L - X_c \tag{A54}$$

$$Z = \text{magnitude of impedance} = \sqrt{R^2 + X^2} \tag{A55}$$

$$\phi = \text{phase angle} = \arctan \frac{X}{R} \tag{A56}$$

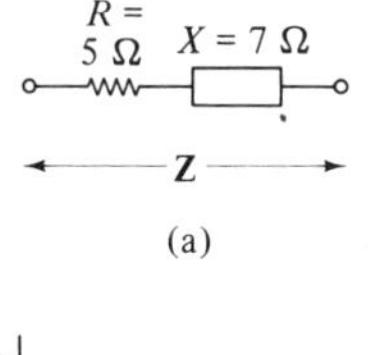

(a)

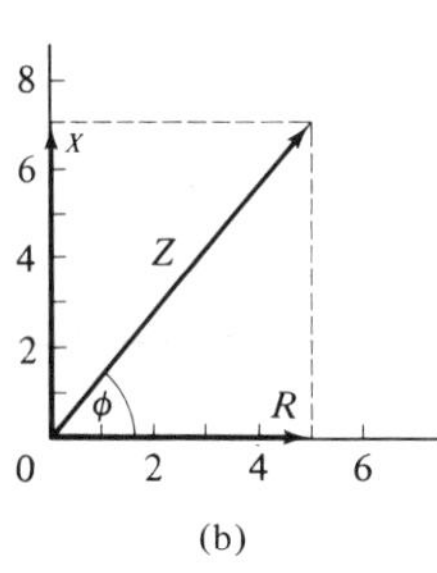

(b)

FIGURE 16-8

Example: A circuit has a resistance of 5 ohms in series with a reactance of 7 ohms (Fig. 16-8a). Represent the impedance by a complex number.

Solution: We draw the vector impedance diagram (Fig. 16-8b). The impedance, in rectangular form, is

$$Z = 5 + j7$$

The magnitude of the impedance is

$$\sqrt{5^2 + 7^2} = 8.49$$

The phase angle is, by Eq. A56,

$$\arctan \frac{7}{5} = 54.5°$$

So we can write the impedance in polar form,

$$\mathbf{Z} = 8.49\underline{/54.5°}$$

"Ohm's Law for AC": We stated at the beginning of this section that the use of complex numbers would make calculation with alternating current almost as easy as for direct current. We do this by means of

"Ohm's Law for AC"	$\mathbf{V} = \mathbf{ZI}$	**A62**

Note the similarity between this equation and Ohm's law, Eq. A40. It is used in the same way.

Example: A voltage of $v = 100 \sin \omega t$ is applied to the circuit of Fig. 16-8. Write a sinusoidal expression for the current i.

Solution:

1. Write the voltage in complex form. By Eq. A58,
$$\mathbf{V} = 100\underline{/0^\circ}$$
2. Find the complex impedance Z. From the example in the preceding section,
$$\mathbf{Z} = 8.49\underline{/54.5^\circ}$$
3. Find the complex current I. By Ohm's law for AC,
$$\mathbf{I} = \frac{\mathbf{V}}{\mathbf{Z}} = \frac{100\underline{/0^\circ}}{8.49\underline{/54.5^\circ}} = 11.8\underline{/-54.5^\circ}$$
4. Write the current in sinusoidal form. By Eq. A59,
$$i = 11.8 \sin (\omega t - 54.5^\circ)$$

EXERCISE 5

AC Calculations

Express each current or voltage in complex form.

1. $i = 250 \sin (\omega t + 25^\circ)$ **2.** $v = 1.5 \sin (112t - 30^\circ)$

3. $v = 57 \sin (\omega t - 90^\circ)$ **4.** $i = 2.7 \sin \omega t$

5. $v = 144 \sin 160t$ **6.** $i = 2.7 \sin (275t - 15^\circ)$

Express each current or voltage in sinusoidal form.

7. $\mathbf{V} = 150\underline{/0^\circ}$ **8.** $\mathbf{V} = 1.75\underline{/70^\circ}$ **9.** $\mathbf{V} = 300\underline{/-90^\circ}$

10. $\mathbf{I} = 25\underline{/30^\circ}$ **11.** $\mathbf{I} = 7.5\underline{/0^\circ}$ **12.** $\mathbf{I} = 15\underline{/-130^\circ}$

Express the impedance of each circuit as a complex number in rectangular and in polar form.

13. Figure 16-9a. **14.** Figure 16-9b.

15. Figure 16-9c. **16.** Figure 16-9d.

17. Figure 16-9e.

18. Figure 16-9f.

19. Figure 16-9g.

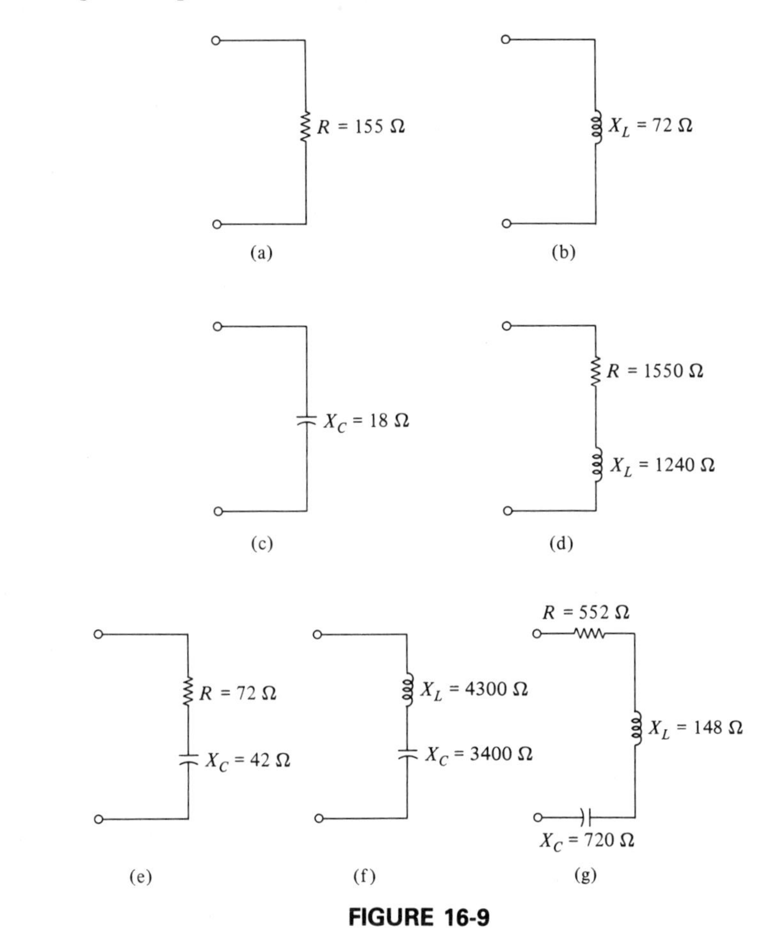

FIGURE 16-9

20. Write a sinusoidal expression for the current i in each part of Fig. 16-10.

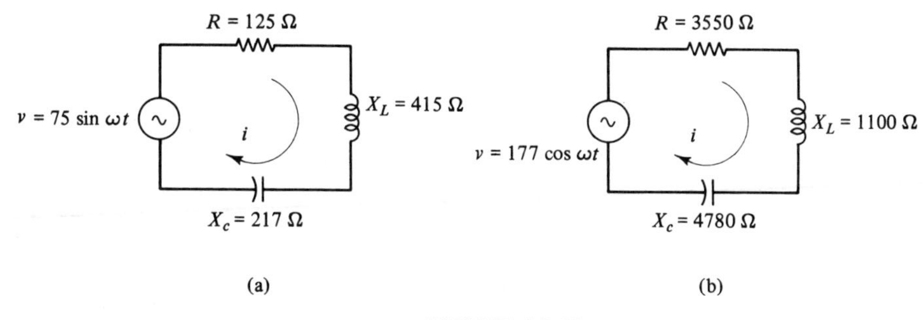

FIGURE 16-10

21. Write a sinusoidal expression for the voltage v in each part of Fig. 16-11.

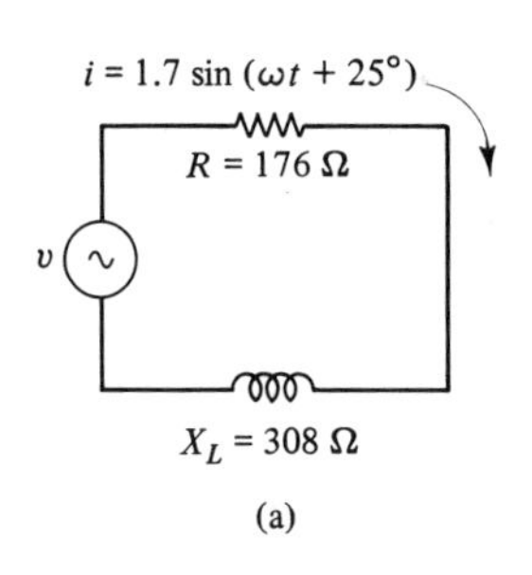

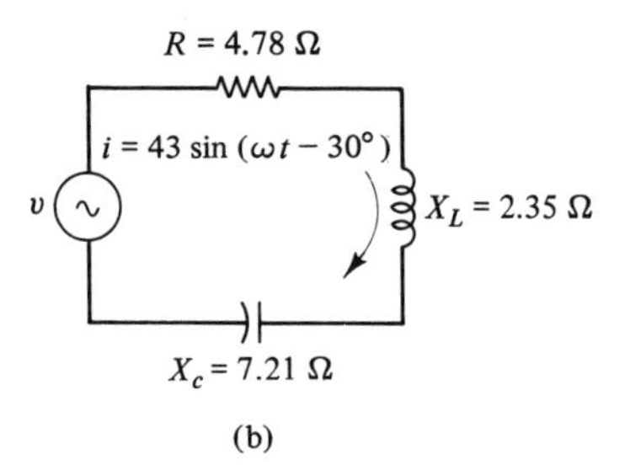

FIGURE 16-11

22. Find the complex impedance **Z** in each part of Fig. 16-12.

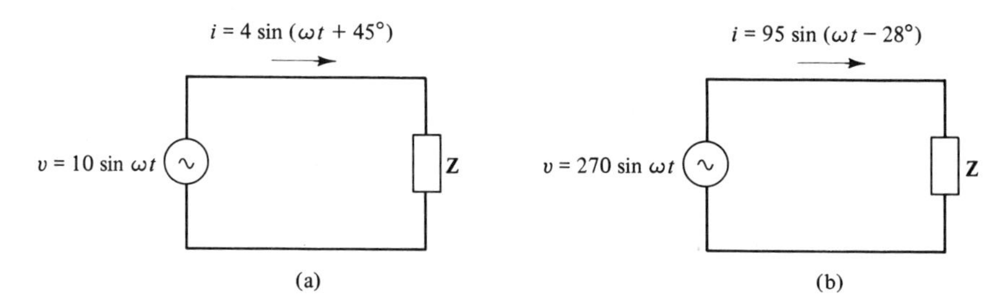

FIGURE 16-12

CHAPTER TEST

Express the following as complex numbers in rectangular, polar, and trigonometric forms.

1. $3 + \sqrt{-4}$ **2.** $-\sqrt{-9} - 5$ **3.** $-2 + \sqrt{-49}$ **4.** $3.25 + \sqrt{-11.6}$

Evaluate.

5. j^{17} **6.** j^{25}

Combine the complex numbers. Leave your answer in rectangular form.

7. $(7 - j3) + (2 + j5)$ **8.** $4.8\angle 28^\circ - 2.4\angle 72^\circ$

9. $52(\cos 50^\circ + j \sin 50^\circ) + 28(\cos 12^\circ + j \sin 12^\circ)$

Multiply and leave your answer in the same form as the complex numbers.

10. $(2 - j)(3 + j5)$ **11.** $(7.3\angle 21^\circ)(2.1\angle 156^\circ)$

12. $2(\cos 20^\circ + j \sin 20^\circ) \cdot 6(\cos 18^\circ + j \sin 18^\circ)$

Divide and leave your answer in the same form as the complex numbers.

13. $(9 - j3) \div (4 + j)$ **14.** $(18\angle 72^\circ) \div (6\angle 22^\circ)$

15. $16(\cos 85^\circ + j \sin 85^\circ) \div 8(\cos 40^\circ + j \sin 40^\circ)$

Graph each complex number.

16. $7 + j4$

17. $2.75\angle 44^\circ$

18. $6(\cos 135^\circ + j \sin 135^\circ)$

Evaluate each power.

19. $(-4 + j3)^2$

20. $(5\angle 12^\circ)^3$

21. $[5(\cos 10^\circ + j \sin 10^\circ)]^3$

22. Express in complex form $i = 45 \sin(\omega t + 32^\circ)$.

23. Express in sinusoidal form $\mathbf{V} = 283\angle -22^\circ$.

24. Write a complex expression for the current i in Fig. 16-13.

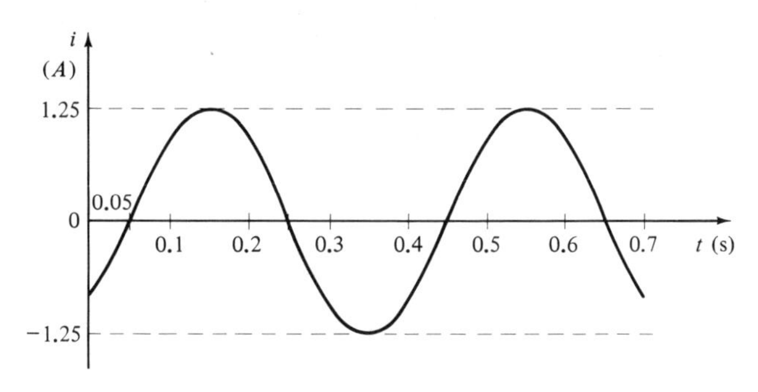

FIGURE 16-13

Inequalities and Linear Programming

So far we have dealt only with equations, where one expression is equal to another, and our aim was to find the value of the variables that makes the two sides equal. In this chapter we work with *inequalities,* in which one expression is not necessarily equal to another, but may be greater than or less than that expression. We will graph inequalities, and also solve them. Here we seek a whole range of values of the variable to satisfy the inequality.

We then introduce *linear programming* (which has nothing to do with *computer* progamming), which plays an important part in certain business decisions. It helps us decide how to allocate resources in order to reduce costs and maximize profit—a concern of technical people as well as management. We then write a computer program to do linear programming.

17-1. DEFINITIONS

Inequalities: An *inequality* is a statement that one quantity is greater than, or less than, another quantity.

Example: The statement $a > b$ means "a is greater than b."

Parts of an Inequality:

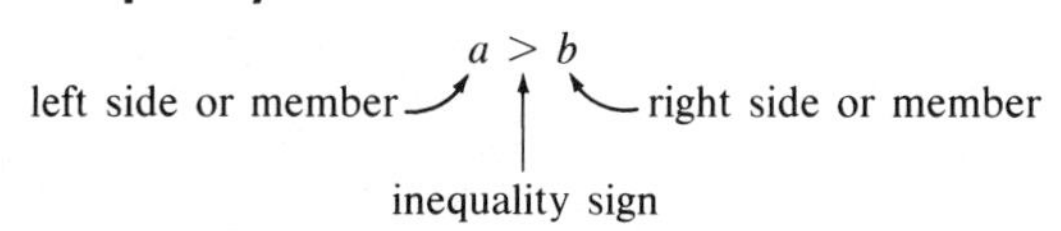

Inequality Symbols: In Sec. 1-1 we introduced the inequality symbols $>$ and $<$. These can be combined with the equals sign to produce the compound symbols $\geqq$ and $\leqq$, which mean "greater than or equal to" and "less than or equal to." Further, a slash through any of these symbols indicates the negation of the inequality.

Example: The statement $a \ngtr b$ means "a is not greater than b."

Sense of an Inequality: The *sense* of an inequality refers to the *direction* in which the inequality sign points. We use this term mostly when operating on an inequality, where certain operations will *change the sense,* whereas others will not.

Inequalities with Three Members: The inequality $3 < x$ has two members, left and right. However, many inequalities have *three* members.

Example: The statement $3 < x < 9$ means "x is greater than 3 and less than 9."

Conditional and Unconditional Inequalities: Just as we have conditional equations and identities, we also have conditional and unconditional inequalities. As with equations, a conditional inequality is one that is satisfied only by certain values of the unknown, whereas unconditional equalities are true for any values of the unknown.

Example: The statement $x - 2 > 5$ is a conditional inequality because it is true only for values of x greater than 7.

In this chapter we limit ourselves to the real *numbers.*

Example: The statement $x^2 + 5 > 0$ is an unconditional inequality (also called an *absolute* inequality) because for any value of x, positive, negative, or zero, x^2 cannot be negative.

Graphical Representation: An inequality having *two members* can be represented on the number line by a *ray,* as in Fig. 17-1. The ray

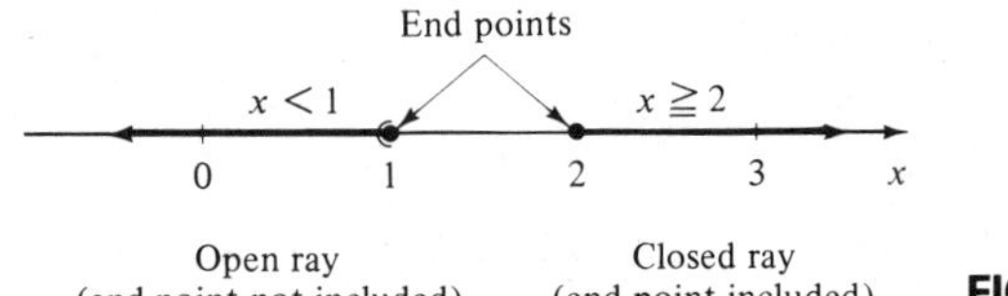

FIGURE 17-1. Rays.

is called *open* or *closed,* depending on whether or not the *endpoint* is included.

An inequality having *three members* can be represented by an *interval* on the number line. The interval is called open, closed, or half-open, depending on how many of the endpoints are included, as shown in Fig. 17-2.

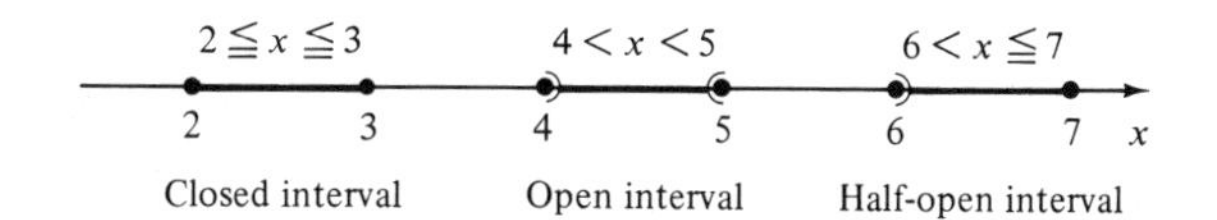

FIGURE 17-2. Intervals.

An inequality having *two variables* can be represented by a *region,* in the *xy* plane.

Example: Graph the inequality

$$y > x + 1$$

Solution: We graph the line $y = x + 1$ (Fig. 17-3). All points in the region *above* the line satisfy the inequality

$$y > x + 1$$

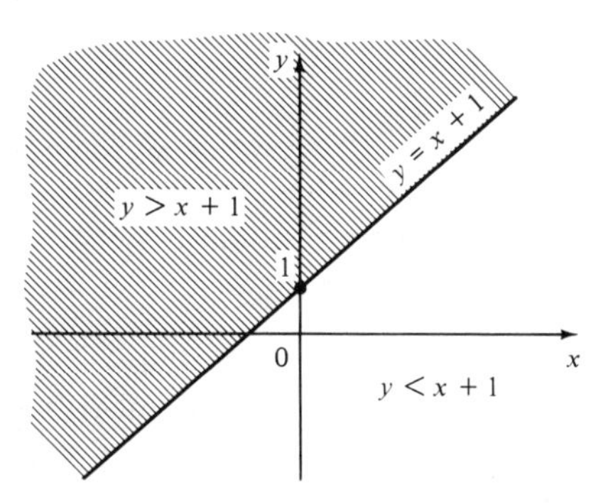

FIGURE 17-3. Points within the shaded region satisfy the inequality $y > x + 1$.

For example, if we substitute the point (1, 3) into our inequality, we get

$$3 > 1 + 1$$

which is true.

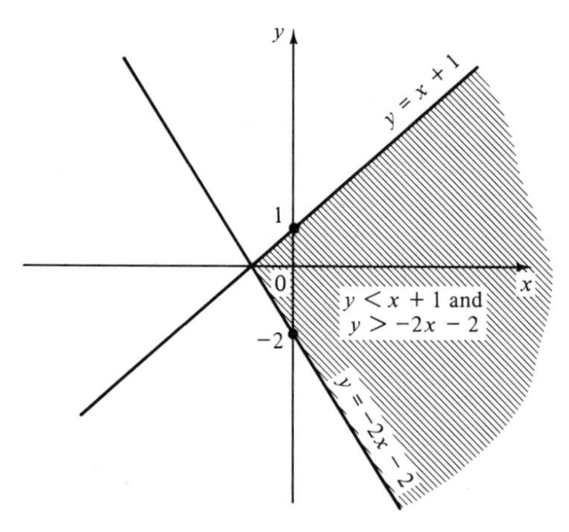

FIGURE 17-4. Points within the shaded region satisfy two inequalities.

All points in the shaded region in Fig. 17-4 satisfy the *two* inequalities

$$y < x + 1 \quad \text{and} \quad y > -2x - 2$$

If we substitute a point, say, (5, 2) into both our inequalities we get

$$2 < 5 + 1$$

which is true, and

$$2 > -2(5) - 2$$

or

$$2 > -12$$

which is also true.

Example: A certain automobile company can make a sedan in 3 h and a hatchback in 2 h. How many of each type of car can be made if the working time is not to exceed 12 h?

Solution: We let

$$x = \text{number of sedans made in 12 h}$$

and $$y = \text{number of hatchbacks made in 12 h}$$

Since each sedan takes 3 h to make, it will take $3x$ hours to make the x sedans. Similarly, it takes $2y$ hours for the hatchbacks. The total time for both is to be 12 h or less, so we get the inequality

$$3x + 2y \leqq 12$$

We plot the equation $3x + 2y = 12$ in Fig. 17-5. Any point on the line or in the shaded region satisfies the inequality. For example, the company could make two sedans and three hatchbacks in the 12-h period, or it could make no cars at all, or it could make four sedans and no hatchbacks, and so forth. It could not, however, make three sedans and three hatchbacks in the allotted 12 h.

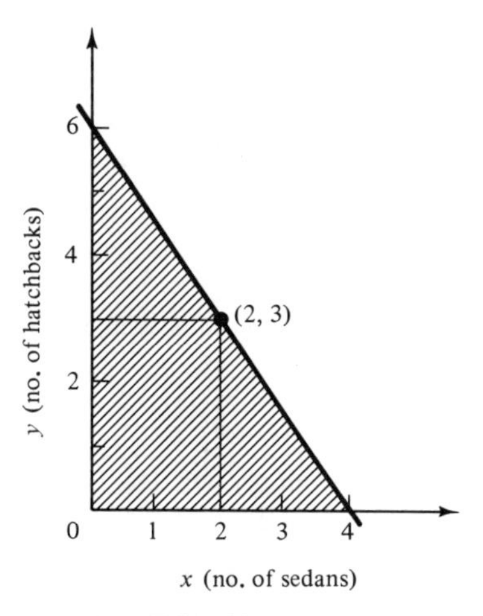

FIGURE 17-5

Inequalities on the Computer: Most computer languages allow us to use the inequality symbols when comparing two quantities.

Example: In BASIC, the instruction

```
100  IF X + Y =< 5 THEN 200
```

will send us to line 200 if the sum of x and y is equal to or less than 5.

Example: For the automobile problem in an earlier example, we could test if x and y gave us a point within the shaded region of Fig. 17-5 with the program statements

```
100  IF X < 0 THEN 200
110  IF Y < 0 THEN 200
120  IF 3 * X + 2 * Y > 12 THEN 200
130
```

We will make use of statements such as these in Sec. 17-3, where we use the computer to do linear programming.

Here we test whether the point (x, y) is within or on the edge of the shaded region with statements which eliminate those points that are *outside* the region. If a point passes the test, control passes to line 130. Otherwise, we go to line 200.

EXERCISE 1

Graphing Inequalities

Graph each inequality as a ray or interval on the number line.

1. $x > 5$ **2.** $x < 3$ **3.** $x \geqq -2$

4. $x \leqq 1$ **5.** $2 < x < 5$ **6.** $-1 \leqq x \leqq 3$

7. $4 \leqq x < 7$ **8.** $-4 < x \leqq 0$

Graph each inequality as a region on the coordinate axes.

9. $y < 3x - 2$ **10.** $x + y > 1$ **11.** $2x - y < 3$

12. $y > -x^2 + 4$

Graph the region in which both inequalities are satisfied.

13. $y > -x + 3$, $y < 2x - 2$ **14.** $y < x^2 - 2$, $y > x$ **15.** $y > x^2 - 4$, $y < 4 - x^2$

16. The supply voltage V to a certain device is required to be over 100 V, but not over 150 V. Express this as an inequality.

17. A manufacturer makes skis and snowshoes. A pair of skis takes 2.5 h to make and a pair of snowshoes takes 3.1 h. Write and graph an inequality to represent the number of pairs of skis and snowshoes that can be made in 8 h.

Computer

18. Write a program that will verify whether a value or values satisfy one or more inequalities. Use your program to check whether selected values of x and y satisfy any of the inequalities in this exercise.

17-2. SOLVING CONDITIONAL INEQUALITIES

Graphical Solution: First transpose all terms to one side of the inequality to get it into the form

$$f(x) > 0 \qquad [\text{or } f(x) < 0]$$

Then plot $y = f(x)$ in the usual way. The solution will then be all values of x for which the curve is above (or, alternatively, below) the x axis.

Example: Solve the inequality

$$x^2 - 3 > 2x$$

Solution: Transposing, we get

$$x^2 - 2x - 3 > 0$$

We now plot the function

$$y = x^2 - 2x - 3$$

as shown in Fig. 17-6. The curve (a parabola) is above the x axis when

$$x < -1 \qquad \text{and} \qquad x > 3$$

These values of x are the solutions to our inequality, because, for these values, $x^2 - 2x - 3 > 0$. For example, substituting the value

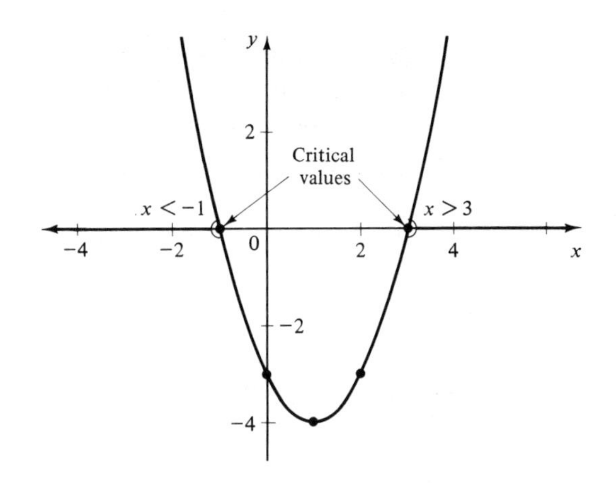

FIGURE 17-6. The points $x = -1$ and $x = 3$ where the curve crosses the x axis are called *critical values.*

$x = 4$ into our original inequality gives

$$4^2 - 3 > 2(4)$$

or

$$16 - 3 > 8$$

which is true. Similarly, the value $x = -2$ gives

$$(-2)^2 - 3 > 2(-2)$$

or

$$4 - 3 > -4$$

which is also true. On the other hand, a value between -1 and 3, such as $x = 1$, gives

$$1^2 - 3 > 2(1)$$

or

$$-2 \ngtr 2$$

which is not true.

Also note that our solutions are all points *on the x axis,* and are not points in some region in the plane.

Algebraic Solution: The object of an *algebraic solution* to a conditional inequality is to locate the *critical values* (the points such as in Fig. 17-6 where the curve crossed the x axis). To do this it is usually a good idea to transpose all terms to one side of the inequality and simplify the expression as much as possible. We make use of the following principles for operations with inequalities:

1. *Addition or subtraction:* The sense of an inequality is not changed if the same quantity is added to or subtracted from both sides.

2. *Multiplication or division by a positive quantity:* The sense of an inequality is not changed if both sides are multiplied or divided by the same *positive* quantity.
3. *Multiplication or division by a negative quantity:* The sense of an inequality *is* changed if both sides are multiplied or divided by the same *negative* quantity.

Example: Solve the inequality

$$21x - 23 < 2x + 15$$

Solution: Transposing,

$$19x < 38$$

Dividing by 19,

$$x < 2$$

Example: Solve the inequality

$$2x - \frac{x^2}{2} > -\frac{21}{2}$$

Solution: Multiplying by 2,

$$4x - x^2 > -21$$

Transposing,

$$4x - x^2 + 21 > 0$$

or

$$-x^2 + 4x + 21 > 0$$

Multiplying by -1 and reversing the sense,

$$x^2 - 4x - 21 < 0$$

Factoring,

$$(x - 7)(x + 3) < 0$$

In order for the product $(x - 7)(x + 3)$ to be negative, one of the factors must be negative and the other positive. But which is which? If we assume that

$$x + 3 < 0 \qquad \text{and} \qquad x - 7 > 0$$

then each x would have to satisfy both the inequalities

$$x < -3 \qquad \text{and} \qquad x > 7$$

But x cannot be greater than 7 and less than -3 at the same time, so this answer is not possible. We then try

$$x + 3 > 0 \qquad \text{and} \qquad x - 7 < 0$$

from which

$$x > -3 \qquad \text{and} \qquad x < 7$$

This answer is possible, and is our solution.

Inequalities Containing Absolute Value Signs: Recall from Sec. 1-1 that the *absolute value* of a quantity is its *magnitude* and is always positive.

Examples:

(a) $|5| = 5$ (b) $|-5| = 5$

It would thus be correct to say that

$$|-5| > 2$$

So if we saw the expression $|x| > 2$, we realize that one permissible value for x would be -5. In fact, any x less than -2 would also be permissible. Of course, any x greater than 2 would also satisfy the expression $|x| > 2$. We can then replace the inequality $|x| > 2$ with the *two* inequalities

$$x < -2 \qquad \text{or} \qquad x > 2$$

To solve an inequality containing an absolute value sign, replace the inequality with two inequalities having no absolute value sign and proceed as before.

Example: Solve the inequality

$$|7 - 3x| > 4$$

Solution: We replace this inequality with

$$(7 - 3x) > 4 \qquad \text{or} \qquad (7 - 3x) < -4$$

Subtracting 7,

$$-3x > -3 \qquad \text{or} \qquad -3x < -11$$

Dividing by -3 and reversing the sense,

$$x < 1 \qquad \text{or} \qquad x > \frac{11}{3}$$

So x may have values less than 1 or greater than $3\frac{2}{3}$.

EXERCISE 2

Solving Inequalities

Solve each inequality.

Solve some of these graphically, as directed by your instructor.

1. $2x - 5 > x + 4$

2. $x^2 + 2x < 3$

3. $x^2 + 4 \geqq 2x^2 - 5$

4. $2x^2 + 15x \leqq -25$

5. $2x^2 - 5x + 3 > 0$

6. $6x < 19x - 10$

7. $\frac{x}{5} + \frac{x}{2} \geqq 4$

8. $\frac{x + 1}{2} - \frac{x - 2}{3} \leqq 4$

Solve each inequality having absolute value signs.

9. $|3x| > 9$ **10.** $2|x + 3| < 8$ **11.** $5|x| > 8$

12. $|3x + 2| \leqq 11$ **13.** $|2 - 3x| \geqq 8$ **14.** $3|x + 2| > 21$

15. A certain machine is rented for \$155 per day, and it costs \$25 per hour to operate. The total cost per day for this machine is not to exceed \$350. Write an inequality for this situation, and find the maximum permissible hours of operation.

17-3. LINEAR PROGRAMMING

We will do problems with only two variables here. For problems with more than two variables, look up the simplex method *in a text on linear programming.*

Linear programming is a method for finding the maximum value of some function (usually representing *profit*) when the variables it contains are themselves restricted to within certain limits. The function that is maximized is called the *objective function,* and the limits on the variables are called *constraints*. The constraints will be in the form of inequalities.

We show the method by means of examples.

Example: Find the values of x and y that will make z a maximum, where

$$z = 5x + 10y$$

In applications, x and y will usually represent amounts of a certain resource, or the number of items produced, and will then be restricted to positive *values.*

and x and y are positive and subject to the constraints

$$x + y \leqq 5$$

and

$$2y - x \leqq 4$$

Solution: We plot the two inequalities (Fig. 17-7). The only permissible values for x and y are the coordinates of points on the edges of or within the shaded region. These are called *feasible solutions*.

But which (x, y) pair selected from all those in the shaded region will give the greatest value of z? In other words, which of the infinite number of feasible solutions is the *optimum* solution?

To help us answer this, let us plot the equation $z = 5x + 10y$ for different values of z. For example:

When $z = 0$:

$$5x + 10y = 0$$

$$y = -\frac{x}{2}$$

When $z = 10$:

$$5x + 10y = 10$$

$$y = -\frac{x}{2} + 1$$

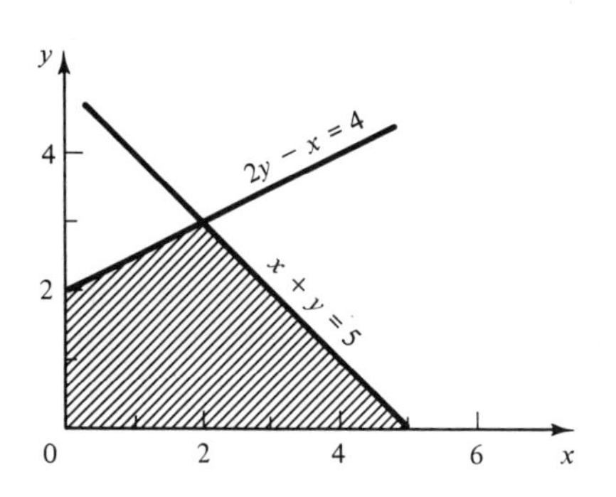

FIGURE 17-7. Feasible solutions.

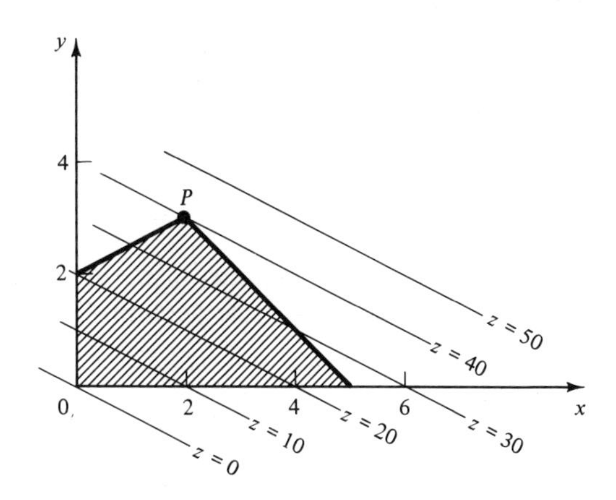

FIGURE 17-8. Optimum solution at P.

and so on. We get a *family* of parallel straight lines, each with a slope of $-\frac{1}{2}$, Fig. 17-8. The point P at which the line having the greatest z value intersects the region gives us the x and y values we seek. Such a point will always occur at a *corner* or *vertex* of the region. Thus, instead of plotting the family of lines, it is only necessary to compute z at the vertices of the region. The coordinates of the vertices are called the *basic feasible solutions*. Of these, the one that gives the greatest z is the *optimum solution*.

In our problem, the vertices are (0, 0), (0, 2), (5, 0), and P, whose coordinates we get by simultaneous solution of the equations of the boundary lines,

$$x + y = 5$$

and

$$x - 2y = -4$$

Subtracting,

$$3y = 9$$

$$y = 3$$

Substituting back,

$$x = 2$$

So P has the coordinates (2, 3). Now computing z at each vertex, we get:

At (0, 0): $z = 5(0) + 10(0) = 0$

At (0, 2): $z = 5(0) + 10(2) = 20$

At (5, 0): $z = 5(5) + 10(0) = 25$

At (2, 3): $z = 5(2) + 10(3) = 40$

So $x = 2$, $y = 3$, which gives the largest z, is our optimum solution.

Example: Find the nonnegative values of x and y that will maximize the function

$$z = 2x + y$$

with the *four* constraints

$$y - x \leqq 6$$
$$x + 4y \leqq 40$$
$$x + y \leqq 16$$
$$2x - y \leqq 20$$

The points of intersection are found by simultaneous solution of the equations of the intersecting lines, and is not shown here.

Solution: We graph the inequalities (Fig. 17-9) and compute the points of intersection. This gives the six basic feasible solutions

$$(0, 0) \quad (0, 6) \quad (3.2, 9.2) \quad (8, 8) \quad (12, 4) \quad \text{and} \quad (10, 0)$$

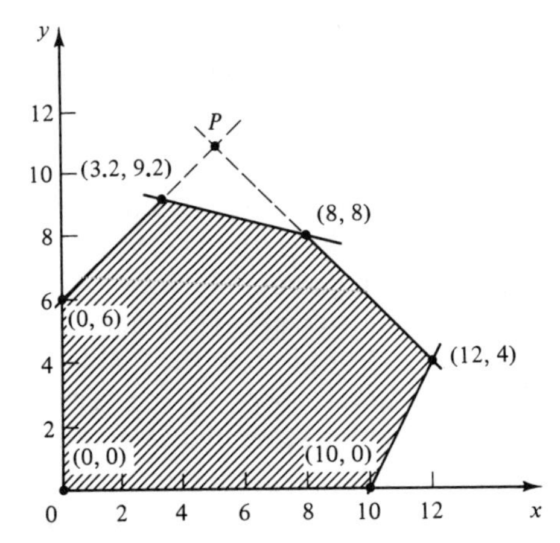

FIGURE 17-9. The coordinates of the corners of the shaded region are the basic feasible solutions. One of these is the optimum solution.

We now compute z at each point of intersection.

Point	z
(0, 0)	$z = 2(0) + 0 = 0$
(0, 6)	$z = 2(0) + 6 = 6$
(3.2, 9.2)	$z = 2(3.2) + 9.2 = 15.6$
(8, 8)	$z = 2(8) + 8 = 24$
(12, 4)	$z = 2(12) + 4 = 28$
(10, 0)	$z = 2(10) + 0 = 20$

The greatest z (28) occurs at $x = 12$, $y = 4$, which is thus our optimum solution.

Applications: Our applications will usually be business problems, where our objective function z (the one we maximize) represents profit, and our variables x and y represent the quantities of two products to be made. A typical problem is to determine how much of each product should be manufactured so that the profit is a maximum.

Example: Each month a company makes x stereo sets and y television sets. The manufacturing hours, material costs, testing time, and profit for each is as follows:

	Manufacturing time (h)	*Material costs*	*Test time (h)*	*Profit*
Stereo	4	$ 95	1	$45
Television	5	48	2	38
Available	180	3600	60	

Also shown in the table is the maximum amount of manufacturing and testing time, and the cost of material available per month. How many of each item should be made for maximum profit?

Solution: First, why should we make any TV sets at all, if we get more profit on stereos? The answer is that we are limited by the money available. In 180 h we could make

$$180 \div 4 = 45 \text{ stereo sets}$$

But 45 stereos, at $95 each, would require $4275, but there is only $3600 available. Then why not make

$$\$3600 \div 95 = 37 \text{ stereo sets}$$

using all our money on the most profitable item? But this would leave us with manufacturing time unused but no money left to make any TV sets.

Thus we wonder if we could do better by making fewer than 37 stereo sets, and using the extra time and money to make some TV sets. We seek the *optimum* solution. We start by writing the equations and inequalities to describe our situation.

Our total profit z will be the profit per item times the number of items made.

$$z = 45x + 38y$$

The manufacturing time must not exceed 180 h, so

$$4x + 5y \leqq 180$$

The materials cost must not exceed $3600, so

$$95x + 48y \leqq 3600$$

And the test time must not exceed 60 h, so

$$x + 2y \leqq 60$$

This gives us our objective function and three constraints. We plot the constraints in Fig. 17-10. We find the vertices of the region by simultaneously solving pairs of equations. To find point A, we take

$$x + 2y = 60$$

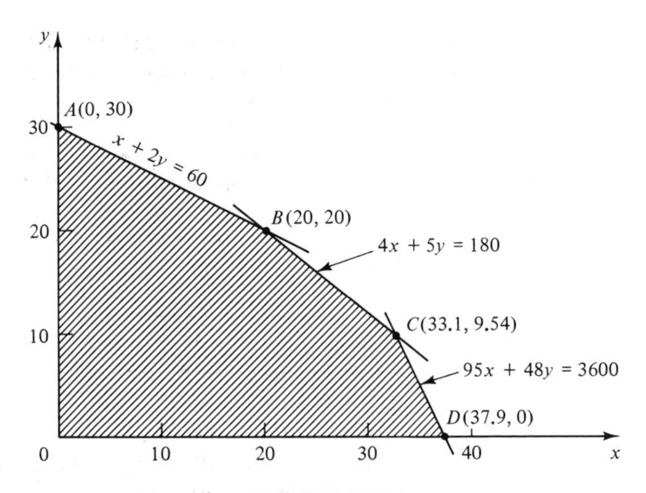

FIGURE 17-10

When $x = 0$,

$$y = 30$$

To find point B, we solve simultaneously

$$x + 2y = 60$$
$$4x + 5y = 180$$

and get

$$B(20, 20)$$

Again, the actual work is omitted here.

To find point C, we solve the equations

$$4x + 5y = 180$$
$$95x + 48y = 3600$$

and get

$$x = 33.1 \qquad y = 9.54$$

Since x and y must be integers, we round and get

$$C(33, 10)$$

To find point D, we use the equation

$$95x + 48y = 3600$$

Letting $y = 0$,

$$x = 38 \quad \text{(rounded)}$$

We then evaluate the profit z at each vertex (omitting the point 0, 0 at which $z = 0$).

At $A(0, 30)$: $\quad z = 45(0) + 38(30) = \$1140$

At $B(20, 20)$: $\quad z = 45(20) + 38(20) = \$1660$

At $C(33, 10)$: $\quad z = 45(33) + 38(10) = \$1865$

At $D(38, 0)$: $\quad z = 45(38) + 38(0) = \$1710$

Thus our maximum profit, \$1865, is obtained when we make 33 stereo sets and 10 television sets.

Linear Programming on the Computer: The computer can reduce the work of linear programming, especially when the number of constraints gets large. The following algorithm is for a problem with two variables and any number of constraints.

1. Using Eq. 63, find the point of intersection of any two of the given boundary lines.
2. If the point of intersection is within the region of permissible solutions, go to step 3. If not, go to step 5. (This will eliminate points such as P in Fig. 17-9.)
3. Compute z at the point of intersection.
4. If z is the largest so far, save it and the coordinates of the point. If not, discard it.
5. If there are combinations of equations yet unused, go to step 1.
6. Print the largest z and the coordinates of the point at which it occurs.

A flowchart for this computation is shown in Fig. 17-11 on page 546.

EXERCISE 3

1. Maximize the function $z = x + 2y$ under the constraints

$$x + y \leqq 5 \quad \text{and} \quad x - y \leqq -2$$

2. Maximize the function $z = 4x + 3y$ under the constraints

$$x - 2y \leqq -2 \quad \text{and} \quad x + 2y \leqq 4$$

3. Maximize the function $z = 5x + 4y$ under the constraints

$$7x + 4y \leqq 35 \quad 2x - 3y \geqq 7 \quad \text{and} \quad x - 2y \leqq -5$$

4. Maximize the function $z = x + 2y$ under the constraints

$$x - y \leqq -5 \quad x + 5y \leqq 45 \quad \text{and} \quad x + y \leqq 15$$

5. A company makes pulleys and sprockets. A pulley takes 3 min to make, requires \$1.25 in materials, and gives a profit of \$2.10. A sprocket takes 4 min to make, requires \$1.30 in materials, and gives a profit of \$2.35. The time to make x pulleys and y sprockets is not to exceed 8 h, and the cost for materials must not exceed \$175. How many of each item should be made for maximum profit?

6. A company makes two computers, a mini and a micro, in plants in Boston and Seattle. The Boston plant operates for 180 h per month and the Seattle plant for 210 h per month. The Boston plant can make a mini in 6 h and a micro in 4 h, while the Seattle plant takes 5 h for either a mini or a micro. The profit is \$1000 for a mini and \$900 for a micro. How many of each computer should be made to give maximum profit?

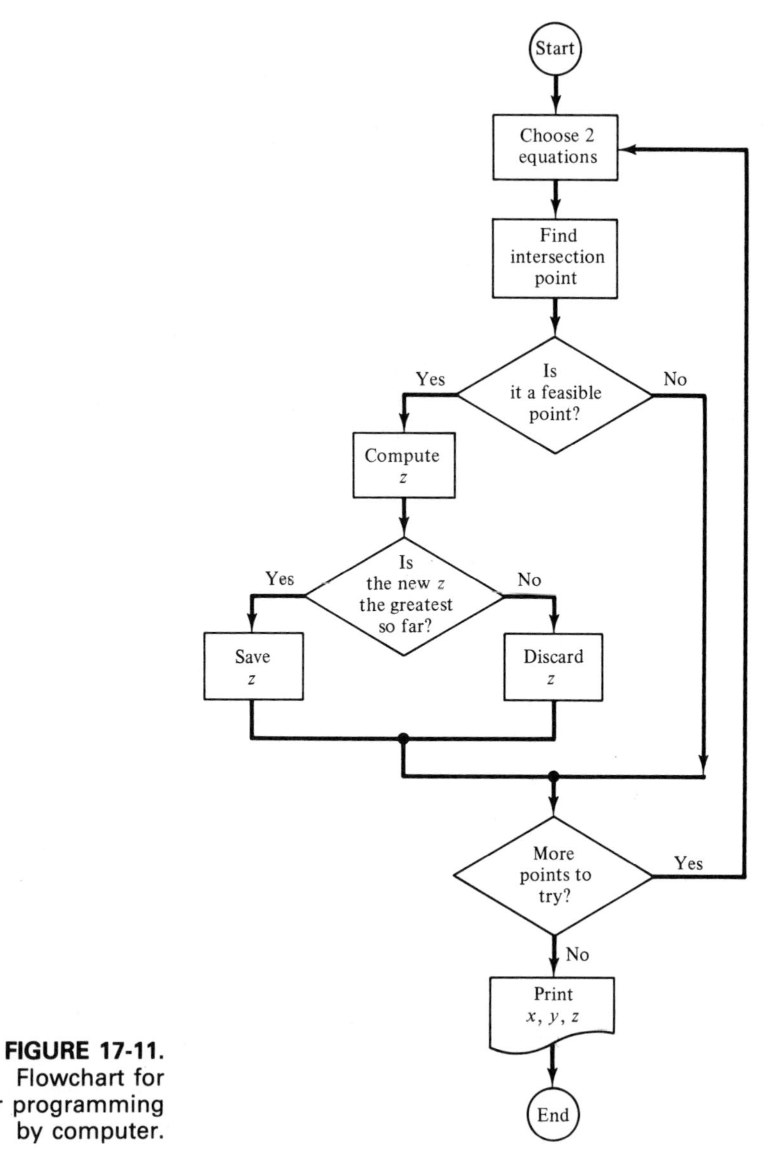

FIGURE 17-11. Flowchart for linear programming by computer.

7. Each week a company makes x disk drives and y monitors. The manufacturing hours, material costs, inspection time, and profit for each is as follows:

	Manufacturing time (h)	*Material costs*	*Inspection time (h)*	*Profit*
Disk drive	5	\$ 21	4.5	\$34
Monitor	3	28	4	22
Available	50	320	50	

Also shown in the table is the maximum amount of manufacturing and inspection time, and the cost of material available per week. How many of each item should be made for maximum profit?

Computer

8. Using the flowchart (Fig. 17-11) as a guide, write a computer program to do linear programming. Try it on any of the problems in this exercise.

CHAPTER TEST

Represent each inequality graphically.

1. $x < 8$

2. $y > 3x - 4$

3. $2 \leqq x < 6$

4. $2x + y > 3$

5. $y > 2x^2 - 3$

6. $y < -2x - 3$ and $y > x + 2$

Solve each inequality.

7. $2x - 5 > x + 3$

8. $|2x| < 8$

9. $|x - 4| > 7$

10. $x^2 > x + 6$

11. Maximize the function $z = 7x + y$ under the constraints

$$2x + 3y \leqq 5 \quad \text{and} \quad x - 4y \leqq -4$$

12. A company makes four-cylinder engine blocks and six-cylinder engine blocks. A "four" requires 4 h of machine time and 5 h of labor, and gives a profit of \$172. A "six" needs 2 h on the machine and 3 h of labor, and yields \$195 in profit. If 130 h of labor and 155 h of machine time are available, how many of each engine block should be made for maximum profit?

Boolean Algebra

Can you imagine an algebra in which the variables can take on *only two values,* 0 and 1, instead of the whole range of real numbers? Such an algebra is not only possible, but it turns out to be very useful. After all, if you are dealing with switch positions, you only *need* two numbers, a 0 to represent the open position and a 1 to represent the closed position. And when Boole applied algebra to the English language, he needed only two numbers, a 0 to represent "false" and a 1 to represent "true." Further, we need only two values when working with binary digits, which, as we saw in Chapter 2, have only two possible values.

Named for the English mathematician George Boole (1815–1864), who wrote "An Investigation of the Laws of Thought On Which Are Founded the Mathematical Theories of Logic and Probabilities" in 1854.

As we develop the Boolean algebra here, we illustrate each new idea in terms of electrical switches and, because they are easy to visualize, in term of sets. We then go on to apply Boolean algebra to more complex circuits, and then to the logic gates which do the arithmetic computations in a computer.

18-1. BOOLEAN VARIABLES AND OPERATIONS

Boolean Variables: We will represent Boolean variables by the capital letters A, B, C, . . ., X, Y, Z. In *ordinary algebra,* a variable such as x will usually have a value that is one of the real numbers. For example, the solution to the equation

$$2x - 8 = 0$$

is

$$x = 4$$

In *Boolean* algebra a variable can never have the value 4. In fact, Boolean variables can have only one of the values

$$0 \quad \text{or} \quad 1$$

These variables should not be thought of as numerical values, but rather as *binary states,* such as TRUE/FALSE, or YES/NO.

When applying Boolean algebra to *sets,* we let

This might be a good time to thumb through Sec. 6-1 to refresh your memory on sets.

0 = the empty set

1 = the univeral set

When applying Boolean algebra to *switching circuits,* we let

0 = off

1 = on

Examples:

(a) If our universal set U is the set of all students at a certain school, and B represents the set of all boys at that school, then

$$B = 0$$

means that there are no boys at that school, and

$$B = 1$$

means that all the students are boys.

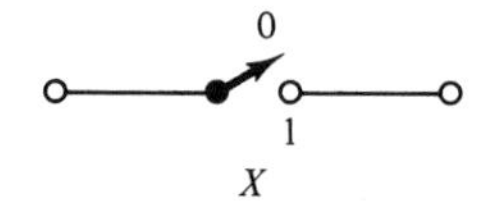

FIGURE 18-1. The two positions of this switch can represent the Boolean variables 0 and 1.

(b) If X represents a certain switch (Fig. 18-1) then

$$X = 0$$

means that the switch is open (no current can flow), and

$$X = 1$$

means that the switch is closed (a current can flow).

Boolean Operations: In ordinary algebra we had the operations of addition, subtraction, and so forth. In Boolean algebra we have the three basic operations

AND OR NOT

which are represented by the multiplication dot (·), the plus sign (+), and the apostrophe (') or overbar ($\overline{\ }$).

Example:

A AND B	A OR B	NOT A
$A \cdot B$	$A + B$	A' or $\overline{A}$

The AND Operator: We represent the AND operation by $A \cdot B$, or simply by AB.

When talking about sets, the AND operator represents the *intersection* of two sets A and B, which, we said in Chapter 6, consists of those elements that are in both A AND B. Recall that in Chapter 6 we used the symbol $\cap$ for intersection.

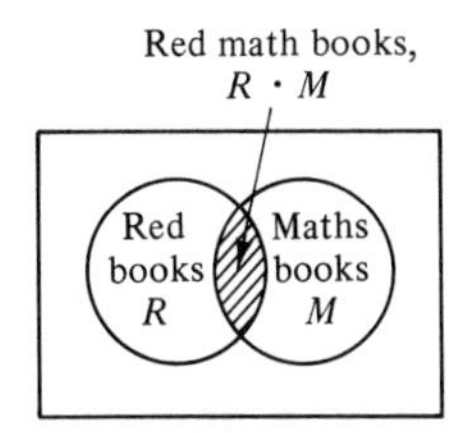

FIGURE 18-2

Example: If R is the set of all red books in a library and M is the set of all math books in the library, then

$$R \cdot M$$

is the set of all red math books in the library, as in the Venn diagram (Fig. 18-2).

When talking about switches, the AND operator represents two switches X and Y *in series,* as in Fig. 18-3.

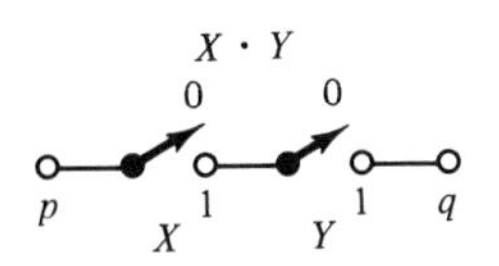

FIGURE 18-3. The AND operator represented by two switches in series.

Example: The expression

$$X \cdot Y = 1$$

means that two switches X and Y in series are both closed. Thus, both X AND Y must be closed in order for current to flow from p to q.

Postulates for the AND Operator: The postulates of Boolean algebra, those ideas that are stated without proof and from which all other theorems are derived, are given here. For each, we give an example for sets and for switching circuits.

Our first postulate is

$$\boxed{0 \cdot 0 = 0}$$

Examples:

(a) The intersection of two empty sets is an empty set.
(b) Two open switches in series is an open circuit.

Our next postulate is

$$\boxed{0 \cdot 1 = 1 \cdot 0 = 0}$$

Examples:

(a) The intersection of the empty set with the universal set is the empty set.
(b) An open switch in series with a closed switch is an open circuit.

Our last postulate for the AND operator is

$$\boxed{1 \cdot 1 = 1}$$

Examples:

(a) The intersection of the universal set and itself is the universal set.

(b) Two closed switches in series is a closed circuit.

Truth Tables: A *truth table* is a convenient way of defining the Boolean operators, for all possible states of the variables. Thus A can be either 0 or 1, and for each of these, B can be either 0 or 1. For each combination, the value of $A \cdot B$, determined from the postulates, is given.

Truth Table for the AND Operator	A	B	$A \cdot B$	224
	0	0	0	
	0	1	0	
	1	0	0	
	1	1	1	

Example: If $A = 1$, $B = 0$, and $C = 1$, evaluate

$$(A \cdot B) \cdot (B \cdot C) \cdot (A \cdot C)$$

Solution: Substituting the values for the variables,

$$(A \cdot B) \cdot (B \cdot C) \cdot (A \cdot C) = (1 \cdot 0) \cdot (0 \cdot 1) \cdot (1 \cdot 1)$$

Then using the truth table or the postulates gives

$$\begin{aligned}(1 \cdot 0) \cdot (0 \cdot 1) \cdot (1 \cdot 1) &= 0 \cdot 0 \cdot 1 \\ &= 0\end{aligned}$$

The OR Operator: When discussing two sets A and B, the OR operator gives us the *union* of A and B. Thus, an element of $A + B$ is a member of sets A OR B. In Chapter 6 we used the symbol ∪ for the union of two sets.

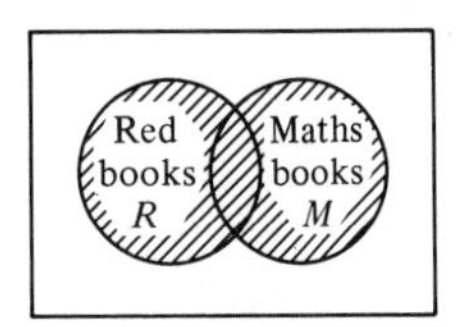

FIGURE 18-4. The shaded region represents $R + M$: those books that are either red or mathematical.

Example: If R is the set of all red books and M is the set of all math books, as before, then

$$R + M$$

is the set of all books that are either red or mathematical (Fig. 18-4).

When used for switching circuits, the OR operator represents two switches *in parallel,* as in Fig. 18-5.

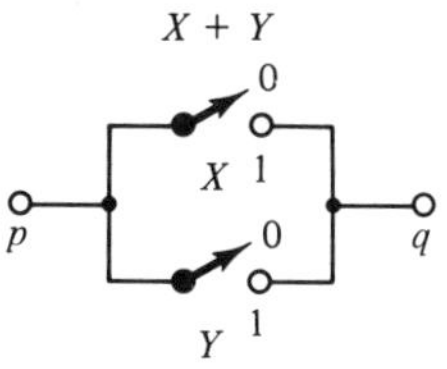

X + Y
0
X 1
p 0 q
Y 1

FIGURE 18-5. The OR operator represented by two switches in parallel.

Example: The expression

$$X + Y = 1$$

means that for two switches X and Y in parallel, current flows from p to q when either X OR Y is closed.

Inclusive and Exclusive OR: In everyday speech the word "or" is used in two different ways. It sometimes means "*A* or *B or both,*" as in

I enjoy jazz or rock.

but at other times "*A* or *B but not both,*" as in

Shall I play jazz or rock on my stereo set?

We give a truth table for the exclusive OR in Sec. 18-5.

The first is called the *inclusive OR* and the second is called the *exclusive OR*. We will take the symbol (+) to stand for the inclusive OR.

Postulates for the OR Operator: The three postulates that apply to the OR operator are given here. The first is

$$\boxed{0 + 0 = 0}$$

Examples:

(a) The union of two empty sets is an empty set.
(b) Two open switches in parallel is an open circuit.

The second postulate is

$$\boxed{1 + 0 = 0 + 1 = 1}$$

Examples:

(a) The union of the universal set and the empty set, in either order, is the universal set.
(b) A closed switch in parallel with an open switch, in either order, is a closed circuit.

Notice that the five preceding postulates were no different from ordinary arithmetic. This is the first one that is different.

The last postulate for the OR operator is

$$\boxed{1 + 1 = 1}$$

Examples:

(a) The union of the universal set with itself is the universal set.

(b) Two closed switches in parallel is a closed circuit.

As with the AND operator, the postulates for the OR operator can be combined into a truth table.

Truth Table for the OR Operator	A	B	$A + B$	**225**
	0	0	0	
	0	1	1	
	1	0	1	
	1	1	1	

Example: If $A = 1$, $B = 0$, and $C = 0$, evaluate

$$(A + B) + (B + C)$$

Solution: Substituting and using the truth table,

$$\begin{aligned}(A + B) + (B + C) &= (1 + 0) + (0 + 0) \\ &= \quad 1 \quad + \quad 0 \\ &= 1\end{aligned}$$

The NOT Operator: We represent the NOT operator with a prime or an overbar:

$$A' \quad \text{or} \quad \overline{A}$$

The NOT operator is defined by the simple truth table;

Truth Table for the NOT Operator	A	$\overline{A}$	**226**
	0	1	
	1	0	

Example: If $A = B = 1$, and $C = D = 0$, evaluate

$$(A + \overline{B}) \cdot (\overline{C} + D)$$

Solution: Substituting, and using the truth tables for AND, OR, and NOT, gives

$$\begin{aligned}(A + \overline{B}) \cdot (\overline{C} + D) &= (1 + 0) \cdot (1 + 0) \\ &= \quad 1 \quad \cdot \quad 1 \\ &= \quad 1\end{aligned}$$

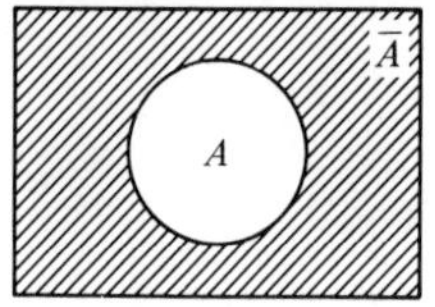

FIGURE 18-6. A Venn diagram representing the NOT operator.

When applied to a set A, the NOT operator will give us the *complement* $\overline{A}$ of that set, such as the shaded region in the Venn diagram (Fig. 18-6). Recall that in Chapter 6 we used the symbol A^c to represent the complement of A.

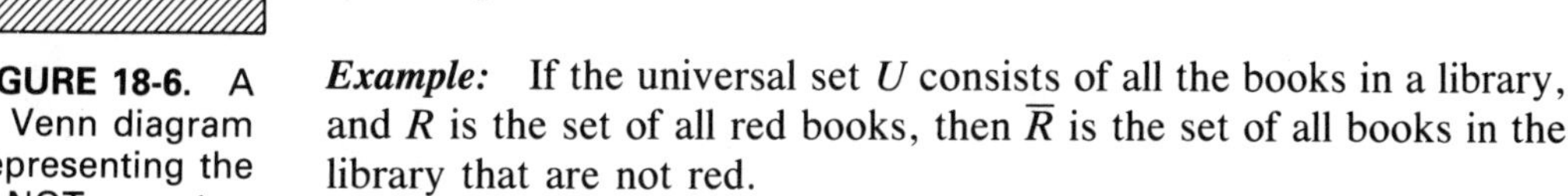

Example: If the universal set U consists of all the books in a library, and R is the set of all red books, then $\overline{R}$ is the set of all books in the library that are not red.

The interpretation of the NOT operator in a switching circuit can be seen in Fig. 18-7, which shows a *double-throw* switch. If we label one position X, then the other position is $\overline{X}$. Thus, when circuit pq is continuous, pr is not, and vice versa.

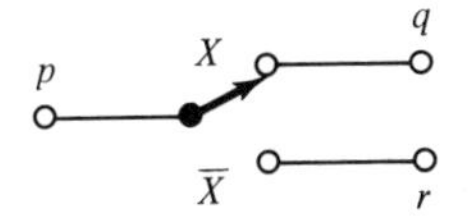

FIGURE 18-7. Switch representation of the NOT operator.

EXERCISE 1

If $A = 1$, $B = 0$, and $C = 1$, evaluate:

1. $(AB)(B + C)$

2. $A + BC + AC$

If $A = 1$, $B = 0$, and $C = 0$, evaluate:

3. $(A + B)(B + C)$

4. $AB + BC + AC$

5. If $A = B = 1$. and $C = D = 0$, evaluate: $(A + \overline{B})(\overline{C} + D)$.

Draw a Venn diagram for three sets A, B, and C, and shade the area representing:

6. AB **7.** $A + B$ •**8.** ABC

9. $A + B + C$ **10.** $\overline{A}$ **11.** $\overline{A}B$

12. $\overline{A}\,\overline{B}$ **13.** $\overline{AB}$

Write a Boolean expression to represent each circuit (see page 556).

14. Figure 18-8a. **15.** Figure 18-8b.

16. Figure 18-8c. **17.** Figure 18-8d.

Use a Venn diagram to decide which of the following expressions is valid.

18. $\overline{A + B} = \overline{A} + \overline{B}$ **19.** $\overline{A + B} = \overline{A}\,\overline{B}$

Use a Venn diagram to simplify each expression.

20. $AB + BC + C\overline{A}$ **21.** $(A + \overline{B})(\overline{A} + C)$

Draw a switch circuit to implement each expression.

22. $X + YZ$ **23.** $X\overline{Y}Z$

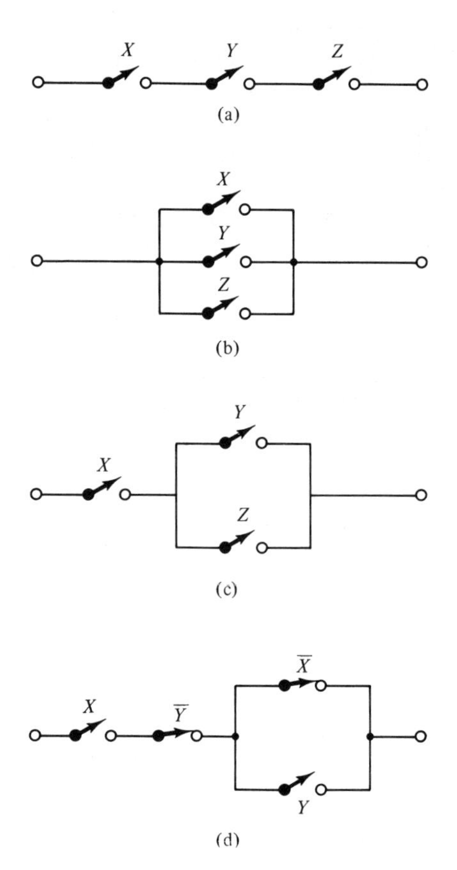

FIGURE 18-8

18-2. LAWS OF BOOLEAN ALGEBRA

Commutative Laws: Let us now combine the truth tables for AND, OR, and NOT into a single table:

Truth Table for AND, OR, NOT	A	B	AND $A \cdot B$	OR $A + B$	NOT $\overline{A}$
	0	0	0	0	1
	0	1	0	1	1
	1	0	0	1	0
	1	1	1	1	0

and see what laws we can deduce. Look first at the AND and OR columns in the combined truth table. We notice that the values for A and B can be interchanged without changing the value of AB or $A + B$.

Note the similarity with the commutative laws for ordinary algebra (Eqs. 1 and 2).

In other words,

$AB \equiv BA$	**228a**
and $A + B \equiv B + A$	**228b**

The value of one variable AND another, or one variable OR another does not depend on the order in which the variables are written.

Boundedness Laws: We next notice that whenever either A or B is 0, then AB is 0. This gives

$A \cdot 0 \equiv 0$	**229a**

The value of a variable AND zero is always zero.

Examples:

(a) The intersection of any set with an empty set is equal to an empty set.

(b) If switch X is in series with an open switch, the circuit remains open regardless of the position of switch X.

Further, whenever either A or B is 1, then $A + B$ is 1.

$A + 1 \equiv 1$	**229b**

The value of any variable OR 1 always equals 1.

Examples:

(a) The union of any set with the universal set is equal to the universal set.

(b) If a switch X is in parallel with a switch that is always closed, the circuit remains closed regardless of the position of switch X.

Identity Laws: Next we see that whenever one variable is 1, then AB takes the value of the other variable. In symbols,

$A \cdot 1 \equiv A$	**230a**

The value of a variable AND 1 is the same as that of the original variable.

Examples:

(a) The intersection of any set A and the universal set is equal to set A.
(b) A switch that remains closed has no effect when connected in series.

Further, when one variable is 0, then $A + B$ takes the value of the other variable.

$A + 0 \equiv A$	**230b**

The value of a variable OR zero is the same as that of the original variable.

Examples:

(a) The union of any set A and an empty set is equal simply to set A.
(b) A switch that is always open has no effect when connected in parallel with any circuit.

An idempotent *quantity is one that remains unchanged when multiplied by or added to itself.*

Idempotent Laws: Next we see that when A and B have the same value, that AB has that same value.

$AA \equiv A$	**231a**

The value of a variable AND itself is the same as the value of the original variable.

Also, when A and B have the same value, then $A + B$ has that same value.

$A + A \equiv A$	**231b**

The value of a variable OR itself is the same as the value of the original variable.

Examples:

(a) The intersection or union of a set A and itself is simply A.
(b) When two switches that operate in unison (both open or both closed) are connected in series or in parallel, one of the switches can be eliminated.

Complement Laws: If $A = 1$, then $\overline{A} = 0$, and if $A = 0$, then $\overline{A} = 1$. Either way, we have

$$A \cdot \overline{A} \equiv 1 \cdot 0$$

which, by law 229a, is equal to 0. So

$A \cdot \overline{A} \equiv 0$	**232a**

The value of a variable AND its complement is zero.

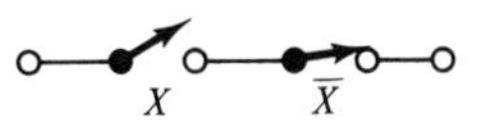

FIGURE 18-9. Switch illustration of law 232a; $A \cdot \overline{A} = 0$.

Examples:

(a) The intersection of a set and the complement of that set is the empty set.

(b) When a switch is in series with another switch which is always in the opposite state (Fig. 18-9) the circuit will always be open.

Similarly, for the OR,

$$A + \overline{A} \equiv 1 + 0$$

which, by law 230b, equals 1. So

$A + \overline{A} \equiv 1$	**232b**

The value of a variable OR its complement is 1.

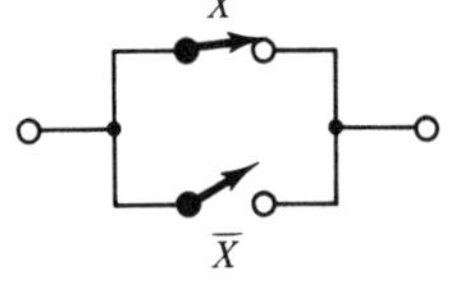

FIGURE 18-10. Switch illustration of law 232b; $A + \overline{A} = 1$.

Examples:

(a) The union of a set and its complement is the universal set.

(b) When a switch is in parallel with another switch which is always in the opposite state (Fig. 18-10), the circuit will always be closed.

Associative Laws: One way to verify that a law of Boolean algebra is true is to see if the truth tables for each side of the equation are identical.

Compare these associative laws with those for ordinary algebra (Eqs. 3 and 4).

Example: Verify the associative law for the AND operator,

$A(BC) \equiv (AB)C$	**233a**

Solution: We make a truth table for $A(BC)$ and for $(AB)C$.

A	B	C	BC	$A(BC)$	AB	$(AB)C$
0	0	0	0	0	0	0
0	0	1	0	0	0	0
0	1	0	0	0	0	0
0	1	1	1	0	0	0
1	0	0	0	0	0	0
1	0	1	0	0	0	0
1	1	0	0	0	1	0
1	1	1	1	1	1	1

identical ($A(BC)$ and $(AB)C$ columns)

Try now to prove the associative law for the OR operator,

$A + (B + C) \equiv (A + B) + C$	**233b**

on your own, again by making a truth table for each side.

Examples:

(a) When finding the union or intersection of three sets, it does not matter in which order they are taken.

(b) When three switches are wired in series or in parallel, the order of the switches does not matter.

Distributive Laws: The first distributive law,

$A(B + C) \equiv AB + AC$	**234a**

is useful because it enables us to *factor* and to *multiply* Boolean expressions in much the same way that Eq. 5 allowed us to factor and to multiply algebraic expressions. Let us verify law 234a by a method different than that we used before. Rather than compare the truth table for each side, we compare the Venn diagram for each, as in Fig. 18-11.

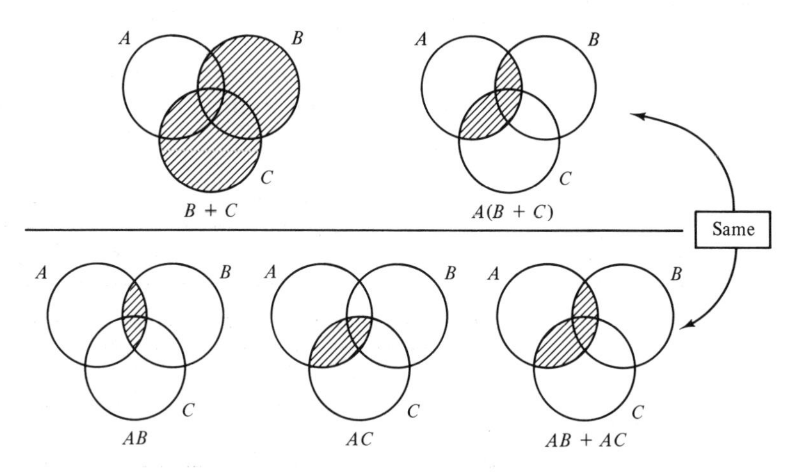

FIGURE 18-11. Use of a Venn diagram to verify law 234a.

Example: Switch Y in parallel with Z, both in series with X (Fig. 18-12a), is equivalent to switch X in series with Y, both in parallel to X in series with Z (Fig. 18-12b).

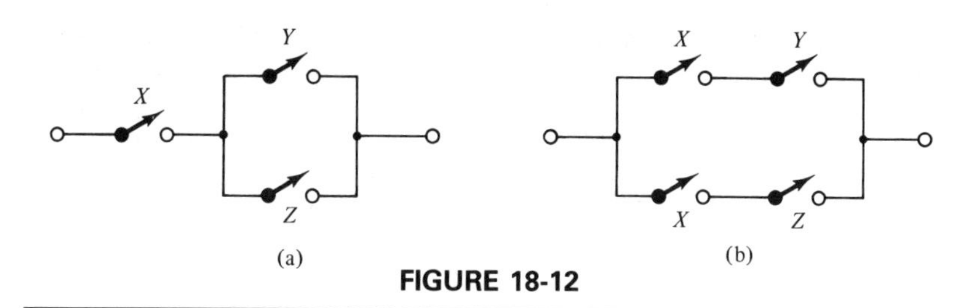

FIGURE 18-12

We now derive the distributive law

$A + BC \equiv (A + B)(A + C)$	**234b**

This is no different from when we used the FOIL rule to multipy two binomials. But the middle terms AB and AC are not present in the final law. Follow the derivation to see how they vanish.

By law 234a, the right side becomes

$$(A + B)(A + C) = (A + B)A + (A + B)C$$
$$= AA + AB + AC + BC$$

But by law 231a, $AA = A$, so

$$(A + B)(A + C) = A + AB + AC + BC$$

Factoring by law 234a,

$$(A + B)(A + C) = A(1 + B) + AC + BC$$

But, by law 229b, $1 + B = 1$, so

$$(A + B)(A + C) = A \cdot 1 + AC + BC$$

and by law 230a,

$$= A + AC + BC$$

Again using law 234a,

$$(A + B)(A + C) = A(1 + C) + BC$$
$$= A \cdot 1 + BC$$
$$= A + BC$$

We now prove the distributive law

$A(\overline{A} + B) \equiv AB$	**235a**

Proof: By law 234a, the left side becomes

$$A(\overline{A} + B) = A\overline{A} + AB$$

Then by law 232a, $= 0 + AB$

and by law 230b, $= AB$

The proof of the distributive law

$A + \overline{A}B \equiv A + B$	**235b**

is left as an exercise.

Absorption Laws: We prove the absorption law

$A(A + B) \equiv A$	**236a**

Proof: By law 234a, the left side becomes

$$A(A + B) = AA + AB$$

Then by law 231a, $= A + AB$

Factoring by law 234a, $= A(1 + B)$

By law 229b, $= A \cdot 1$

and by law 230a, $= A$

Notice that the variable B originally on the left side has vanished, or has been absorbed.

The proof of the other absorption law

$A + AB \equiv A$	**236b**

is left as an exercise.

DeMorgan's Laws:

Augustus DeMorgan (1806–1871).

$\overline{A \cdot B} \equiv \overline{A} + \overline{B}$	**237a**

The complement of one variable AND another is equal to the complement of one of the variables OR the complement of the other variable.

Proof: We make a truth table for each side.

A	B	$A \cdot B$	$\overline{A \cdot B}$
0	0	0	1
0	1	0	1
1	0	0	1
1	1	1	0

A	B	$\overline{A}$	$\overline{B}$	$\overline{A} + \overline{B}$
0	0	1	1	1
0	1	1	0	1
1	0	0	1	1
1	1	0	0	0

↑ identical ↑

The identical last columns of each truth table proves DeMorgan's law.

We leave the proof of the other of DeMorgan's laws,

$\overline{A + B} \equiv \overline{A} \cdot \overline{B}$	**237b**

as an exercise.

The Involution Law: Our last law is

$\overline{\overline{A}} \equiv A$	**238**

The complement of a complement of a variable is equal to that variable.

Proof: We verify it with the following truth table:

A	$\bar{A}$	$\bar{\bar{A}}$
0	1	0
1	0	1

↑ (column A) ——— ↑ (column $\bar{\bar{A}}$) identical

Duality: The 21 laws are listed in Table 18-1. The first 12 can be divided into six that apply to the AND operator, and six to the OR. The other laws cannot be separated in that way.

Now compare each law in the left column of Table 18-1 with the one alongside it in the right column. Do you see that each of the pair could be obtained from the other by

1. Interchanging AND and OR operators, and
2. Interchanging 1 and 0?

One such expression is said to be the *dual* of the other. The *principle of duality* in Boolean algebra states that *the dual of any law is also a law.*

TABLE 18-1 Laws of Boolean Algebra

		AND	*OR*
Commutative laws	228	(a) $AB \equiv BA$	(b) $A + B \equiv B + A$
Boundedness laws	229	(a) $A \cdot 0 \equiv 0$	(b) $A + 1 \equiv 1$
Identity laws	230	(a) $A \cdot 1 \equiv A$	(b) $A + 0 \equiv A$
Idempotent laws	231	(a) $AA \equiv A$	(b) $A + A \equiv A$
Complement laws	232	(a) $A \cdot \bar{A} \equiv 0$	(b) $A + \bar{A} \equiv 1$
Associative laws	233	(a) $A(BC) \equiv (AB)C$	(b) $A + (B + C) \equiv (A + B) + C$
Distributive laws	234	(a) $A(B + C) \equiv AB + AC$	(b) $A + BC \equiv (A + B)(A + C)$
	235	(a) $A(\bar{A} + B) \equiv AB$	(b) $A + \bar{A} \cdot B \equiv A + B$
Absorption laws	236	(a) $A(A + B) \equiv A$	(b) $A + (AB) \equiv A$
DeMorgan's laws	237	(a) $\overline{A \cdot B} \equiv \bar{A} + \bar{B}$	(b) $\overline{A + B} \equiv \bar{A} \cdot \bar{B}$
Involution law	238	$\bar{\bar{A}} \equiv A$	

Example: Write the dual of the Boolean expression

$$(1 + BC)D + 0$$

Solution: We interchange AND and OR operators

$$\{[1 \cdot (B + C)] + D\} \cdot 0$$

and interchange 1 and 0:

$$\{[0 \cdot (B + C)] + D\} \cdot 1$$

This expression is the dual of the original.

EXERCISE 2

Verify each law. Use a truth table, a Venn diagram, or any Boolean law having a number lower than the one you are proving.

1. The associative law for the AND operator,

$$A(BC) \equiv (AB)\,C \qquad (233a)$$

2. The associative law for the OR operator,

$$A + (B + C) \equiv (A + B) + C \qquad (233b)$$

3. The absorption law,

$$A + AB \equiv A \qquad (236b)$$

4. DeMorgan's law,

$$\overline{A + B} \equiv \overline{A} \cdot \overline{B} \qquad (237b)$$

18-3. SIMPLIFYING BOOLEAN EXPRESSIONS

In Secs. 18-4 and 18-5 we will write Boolean expressions to represent electrical hardware. Thus if we can replace a Boolean expression with one that is exactly equivalent but *simpler,* we can replace the hardware with a simpler version that does the same thing.

Rearrange Terms: Laws 228 allow us to rearrange the variables in an AND or an OR expression.

Example:

$$A + B + C + B + C + A = A + A + B + B + C + C$$

This rearrangement allows us to simplify further, as we see in the next example.

Identical Variables: Look for identical variables. By laws 231, one of the variables can be dropped.

Example: Simplify

$$A + A + B + B + C + C$$

Solution: By law 231b,

$$A + A + B + B + C + C = A + B + C$$

Example: Simplify

$$A + B \cdot B + C + A$$

Solution: By law 228b,

$$A + B \cdot B + C + A = A + A + B \cdot B + C$$

and by laws 231,

$$= A + B + C$$

Expressions Containing 1 or 0: Expressions containing 1 or 0 can be simplified using laws 229 and 230.

Example: Simplify

$$A + 1 + B \cdot 0 + C + 0 + D \cdot 1$$

Solution: By laws 229 and 230,

$$A + 1 + B \cdot 0 + C + 0 + D \cdot 1 = 1 + 0 + C + D$$

Regrouping,

$$= (C + 1) + (D + 0)$$

By laws 229b and 230b,

$$= 1 + D$$

and by law 229b,

$$= 1$$

Removing Parentheses: When possible, remove parentheses by using the distributive law (234a).

Example: Simplify the expression

$$A(\overline{A} + 1)$$

Solution: By law 234a,

$$A(\overline{A} + 1) = A\overline{A} + A \cdot 1$$

By laws 230a and 232a,

$$= 0 + A$$

and by law 230b,

$$= A$$

A Variable and Its Complement: When an expression contains a variable and its complement, it can often be simplified using laws 232.

Example: Simplify

$$A + B\overline{B} + C + \overline{A}$$

Solution: By law 228b,

$$A + B\overline{B} + C + \overline{A} = (A + \overline{A}) + B\overline{B} + C$$

By laws 232, $= 1 + 0 + C$

By law 230b, $= 1 + C$

and by law 229b, $= 1$

Factoring: As in ordinary algebra, factoring will sometimes help us simplify an expression.

Example: Simplify $ABC + ABCD$.

Solution: By law 230a,

$$ABC + ABCD = ABC \cdot 1 + ABCD$$

By law 234a, $= ABC(1 + D)$

By law 229b, $= ABC(1)$

and by law 229a, $= ABC$

Notice that the variable D is not present in the final expression. Thus, if D represented a switch in a circuit, we see here that it is *not needed*.

Example: Simplify $AB\overline{C}D + ABCD$.

Solution: Factoring using law 234a,

$$AB\overline{C}D + ABCD = ABD(\overline{C} + C)$$

By law 232b, $= ABD(1)$

and by law 230a, $= ABD$

As in the preceding example, one variable has vanished.

DeMorgan's Laws: When we have the complement of two variables joined by an AND operator, such as

$$(A \cdot B)' \qquad \text{also written} \qquad \overline{A \cdot B}$$

DeMorgan's laws enable us to *separate* the variables, provided that we change the operator to OR.

$$\overline{A \cdot B} = \overline{A} + \overline{B}$$

If we use the overbar to represent a complement, we can make use of the rhyme *Break the line, change the sign.*

Break the line,

$$\overline{A \cdot B} \qquad = \qquad \overline{A} + \overline{B}$$

change the sign.

The same rule holds if we start with the complement of A OR B. When we break the overbar, we change the sign from + to (·).

Example: Simplify

$$A(\overline{B + \overline{CD}})$$

Solution: We first break the line between C and D,

$$A(\overline{B + \overline{CD}}) = A(\overline{B + \overline{C} + \overline{D}})$$

Then between B and C, $= A(\overline{B}\,\overline{\overline{C} + \overline{D}})$

and between $\overline{C}$ and $\overline{D}$, $= A(\overline{B}\,\overline{\overline{C}}\,\overline{\overline{D}})$

Then by law 238, $= A\overline{B}CD$

EXERCISE 3

Expand each expression and simplify.

1. $(A + \overline{B})(\overline{A} + \overline{B})$
2. $(A + BC)(A + A\overline{B})$
3. $(A + B)(\overline{A} + B)$
4. $(A + \overline{B})(\overline{A} + \overline{B})(A + B)(\overline{A} + B)$

Factor each expression and simplify.

5. $XYZ + XZ$
6. $\overline{A}B + AB$
7. $X\overline{Z} + XYZ + XZ$
8. $XY + X\overline{Y} + \overline{X}Y + \overline{X}\,\overline{Y}$

Simplify each expression.

9. $(\overline{X} + Y)(X + Y)$
10. $\overline{X}Y\overline{Y}\overline{X}$
11. $XY + XYZ + X\overline{Y}$
12. $\overline{X} + \overline{Y} + \overline{Z} + XYZ$
13. $X + Z + X + Y \cdot Y$
14. $X + Y \cdot 0 + Z + 0 + W \cdot 1 + 1$
15. $(\overline{X} + 1)X$
16. $\overline{X} + Z + X + Y\overline{Y}$
17. $WXYZ + WYZ$
18. $WXY\overline{Z} + WXYZ$
19. $(\overline{Y + \overline{WZ}})\,X$
20. $\overline{XY + X\overline{Y} + \overline{X}\,\overline{Y}}$
21. $(\overline{\overline{X}\,Y\overline{Z}})(\overline{X\overline{Y}\,\overline{Z}})$
22. $XYZ + X\overline{Y}\,Z + XYZ\overline{Z}$

18-4. SWITCHING CIRCUITS

A Switch as a Boolean Variable: As we did in Sec. 18-1, we represent the closed position of a switch by 1 and the open position by 0. As before, we designate a switch by a capital letter, X, Y, and so on. If

George Boole's book on the algebra of propositions was published in 1854, before the invention of electricity. Eighty-four years later, Claude E. Shannon wrote a paper showing how Boole's algebra could be applied to relay and switching circuits. Boolean algebra is now used more in this application than for the analysis of propositions or the algebra of sets.

The symbols now in current use for AND and OR, as well as the meanings of 0 and 1, are the reverse of those used by Shannon.

two switches operate so that they open and close simultaneously, we will give them the same letter. If they operate so that one is always open when the other is closed, and vice versa, we label one switch as the complement of the other, say, X and $\overline{X}$.

The AND, OR, and NOT Operators: As in Sec. 18-1, two switches X and Y in *series* are denoted by XY, and in parallel are denoted by $X + Y$ (Fig. 18-13a and b). In Fig. 18-13c, the switch position $\overline{X}$ is always opposite to that of X, and hence represents the NOT operator.

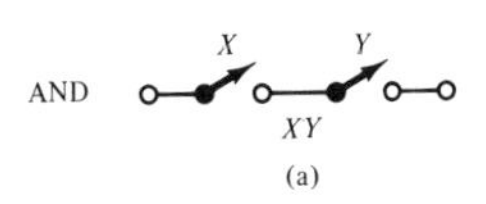

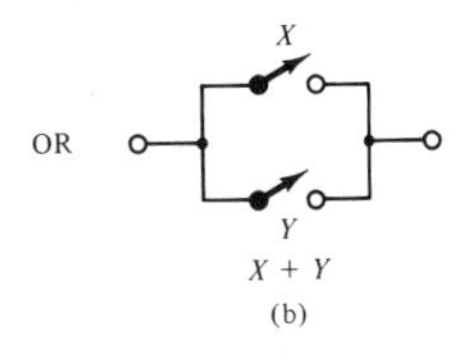

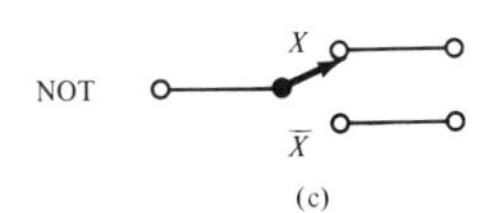

FIGURE 18-13. Switch representation of the AND, OR, and NOT operators.

Representing a Switch Circuit by a Boolean Expression: Now that we can represent a pair of switches by a Boolean expression, we go on to represent circuits with many switches in combination by more complex Boolean expressions.

Example: Write a Boolean expression for the circuit of Fig. 18-14a.

Solution: Note that switch C appears twice. This means that they work in unison, and that they are always in the same state. Also, switches C and $\overline{C}$ work in unison, but are always in opposite states.

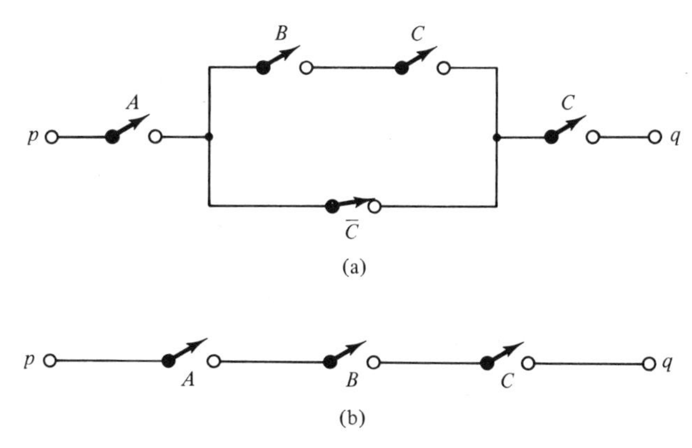

FIGURE 18-14

We first combine switches B and C in the upper branch. Since they are in series, we write $B \cdot C$ or simply BC. This pair is in parallel with switch $\overline{C}$ in the lower branch so we write

$$BC + \overline{C}$$

This entire combination is in series with switches A and C, so we write

$$A(BC + \overline{C})C$$

or

$$AC(BC + \overline{C})$$

Simplifying Circuits with Boolean Algebra: Once a circuit has been represented by a Boolean expression, simplification of the expression may lead to simplification of the circuit.

Example: Simplify the circuit of Fig. 18-14a.

Solution: Starting with the Boolean expression from the preceding problem, we use law 234a to get

$$AC(BC + \overline{C}) = A(BCC + C\overline{C})$$

But by laws 231a and 232a,

$$CC = C \qquad \text{and} \qquad C\overline{C} = 0$$

so

$$\begin{aligned} A(BCC + C\overline{C}) &= A(BC + 0) \\ &= A(BC) \\ &= ABC \end{aligned}$$

But ABC is the Boolean expression for three switches in series (Fig. 18-14b). This simpler circuit will perform identically to the one in Fig. 18-14a.

Drawing a Circuit from a Boolean Expression: Earlier we wrote a Boolean expression for a given switch circuit, and now we do the opposite. Starting with a Boolean expression, we draw a corresponding switch circuit. Any variables connected by the AND operator are placed in series, and those connected by the OR operator are placed in parallel.

Example: Draw the circuit represented by the expression

$$X(\overline{Y} + Z) + \overline{X}\,YZ$$

Solution: The first term has switch $\overline{Y}$ in parallel with Z, and both of these in series with X (Fig. 18-15a). The second term, $\overline{X}\,YZ$, represents three switches $\overline{X}$, Y, and Z in series (Fig. 18-15b). Then both these groups are connected in parallel with each other (Fig. 18-15c).

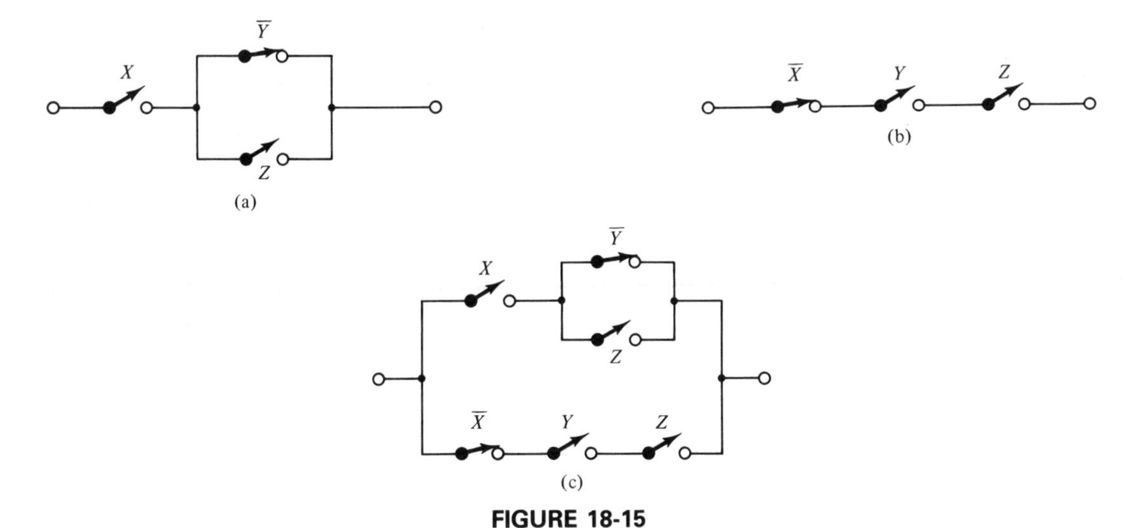

FIGURE 18-15

Designing a Switch Circuit to Perform a Given Function:

1. Make a truth table for the circuit.
2. Write a Boolean expression from the truth table.
3. Simplify the expression.
4. Draw a circuit to match the expression.

Example: A stairway (Fig. 18-16) is to have a light which can be turned on or off by the switch at the head or at the foot of the stairs. Design the circuit.

FIGURE 18-16. A lamp which can be turned on or off by either switch.

Solution:

1. *Make a truth table:* Let us start with both switches in position 0, with the light off. Then if we put switch B into position 1 the light should come on.

Switch A	*Switch B*	*Light*
0	0	Off
0	1	On

Putting switch A into position 1 should now turn the light off.

Switch A	*Switch B*	*Light*
0	0	Off
0	1	On
1	1	Off

Then putting switch B into position 0 should turn the light back on.

Switch A	*Switch B*	*Light*
0	0	Off
0	1	On
1	1	Off
1	0	On

We will see in Sec. 18-5 that this is the truth table for the exclusive OR.

If we represent "light on" by 1 and "light off" by 0, we get the truth table,

A	B	*Light*
0	0	0
0	1	1
1	1	0
1	0	1

2. *Write the Boolean expression:* If we now represent the 0 positions of the switches by $\overline{A}$ and $\overline{B}$, and their 1 positions by A and B, the second row of the truth table shows that the switch combination $\overline{A}B$ will turn the light on. Further, the fourth row shows that the combination $A\overline{B}$ will also turn the light on. Since either combination will work, we connect them with the OR operator. Thus the truth table for the expression

$$\overline{A}B + A\overline{B}$$

will be identical to the one for our switch problem.
3. *Simplify:* The expression is already as simple as possible.
4. *Draw the circuit:* Our Boolean expression calls for switches $\overline{A}$ and B in series, and A and $\overline{B}$ in series, with each of these pairs in parallel (Fig. 18-17a). Then switches $\overline{A}$ and A can be combined into a single switch, and B and $\overline{B}$ into a single switch (Fig. 18-17b).

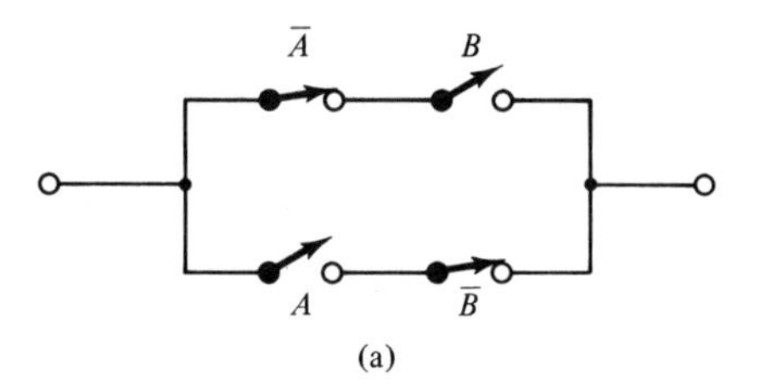

(a)

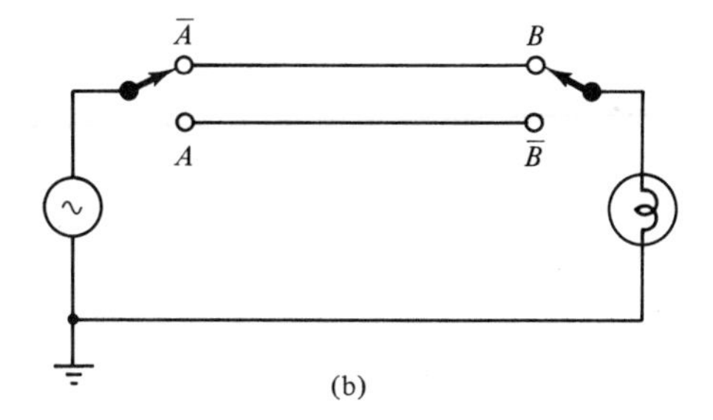

(b)

FIGURE 18-17. Stairway light circuit.

EXERCISE 4

Write a Boolean expression for each circuit.

1. Figure 18-18a.
2. Figure 18-18b.
3. Figure 18-18c.
4. Figure 18-18d.

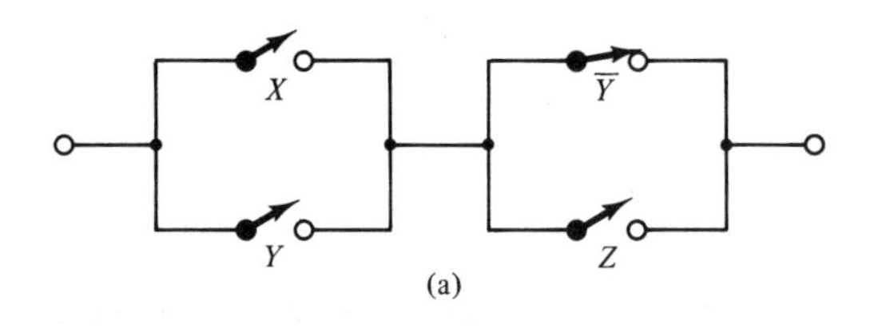

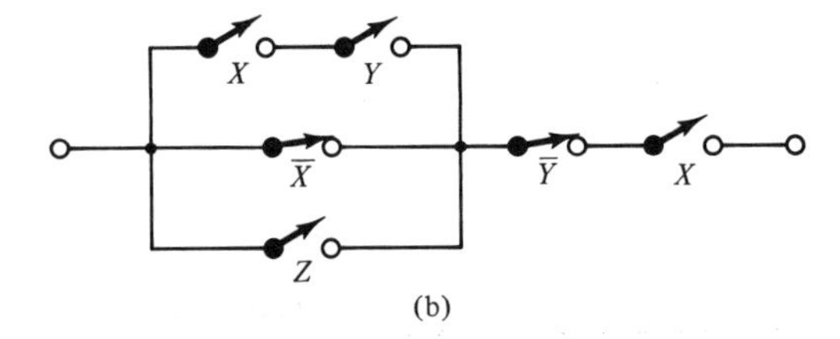

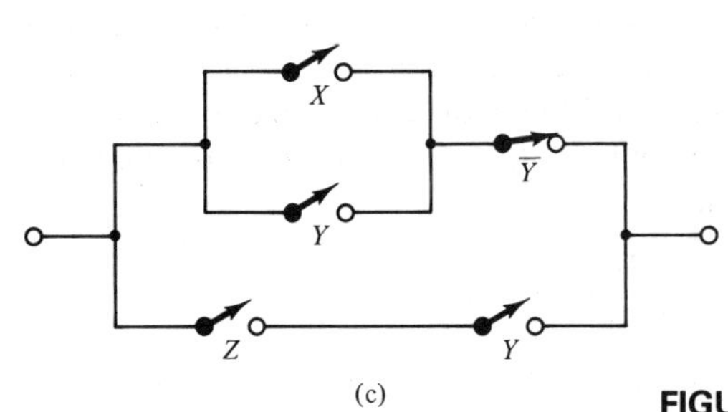

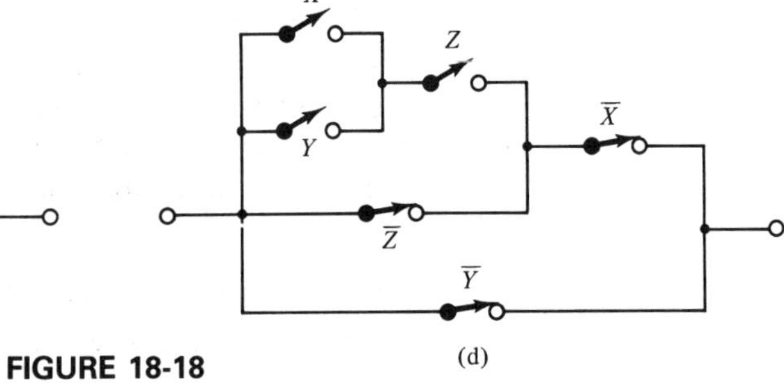

FIGURE 18-18

Simplify each circuit.

5. Figure 18-19a.
6. Figure 18-19b.
7. Figure 18-19c.
8. Figure 18-19d.

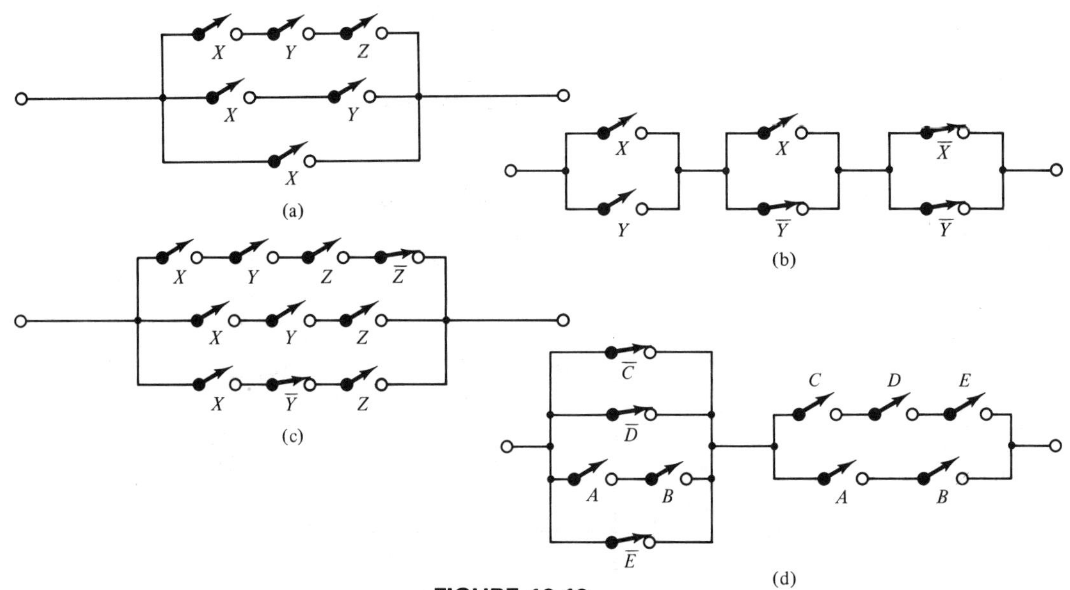

FIGURE 18-19

Without simplifying, draw the circuit represented by each expression.

9. $X(Y + Z) + XYZ$ **10.** $XY(ZW + AB) + AC$

11. $X + Y\overline{Z} + X\overline{Z}$ **12.** $(X + \overline{Y})(\overline{X} + Y)Z$

Draw a truth table for each Boolean expression.

13. $XY + XZ$ **14.** $X(\overline{Y} + \overline{Z})$

Draw a truth table for each circuit.

15. Figure 18-18a. **16.** Figure 18-18b.

17. Figure 18-18c. **18.** Figure 18-18d.

18-5. LOGIC GATES

In Chapter 2 we said that computers perform arithmetic using binary rather than decimal numbers, and then we did binary addition, subtraction, and so forth. But how does a computer actually add or subtract binary numbers? We will see how in this section. First we describe the operation of four types of devices called *logic gates*.

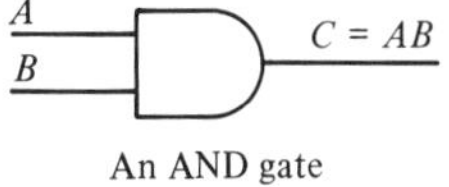

An AND gate

FIGURE 18-20. An AND gate.

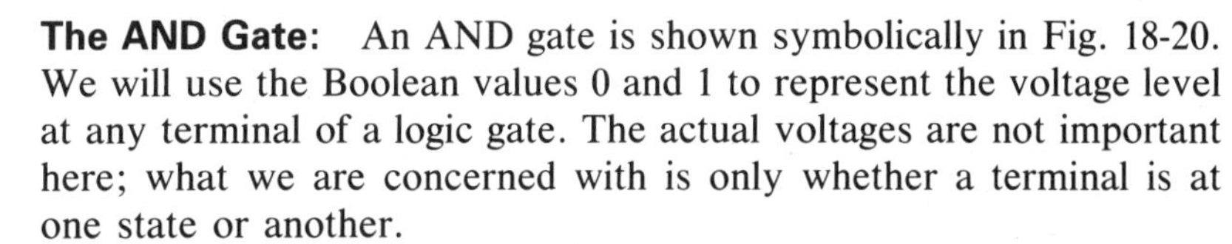

The AND Gate: An AND gate is shown symbolically in Fig. 18-20. We will use the Boolean values 0 and 1 to represent the voltage level at any terminal of a logic gate. The actual voltages are not important here; what we are concerned with is only whether a terminal is at one state or another.

An AND gate can have any number of input terminals. The output C is 1 only when *all* the inputs are 1.

The actual voltages for currently available devices is 0 V for a logical 0, and 5 V for a logical 1.

Actual AND gates may be constructed using relays, diodes, transistors, and so forth. We will not be concerned here with the actual circuitry used for the various gates, only with the logic of their operation.

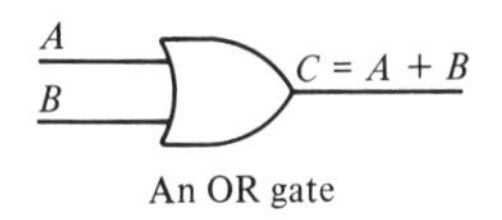

An OR gate

FIGURE 18-21. An OR gate.

The OR Gate: An OR gate (Fig. 18-21) is one whose output is 1 if either of the inputs is 1. As with the AND gate, an OR gate can have any number of inputs. The output is 1 if *any one* of the inputs is 1.

Exclusive OR Gate: The output of the exclusive OR gate (Fig. 18-22) is 1 when either input is 1, as for the ordinary OR gate. This gate differs, however, in that *the output is 0 when* both *inputs are 1,* as in the following truth table.

Truth Table for the exclusive OR

A	B	$A \oplus B$
0	0	0
0	1	1
1	0	1
1	1	0

227

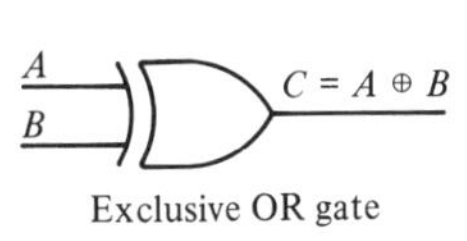

Exclusive OR gate

FIGURE 18-22. An exclusive OR gate.

Note the symbol $\oplus$ for the exclusive OR.

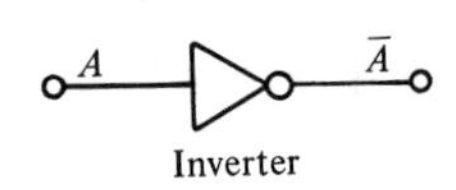

FIGURE 18-23. An inverter.

The NOT Gate, or Inverter: The NOT gate, also called an *inverter,* is usually represented by a triangle with a circle at its vertex (Fig. 18-23). The output of an inverter always has the opposite state as the input. Inversion is often shown by placing a circle at a terminal of another logic gate, as in Fig. 18-24.

Converting a Boolean Expression to a Logic Circuit: Suppose that we have a Boolean expression which describes some operation we wish to carry out, and we need to design the circuit to do it.

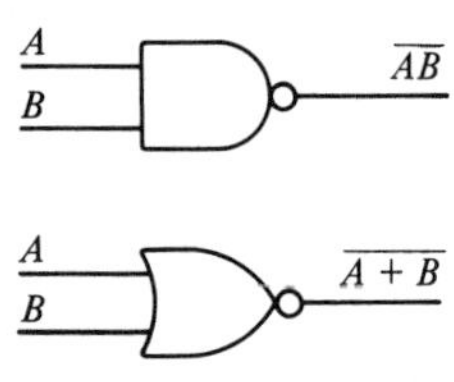

FIGURE 18-24. An AND gate with an inverted output is called a NAND gate, and an OR gate with an inverted output is called a NOR gate. NAND and NOR logic is currently used more than AND and OR logic.

Example: Design the circuit to implement the Boolean expression

$$\overline{A}B + \overline{CD}$$

Solution: We start at the output and work backward. Our expression has two terms joined by an OR operator, so we draw an OR gate with inputs $\overline{A}B$ and $\overline{CD}$ (Fig. 18-25a). We next show $\overline{A}B$ as the output of an AND gate with inputs $\overline{A}$ and B, and $\overline{CD}$ as the inverted output of an AND gate with inputs C and D (Fig. 18-25b). We then use an inverter to convert A to $\overline{A}$ (Fig. 18-25c). Finally, we draw the circuit more simply by showing each inversion as a circle at the input or output of the other devices (Fig. 18-26).

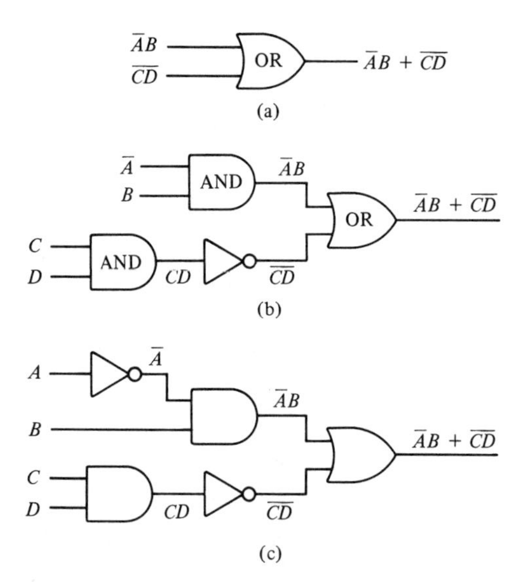

FIGURE 18-25. Designing a logic circuit to implement a given Boolean expression.

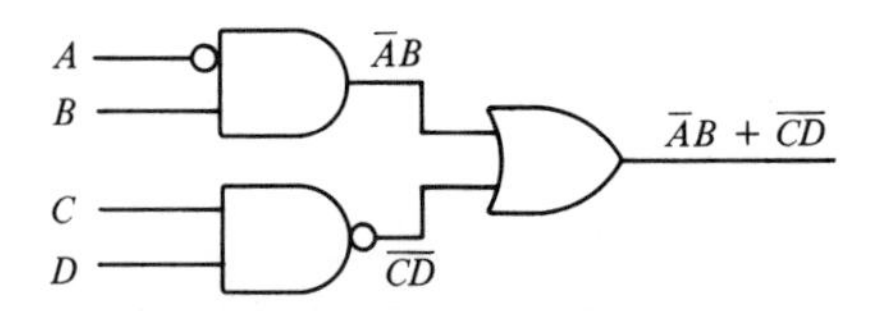

FIGURE 18-26

Writing a Boolean Expression for a Given Logic Circuit: Suppose now that we have a logic circuit and wish to write a Boolean expression for the output. We proceed as in the following example.

Example: Write a Boolean expression for the output of the logic circuit in Fig. 18-27.

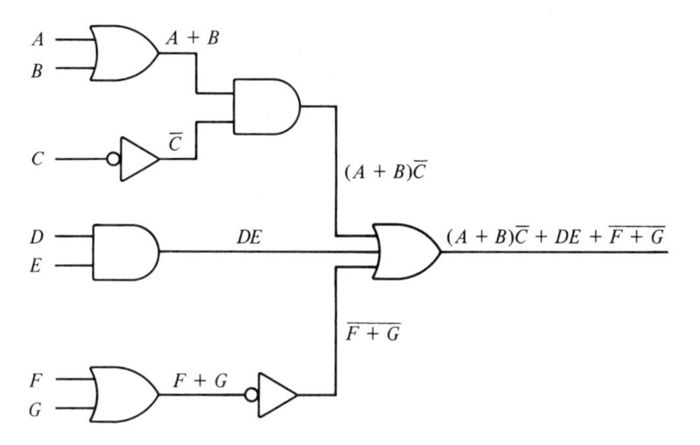

FIGURE 18-27. Finding the Boolean expression implemented by a given logic circuit.

Solution: We start by labeling each input. Then we proceed from left to right, putting the output of each gate right on the circuit diagram until the output of the last gate is reached. In this example, the output is

$$(A + B)\overline{C} + DE + \overline{F + G}$$

Binary Addition: There are two general methods for binary addition, *serial* and *parallel*.

Serial addition requires just one input wire for each number, with the digits being represented by a pulse train, usually with the least significant bit first.

Parallel addition requires one input wire for each binary digit. Thus to add two 8-bit binary numbers would require 16 input wires. Each input would be high (1) or low (0) depending on whether the digit is 1 or 0. We will show only parallel addition here.

This is a good time to review binary arithmetic in Chapter 2, before continuing.

Parallel Addition of Two Binary Digits: Binary addition of two bits is shown in Chapter 2. When both bits are 0 the sum is zero, when one bit is 1 the sum is 1, and when both bits are 1, the sum is 0 with a carry of 1. The circuit used to accomplish this addition is called a *half-adder* (Fig. 18-28). The two input lines carry the two bits to be added, and the two outputs are the sum and the carry.

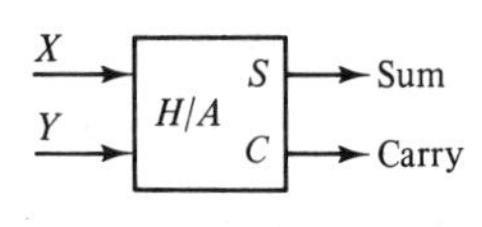

FIGURE 18-28. A half-adder.

Notice now that the carry is 1 only when both X and Y are 1. Thus the carry can be obtained by using an AND gate.

Further, the sum is 1 if either X OR Y is 1, but not when both are 1. This is exactly the function of the exclusive OR gate. Thus the half-adder can be made by combining two basic logic gates, the AND gate for the carry and the exclusive OR gate for the sum, as in Fig. 18-29.

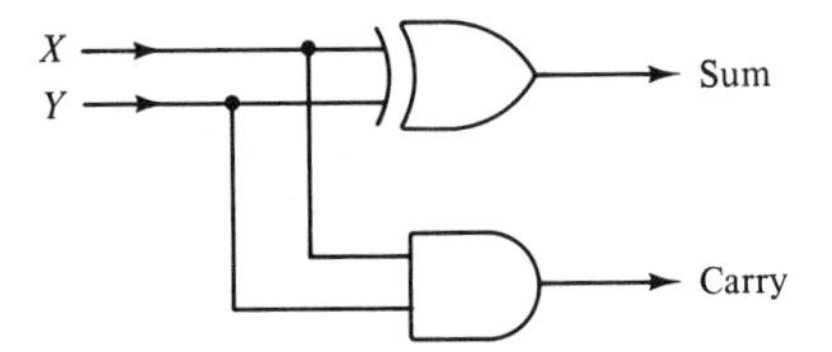

FIGURE 18-29. A half-adder made by combining an AND gate and an exclusive OR gate.

Addition of Three Binary Digits: A circuit called a *full adder* (Fig. 18-30) is used to add three 1-bit binary numbers. The three inputs are usually the bits X and Y from two numbers to be added, and a carry from a preceding addition. The outputs are a sum and a carry, as with the half-adder.

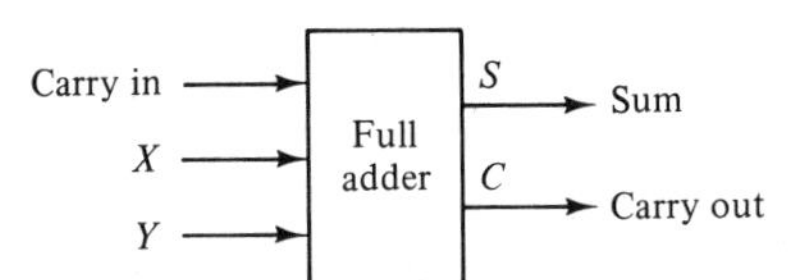

FIGURE 18-30. A full adder.

We want sum and carry = 0 when all inputs are 0, sum = 1 and carry = 0 when any one input is 1, sum = 0 and carry = 1 when any two inputs are 1, and sum = 1 and carry = 1 when all three inputs are 1.

Carry-in	*X*	*Y*	*Sum*	*Carry-out*
0	0	0	0	0
1	0	0	1	0
0	1	0	1	0
0	0	1	1	0
1	1	0	0	1
1	0	1	0	1
0	1	1	0	1
1	1	1	1	1

To add three numbers, we can add any two of them, then add their sum to the third. We do that here, using a half-adder to add X and Y, and another half-adder to add the carry-in to the sum of X and Y, as in Fig. 18-31.

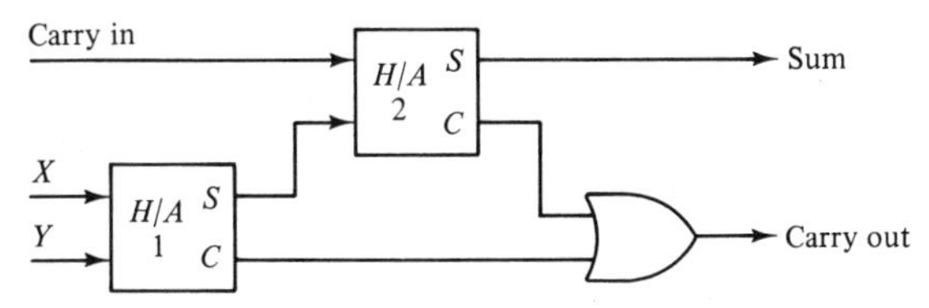

FIGURE 18-31. A full adder made from two half-adders and an OR gate.

When either half-adder has a carry, the OR gate passes it along to the carry-out terminal. Note that *both* half-adders cannot have a carry, for if half-adder 1 had a carry it must have had a sum of 0, thus making it impossible for half-adder 2 to have a carry.

Addition of Binary Numbers of Any Length: The next step is to add two binary numbers X and Y, each of which has many bits. Let us assume numbers having a length of four bits.

Let us indicate the bit position of each number with a subscript. Thus, X_1 is the least significant bit in the number X and Y_4 is the most significant bit in the number Y. Our addition problem would then look like this:

$$\begin{array}{rrrrr} & X_4 & X_3 & X_2 & X_1 \\ + & Y_4 & Y_3 & Y_2 & Y_1 \\ \hline S_5 & S_4 & S_3 & S_2 & S_1 \end{array}$$

where each S stands for a sum.

Figure 18-32 shows a circuit to add two 4-bit binary numbers X

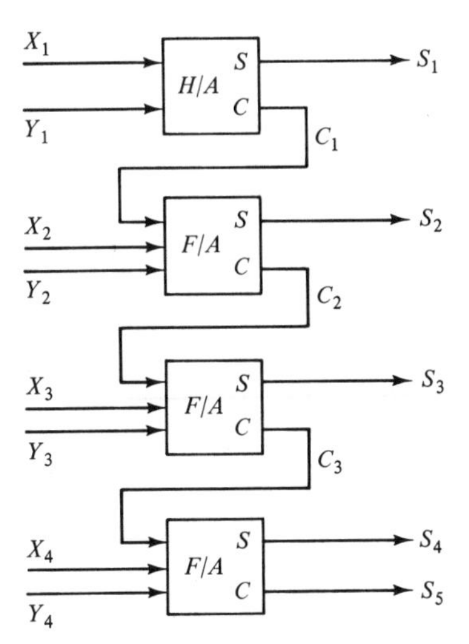

FIGURE 18-32. A circuit for adding two 4-bit binary numbers.

and Y. The least significant bits X_1 and Y_1 are added with a half-adder. The next higher bits X_2 and Y_2 and the carry C_1 from the first addition are combined with a full adder, and so on, the carry from each addition becoming an input for the next addition.

EXERCISE 5

Without simplifying, design the circuit to implement each Boolean expression

1. $A + BC$

2. $(X + Y)Z$

3. $(A + B\overline{C}) + \overline{A}$

4. $X\overline{Y} + \overline{X}Y$

5. $X + XY + X\overline{Y}$

6. $X\overline{Y}Z + \overline{Y}\overline{Z} + X\overline{Y}Z$

Simplify each expression and draw a logic circuit that will implement it.

7. $XY + X\overline{Y} + \overline{X}Y + \overline{X}\,\overline{Y}$

8. $(XY + Z)(W + XY)$

9. $\overline{X}\,\overline{Y}Z + X\overline{Y}Z$

10. $XYZ + \overline{X}YZ + X\overline{Y}Z$

Write a Boolean expression for the output of each logic circuit.

11. Figure 18-33a.

12. Figure 18-33b.

13. Figure 18-33c.

14. Figure 18-33d.

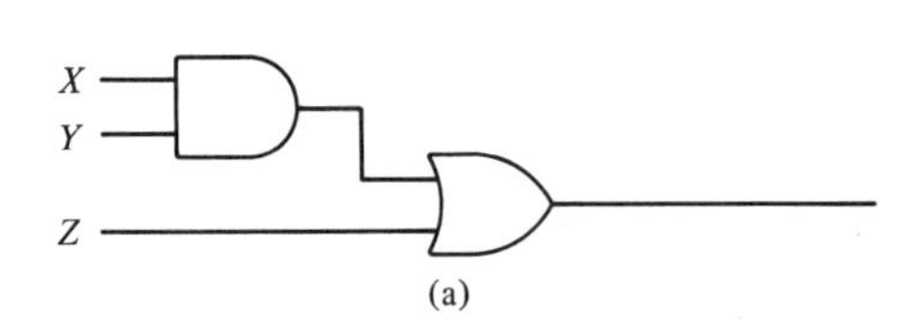

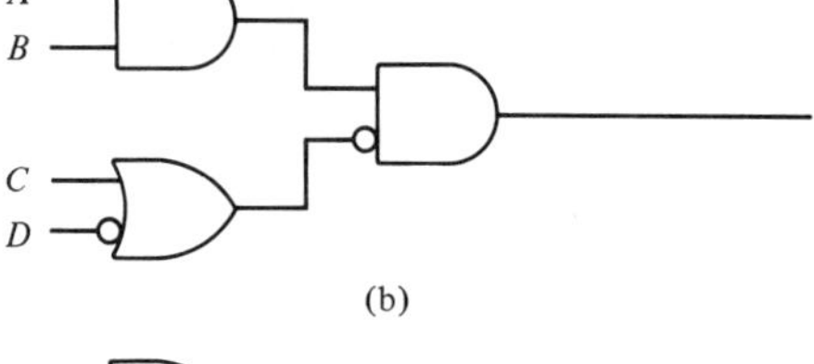

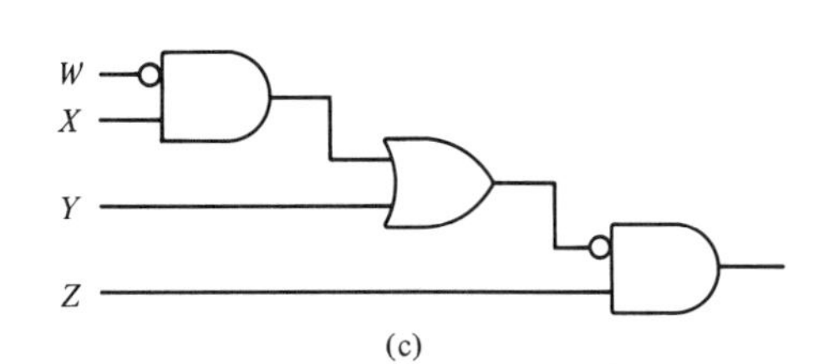

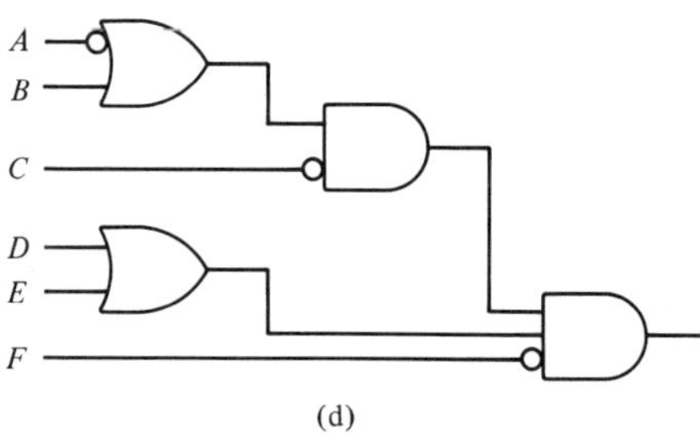

FIGURE 18-33

CHAPTER TEST

1. If $X = 1$, $Y = 0$, and $Z = 1$, evaluate $(X \cdot Y) \cdot (Y \cdot Z) \cdot (X \cdot Z)$

2. Simplify: $X + Z + X + Y \cdot Y$

3. Draw the switch circuit represented by the expression $\overline{A}BC + A(B + \overline{C})$

4. Simplify: $XYZ + WXYZ$

5. Write a Boolean expression for the circuit of Fig. 18-34.

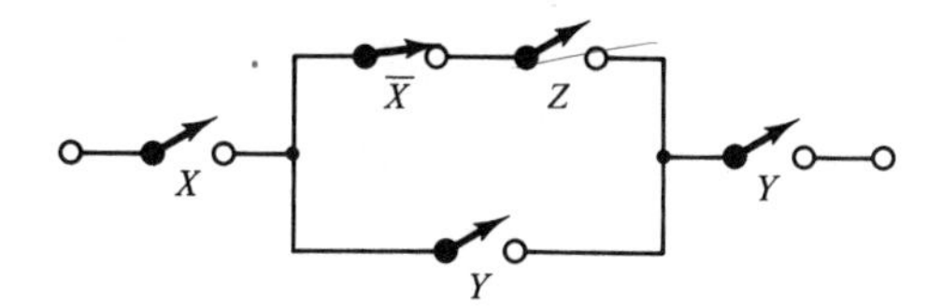

FIGURE 18-34

6. Simplify: $XY\overline{Z}W + WXYZ$
7. If $A = 1$, $B = 1$, and $C = 0$, evaluate $\overline{(A + C) + AB}$.
8. Simplify: $(Y + \overline{Z}\overline{W})X$
9. Simplify the circuit of Fig. 18-35.

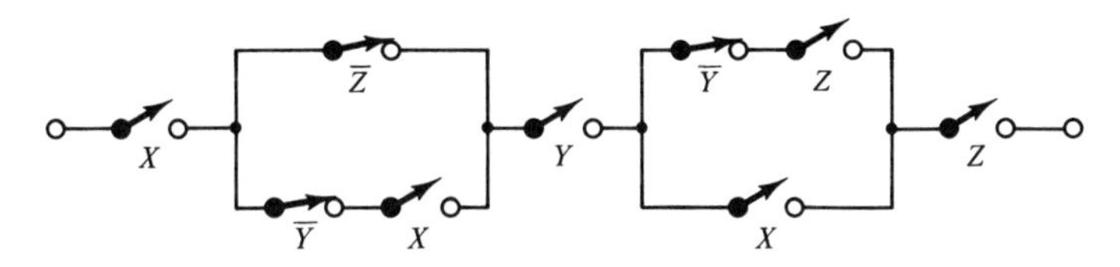

FIGURE 18-35

10. Simplify: $Y \cdot 0 + Z + 0 + W \cdot 1 + 1 + X$
11. Design the switch circuit to implement the Boolean expression $\overline{WZ} + X\overline{Y}$.
12. Design a logic circuit to implement the expression in Problem 11.
13. Simplify the expression: $\overline{X}\,(1 + X)$
14. Write a Boolean expression for the output of the logic circuit in Fig. 18-36.
15. Simplify: $\overline{Z} + X + Y\overline{Y} + Z + \overline{X}$
16. If $X = Y = 0$, and $Z = W = 1$, evaluate

$$(\overline{X} + \overline{Y}) \cdot (\overline{Z} + W)$$

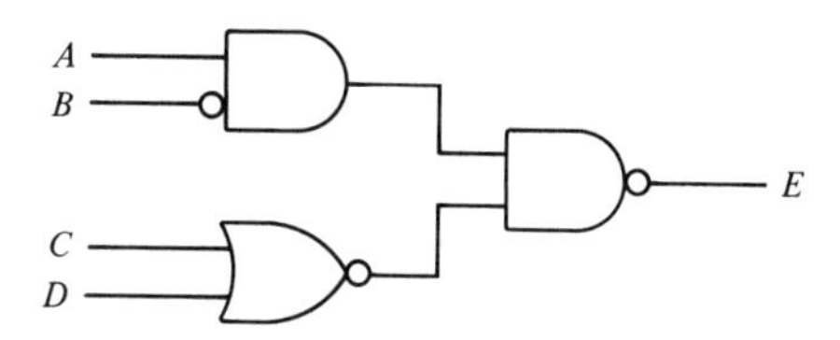

FIGURE 18-36

Appendices

Summary of Facts and Formulas

					Page
ALGEBRAIC LAWS	1	Commutative Law	Addition	$a + b = b + a$	9
	2		Multiplication	$ab = ba$	12, 117
	3	Associative Law	Addition	$a + (b + c) = (a + b) + c = (a + c) + b$	9
	4		Multiplication	$a(bc) = (ab)c = (ac)b = abc$	13
	5	Distributive Law		$a(b + c) = ab + ac$	13, 118, 193
RULES OF SIGNS	6	Addition and Subtraction		$a + (-b) = a - (+b) = a - b$	8
	7			$a + (+b) = a - (-b) = a + b$	8
	8	Multiplication		$(+a)(+b) = (-a)(-b) = +ab$	13, 117
	9			$(+a)(-b) = (-a)(+b) = -(+a)(+b) = -ab$	13, 117
	10	Division		$\frac{+a}{+b} = \frac{-a}{-b} = -\frac{-a}{+b} = -\frac{+a}{-b} = \frac{a}{b}$	17, 126, 206
	11			$\frac{+a}{-b} = \frac{-a}{+b} = -\frac{-a}{-b} = -\frac{a}{b}$	17, 126, 206
PERCENTAGE	12	Amount = base × rate $A = BP$			40
	13	Percent change = $\frac{\text{new value} - \text{original value}}{\text{original value}} \times 100\%$			44
	14	Percent error = $\frac{\text{relative error}}{\text{true value}} \times 100\%$			97
	15	Percent concentration of ingredient A = $\frac{\text{amount of } A}{\text{amount of mixture}} \times 100\%$			148
	16	Percent efficiency = $\frac{\text{output}}{\text{input}} \times 100\%$			45

					Page
BINARY NUMBERS	17	Largest n-Bit Binary Number		$2^n - 1$	62
	18	Addition		$0 + 0 = 0$ $0 + 1 = 1$ $1 + 0 = 1$ $1 + 1 = 0$ with carry to next column	68
	19	Subtraction		$0 - 0 = 0$ $0 - 1 = 1$ with borrow from next column $1 - 0 = 1$ $1 - 1 = 0$ $10 - 1 = 1$	71
	20			Subtracting B from A is equivalent to adding the two's complement of B to A	80
	21	Multiplication		$0 \times 0 = 0$ $0 \times 1 = 0$ $1 \times 0 = 0$ $1 \times 1 = 1$	69
	22	Division		$0 \div 0$ is not defined $1 \div 0$ is not defined $0 \div 1 = 0$ $1 \div 1 = 1$	73
	23	Complements	n-Bit One's Complement of x	$(2^n - 1) - x$	76
	24			The one's complement of a number can be written by changing the 1's to 0's and the 0's to 1's	77
	25		n-Bit Two's Complement of x	$2^n - x$	77
	26			To find the two's complement of a number first write the one's complement, and add 1	77
	27	Negative Binary Numbers		If M is a positive binary number, then $-M$ is the two's complement of M	79

	No.				Page
EXPONENTS	28	Definition		$a^n = \underbrace{a \cdot a \cdot a \cdot \ldots \cdot a}_{n \text{ factors}}$	111
	29	Laws of Exponents	Products	$x^a \cdot x^b = x^{a+b}$	112
	30		Quotients	$\frac{x^a}{x^b} = x^{a-b}$	112, 126
	31		Powers	$(x^a)^b = x^{ab} = (x^b)^a$	113
	32		Product Raised to a Power	$(xy)^n = x^n \cdot y^n$	113
	33		Quotient Raised to a Power	$\left(\frac{x}{y}\right)^n = \frac{x^n}{y^n}$	114
	34		Zero Exponent	$x^0 = 1$	114
	35		Negative Exponent	$x^{-a} = \frac{1}{x^a}$	115
	36	Fractional Exponents		$a^{1/n} = \sqrt[n]{a}$	22, 308
	37			$a^{m/n} = \sqrt[n]{a^m} = (\sqrt[n]{a})^m$	308
RADICALS	38	Rules of Radicals	Root of a Product	$\sqrt[n]{ab} = \sqrt[n]{a}\,\sqrt[n]{b}$	309
	39		Root of a Quotient	$\sqrt[n]{\frac{a}{b}} = \frac{\sqrt[n]{a}}{\sqrt[n]{b}}$	309
	40		Root of a Power	$\sqrt[n]{a^m} = (\sqrt[n]{a})^m$	310
SPECIAL PRODUCTS AND FACTORING	41	Binomials	Difference of Two Squares	$a^2 - b^2 = (a - b)(a + b)$	121, 196
	42		Sum of Two Cubes	$a^3 + b^3 = (a + b)(a^2 - ab + b^2)$	
	43		Difference of Two Cubes	$a^3 - b^3 = (a - b)(a^2 + ab + b^2)$	
	44	Trinomials	Test for Factorability	$ax^2 + bx + c$ is factorable if $b^2 - 4ac$ is a perfect square	198
	45		Leading Coefficient = 1	$x^2 + (a + b)x + ab = (x + a)(x + b)$	121, 199
	46		General Quadratic Trinomial	$acx^2 + (ad + bc)x + bd = (ax + b)(cx + d)$	121, 201
	47		Perfect Square Trinomials	$a^2 + 2ab + b^2 = (a + b)^2$	121, 203
	48			$a^2 - 2ab + b^2 = (a - b)^2$	122, 203
	49	Factoring by Grouping		$ac + ad + bc + bd = (a + b)(c + d)$	

					Page
FRACTIONS	50	Simplifying		$\frac{ad}{bd} = \frac{a}{b}$	207
	51	Multiplication		$\frac{a}{b} \cdot \frac{c}{d} = \frac{ac}{bd}$	210
	52	Division		$\frac{a}{b} \div \frac{c}{d} = \frac{a}{b} \cdot \frac{d}{c} = \frac{ad}{bc}$	211
	53	Addition and Subtraction	Same Denominators	$\frac{a}{b} \pm \frac{c}{b} = \frac{a \pm c}{b}$	213
	54		Different Denominators	$\frac{a}{b} \pm \frac{c}{d} = \frac{ad}{bd} \pm \frac{bc}{bd} = \frac{ad \pm bc}{bd}$	215
PROPORTION	55	In the Proportion $a:b = c:d$	The product of the means equals the product of the extremes	$ad = bc$	
	56		The extremes may be interchanged	$d:b = c:a$	
	57		The means may be interchanged	$a:c = b:d$	
	58		The means may be interchanged with the extremes	$b:a = d:c$	
	59	Mean Proportional	In the Proportion	$a:b = b:c$	
			Geometric Mean	$b = \pm\sqrt{ac}$	
VARIATION	60	k = Constant of Proportionality	Direct	$y \propto x$ or $y = kx$	
	61		Inverse	$y \propto \frac{1}{x}$ or $y = \frac{k}{x}$	
	62		Joint	$y \propto xw$ or $y = kxw$	

Topic	No.				Formula	Page
SYSTEMS OF LINEAR EQUATIONS	63	Algebraic solution			$a_1x + b_1y = c_1$ $a_2x + b_2y = c_2$ $x = \dfrac{b_2c_1 - b_1c_2}{a_1b_2 - a_2b_1}$, and $y = \dfrac{a_1c_2 - a_2c_1}{a_1b_2 - a_2b_1}$ where $a_1b_2 - a_2b_1 \neq 0$	246, 263, 297, 302
SYSTEMS OF LINEAR EQUATIONS	64	Algebraic solution			$a_1x + b_1y + c_1z = k_1$ $a_2x + b_2y + c_2z = k_2$ $a_3x + b_3y + c_3z = k_3$ $x = \dfrac{b_2c_3k_1 + b_1c_2k_3 + b_3c_1k_2 - b_2c_1k_3 - b_3c_2k_1 - b_1c_3k_2}{a_1b_2c_3 + a_3b_1c_2 + a_2b_3c_1 - a_3b_2c_1 - a_1b_3c_2 - a_2b_1c_3}$ $y = \dfrac{a_1c_3k_2 + a_3c_2k_1 + a_2c_1k_3 - a_3c_1k_2 - a_1c_2k_3 - a_2c_3k_1}{a_1b_2c_3 + a_3b_1c_2 + a_2b_3c_1 - a_3b_2c_1 - a_1b_3c_2 - a_2b_1c_3}$ $z = \dfrac{a_1b_2k_3 + a_3b_1k_2 + a_2b_3k_1 - a_3b_2k_1 - a_1b_3k_2 - a_2b_1k_3}{a_1b_2c_3 + a_3b_1c_2 + a_2b_3c_1 - a_3b_2c_1 - a_1b_3c_2 - a_2b_1c_3}$	267
SYSTEMS OF LINEAR EQUATIONS	65	Determinants	Value of a Determinant	Second Order	$\begin{vmatrix} a_1 & b_1 \\ a_2 & b_2 \end{vmatrix} = a_1b_2 - a_2b_1$	262
SYSTEMS OF LINEAR EQUATIONS	66	Determinants	Value of a Determinant	Third Order	$\begin{vmatrix} a_1 & b_1 & c_1 \\ a_2 & b_2 & c_2 \\ a_3 & b_3 & c_3 \end{vmatrix} = a_1b_2c_3 + a_3b_1c_2 + a_2b_3c_1 - (a_3b_2c_1 + a_1b_3c_2 + a_2b_1c_3)$	267
SYSTEMS OF LINEAR EQUATIONS	67	Determinants	Value of a Determinant	Minors	The signed minor of element b in the determinant $\begin{vmatrix} a & b & c \\ d & e & f \\ g & h & i \end{vmatrix}$ is $-\begin{vmatrix} d & f \\ g & i \end{vmatrix}$	265
SYSTEMS OF LINEAR EQUATIONS	68	Determinants	Value of a Determinant	Minors	To find the value of a determinant: 1. Choose any row or any column to develop by minors. 2. Write the product of every element in that row or column and its signed minor. 3. Add these products to get the value of the determinant.	266
SYSTEMS OF LINEAR EQUATIONS	69	Determinants	Cramer's Rule		The solution for any variable is a fraction whose denominator is the determinant of the coefficients, and whose numerator is also the determinant of the coefficients, except that the column of coefficients of the variable being solved for is replaced by the column of constants	268
SYSTEMS OF LINEAR EQUATIONS	70	Determinants	Cramer's Rule	Two Equations	$x = \dfrac{\begin{vmatrix} c_1 & b_1 \\ c_2 & b_2 \end{vmatrix}}{\begin{vmatrix} a_1 & b_1 \\ a_2 & b_2 \end{vmatrix}}$ and $y = \dfrac{\begin{vmatrix} a_1 & c_1 \\ a_2 & c_2 \end{vmatrix}}{\begin{vmatrix} a_1 & b_1 \\ a_2 & b_2 \end{vmatrix}}$	263
SYSTEMS OF LINEAR EQUATIONS	71	Determinants	Cramer's Rule	Three Equations	$x = \dfrac{\begin{vmatrix} k_1 & b_1 & c_1 \\ k_2 & b_2 & c_2 \\ k_3 & b_3 & c_3 \end{vmatrix}}{\Delta}$, $y = \dfrac{\begin{vmatrix} a_1 & k_1 & c_1 \\ a_2 & k_2 & c_2 \\ a_3 & k_3 & c_3 \end{vmatrix}}{\Delta}$, $z = \dfrac{\begin{vmatrix} a_1 & b_1 & k_1 \\ a_2 & b_2 & k_2 \\ a_3 & b_3 & k_3 \end{vmatrix}}{\Delta}$ Where $\Delta = \begin{vmatrix} a_1 & b_1 & c_1 \\ a_2 & b_2 & c_2 \\ a_3 & b_3 & c_3 \end{vmatrix} \neq 0$	268

SYSTEMS OF LINEAR EQUATIONS (Continued)					Page
	72	Properties of Determinants	Zero Row or Column	If all elements in a row (or column) are zero, the value of the determinant is zero	272
	73		Identical Rows or Columns	The value of a determinant is zero if two rows (or columns) are identical	272
	74		Zeros below the Principal Diagonal	If all elements below the principal diagonal are zeros, then the value of the determinant is the product of the elements along the principal diagonal	273
	75		Interchanging Rows with Columns	The value of a determinant is unchanged if we change the rows to columns and the columns to rows	273
	76		Interchange of Rows or Columns	A determinant will change sign when we interchange two rows (or columns)	273
	77		Multiplying by a Constant	If each element in a row (or column) is multiplied by some constant, the value of the determinant is multiplied by that constant	274
	78		Multiples of One Row or Column Added to Another	The value of a determinant is unchanged when the elements of a row (or column) are multiplied by some factor, and then added to the corresponding elements of another row or column	275

	No.		Topic	Formula	Page
MATRICES	79	Addition	Commutative Law	$\mathbf{A} + \mathbf{B} = \mathbf{B} + \mathbf{A}$	281
	80		Associative Law	$\mathbf{A} + (\mathbf{B} + \mathbf{C}) = (\mathbf{A} + \mathbf{B}) + \mathbf{C} = (\mathbf{A} + \mathbf{C}) + \mathbf{B}$	281
	81		Addition and Subtraction	$\begin{pmatrix} a & b \\ c & d \end{pmatrix} + \begin{pmatrix} w & x \\ y & z \end{pmatrix} = \begin{pmatrix} a+w & b+x \\ c+y & d+z \end{pmatrix}$	281
	82	Multiplication	Conformable Matrices	The product **AB** of two matrices **A** and **B** is defined only when the number of columns in **A** equals the number of rows in **B**	281
	83		Commutative Law	$\mathbf{AB} \neq \mathbf{BA}$ Matrix multiplication is not commutative	286
	84		Associative Law	$\mathbf{A}(\mathbf{BC}) = (\mathbf{AB})\mathbf{C} = \mathbf{ABC}$	282
	85		Distributive Law	$\mathbf{A}(\mathbf{B} + \mathbf{C}) = \mathbf{AB} + \mathbf{AC}$	282
	86		Dimensions of the Product	$(m \times p)(p \times n) = (m \times n)$	282
	87		Product of a Scalar and a Matrix	$k \begin{pmatrix} a & b \\ c & d \end{pmatrix} = \begin{pmatrix} ka & kb \\ kc & kd \end{pmatrix}$	281
	88		Scalar Product of a Row Vector and a Column Vector	$(1 \times 2)\,(2 \times 1) \quad (1 \times 1)$ $(a \quad b) \begin{pmatrix} x \\ y \end{pmatrix} = (ax + by)$	282
	89		Tensor Product of a Column Vector and a Row Vector	$(2 \times 1)\,(1 \times 2) \quad (2 \times 2)$ $\begin{pmatrix} a \\ b \end{pmatrix} (x \quad y) = \begin{pmatrix} ax & ay \\ bx & by \end{pmatrix}$	288
	90		Product of a Row Vector and a Matrix	$(1 \times 2) \quad (2 \times 3) \quad (1 \times 3)$ $(a \quad b) \begin{pmatrix} u & v & w \\ x & y & z \end{pmatrix} = (au + bx \quad av + by \quad aw + bz)$	283

					Page
MATRICES (Continued)	91	Matrix Multiplication (Continued)	Product of a Matrix and a Column Vector	$\underset{(2 \times 2)}{\begin{pmatrix} a & b \\ c & d \end{pmatrix}} \underset{(2 \times 1)}{\begin{pmatrix} x \\ y \end{pmatrix}} = \underset{(2 \times 1)}{\begin{pmatrix} ax + by \\ cx + dy \end{pmatrix}}$	287
	92		Product of Two Matrices	$\underset{(2 \times 3)}{\begin{pmatrix} a & b & c \\ d & e & f \end{pmatrix}} \underset{(3 \times 2)}{\begin{pmatrix} u & x \\ v & y \\ w & z \end{pmatrix}} = \underset{(2 \times 2)}{\begin{pmatrix} au + bv + cw & ax + by + cz \\ du + ev + fw & dx + ey + fz \end{pmatrix}}$	286
	93		Product of a Matrix and Its Inverse	$\mathbf{AA}^{-1} = \mathbf{A}^{-1}\mathbf{A} = \mathbf{I}$	298
	94		Multiplying by the Unit Matrix	$\mathbf{AI} = \mathbf{IA} = \mathbf{A}$	292
	95	Solving Systems of Equations	Matrix Form for a System of Equation	$\mathbf{AX} = \mathbf{B}$	293
	96		Elementary Transformations of a Matrix	1. Interchange any rows 2. Multiply a row by a nonzero constant 3. Add a constant multiple of one row to another row	293
	97		Gauss Elimination	When we transform the coefficient matrix **A** into the unit matrix **I**, the column vector **B** gets transformed into the solution set	296
	98		Solving a Set of Equations Using the Inverse	$\mathbf{X} = \mathbf{A}^{-1}\mathbf{B}$	301

					Page
QUADRATICS	99	General Form	$ax^2 + bx + c = 0$		324
	100	Quadratic Formula	$x = \dfrac{-b \pm \sqrt{b^2 - 4ac}}{2a}$		332
	101	Nature of the Roots	If a, b, and c are real, and	if $b^2 - 4ac > 0$ the roots are real and unequal if $b^2 - 4ac = 0$ the roots are real and equal if $b^2 - 4ac < 0$ the roots are not real	334
	102	Polynomial of Degree n	$a_0x^n + a_1x^{n-1} + \cdots + a_{n-1}x + a_n$		346
	103	Factor Theorem	If a polynomial equation $f(x) = 0$ has a root r, then $(x - r)$ is a factor of the polynomial $f(x)$; conversely, if $(x - r)$ is a factor of a polynomial $f(x)$, then r is a root of $f(x) = 0$		347

	No.	Figure	Description	Formula	Page
INTERSECTING LINES	104	Opposite angles of two intersecting straight lines are equal			
	105	If two parallel straight lines are cut by a transversal, corresponding angles are equal and alternate interior angles are equal			
	106	If two lines are cut by a number of parallels, the corresponding segments are proportional			
QUADRILATERALS	107	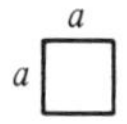 a, a	Square	Area = a^2	49
	108	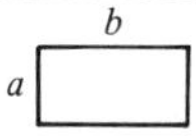 a, b	Rectangle	Area = ab	49
	109	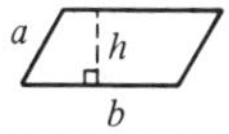 a, h, b	Parallelogram: Diagonals bisect each other	Area = bh	49
	110	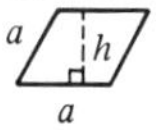 a, h, a	Rhombus: Diagonals intersect at right angles	Area = ah	49
	111	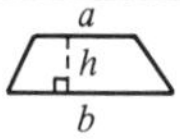 a, h, b	Trapezoid	Area = $\frac{(a+b)h}{2}$	49
POLYGON	112	 n sides	Sum of Angles = $(n-2)\ 180°$		
CIRCLES	113	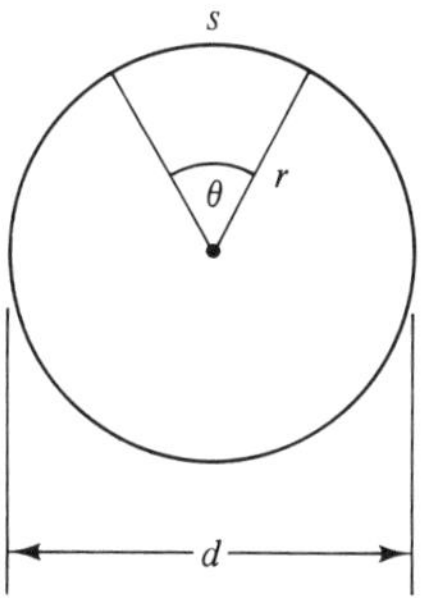 s, θ, r, d	Circumference = $2\pi r = \pi d$		49
	114		Area = $\pi r^2 = \frac{\pi d^2}{4}$		28, 49
	115		Central angle θ (radians) = $\frac{s}{r}$		424
	116		Area of sector = $\frac{rs}{2} = \frac{r^2\theta}{2}$		425
	117		1 revolution = 2π radians = 360°		356
	118		Any angle inscribed in a semicircle is a right angle		
	119	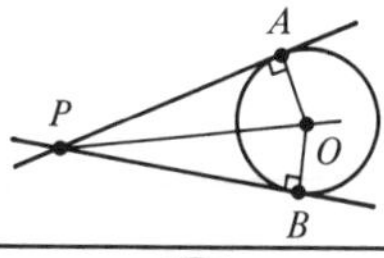 A, P, O, B	Tangents to a Circle	Tangent AP is perpendicular to radius OA	
	120			Tangent AP = tangent BP OP bisects angle APB	
	121	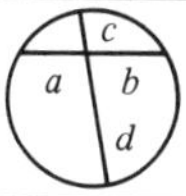 c, a, b, d	Intersecting Chords	$ab = cd$	

	No.	Figure	Formula	Page
SOLIDS	122	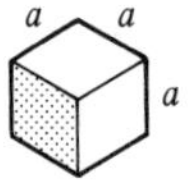 Cube	Volume = a^3	51
	123		Surface area = $6a^2$	51
	124	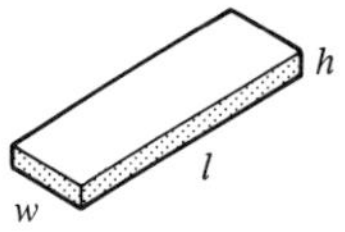 Rectangular Parallel-epiped	Volume = lwh	51
	125		Surface area = $2(lw + hw + lh)$	51
	126	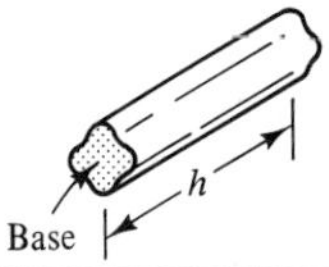 Any Cylinder or Prism	Volume = (area of base)(altitude)	51
	127	Right Cylinder or Prism	Lateral area (not incl. bases) = (perimeter of base)(altitude)	51
	128	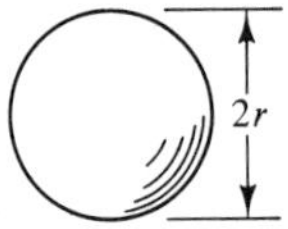 Sphere	Volume = $\frac{4}{3}\pi r^3$	51
	129		Surface area = $4\pi r^2$	51
	130	 Any Cone or Pyramid	Volume = $\frac{1}{3}$(area of base)(altitude)	51
	131	Right Circular Cone or Regular Pyramid	Lateral area = $\frac{1}{2}$(perimeter of base) $\times$ (slant height)	51
	132	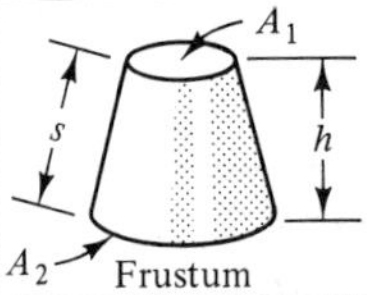 Frustum: Any Cone or Pyramid	Volume = $\frac{h}{3}(A_1 + A_2 + \sqrt{A_1A_2})$	51
	133	Frustum: Right Circular Cone or Regular Pyramid	Lateral area = $\frac{s}{2}$(sum of base perimeters) = $\frac{s}{2}(P_1 + P_2)$	51
SIMILAR FIGURES	134	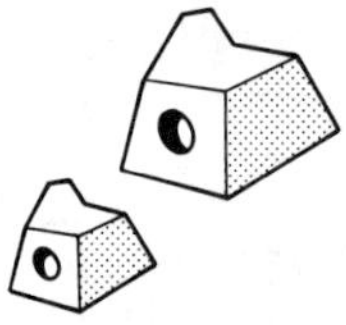	Corresponding dimensions of plane or solid similar figures are in proportion	
	135		Areas of similar plane or solid figures are proportional to the squares of any two corresponding dimensions	
	136		Volumes of similar solid figures are proportional to the cubes of any two corresponding dimensions	

Section	No.	Figure	Item	Formula		Page
ANY TRIANGLE	137	Areas		Area = $\frac{1}{2}bh$		47
	138		Hero's Formula:	Area = $\sqrt{s(s-a)(s-b)(s-c)}$ where $s = \frac{1}{2}(a+b+c)$		47
	139		Sum of the Angles	$A + B + C = 180°$		48
	140		Law of Sines	$\frac{a}{\sin A} = \frac{b}{\sin B} = \frac{c}{\sin C}$		398
	141		Law of Cosines	$a^2 = b^2 + c^2 - 2bc \cos A$ $b^2 = a^2 + c^2 - 2ac \cos B$ $c^2 = a^2 + b^2 - 2ab \cos C$.		403
	142		Exterior Angle	$\theta = A + B$		
SIMILAR TRIANGLES	143	If two angles of a triangle equal two angles of another triangle, the triangles are similar				
	144	Corresponding sides of similar triangles are in proportion				
RIGHT TRIANGLES	145		Pythagorean Theorem	$a^2 + b^2 = c^2$		48, 367
	146	Trigonometric Ratios	Sine	$\sin\theta = \frac{y}{r} = \frac{\text{opposite side}}{\text{hypotenuse}}$		359, 367
	147		Cosine	$\cos\theta = \frac{x}{r} = \frac{\text{adjacent side}}{\text{hypotenuse}}$		359, 367
	148		Tangent	$\tan\theta = \frac{y}{x} = \frac{\text{opposite side}}{\text{adjacent side}}$		359, 367
	149		Cotangent	$\cot\theta = \frac{x}{y} = \frac{\text{adjacent side}}{\text{opposite side}}$		359
	150		Secant	$\sec\theta = \frac{r}{x} = \frac{\text{hypotenuse}}{\text{adjacent side}}$		359
	151		Cosecant	$\csc\theta = \frac{r}{y} = \frac{\text{hypotenuse}}{\text{opposite side}}$		359
	152	Reciprocal Relationships	(a) $\sin\theta = \frac{1}{\csc\theta}$	(b) $\cos\theta = \frac{1}{\sec\theta}$	(c) $\tan\theta = \frac{1}{\cot\theta}$	360, 452
	153		In a right triangle, the altitude to the hypotenuse forms two right triangles which are similar to each other and to the original triangle			
	154	A and B Are Complementary Angles	Cofunctions	(a) $\sin A = \cos B$ (b) $\cos A = \sin B$ (c) $\tan A = \cot B$	(d) $\cot A = \tan B$ (e) $\sec A = \csc B$ (f) $\csc A = \sec B$	370
CONGRUENT TRIANGLES	155	Two Triangles Are Congruent If	Two angles and a side of one are equal to two angles and the corresponding side of the other (ASA), (AAS)			
	156		Two sides and the included angle of one are equal, respectively to two sides and the included angle of the other (SAS)			
	157		Three sides of one are equal to the three sides of the other (SSS)			

				Page
COORDINATE SYSTEMS	158	Rectangular	$x = r \cos \theta$	450
	159		$y = r \sin \theta$	450
	160	Polar	$r = \sqrt{x^2 + y^2}$	450
	161		$\theta = \arctan \dfrac{y}{x}$	450
TRIGONOMETRIC IDENTITIES	162	Quotient Relations	$\tan \theta = \dfrac{\sin \alpha}{\cos \alpha}$	452
	163		$\cot \alpha = \dfrac{\cos \alpha}{\sin \alpha}$	452
	164	Pythagorean Relations	$\sin^2 \alpha + \cos^2 \alpha = 1$	453
	165		$1 + \tan^2 \alpha - \sec^2 \alpha$	453
	166		$1 + \cot^2 \alpha = \csc^2 \alpha$	453
	167	Sum or Difference of Two Angles	$\sin (\alpha \pm \beta) = \sin \alpha \cos \beta \pm \cos \alpha \sin \beta$	
	168		$\cos (\alpha \pm \beta) = \cos \alpha \cos \beta \mp \sin \alpha \sin \beta$	
	169		$\tan (\alpha \pm \beta) = \dfrac{\tan \alpha \pm \tan \beta}{1 \mp \tan \alpha \tan \beta}$	
	170	Double-Angle Relations	$\sin 2\alpha = 2 \sin \alpha \cos \alpha$	
	171		(a) $\cos 2\alpha = \cos^2 \alpha - \sin^2 \alpha$ (b) $\cos 2\alpha = 1 - 2 \sin^2 \alpha$ (c) $\cos 2\alpha = 2 \cos^2 \alpha - 1$	
	172		$\tan 2\alpha = \dfrac{2 \tan \alpha}{1 - \tan^2 \alpha}$	
	173	Half-Angle Relations	$\sin \dfrac{\alpha}{2} = \pm \sqrt{\dfrac{1 - \cos \alpha}{2}}$	
	174		$\cos \dfrac{\alpha}{2} = \mp \sqrt{\dfrac{1 + \cos \alpha}{2}}$	
	175		(a) $\tan \dfrac{\alpha}{2} = \dfrac{1 - \cos \alpha}{\sin \alpha}$ (b) $\tan \dfrac{\alpha}{2} = \dfrac{\sin \alpha}{1 + \cos \alpha}$ (c) $\tan \dfrac{\alpha}{2} = \pm \sqrt{\dfrac{1 - \cos \alpha}{1 + \cos \alpha}}$	
	176	Inverse Trigonometric Functions	$\text{arcsine } \theta = \arctan \dfrac{\theta}{\sqrt{1 - \theta^2}}$	365
	177		$\arccos \theta = \arctan \dfrac{\sqrt{1 - \theta^2}}{\theta}$	365

					Page
LOGARITHMS	178	Exponential to Logarithmic Form		If $b^x = y$ then $x = \log_b y$	474
	179	Laws of Logarithms	Products	$\log_b MN = \log_b M + \log_b N$	478
	180		Quotients	$\log_b \dfrac{M}{N} = \log_b M - \log_b N$	479
	181		Powers	$\log_b M^p = p \log_b M$	480
	182		Roots	$\log_b \sqrt[q]{M} = \dfrac{1}{q} \log_b M$	481
	183		Log of 1	$\log_b 1 = 0$	482
	184		Log of the Base	$\log_b b = 1$	482
	185	Log of the Base Raised to a Power		$\log_b b^n = n$	482
	186	Change of Base		$\log N = \dfrac{\ln N}{\ln 10} = \dfrac{\ln N}{2.3026}$	487

SOME USEFUL FUNCTIONS

No.	Name	Graph	Formula	Page
187	Power Function	y, a, $n = \infty$, $n = 1$, $n = 0$, 0, 1, x	$y = ax^n$	499
188	Exponential Function	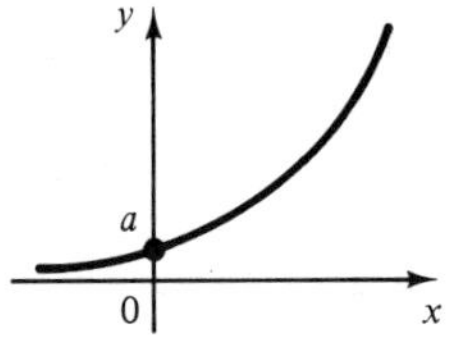 y, a, 0, x	$y = a(b)^{nx}$	500
189	Series Approximation		$b^x = 1 + x \ln b + \frac{(x \ln b)^2}{2!} + \frac{(x \ln b)^3}{3!} + \cdots \quad (b > 0)$	489
190	Exponential Growth	 y, a, t	$y = ae^{nt}$	467, 500
191			Doubling Time or Half-Life: $\frac{\ln 2}{n}$	491, 492
192	Exponential Decay	 y, a, t	$y = ae^{-nt}$	468, 500
193	Exponential Growth to an Upper Limit	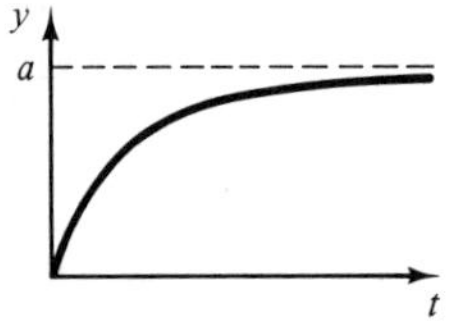 y, a, t	$y = a(1 - e^{-nt})$	468
194	Time Constant		$T = \frac{1}{\text{growth rate } n}$	469
195	Series Approximations		$e = 2 + \frac{1}{2!} + \frac{1}{3!} + \frac{1}{4!} + \cdots$	473
196			$e^x = 1 + x + \frac{x^2}{2!} + \frac{x^3}{3!} + \frac{x^4}{4!} + \cdots$	473

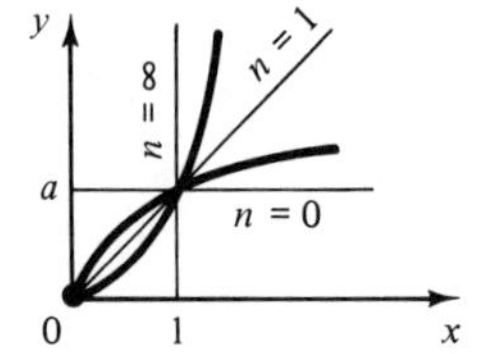

SOME USEFUL FUNCTIONS (Continued)

No.	Function			Page
197	Logarithmic Function	y, x, 1	$y = \log_b x$	475
198	Series Approximation	$\ln x = 2a + \frac{2a^3}{3} + \frac{2a^5}{5} + \frac{2a^7}{7} + \cdots$ where $a = \frac{x-1}{x+1}$		488
199	Sine Wave of Amplitude a	y, $\frac{2\pi}{b}$, a, 0, $-a$, x, $\frac{c}{b}$	$y = a \sin (bx + c)$	432
200			Period $= \frac{360}{b}$ deg/cycle $= \frac{2\pi}{b}$ rad/cycle	435
201			Frequency $= \frac{b}{360}$ cycle/deg $= \frac{b}{2\pi}$ cycle/rad	436
202			Phase Displacement $= \frac{c}{b}$	437
203	Series Approximations	$\sin x = x - \frac{x^3}{3!} + \frac{x^5}{5!} - \frac{x^7}{7!} + \cdots$		366
204		$\cos x = 1 - \frac{x^2}{2!} + \frac{x^4}{4!} - \frac{x^6}{6!} + \cdots$		367

	No.	Topic	Form	Subtopic	Formula	Page
COMPLEX NUMBERS	205	Powers of j			$j = \sqrt{-1}, j^2 = -1, j^3 = -j, j^4 = 1, j^5 = i$, etc.	511
	206		Rectangular Form		$a + jb$	511
	207			Sums	$(a + jb) + (c + jd) = (a + c) + j(b + d)$	511
	208			Differences	$(a + jb) - (c + jd) = (a - c) + j(b - d)$	511
	209			Products	$(a + jb)(c + jd) = (ac - bd) + j(ad + bc)$	512
	210			Quotients	$\dfrac{a + jb}{c + jd} = \dfrac{ac + bd}{c^2 + d^2} + j\,\dfrac{bc - ad}{c^2 + d^2}$	513
	211		Trigonometric Form		$a + jb = r(\cos\theta + j\sin\theta)$	517
	212			where	$a = r\cos\theta$	516
	213				$b = r\sin\theta$	516
	214	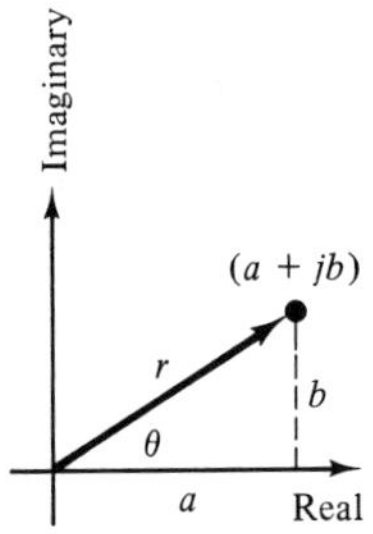			$r = \sqrt{a^2 + b^2}$	516
	215				$\theta = \arctan\dfrac{b}{a}$	517
	216		Polar Form		$r\angle\theta = a + jb$	516
	217			Products	$r\angle\theta \cdot r'\angle\theta' = rr'\angle\theta + \theta'$	518
	218			Quotients	$r\angle\theta \div r'\angle\theta' = \dfrac{r}{r'}\angle\theta - \theta'$	518
	219			Roots and Powers	DeMoivre's Theorem: $(r\angle\theta)^n = r^n\angle n\theta$	519
	220		Exponential Form	Euler's Formula	$re^{j\theta} = r(\cos\theta + j\sin\theta)$	
	221			Products	$r_1e^{j\theta_1} \cdot r_2e^{j\theta_2} = r_1r_2e^{j(\theta_1 + \theta_2)}$	
	222			Quotients	$\dfrac{r_1e^{j\theta_1}}{r_2e^{j\theta_2}} = \dfrac{r_1}{r_2}\,e^{j(\theta_1 - \theta_2)}$	
	223			Powers and Roots	$(re^{j\theta})^n = r^ne^{jn\theta}$	

BOOLEAN ALGEBRA AND SETS		Boolean Operators		Truth Table	Venn Diagram	Switch Diagram	Logic Gate	Page
	224		AND	A B $A \cdot B$ 0 0 0 0 1 0 1 0 0 1 1 1	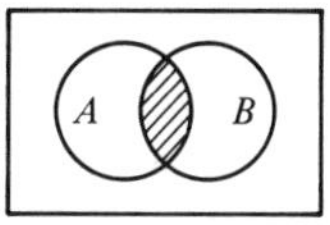 Intersection $A \cap B = \{x : x \in A, x \in B\}$	 In Series	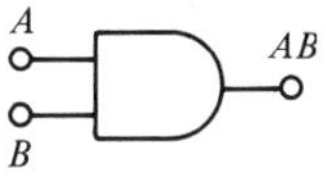 AND Gate	158, 552
	225		OR	A B $A + B$ 0 0 0 0 1 1 1 0 1 1 1 1	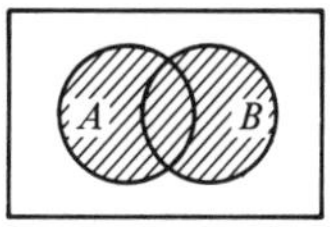 Union $A \cup B = \{x : x \in A \text{ or } x \in B\}$	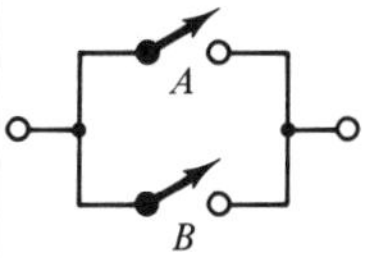 In Parallel	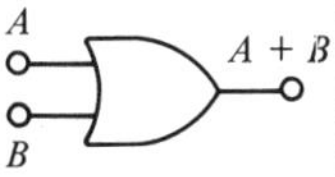 OR Gate	159, 554
	226		NOT	A $\bar{A}$ 0 1 1 0	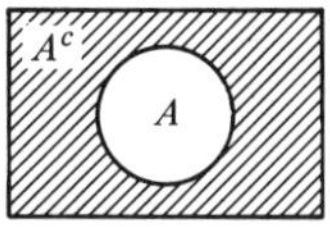 Complement $A^c = \{x : x \in U, x \notin A\}$	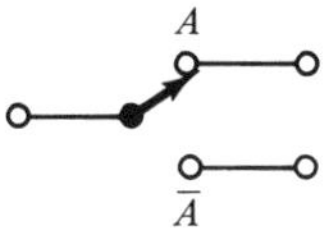	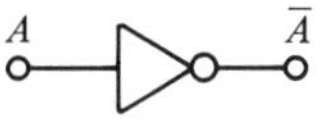 Inverter	159, 554
	227		Exclusive OR	A B $A \oplus B$ 0 0 0 0 1 1 1 0 1 1 1 0	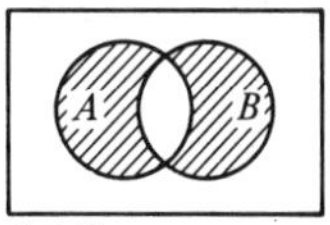 $A \oplus B = \{x : x \in A \text{ or } x \in B, x \notin A \cap B\}$	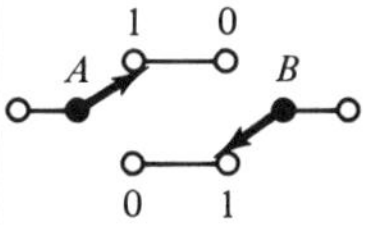	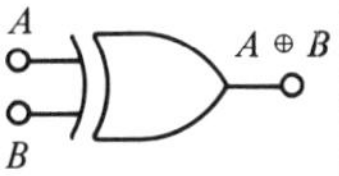 Exclusive OR Gate	574

BOOLEAN ALGEBRA (Continued)	No.	Laws of Boolean Algebra	Law	AND	OF	Page
	228		Cummutative Laws	(a) $AB \equiv BA$	(b) $A + B \equiv B + A$	557, 563
	229		Boundedness Laws	(a) $A \cdot 0 \equiv 0$	(b) $A + 1 \equiv 1$	557, 563
	230		Identity Laws	(a) $A \cdot 1 \equiv A$	(b) $A + 0 \equiv A$	557, 563
	231		Idempotent Laws	(a) $AA \equiv A$	(b) $A + A \equiv A$	558, 563
	232		Complement Laws	(a) $A \cdot \overline{A} \equiv 0$	(b) $A + \overline{A} \equiv 1$	559, 563
	233		Associative Laws	(a) $A(BC) \equiv (AB)C$	(b) $A + (B + C) \equiv (A + B) + C$	559, 563
	234		Distributive Laws	(a) $A(B + C) \equiv AB + AC$	(b) $A + BC \equiv (A + B)(A + C)$	560, 563
	235			(a) $A(\overline{A} + B) \equiv AB$	(b) $A + \overline{A} \cdot B \equiv A + B$	561, 563
	236		Absorption Laws	(a) $A(A + B) \equiv A$	(b) $A + (AB) \equiv A$	562, 563
	237		DeMorgan's Laws	(a) $\overline{A \cdot B} \equiv \overline{A} + \overline{B}$	(b) $\overline{A + B} \equiv \overline{A} \cdot \overline{B}$	562, 563
	238		Involution Law	$\overline{\overline{A}} \equiv A$		562, 563

APPLICATIONS

Category	No.	Application	Formula		Page
MIXTURES	A1	Mixture Containing Ingredients $A, B, C, \ldots$	Total amount of mixture = amount of A + amount of B + · · ·		147
	A2		Final amount of each ingredient = initial amount + amount added − amount removed		147
	A3	Combination of Two Mixtures	Final amount of A = amount of A from mixture 1 + amount of A from mixture 2		148
	A4	Fluid Flow	Amount of flow = flow rate × elapsed time $A = QT$		
WORK	A5	Amount done = rate of work × time worked			223
	A6	d, F	Work = force × distance = Fd		
FINANCIAL	A7	Unit Cost	Unit cost = $\dfrac{\text{total cost}}{\text{number of units}}$		
	A8	Interest: Principal a Invested at Rate n for t years Accumulates to Amount y	Simple	$y = a(1 + nt)$	29
	A9		Compounded Annually	$y = a(1 + n)^t$	29, 465
	A10		Compounded m times/yr	$y = a\left(1 + \dfrac{n}{m}\right)^{mt}$	466

						Page
STATICS	A11	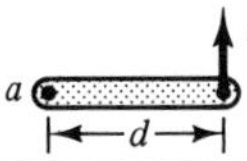		Moment about Point a	$M_a = Fd$	
	A12	Equations of Equilibrium (Newton's First Law)		The sum of all horizontal forces = 0		
	A13			The sum of all vertical forces = 0		
	A14			The sum of all moments about any point = 0		
	A15	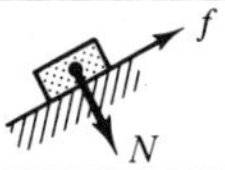		Coefficient of Friction	$\mu = \frac{f}{N}$	
MOTION	A16	Linear Motion	Uniform Motion (Constant Speed)	Distance = rate × time	$D = Rt$	37, 146
	A17		Uniformly Accelerated (Constant Acceleration a, Initial Velocity v_0) For free fall, $a = g = 9.807\ m/s^2 = 32.2\ ft/s^2$	Displacement at Time t	$s = v_0 t + \frac{at^2}{2}$	29, 229
	A18			Velocity at Time t	$v = v_0 + at$	
	A19			Newton's Second Law	$F = ma$	
	A20		Average Speed	Average speed = $\frac{\text{total distance traveled}}{\text{total time elapsed}}$		
	A21	Rotation	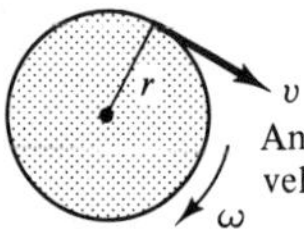 Angular velocity	Linear Speed of Point at Radius r	$v = \omega r$	429
	A22			Angle = (angular velocity)(time)	$\theta = \omega t$	428
MATERIAL PROPERTIES	A23	Density		Density = $\frac{\text{weight}}{\text{volume}}$ or $\frac{\text{mass}}{\text{volume}}$		
	A24	Mass		Mass = $\frac{\text{weight}}{\text{acceleration due to gravity}}$		
	A25	Specific Gravity		SG = $\frac{\text{density of substance}}{\text{density of water}}$		
	A26	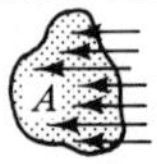 Pressure		Total Force on a Surface	Force = pressure × area	37
	A27	pH		pH = −10 log concentration		498

	No.			Formula	Page
TEMPERATURE	A28	Conversions between Degrees Celsuis (C) and Degrees Fahrenheit (F)		$C = \frac{5}{9}(F - 32)$	
	A29			$F = \frac{9}{5}C + 32$	29
STRENGTH OF MATERIALS	A30	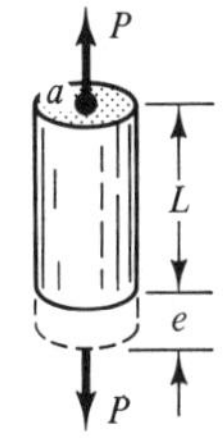 Tension or Compression	Normal Stress	$\sigma = \frac{P}{a}$	37
	A31		Strain	$\epsilon = \frac{e}{L}$	37
	A32		Modulus of Elasticity and Hooke's Law	$E = \frac{PL}{ae}$	28, 29, 37, 229
	A33			$E = \frac{\sigma}{\epsilon}$	
	A34	Thermal Expansion	Elongation	$e = \alpha L \, \Delta t$	
	A35	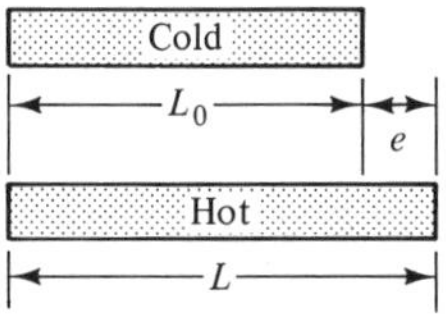	New Length	$L = L_0 (1 + \alpha \Delta t)$	229
	A36		Strain	$\epsilon = \frac{e}{L} = \alpha \Delta t$	
	A37	Temperature change = Δt	Stress, if Restrained	$\sigma = E\epsilon = E\alpha \, \Delta t$	
	A38	Coefficient of thermal expansion = α	Force, if Restrained	$P = a\sigma = aE\alpha \, \Delta t$	
	A39	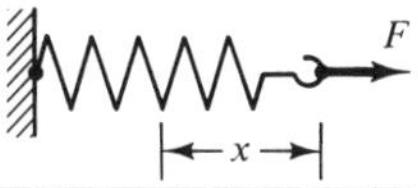	Force needed to Deform a Spring	F = spring constant × distance = kx	
ELECTRICAL TECHNOLOGY	A40	Ohm's Law		Current = $\frac{\text{voltage}}{\text{resistance}}$ $I = \frac{V}{R}$	
	A41	Combinations of Resistors	In Series	$R = R_1 + R_2 + R_3 + \cdots$	37
	A42		In Parallel	$\frac{1}{R} = \frac{1}{R_1} + \frac{1}{R_2} + \frac{1}{R_3} + \cdots$	37, 229
	A43	Power Dissipated in a Resistor		Power = $P = VI$	37
	A44	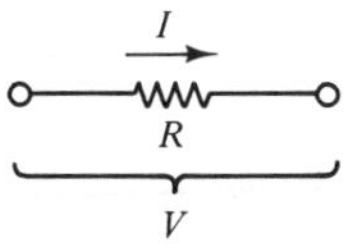		$P = \frac{V^2}{R}$	37
	A45			$P = I^2 R$	
	A46	Kirchhoff's Laws	Loops	The sum of the voltage rises and drops around any closed loop is zero	230, 253
	A47		Nodes	The sum of the currents entering and leaving any node is zero	

ELECTRICAL TECHNOLOGY (Continued)

				Page
A48	Resistance Change with Temperature		$R = R_1[1 + \alpha(t - t_1)]$	30, 230
A49	Resistance of a Wire 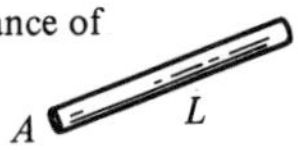		$R = \dfrac{\rho L}{A}$	
A50	Combinations of Capacitors	In Series	$\dfrac{1}{C} = \dfrac{1}{C_1} + \dfrac{1}{C_2} + \dfrac{1}{C_3} + \cdots$	
A51		In Parallel	$C = C_1 + C_2 + C_3 + \cdots$	37
A52	Charge on a Capacitor at Voltage V		$Q = CV$	
A53	Impedance of a Series *RLC* Circuit		$Z = \sqrt{R^2 + \left(\omega L - \dfrac{1}{\omega C}\right)^2}$	320
A54	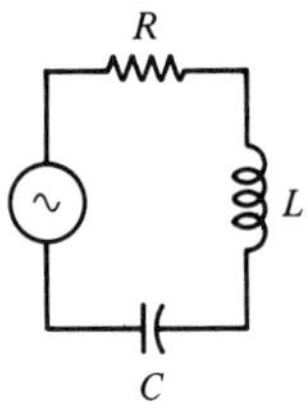	Reactance	$X = X_L - X_C$	380
A55		Magnitude of Impedance	$\lvert Z \rvert = \sqrt{R^2 + X^2}$	380
A56		Phase Angle	$\phi = \arctan \dfrac{X}{R}$	380
A57		Complex Impedance	$\mathbf{Z} = R + jx = Z\angle\phi = Ze^{j\phi}$	525
A58	Alternating Voltage	Sinusoidal Form: $v = V_m \sin(\omega t + \theta_1)$	Complex Form: $\mathbf{V} = V_m \angle\theta_1$	446, 524
A59	Alternating Current	$i = I_m \sin(\omega t + \theta_2)$	$\mathbf{I} = I_m \angle\theta_2$	446, 524
A60	Period		$P = \dfrac{2\pi}{\omega}$ seconds	444
A61	Frequency		$f = \dfrac{1}{P} = \dfrac{\omega}{2\pi}$ hertz	445
A62	Ohm's Law for AC		$\mathbf{V} = \mathbf{ZI}$	526
A63	Decibels Gained or Lost		$G = 10 \log_{10} \dfrac{P_2}{P_1}$ dB	498

Conversion Factors

Unit	Equals	
	Length	
1 angstrom	1×10^{-10}	meter
	1×10^{-4}	micrometer (micron)
1 centimeter	10^{-2}	meter
	0.3937	inch
1 foot	12	inches
	0.3048	meter
1 inch	25.4	millimeters
	2.54	centimeters
1 kilometer	3281	feet
	0.5400	nautical mile
	0.6214	statute mile
	1094	yards
1 light-year	9.461×10^{12}	kilometers
	5.879×10^{12}	statute miles
1 meter	10^{10}	angstroms
	3.281	feet
	39.37	inches
	1.094	yards
1 micron	10^{4}	angstroms
	10^{-4}	centimeter
	10^{-6}	meter
1 nautical mile (International)	8.439	cables
	6076	feet
	1852	meters
	1.151	statute miles
1 statute mile	5280	feet
	8	furlongs
	1.609	kilometers
	0.8690	nautical mile
1 yard	3	feet
	0.9144	meter
	Angles	
1 degree	60	minutes
	0.01745	radian
	3600	seconds
	2.778×10^{-3}	revolution
1 minute of arc	0.01667	degree
	2.909×10^{-4}	radian
	60	seconds
1 radian	0.1592	revolution
	57.296	degrees
	3438	minutes
1 second of arc	2.778×10^{-4}	degree
	0.01667	minute

Unit	Equals	
	Area	
1 acre	4047	square meters
	43 560	square feet
1 are	0.024 71	acre
	1	square dekameter
	100	square meters
1 hectare	2.471	acres
	100	ares
	10 000	square meters
1 square foot	144	square inches
	0.092 90	square meter
1 square inch	6.452	square centimeters
1 square kilometer	247.1	acres
1 square meter	10.76	square feet
1 square mile	640	acres
	2.788×10^7	square feet
	2.590	square kilometers
	Volume	
1 board-foot	144	cubic inches
1 bushel (U.S.)	1.244	cubic feet
	35.24	liters
1 cord	128	cubic feet
	3.625	cubic meters
1 cubic foot	7.481	gallons (U.S. liquid)
	28.32	liters
1 cubic inch	0.01639	liter
	16.39	milliliters
1 cubic meter	35.31	cubic feet
	10^6	cubic centimeter
1 cubic millimeter	6.102×10^{-5}	cubic inch
1 cubic yard	27	cubic feet
	0.7646	cubic meter
1 gallon (imperial)	277.4	cubic inches
	4.546	liters
1 gallon (U.S. liquid)	231	cubic inches
	3.785	liters
1 kiloliter	35.31	cubic feet
	1.000	cubic meter
	1.308	cubic yards
	220	imperial gallons
1 liter	10^3	cubic centimeters
	10^6	cubic millimeters
	10^{-3}	cubic meter
	61.02	cubic inches

Unit	Equals	
Mass		
1 gram	10^{-3}	kilogram
	6.854×10^{-5}	slug
1 kilogram	1000	grams
	0.06854	slug
1 slug	14.59	kilograms
	14,590	grams
1 metric ton	1000	kilograms
Force		
1 dyne	10^{-5}	newton
1 newton	10^{5}	dynes
	0.2248	pound
	3.597	ounces
1 pound	4.448	newtons
	16	ounces
1 ton	2000	pounds
On earth's surface only		
1 pound	2.2046	kilograms
Velocity		
1 foot/minute	0.3048	meter/minute
	0.011 364	mile/hour
1 foot/second	1.097	kilometers/hour
	18.29	meters/minute
	0.6818	mile/hour
1 kilometer/hour	3281	feet/hour
	54.68	feet/minute
	0.6214	mile/hour
1 kilometer/minute	3281	feet/minute
	37.28	miles/hour
1 knot	6076	feet/hour
	101.3	feet/minute
	1.852	kilometers/hour
	30.87	meters/minute
	1.151	miles/hour
1 meter/hour	3.281	feet/hour
1 mile/hour	1.467	feet/second
	1.609	kilometers/hour
Power		
1 British thermal unit/hour	0.2929	watt
1 Btu/pound	2.324	joules/gram

Unit	Equals	
Power (continued)		
1 Btu-second	1.414	horsepower
	1.054	kilowatts
	1054	watts
1 horsepower	42.44	Btu/minute
	550	foot-pounds/second
	746	watts
1 kilowatt	3414	Btu/hour
	737.6	foot-pounds/second
	1.341	horsepower
	10^3	joules/second
	999.8	international watt
1 watt	44.25	foot-pounds/minute
	1	joule/second
Pressure		
1 atmosphere	1.013	bars
	14.70	pounds/square inch
	760	torrs
	101	kilopascals
1 bar	10^6	baryes
	14.50	pounds/square inch
1 barye	10^{-6}	bar
1 inch of mercury	0.033 86	bar
	70.73	pounds/square foot
1 pascal	1	newton/square meter
1 pound/square inch	0.068 03	atmosphere
Energy		
1 British thermal unit	1054	joules
	1054	wattseconds
1 foot-pound	1.356	joules
	1.356	newtonmeters
1 joule	0.7376	foot-pound
	1	wattsecond
1 kilowatthour	3410	British thermal units
	1.341	horsepowerhours
1 newtonmeter	0.7376	foot-pounds
1 watthour	3.414	British thermal units
	2655	foot-pounds
	3600	joules

Summary of BASIC

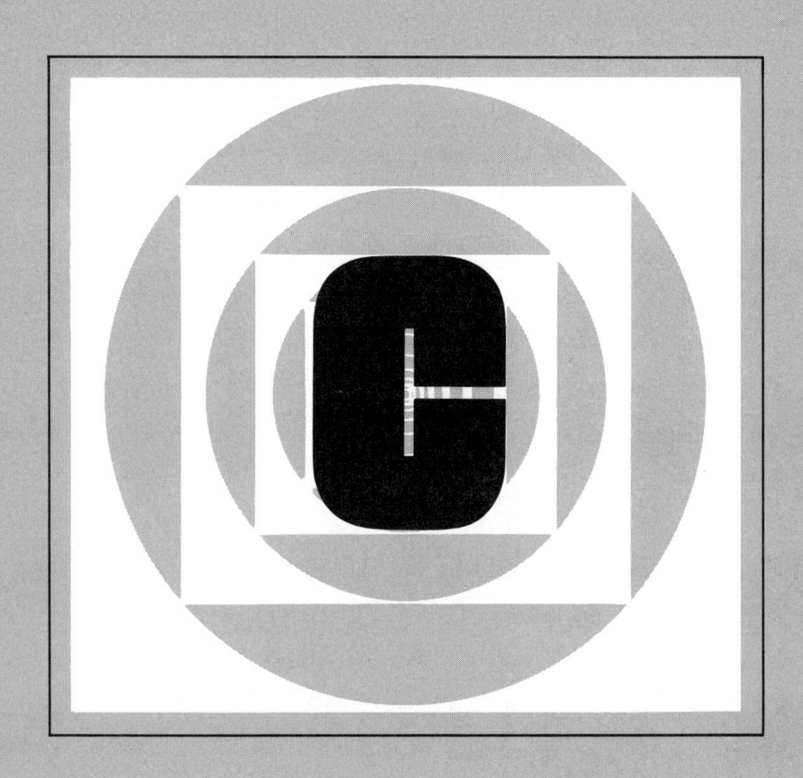

This brief listing contains commands, statements, and functions. *Commands,* such as RUN, or LIST are not part of a program, but tell the computer what to do with a program. Commands are typed without line numbers and are executed immediately. Program *statements,* such as PRINT or GO TO, tell the computer what to do during a run of the program. *Functions,* such as COS( ) or EXP( ), tell the computer what operation to perform on the quantity enclosed in parentheses (called the *argument*).

ABS( )

A function which returns the absolute value of the expression in parentheses.

Example: The lines

```
10 LET X = -5
20 PRINT ABS(X)
```

will cause the value 5 to be printed.

ATN( )

A function which returns the angle (in radians) whose tangent is specified.

Example: The statement PRINT ATN(2) will cause the arctangent of 2 (1.1071 radians) to be printed.

CONT

A command which causes a program to continue running after a STOP or CTRL-C (interrupt) has been executed.

COS( )

A function which returns the cosine of the angle (in radians).

Example: The statement PRINT COS(1.2) will cause the number 0.362358 (the cosine of 1.2 radians) to be printed.

DATA

A statement used to store numbers and strings, for later access by the READ statement.
Example: The statement DATA 5, 2, 9, 3 stores the numbers 5, 2, 9, and 3 for later access, and the statement

DATA JOHN, MARY, BILL

stores the given three names for later access. See READ.

DEF FN

A statement which defines a new function written by the user.
Examples: DEF FNA(X) = 3*X + 2
DEF FNG(X,Y,Z) = (5*X − 2*Y^2)*Z

DIM

A statement which reserves space for lists or tables.
Example: DIM A(50), DIM P$(20), B(25,30)

END

A statement which stops the run and closes all files. It is the last line in a program.

EXP( )

A function which raises e (the base of natural logarithms) to the power specified.
Example: The statement PRINT EXP(2.5) will compute and print the value of $e^{2.5}$, (12.182).

FOR . . . NEXT

A pair of statements used to set up a loop.
Examples: The lines

```
10 FOR N = 1 TO 10
20 PRINT "HELLO"
30 NEXT N
```

will cause the word "HELLO" to be printed ten times.
The lines

```
10 FOR X = 20 TO 30 STEP .5
20 PRINT X,
30 NEXT X
```

will cause the numbers 20, 20.5, 21, . . . 30 to be printed.

GOSUB . . . RETURN

A pair of statements used to branch to and return from a subroutine.
Example: The line GOSUB 500 causes a branching to the subroutine starting on line 500. The RETURN statement placed at the end of the subroutine returns us to the line immediately following the GOSUB statement.

GOTO

Causes a branch to the specified line number.
Example: GOTO 150

IF . . . THEN

A statement which causes a branching to another line if the specified condition is met.
Example: The line

40 IF A > B THEN 200

will cause branching to line 200 if A is greater than B. If not, we go to the next line.

INPUT — A statement which allows input of data from the keyboard during a run.
Example: The statement INPUT X will cause the run to stop, and a question mark will be printed. The operator must then type a number, which will be accepted as X by the program.
Example: The statement
20 INPUT "WHAT ARE THE TWO NAMES"; A$, B$
will print the *prompt string* WHAT ARE THE TWO NAMES, then a question mark, and then stop the run until you enter two strings, separated by a comma.

INT( ) — A function which returns the integer part of a number.
Example: The statement PRINT INT(5.995) will cause the integer 5 to be printed.

KILL — A command which deletes a file from the disk.
Example: KILL "CIRCLE.BAS"

LEFT$( ) — A function which returns the leftmost characters of the specified string.
Example: The line 50 PRINT LEFT$(A$,5) will cause the five leftmost characters of A$ to be printed.

LEN( ) — A function which returns the number of characters, incuding blanks, in the specified string.
Example: 10 PRINT LEN(B$)

LET — A statement used to assign a value to a variable.
Examples: 30 LET X = 5
40 LET Y = 3*X−2
50 LET SUM = X + Y
(The LET statement is optional in many versions of BASIC.)

LIST — A command to list all or part of a program.
Examples: LIST lists the entire program.
LIST 100–150 lists lines 100 to 150 only.
LIST–200 lists up to line 200.
LIST 300– lists from line 300 to the end.

LLIST — A command to list the program on the printer.

LOAD — A command to load a program or file from the disk.
Example: LOAD "B:QUADRATIC" will load the program QUADRATIC from drive B.

LOG( ) — A function which returns the natural logarithm of the argument.

LPRINT — See PRINT.

MID$( ) — A function which returns a string of characters from the middle of the specified string.
Example: In the program
10 A$ = "COMPUTER"
20 PRINT MID$(A$,4,3)
line 20 will extract and print a string of length 3 from A$, starting with the 4th character. Thus, "PUT" will be printed.

NEW — A command used to clear the memory before entering a new program.

ON . . . GO TO — A statement which allows branching to one of several line numbers.
Example: Given the line
10 ON X GO TO 150, 300, 450
we branch to line 150 if X = 1, to line 300 if X = 2, and to line 450 if X = 3.

PRINT — A statement which causes information to be printed.
Examples:
20 PRINT X causes the value of X to be printed.
30 PRINT A, B, C causes the values of A, B, and C to be printed, each in a separate column.
40 PRINT A; B; C causes the values of A, B, and C to be printed with just a single space between them.
50 PRINT "HELLO" causes the word "HELLO" to be printed.
The statement LPRINT will cause the printing to occur at the printer, rather than at the video terminal.

PRINT USING — A statement which specifies a format for the printing of strings or numbers.
Example: 20 PRINT USING "###.##"; X
will cause the number X to be printed with three digits before the decimal point and two digits following the decimal point. There are many other ways to format numbers and strings. Consult your users' manual.

RANDOMIZE — A statement which allows you to obtain a different list of random numbers each time you use RND( ). Put the line 10 RANDOMIZE at the start of each program that uses random numbers. Then when you RUN, you will be asked to enter a number. Each such number will cause a different set of random numbers to be generated.

READ — A statement which reads values from a DATA statement and assigns them to variables.
Example: 10 DATA JOHN, 24
20 READ N$, A

Line 20 will read the string "JOHN" and assign it to N$, and read the number 24 and assign it to N.

REM — This statement allows us to place remarks in a program. REM's are ignored during a RUN.
Example: 10 REM THIS PROGRAM COMPUTES AVERAGES

RENUM — A command to renumber a program.
Example: RENUM renumbers a program with the line numbers, 10, 20, 30, . . .
RENUM 100, 300, 20 will start renumbering your program at the old line 300, which it changes to 100, and continues in increments of 20.

RESTORE — A statement which allows data to be reread.

RETURN — see GOSUB

RIGHT$( ) — A statement which returns the rightmost characters from the specified string. Similar to LEFT$( ).

RND — A function which returns a random number between 0 and 1.
Example: The program

```
10 FOR N = 1 TO 10
20 PRINT RND
30 NEXT N
```

will cause 10 random numbers to be printed.

RUN — A command to run the program in memory.

SAVE — A command to save a program on disk.
Example: SAVE "B:ROOTS"
will save the program "ROOTS" on the disk in drive B.

SGN( ) — A function which returns a value of +1, 0, or −1, depending on whether the argument is positive, zero, or negative, respectively.

SIN( ) — A function which returns the sine of the angle (in radians).

SQR( ) — A function which returns the square root of the argument.

STOP — A statement which stops a run, but does not close files. To resume the run, type CONT.

SYSTEM — A command to exit BASIC and return to the operating system.

TAB( ) — A function which tabs to the print position specified. *Example:* The line 30 PRINT TAB(25); "HELLO" will cause the word HELLO to be printed starting at a position 25 characters from the left edge of the screen or paper.

TAN( ) — A function which returns the tangent of the angle (in radians).

Answers to Selected Problems

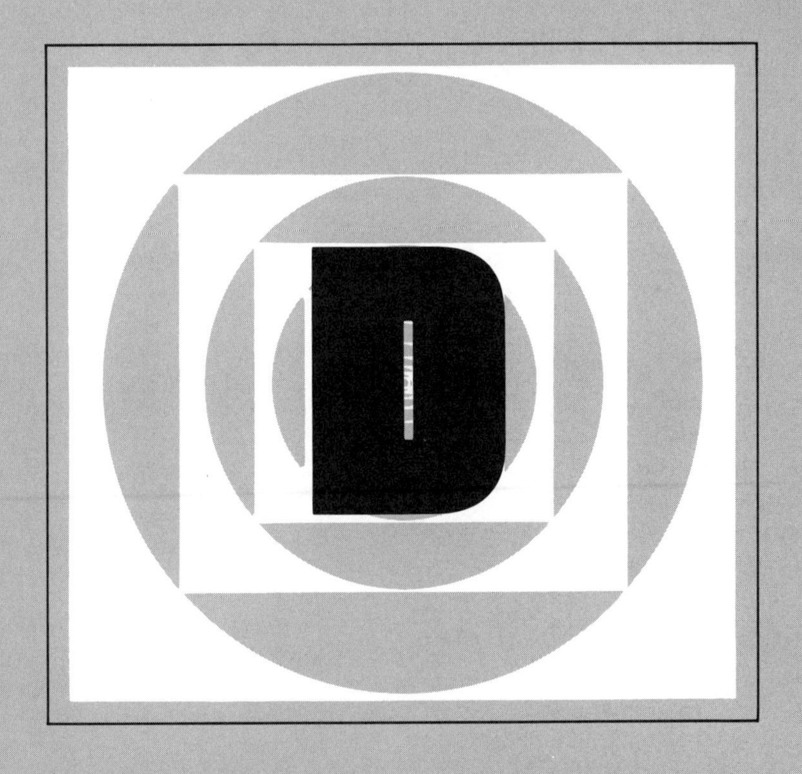

CHAPTER ONE: NUMERICAL COMPUTATION

Exercise 1, Page 6

1. $<$ **3.** $<$ **5.** $=$ **7.** -18
9. 13 **11.** 4 **13.** 2 **15.** 5
17. 5 **19.** 2.00 **21.** 55.86 **23.** 2.96
25. 278.38 **27.** 745.6 **29.** 0.5 **31.** 34.9
33. 0.8 **35.** 7600 **37.** 274,800 **39.** 2860
41. 484,000 **43.** 29.6 **45.** 8370

Exercise 2, Page 11

1. -1090 **3.** -1116 **5.** 105,223 **7.** 1789
9. -1129 **11.** -850 **13.** 1827 **15.** 4931
17. 593.44 **19.** -0.00031 **21.** 78,388 sq mi **23.** 35.0 cm
25. 41.1 Ω

Exercise 3, Page 15

1. $73\bar{0}0$ **3.** 0.525 **5.** $-17,800$ **7.** 22.9
9. $3320 **11.** $1400 **13.** $1,570,000 **15.** 17,180 rev
17. 980.03 cm

Exercise 4, Page 18

1. 163	**3.** -0.347	**5.** 0.7062	**7.** 70,840
9. 17	**11.** 0.00144	**13.** -0.00000253	**15.** -175
17. 0.2003	**19.** 0.0901	**21.** 0.9930	**23.** 313 Ω
25. 0.279			

Exercise 5, Page 22

1. 8	**3.** -8	**5.** 1	**7.** 1
9. 1	**11.** -1	**13.** 100	**15.** 1
17. 1000	**19.** 0.01	**21.** 10,000	**23.** 0.00001
25. 1.035	**27.** -112	**29.** 0.0146	**31.** 0.0279
33. 59.8	**35.** -0.0000193	**37.** 125 W	**39.** 878,000 cm^3
41. 5	**43.** 7	**45.** -2	**47.** 7.01
49. 4.45	**51.** 62.25	**53.** -7.28	**55.** -1.405
57. 4480 Ω			

Exercise 6, Page 27

1. 56.4 m	**3.** 391 in.	**5.** 321,000 m^2	**7.** 13.7 in.^{2}
9. 68.1 slugs	**11.** 278 oz	**13.** 1243 lb	**15.** 43 mi/h
17. 89.1 km/h			

Exercise 7, Page 29

1. 17	**3.** -37	**5.** 14	**7.** 14.1
9. 233	**11.** 8.00	**13.** $3975	**15.** 53°C
17. 13 ft lb	**19.** $13,266	**21.** 958 Ω	

Exercise 8, Page 35

1. 10^2	**3.** 10^{-4}	**5.** 10^8	**7.** 0.01
9. 0.1	**11.** 1.86×10^5	**13.** 2.5742×10^4	**15.** 9.83×10^4
17. 1.15×10^2	**19.** 1.452×10^2	**21.** 2850	**23.** 90,000
25. 0.003667	**27.** 10	**29.** 10^3	**31.** 10^3
33. 10^7	**35.** 10^2	**37.** 4×10^2	**39.** 2.1×10^9
41. 1.2×10^{-7}	**43.** 3×10^2	**45.** 2×10^6	**47.** 2×10^{-9}
49. 7×10^5	**51.** 3×10^{-2}	**53.** 11×10^3	**55.** 3×10^{-2}
57. 2.5×10^4	**59.** 9×10^{-5}	**61.** 3.55×10^2	**63.** 3.00×10^{-3}
65. 2.06×10^{17}	**67.** 9.19×10^2	**69.** 2.26×10^2	**71.** -3.04×10^{-5}
73. 2.43	**75.** 2.35×10^4 Ω	**77.** 1.04×10^{-11} W	**79.** 153 lb/in.^{2}
81. 2.28×10^7 lb/in.^{2}	**83.** 120 h	**85.** 22 years	

Exercise 9, Page 42

1. 372%	**3.** 0.55%	**5.** 23.4%	**7.** 40%
9. 70%	**11.** 87.5%	**13.** 0.23	**15.** 2.875
17. 0.0075	**19.** 3/8	**21.** 1 1/2	**23.** 1/6
25. 100 tons	**27.** 220 kg	**29.** 120 liters	**31.** 1090 Ω

33. $1400
35. 39.4 tons
37. 3000
39. 200
41. 80
43. $3000
45. 19.0 billion bbl
47. 33 1/3%
49. 50%
51. 66 2/3%
53. 87.5%
55. 53.6%
57. 64.5%

Exercise 10, Page 45

1. 500
3. 600
5. 100
7. 640
9. 96.6%
11. 31.3%
13. 11%
15. 5.99%
17. $88.56
19. 7216
21. 28.1 mi/h
23. 1.42 hp
25. 77%

Exercise 11, Page 51

1. $4405
3. 43.4 acres
5. 8660
7. $2768
9. 1024
11. 13.2
13. $162.93
15. 15.8 in.
17. 104 m
19. 319 acres
21. 6.06 in.
23. 114 cm^2
25. 247 cm
27. 47 cuts
29. 428 loads
31. 0.64 in.
33. 1.0 in.
35. 760 in^2
37. 3800 lb
39. 38 ft^3
41. $V = 1.72 \times 10^9$, S.A. $= 6.96 \times 10^6$
43. $r = 1.12$, SA $= 15.8$
45. 213 lb

47.
```
10 '     HERO
20 '
30 '     This program finds the area of a triangle using Hero's formula.
40 '
50 '     Run the program and enter the three sides in the input statements.
60 '
70 INPUT "What is side 1"; A
80 INPUT "What is side 2"; B
90 INPUT "What is side 3"; C
100 LET S = (A + B + C) / 2
110 LET AA = (S * (S - A) * (S - B) * (S - C))^.5
120 PRINT "Side 1", "Side 2", "Side 3", "Area"
130 PRINT A, B, C, AA
140 END

What is side 1; 15
What is side 2; 12
What is side 3; 13
Side 1        Side 2        Side 3        Area
 15            12            13            74.83315
```

49.
```
10 '     OPTICAL
20 '
30 '     This program calculates the diameter
40 '     of a circular stop placed
50 '     front of a lens that reduces the
60 '     area of the lens by 15%.
70 '
80  LPRINT "Diameter (mm)", "Area (Sq.mm)"
90  LET C = 75
```

Diameter (mm)	Area (Sq.mm)
0	4417.875
69.14658	3755.194
63.74999	3191.915
58.77461	2713.128
54.1875	2306.159
⋮	⋮
22.16682	385.9202
20.43679	328.0322

```
100 LET P = 3.1416
110 LET A = P * (C / 2)^2
120 LPRINT D, A
130 A = A * .85
140 LET D = 2 * (A / P)^.5
150 IF D < 20 THEN 170
160 GOTO 120
170 END
```

Chapter Test, Page 55

1. 83.35	**3.** 88.1	**5.** 17.5 m	**7.** 7.63%
9. 1440 mi/h	**11.** 9.64	**13.** −13	**15.** 2.9 m
17. 255 ft	**19.** 5.46	**21.** 6.76	**23.** 13 m
25. 6.27	**27.** −64	**29.** **(a)** 7.98	**29.** **(b)** 4.66
29. **(c)** 11.84	**29.** **(d)** 1.00	**31.** 43 in.	**33.** 470 kWh
35. 54.1 lb/in.^{2}	**37.** 10,200	**39.** 4.16×10^{-10}	**41.** 2.42
43. $-2/3 < -0.660$	**45.** 2375	**47.** −1.72	**49.** 47.1 cm, 534 cm^2
51. 219 N	**53.** 5340 ft	**55.** 22.00%	

CHAPTER TWO: BINARY NUMBERS

Exercise 1, Page 66

1. 2	**3.** 3	**5.** 6	**7.** 13
9. 12	**11.** 5	**13.** 103	**15.** 119
17. 101	**19.** 10	**21.** 0100 1000	**23.** 0101 1101
25. 1 0001 0010	**27.** 0111 0110	**29.** 10 0000 1011 0111	**31.** 1 0100 0011 0011 0100
33. 0.1	**35.** 0.11	**37.** 0.0100 1100	**39.** 0.1000 1100
41. 0.111	**43.** 0.0110 0011	**45.** 0.5	**47.** 0.25
49. 0.5625	**51.** 101.1	**53.** 100.011	**55.** 11 1011 0100.0111 1000
57. 1.5	**59.** 2.25	**61.** 25.40625	

Exercise 2, Page 75

1. 1	**3.** 101	**5.** 110	**7.** 1 0011
9. 1 1110	**11.** 1 0011 0100	**13.** 1 0110 1101 1111 1001	**15.** 1.01
17. 1.011	**19.** 10.00	**21.** 1 0001.0011	**23.** 1 1010
25. 111	**27.** 0	**29.** 100	**31.** 110
33. 0110 1100	**35.** 0.011	**37.** 111.1	**39.** 1 0101.0111
41. 1	**43.** 10	**45.** 10 0001	**47.** 0.01
49. 1.1	**51.** 1 0100	**53.** 101	**55.** 11 1101
57. 11 1101	**59.** 100.0101 0101	**61.** 1 0111.1100 11	

Exercise 3, Page 81

1. 1111 0010	**3.** 1101 1001	**5.** 1111 0010	**7.** 1100 0101
9. 1111 1001	**11.** 1101 0000	**13.** 1000 1000	**15.** 1100 0111
17. 0111	**19.** 0010 0001	**21.** 1100	**23.** 1110

Exercise 4, Page 83

1.

0 111 1000 1000 0000 0000 0000	0 000 1001

3.

0 111 1000 0000 0000 0000 0000	0 000 0011

5.

0 100 1001 1010 0000 1100 0101	0 001 0111

7. 0.59375 **9.** 0.04296875

Chapter Test, Page 84

1. 11 0010.01 **3.** 1001 1010 **5.** 1 0000.001 **7.** 1011 0101
9. 1 1010.111 **11.** −50.75 **13.** 1 0110.0111 0001 **15.** 6.125

CHAPTER THREE: HEXADECIMAL, OCTAL, AND BCD NUMBERS. ERRORS IN COMPUTATION

Exercise 1, Page 92

1. D **3.** 9 **5.** 93 **7.** D8
9. 92A6 **11.** 9.3 **13.** 1.38 **15.** 0110 1111
17. 0100 1010 **19.** 0010 1111 0011 0101 **21.** 0100 0111 1010 0010 **23.** 0101.1111
25. 1001.1010 1010 **27.** 242 **29.** 51 **31.** 14,244
33. 62,068 **35.** 3.9375 **37.** 2748.8671875 **39.** 27
41. 399 **43.** AB5 **45.** 6C8 **47.** 11
49. 19 **51.** DB **53.** F7 **55.** 184B7
57. B **59.** 5 **61.** C4 **63.** 33
65. 683 **67.** 63EF **69.** 2A **71.** 75
73. 1FB **75.** 648C **77.** 53858 **79.** 5
81. 2 **83.** 33

Exercise 2, Page 94

1. 6 **3.** 7 **5.** 63 **7.** 155
9. 111 **11.** 01 0110 **13.** 1 1001 0011 **15.** 1010 1010 0011
17. 0110 0110 0110 1000

Exercise 3, Page 96

1. 0110 0010 **3.** 0010 0111 0100 **5.** 0100 0010.1001 0001 **7.** 9
9. 61 **11.** 36.8 **13.** 0110 0001 0010 **15.** 0100 0001 0100

Exercise 4, Page 100

1. 2.34% **3.** 112.5 V to 312.5 V **5.** 0.108%

Chapter Test, Page 101

1. 67ED **3.** 0010 1101 **5.** 1264 **7.** 0.32% high
9. 1FCDFA **11.** BE **13.** 20AA **15.** 134
17. 0100 1110 0101 1101 **19.** 0100 1010 **21.** 0001 0110 0010 **23.** 0001 0011 0010 0011

CHAPTER FOUR: INTRODUCTION TO ALGEBRA

Exercise 1, Page 109

1. 2 **3.** 1 **5.** 5
7. $b/4$ **9.** $4(a + 4)$ **11.** $12x$
13. $-10ab$ **15.** $2m - c$ **17.** $4a - 2b - 3c - d - 17$
19. $11x + 10ax$ **21.** $-26x^3 + 2x^2 + x - 31$ **23.** 0
25. $-2ay^3 + 20$ **27.** $18b^2 - 3ac - 2d$
29. $-5m^3 + 7m^2 + 15ab - 4q - z$ **31.** $5.83(a + b)$
33. $3a^2 - 3a + x - b$ **35.** $-1.1 - 4.4x$ **37.** $4a^2 - 5a + y + 1$
39. $2a + 2b - 2m$ **41.** $6m - 3z$ **43.** $5w + 6x + 2z$
45. $a + 12$

Exercise 2, Page 115

1. 16 **3.** -27 **5.** 0.0001 **7.** a^8
9. y^{2a-2} **11.** 10^{a+b} **13.** y^3 **15.** x
17. 100 **19.** x **21.** x^{12} **23.** a^{xy}
25. x^{2a+2} **27.** $8x^3$ **29.** $27a^3b^3c^3$ **31.** $-1/27$
33. $\dfrac{8a^3}{27b^6}$ **35.** $1/a^2$ **37.** $y^3/27$ **39.** $9b^6/4a^2$
41. $2/x^2 + 3/y^3$ **43.** x^{-1} **45.** x^2y^{-2} **47.** $a^{-3}b^{-2}$
49. 1 **51.** $1/9$ **53.** 1

Exercise 3, Page 122

1. x^5 **3.** $10a^3b^4$ **5.** $36a^6b^5$
7. $-8p^4q^4$ **9.** $2a^2 - 10a$ **11.** $x^2y - xy^2 + x^2y^2$
13. $-12p^3q - 8p^2q^2 + 12pq^3$ **15.** $6a^3b^3 - 12a^3b^2 + 9a^2b^3 + 3a^2b^2$
17. $a^2 + ac + ab + bc$ **19.** $x^3 - xy + 3x^2y - 3y^2$ **21.** $9m^2 - 9n^2$
23. $49c^2d^4 - 16y^6z^2$ **25.** $x^4 - y^4$ **27.** $a^2 - ac - c - 1$
29. $m^3 - 5m^2 - m + 14$ **31.** $x^8 - x^2$ **33.** $-z^3 + 2z^2 - 3z + 2$
35. $a^3 + y^3$ **37.** $a^3 - y^3$
39. $x^3 - px^2 - mx^2 + mpx + nx^2 - npx - mnx + mnp$ **41.** $m^5 - 1$
43. $9a^2x - 6abx + b^2x$ **45.** $x^3 - 3x - 2$
47. $x^3 - cx^2 + bx^2 - bcx + ax^2 - acx + abx - abc$ **49.** $-c^5 - c^4 + c + 1$
51. $4a^6 - 10a^4c^2 + 14a^4 - 25a^2c^2 + 10a^2$
53. $a^3 + a^2c - ab^2 + b^2c - b^3 + 2abc - ac^2 + bc^2 - c^3 + a^2b$
55. $n^5 - 34n^3 + 57n^2 - 20$ **57.** $a^2 + 2ac + c^2$ **59.** $m^2 - 2mn + n^2$
61. $A^2 + 2AB + B^2$ **63.** $14.4a^2 - 16.7ax + 4.84x^2$ **65.** $4c^2 - 12cd + 9d^2$
67. $a^2 - 2a + 1$ **69.** $y^4 - 40y^2 + 400$
71. $a^2 + 2ab + b^2 + 2ac + 2bc + c^2$
73. $x^2 + 2x - 2xy - 2y + y^2 + 1$
75. $x^4 + 2x^3y + 3x^2y^2 + 2xy^3 + y^4$
77. $a^3 + 3a^2b + 3ab^2 + b^3$ **79.** $p^3 - 9p^2q + 27pq^2 - 27q^3$

81. $a^3 - 3a^2b + 3ab^2 - b^3$ 83. $2b - d$ 85. $2N - M$
87. $31y - 5z$ 89. $2x - 4$ 91. $7 - 5x$

Exercise 4, Page 131

1. z^2 3. $-5xyz$ 5. $2f$ 7. $-2xz$
9. $-5xy^2$ 11. $3c$ 13. $-5y$ 15. $-3w$
17. $7n$ 19. $4b$ 21. $2a^2 - a$ 23. $7x^2 - 1$
25. $3x^4 - 5x^2$ 27. $2x^2 - 3x$ 29. $a + 2c$ 31. $ax - 2y$
33. $2x + y$ 35. $ab - c$ 37. $xy - x^2 - y^2$ 39. $1 - a - b$
41. $x + 8$ 43. $x + 8$ 45. $a + 5$ 47. $-3a - 2$

Chapter Test, Page 132

1. $b^6 - b^4x^2 + b^4x^3 - b^2x^5 + b^2x^4 - x^6$ 3. 3.86×10^{14}
5. $9x^4 - 6mx^3 - 6m^3x + 10m^2x^2 + m^4$ 7. $8x^3 + 12x^2 + 6x + 1$
9. $(a - c)^{m-2}$ 11. $16a^2 - 24ab + 9b^2$ 13. $17x^2y^2 - 35xy + 8.8$
15. $4a^2 - 12ab + 9b^2$ 17. $ab - b^4 - a^2b$ 19. $16m^4 - c^4$
21. $2a^7$ 23. $24 - 8y$ 25. $6x^3 + 9x^2y - 3xy^2 - 6y^3$
27. $13w - 6$

CHAPTER FIVE: SIMPLE EQUATIONS AND WORD PROBLEMS

Exercise 1, Page 141

1. 33 3. 4/5 5. 1/2 7. −4
9. 4 11. 3 13. −1.4 15. 1
17. 4 19. 2 21. 5 23. 28
25. 3 27. 3 29. 35 31. −5/23
33. 5/6 35. 19/5

Exercise 2, Page 145

1. $5x + 8$ 3. x, $83 - x$
5. x = gallons of antifreeze, $6 - x$ = gallons of water
7. A, $2A$, $180 - 3A$ 9. $82x$ 11. 5 13. 57, 58, 59
15. 14 17. 6 19. 4 21. 20

Exercise 3, Page 146

1. 113 km/h 3. 5.36 days 5. 4:40 PM, 1400 km 7. 137 km
9. 170 mi

Exercise 4, Page 150

1. 333 gal 3. 485 kg, 215 kg 5. 1375 kg 7. 1.3 liters
9. 0.37 liters

Exercise 5, Page 151

1. 8 3. $130,000 5. 57,000 gal 7. $64.29, $128.57, $257.14
9. $94,000 11. $12,000

Chapter Test, Page 152

1. 12 **3.** 5 **5.** 0 **7.** $10\frac{1}{2}$
9. 2 **11.** 4 **13.** 4 **15.** 6
17. $-5/7$ **19.** -6 **21.** 9.3 **23.** 47 cm, 53 cm
25. 10 m **27.** $30

CHAPTER SIX: SETS, FUNCTIONS, AND GRAPHS

Exercise 1, Page 160

1. $A = C$, and $B = D$ **3.** $B = \{2, 4, 6\}$ **5.** $N = \emptyset$

7. (a) $A \cap B = \{2, 4, 6\}$
(b) $A \cup B = \{1, 2, 3, 4, 5, 6, 8, 10\}$
(c) $n(A \cup B) = 8$
(d) $n(A) = 5$

9. (a) $P - Q = \{a, b, c\}$
(b) $Q - P = \{f, g, h\}$
(c) $R - Q = 0$
(d) $n(R - Q) = 0$

Exercise 2, Page 168

1. Is a function. **3.** Is not a function. **5.** Not a function.
7. Is a function. There is only one y for each x. **9.** $y = x^3$
11. $y = 2x^2 + x$ **13.** $\frac{2}{3}(x - 4)$
15. The amount by which 5 exceeds x. **17.** Twice the cube root of x.
19. $A = \frac{1}{2}bh$ **21.** $V = \frac{4}{3}\pi r^3$ **23.** $d = 55t$
25. $H = 125t - 4.9t^2$ **27.** Domain = $-10, 5, -7, 10, 0$. Range = 20, 7, 10, 20, 3.
29. Domain: $x \geqq 7$. Range: $y \geqq 0$ **31.** All x but 0. All y.

Exercise 3, Page 176

1. explicit **3.** implicit
5. x is independent, y is dependent
7. x and y are independent, w is dependent
9. x and y are independent, r is dependent
11. $y = 2/x + 3$ **13.** $y = x(2x + 1)/3$ **15.** $R_2 = RR_1/(R_1 - R)$
17. $x = 4z - (2w + 5)^2$ **19.** $Z = \sqrt{R^2 - (\omega L - 1/\omega C)^2}$
21. (a) $6x - 13$
(b) $6u - 1$
23. 6 **25.** -21
27. 12.5 **29.** -5 **31.** -15
33. 5/4 **35.** $2a^2 + 4$ **37.** $5a + 5b + 1$
39. 9.5 **41.** 142
43. (a) 9/5
(b) 18
(c) 16
45. 10 **47.** $5a^2$ **49.** 18
51. $(x + 15)/14$ **53.** $1.063 \times 10^4\ \Omega$, $1.084 \times 10^4\ \Omega$, $1.105 \times 10^4\ \Omega$

55.
```
10 '  FUNCTION
20 '
30 '     This program shows how functions
40 '     can be manipulated by computer.
50 '
```

```
60  LPRINT " X", " Y"
70  DEF FNA(X) = 2 * X - 5
80  DEF FNB(X) = 3 * X^2 + 2 * X - 3
90  FOR X = 1 TO 10
100    LET Y = (4 * FNA(X)) / (FNB(X) + 3)
110    LPRINT X, Y
120 NEXT X
130 END
```

X	Y
1	−2.4
2	−.25
3	.1212121
⋮	⋮
9	.1992337
10	.1875

Exercise 4, Page 185

1. fourth
3. second
5. fourth
7. first and fourth
9. 7
11. E(−1.8, −0.7) F(−1.4, −1.4) G(1.4, −0.6) H(2.5, −1.8)

13.

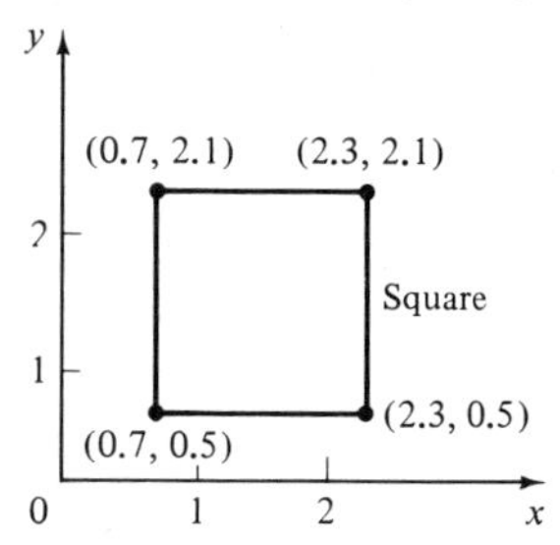

15.

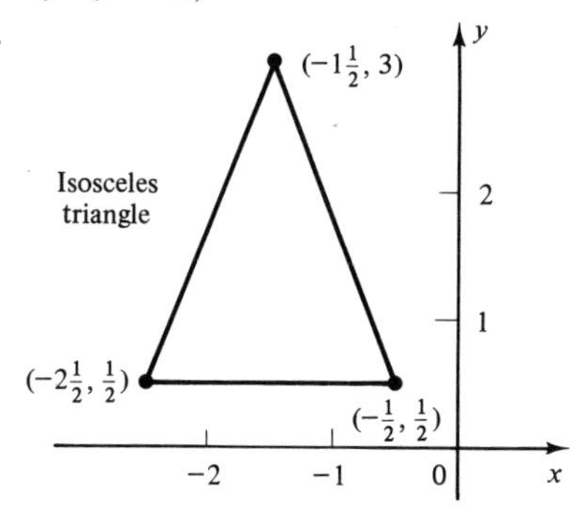

17. (4, −5)

19. 

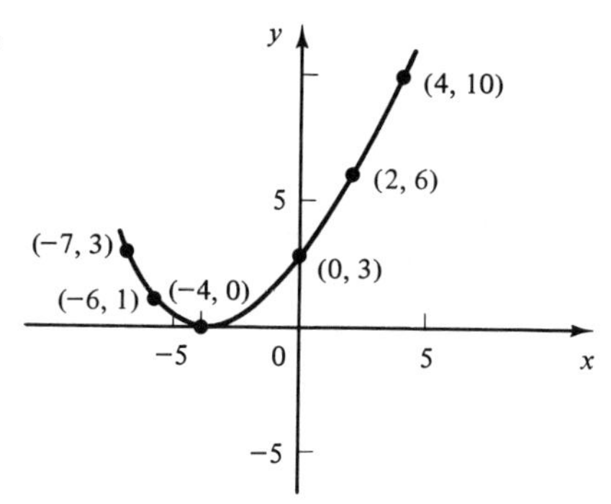

21.

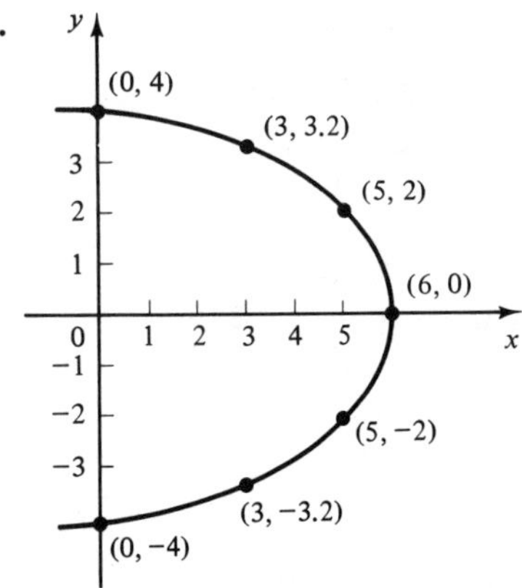

23.

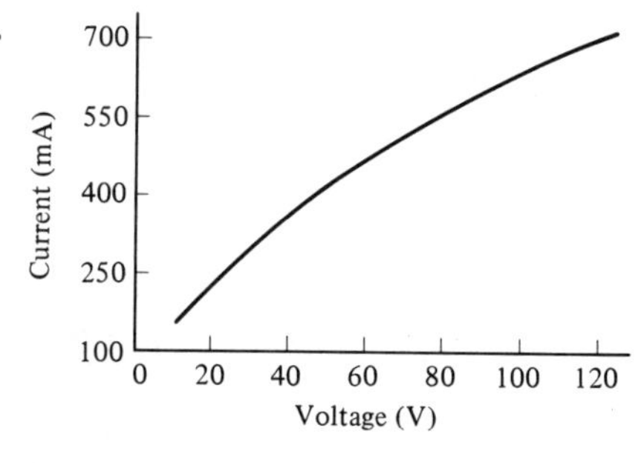

25. **(a)** Not a function
(b) a function
(c) a function

27.

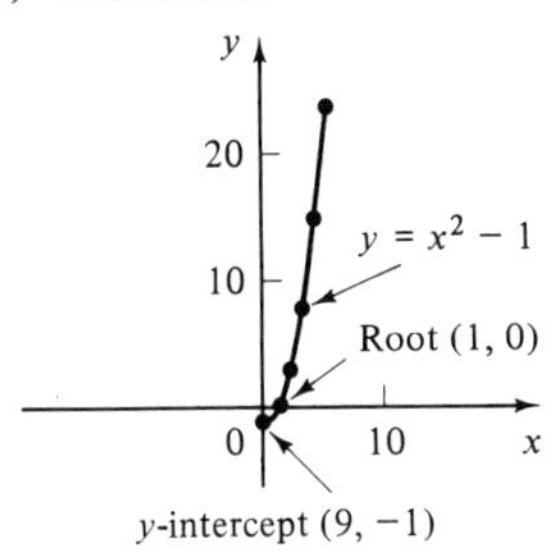

x	0	1	2	3	4	5
y	−1	0	3	8	15	24

29.

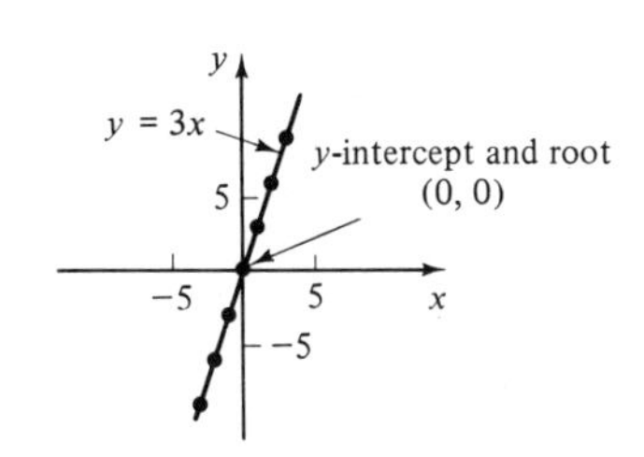

x	−3	−2	−1	0	1	2	3
y	−9	−6	−3	0	3	6	9

31.

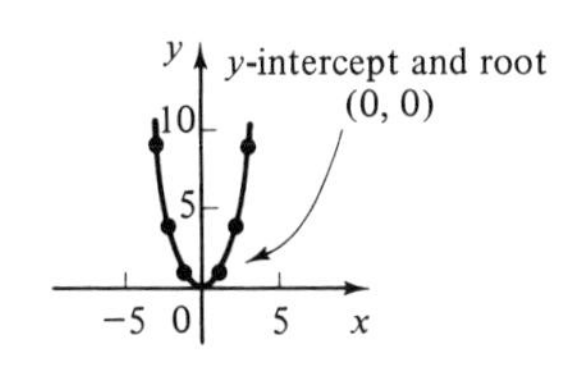

x	−3	−2	−1	0	1	2	3
y	9	4	1	0	1	4	9

33.

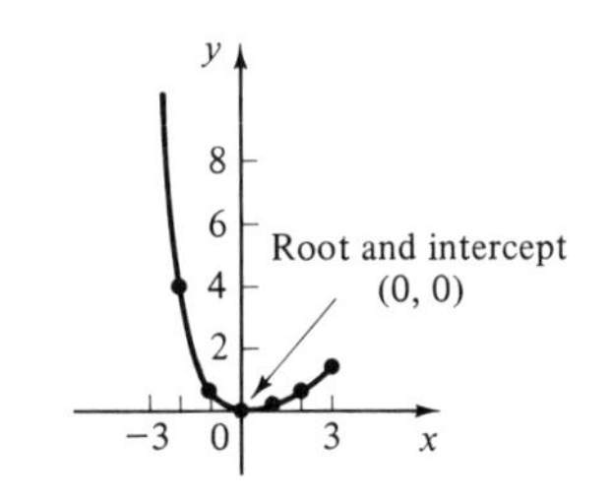

x	−3	−2	−1	0	1	2	3
y	—	4	$\frac{1}{2}$	0	$\frac{1}{4}$	$\frac{4}{5}$	$\frac{3}{2}$

35.

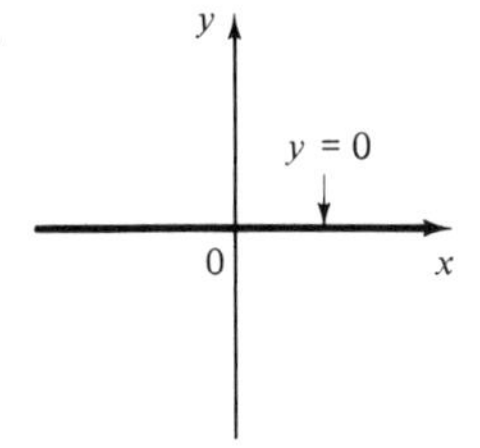

37.

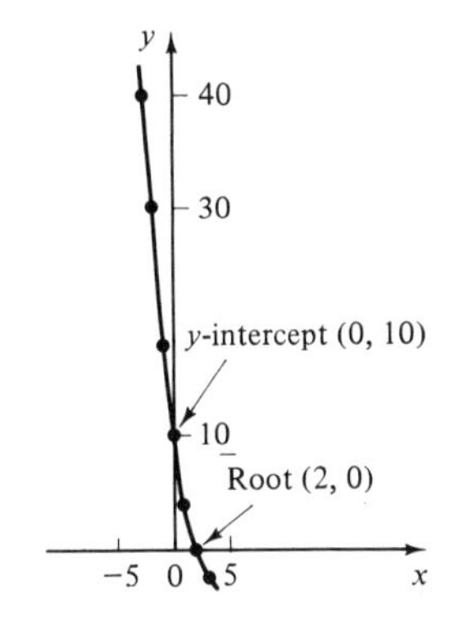

x	−3	−2	−1	0	1	2	3
y	40	28	18	10	4	0	−2

39. **(c)** Domain: $-6 \leqq x \leqq 6$, Range: $-6 \leqq x \leqq 6$
(d) Domain: $-10 \leqq x \leqq 10$, Range: $-6 \leqq x \leqq 6$

41. 3.2

43. −0.4, 2.5

45. −0.8, 1.1

47. (0.33, 1.67)

49. (2.25, −1.75)

51. 10 ' MIDPOINT
20 '

```
30 '  This program finds the roots of an equation by the midpoint method.
40 '
50 '  Define the equation on line 80.
60 '
70 INPUT "WHAT ARE YOUR LEFT AND RIGHT LIMITS"; X1, X2
80 DEF FNA(X) = 2.4 * X^3 - 7.2 * X^2 - 3.3
90 Y1 = FNA(X1) : Y2 = FNA(X2)
100 IF Y1 < 0 AND Y2 > 0 THEN GOTO 140
110 IF Y2 < 0 AND Y1 > 0 THEN GOTO 140
120 PRINT "Your endpoints are not on different sides of the root."
130 GOTO 70
140 X3 = (X1 + X2) / 2
150 Y3 = FNA(X3)
160 IF Y3 > 0 AND Y1 > 0 THEN X1 = X3
170 IF Y3 < 0 AND Y1 < 0 THEN X1 = X3
180 IF Y3 > 0 AND Y2 > 0 THEN X2 = X3
190 IF Y3 < 0 AND Y2 < 0 THEN X2 = X3
200 IF ABS(Y3) < .00001 THEN GOTO 220
210 GOTO 90
220 LPRINT "The root of this equation is "; X3
230 END
```

Chapter Test, Page 189

1. Is a function.
3. Is not.
5. $(A \cup B) - C = \{b, p, r, s\}$
7. $y = x^3/(2 - 2x)$
9. $x < 3, y > 0$
11. yes
13. Implicit
15. $\dfrac{3 - x^2 - y^2}{2}$
17. $A \cap B = \{a\}$
19. $A^c = \{b, c, d, f, g, h, j, k, l, m, n, p, q, r, s, t, v, w, x, y, z\}$
21. $A - B = \{e, i, o, u\}$
23. $y = 7x^2$
25. 6

27.

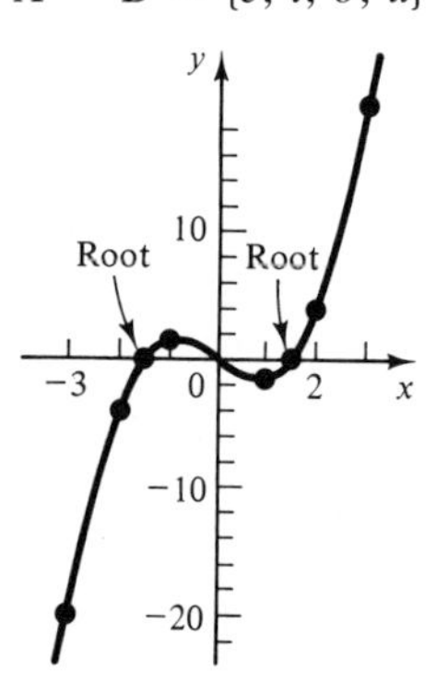

29.

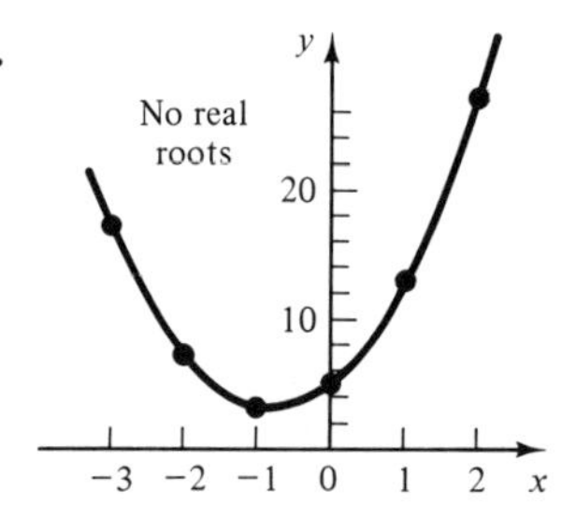

31. true
33. false
35. true
37. false
39. true
41. y equals five times the cube of x, diminished by seven.
43. $x = \pm\sqrt{\dfrac{6 - y}{3}}$
45. $8u^2 - 29$
47. 28
49. $x = (y + 5)/9$

CHAPTER SEVEN: FACTORING AND FRACTIONS

Exercise 1, Page 194

1. $y^2(3 + y)$ **3.** $x^3(x^2 - 2x + 3)$ **5.** $a(3 + a - 3a^2)$
7. $5(x + y)[1 + 3(x + y)]$ **9.** $\frac{1}{x}\left(3 + \frac{2}{x} - \frac{5}{x^2}\right)$ **11.** $a^2(5b + 6c)$
13. $xy(4x + cy + 3y^2)$ **15.** $3ay(a^2 - 2ay + 3y^2)$ **17.** $cd(5a - 2cd + b)$
19. $4x^2(2y^2 + 3z^2)$ **21.** $ab(3a + c - d)$ **23.** $L_0(1 + \alpha t)$
25. $R_1[1 + \alpha(t - t_1)]$ **27.** $t\left(v_0 + \frac{a}{2}t\right)$ **29.** $\frac{4\pi D}{3}(r_2^3 - r_1^3)$

Exercise 2, Page 197

1. $(2 - x)(2 + x)$ **3.** $(3a - x)(3a + x)$ **5.** $4(x - y)(x + y)$
7. $(x - 3y)(x + 3y)$ **9.** $(3c - 4d)(3c + 4d)$ **11.** $(3y - 1)(3y + 1)$
13. $y^2(x - 2z)(x + 2z)$ **15.** $(9xy - 1)(9xy + 1)$ **17.** $(5x - a)(5x + a)$
19. $(1 - 4b)(1 + 4b)$ **21.** $(7xy - 4ab)(7xy + 4ab)$ **23.** $4\pi(r_1 - r_2)(r_1 + r_2)$
25. $\frac{g}{2}(t_2 - t_1)(t_2 + t_1)$ **27.** $\frac{k}{2}(x_2 - x_1)(x_2 + x_1)$ **29.** $\pi h(R - r)(R + r)$

Exercise 3, Page 200

1. not factorable **3.** factorable **5.** factorable
7. $(x - 7)(x - 3)$ **9.** $(x - 9)(x - 1)$ **11.** $(x + 10)(x - 3)$
13. $(x + 4)(x + 3)$ **15.** $(x - 7)(x + 3)$ **17.** $(x + 4)(x + 2)$
19. $(b - 5)(b - 3)$ **21.** $(b - 4)(b + 3)$ **23.** $(x + 2)(x - 9)$
25. $(x - 4)(x - 3)$ **27.** $(a + 24)(a - 2)$ **29.** $(a + 5)(a + 2)$

Exercise 4, Page 202

1. $(4x - 1)(x - 3)$ **3.** $(5x + 1)(x + 2)$ **5.** $(3b + 2)(4b - 3)$
7. $(2a - 3)(a + 2)$ **9.** $(x - 7)(5x - 3)$ **11.** $3(x + 1)^2$
13. $(3x + 2)(x - 1)$ **15.** $2(2x - 3)(x - 1)$ **17.** $(2a - 1)(2a + 3)$
19. $(3a - 7)(3a + 2)$ **21.** $16(4 - t)(2 + t)$ **23.** $(m - 500)(m - 500)$

Exercise 5, Page 204

1. $(x + 2)^2$ **3.** $2(y - 3)^2$ **5.** $(y - 1)^2$
7. $(3 + x)^2$ **9.** $(3x + 1)^2$ **11.** $9(y - 1)^2$
13. $4(2 + a)^2$ **15.** $(7a - 2)^2$

Exercise 6, Page 209

1. 0 **3.** 5 **5.** 1 or 2 **7.** −1
9. $d - c$ **11.** $\frac{a}{3}$ **13.** $\frac{3m}{4p^2}$ **15.** $\frac{2a - 3b}{2a}$
17. $\frac{x}{x - 1}$ **19.** $\frac{x}{5}$ **21.** $\frac{2}{5xy}$ **23.** $\frac{2(a + 1)}{a - 1}$
25. $\frac{x + z}{x^2 + xz + z^2}$ **27.** 0.5556 **29.** 0.8065 **31.** 2.778

33. $7/11$ 35. $9/32$ 37. $25/64$ 39. $2\frac{1}{5}$

41. $5\frac{2}{3}$ 43. $10\frac{5}{12}$ 45. $\frac{1}{2x} - 2$

Exercise 7, Page 212

1. $\frac{2}{15}$ 3. $\frac{6}{7}$ 5. $\frac{35}{8}$ 7. $\frac{77}{3}$

9. $22\frac{1}{5}$ 11. 1 13. $\frac{x(x^2 - y^2)}{a}$ 15. $\frac{15x^2}{2a}$

17. $\frac{a^4b^4}{2y^{2n}}$ 19. $\frac{7}{32}$ 21. $1\frac{1}{14}$ 23. b^2

25. $\frac{8a^3}{3dxy}$ 27. $\frac{ab + cd}{a^2 - c^2}$ 29. $\frac{3an + cm}{x^4 - y^4}$ 31. $\frac{4}{3ayz}$

33. $\frac{4F}{\pi d^2}$ 35. $\frac{3F}{4\pi r^3 D}$ 37. $\frac{\pi d}{s}$

Exercise 8, Page 216

1. 1 3. $\frac{1}{7}$ 5. $5/3$ 7. $7/6$

9. $19/16$ 11. $7/18$ 13. $\frac{13}{5}$ 15. $-\frac{92}{15}$

17. $\frac{67}{35}$ 19. $\frac{8}{3}$ 21. $\frac{13}{4}$ 23. $\frac{91}{16}$

25. $\frac{23}{3}$ 27. $\frac{4}{x}$ 29. $\frac{x - y + z}{3a}$ 31. $\frac{2}{x + 2}$

33. $\frac{a - b + c}{x^2}$ 35. $\frac{3x - 2}{x^2}$ 37. $\frac{yz + xz + xy}{xyz}$ 39. $\frac{2x + 1}{x^2 + x - 6}$

41. $\frac{mm(3 + 2y)}{4y^2}$ 43. $\frac{5a - b}{a^2 - b^2}$ 45. $\frac{4x}{x^2 - 1}$ 47. $\frac{1 - 2x}{2}$

49. $\frac{a^3 - ab^2 + 1}{a^2 - b^2}$ 51. $\frac{5x - 2x^3 + 3}{x^2}$ 53. $\frac{V_1V_2 d + VV_2d_1 + VV_1d_2}{VV_1V_2}$

Exercise 9, Page 219

1. $\frac{85}{12}$ 3. $\frac{65}{132}$ 5. $\frac{69}{95}$ 7. $\frac{3(4x + y)}{4(3x - y)}$

9. $\frac{y}{y - x}$ 11. $\frac{x + y}{y - x}$

Exercise 10, Page 221

1. 12 3. 20 5. 14 7. 72

9. 24 11. 7 13. $-2/3$ 15. $-11/4$

17. $\frac{1}{2}$ 19. 4

Exercise 11, Page 224

1. 150 **3.** 5 and 6 **5.** 2.0 days **7.** 2.4 h
9. 40 min **11.** 26 h **13.** 240 mi **15.** 7.9 months
17. 6.1 h **19.** $11,360 **21.** $1500, $2000

Exercise 12, Page 228

1. $\dfrac{bc}{2a}$ **3.** $\dfrac{bz - ay}{a - b}$ **5.** $\dfrac{a^2d + 3d^2}{4ac + d^2}$ **7.** $\dfrac{b + cd}{a^2 + a - d}$
9. $2z/a + 3w$ **11.** $\dfrac{b - m}{3}$ **13.** $\dfrac{c - m}{a - b - d}$ **15.** b
17. $\dfrac{2c + bc + 3b}{3 - 2b + c}$ **19.** $\dfrac{c - ab}{a + b}$ **21.** $\dfrac{b + ac}{bc - a}$ **23.** $\dfrac{4c}{5a - b}$
25. $\dfrac{a(c + bd)}{d(a + 1)}$ **27.** $\dfrac{b(a^2 - 1)}{2}$ **29.** $\sqrt[3]{\dfrac{24Cp^2}{w^2}}$ **31.** $\dfrac{RR_1}{R_1 - R}$
33. $\dfrac{2(s - v_0t)}{t^2}$ **35.** $\dfrac{R_1(1 + \alpha t) - R}{R_1\alpha}$ **37.** $\dfrac{E + i_2R_2}{R_1 + R_2}$ **39.** $\dfrac{M - Fx}{R_1 - F}$
41. $\dfrac{R}{1 + \alpha(t - t_1)}$ **43.** $\dfrac{RT}{V^2 - Rg}$

Chapter Test, Page 231

1. $\dfrac{2ab}{a - b}$ **3.** $\dfrac{1}{3m - 1}$ **5.** $a^2 + 1 + \dfrac{1}{a^2}$
7. $\dfrac{a}{c}$ **9.** $(x - 5)(x + 3)$ **11.** $(x^3 - y^2)(x^3 + y^2)$
13. $(2x - y/3)(4x^2 + 2xy/3 + y^2/9)$
15. $2a(xy - 3)(xy + 3)$ **17.** $(3a + 4)(a - 2)$ **19.** $(x - 3)(x - 4)$
21. $4(a - b)(a + b)(a^4 + a^2b^2 + b^4)$
23. $(x - 3)(x + 1)$ **25.** $(a - 4)(a + 2)$ **27.** $(x - 10)(x - 11)$
29. $\dfrac{3a + 2b + c}{a - b}$ **31.** 4 **33.** -16
35. $\dfrac{np - m}{mp - n}$ **37.** $\dfrac{am + bn + cp}{m + n + p}$ **39.** 47/7
41. 19 29/45 **43.** $\dfrac{2ax^2y}{7w}$

CHAPTER EIGHT: SETS OF EQUATIONS

Exercise 1, Page 241

1. $x = 2, y = 1$ **3.** $x = 1, y = 2$ **5.** $x = 2, y = -1$
7. $x = -0.24, y = 0.90$ **9.** $x = 3, y = 5$ **11.** $x = -3, y = 3$
13. $x = 1, y = 2$ **15.** $x = 3, y = 2$ **17.** $x = 15, y = 6$
19. $x = 3, y = 4$ **21.** $x = 2, y = 3$ **23.** $x = 1.63, y = 0.0967$
25. $x = 2, y = 3$ **27.** $x = 6, y = 1$

29.
```
10 '    SEIDEL
20 '
30 '     This program uses the Gauss-Seidel method
40 '     to solve a set of two equations.
50 '     Enter your equations on lines 70 and 80.
60 '
70  DEF FNA(Y) = (Y + 4) / 3
80  DEF FNB(X) = 11 - 2 * X
90  Y = 0
100 X1 = FNA(Y)
110 Y1 = FNB(X1)
120 IF ABS(Y1 - Y) < .0001 THEN 150
130 Y = Y1
140 GOTO 100
150 LPRINT "The roots are "; X1, Y1
160 END
```

The roots are 2.99998 5.00004

Exercise 2, Page 246

1. $x = 60, y = 36$ **3.** $x = 87/7, y = 108/7$ **5.** $x = 15, y = 12$
7. $x = 4, y = 3$ **9.** $x = 4, y = 6$ **11.** $x = 1, y = 3$
13. $x = -2, y = 2$ **15.** $x = 1/3, y = 1/5$ **17.** $x = \frac{a + 2b}{7}, y = \frac{3b - 2a}{7}$

19. $x = \frac{c(n - d)}{an - dm}, y = \frac{c(m - a)}{an - dm}$

21.
```
10  '    TWO-EQ
20  '
30  '    This program solves a set of
40  '    two equations.
50  '
60  '    Enter the coefficients of the
70  '    first equation; A1, B1, C1, in
80  '    the first data statement and
90  '    the coefficients of the second
100 '    equation in the second data
110 '    statement.
120 '
130 DATA 2, -3, -4
140 DATA 1, 1, 3
150 READ A1, B1, C1
160 READ A2, B2, C2
170 PRINT " X", " Y"
180 LET D = A1*B2 - A2*B1
190 LET X = (C1*B2 - C2*B1)/D
200 LET Y = (A1*C2 - A2*C1)/D
210 PRINT X, Y
220 END
```

Exercise 3, Page 248

1. 16 and .8 **3.** 4/21 **5.** 7 and 41
7. 7 ft/s for A. $5\frac{1}{4}$ ft/s for B **9.** \$2500 @ 4%
11. \$6000 @ 4% and \$4000 @ 5% **13.** 2.86 kg lead and 18.6 kg zinc.
15. 25.1 days, 37.6 days **17.** 12,000 gal/h, 18,000 gal/h **19.** 93,000, 162,000
21. $I_1 = 13$ mA, $I_2 = 22$ mA

Exercise 4, Page 252

1. 15, 20, 25 **3.** 1, 2, 3 **5.** 5, 6, 7
7. 1, 2, 3 **9.** 3, 4, 5 **11.** 6, 8, 10
13. 2, 3, 1 **15.** 1/2, 1/3, 1/4 **17.** 1, −1, 1/2
19. 361
21. 20 dollars, 24 half-dollars, 40 quarters **23.** $I_1 = 6.35$ mA, $I_2 = 0.275$ mA, $I_3 = -2.53$ mA

Chapter Test, Page 254

1. $x = 3, y = 5$ **3.** $x = 5, y = -2$ **5.** $x = 2, y = -1, z = 1$

7. $x = 1, y = 2, z = -3$ 9. $x = 8, y = 10$ 11. $x = 2, y = 3$

13. $x = (a + b - c)/2, y = (a - b + c)/2, z = (b - a + c)/2$

15. $x = 7, y = 5$ 17. 2 14/17 days 19. 40 sheep, 8 cows, 6 horses

CHAPTER NINE: MATRICES AND DETERMINANTS

Exercise 1, Page 260

1. A, B, D, E, F, I 3. C, H 5. I

7. B, I 9. F 11. 6

13. 4×3 15. 2×4 17. $\begin{pmatrix} 0 & 0 \\ 0 & 0 \\ 0 & 0 \\ 0 & 0 \end{pmatrix}$

19. $x = 2, y = 6, z = 8, w = 7$

Exercise 2, Page 264

1. -14 3. 15 5. 0

7. $ad - bc$ 9. 0 11. $11 i_1 i_2$

13. $x = 5, y = 1$ 15. $x = 9, y = 6$ 17. $x = 1, y = 1$

Exercise 3, Page 270

1. 11 3. 45 5. 48

7. -28 9. $x = 5, y = 6, z = 7$ 11. $x = 15, y = 20, z = 25$

13. $x = 1, y = 2, z = 3$ 15. $x = 3, y = 4, z = 5$ 17. $x = 6, y = 8, z = 10$

19. $x = 3, y = 6, z = 9$ 21. $x = 2, y = -1, z = 5$

Exercise 4, Page 279

1. 2 3. 18 5. -66

7. $x = 2, y = 3, z = 4, w = 5$

9. $x = -4, y = -3, z = 2, w = 5$

11. $x = a - c, y = b + c, z = 0, w = a - b$

13. $u = 8, w = 7, x = 4, y = 5, z = 6$

15. $w = 2, v = 3, x = 4, y = 5, z = 6$

17. \$40,000, \$30,000, \$24,000, \$26,000

19. 140 kg of alloy 1, 100 kg of alloy 2, 200 kg of alloy 3 and 160 kg of alloy 4.

Exercise 5, Page 288

1. $(12 \quad 13 \quad 5 \quad 8)$ 3. $\begin{pmatrix} 14 \\ -10 \\ 6 \\ 14 \end{pmatrix}$ 5. $\begin{pmatrix} 7 & 1 & -4 & -7 \\ -9 & -4 & 9 & 2 \\ 3 & 7 & -3 & -9 \\ -1 & -4 & -5 & 6 \end{pmatrix}$

7. $(-15 \quad -27 \quad 6 \quad -21 \quad -9)$ 9. $\begin{pmatrix} -12 & -6 & 9 & 0 \\ 3 & -18 & -9 & 12 \end{pmatrix}$

11. $7\begin{pmatrix} 3 & 1 & -2 \\ 7 & 9 & 4 \\ 2 & -3 & 8 \\ -6 & 10 & 12 \end{pmatrix}$ 13. $\begin{pmatrix} 6 & 16 & 3 \\ -16 & 8 & 19 \\ 31 & -14 & 15 \end{pmatrix}$ 17. 35

19. -61

21. $(10 \quad 27 \quad -14 \quad 15 \quad -7)$

23. $(-60 \quad 18 \quad 10)$

25. $(15 \quad 2 \quad 40 \quad 2 \quad 34)$

27. $\begin{pmatrix} 19 & -13 \\ 17 & -43 \end{pmatrix}$

29. $\begin{pmatrix} -20 & 10 \\ 8 & -2 \\ -4 & 16 \end{pmatrix}$

31. $\begin{pmatrix} 6 & -10 \\ 3 & -5 \end{pmatrix}$

33. $\begin{pmatrix} 0 & 2 \\ 2 & 0 \\ -10 & -6 \\ -5 & 0 \end{pmatrix}$

35. $\begin{pmatrix} 2 & 0 & -2 & 1 \\ 8 & 8 & -1 & 2 \\ 2 & 2 & 2 & 6 \end{pmatrix}$

37. $AB = \begin{pmatrix} 1 & 2 & -1 \\ -2 & 9 & 2 \\ -1 & -2 & 1 \end{pmatrix}$ $BA = \begin{pmatrix} 5 & -2 & 2 \\ 4 & -2 & 0 \\ -2 & 3 & 8 \end{pmatrix}$

39. $\begin{pmatrix} 788 & 1050 \\ 1225 & 1653 \end{pmatrix}$

41. $\begin{pmatrix} 5 \\ 1 \end{pmatrix}$

43. $\begin{pmatrix} 13 \\ -1 \\ 10 \end{pmatrix}$

45. $\begin{pmatrix} 6 & 4 & -8 \\ 0 & 0 & 0 \\ 9 & 6 & -12 \\ -3 & -2 & 4 \end{pmatrix}$

49.

```
10 '       SCALAR
20 '
30 '       This program multiplies a
40 '       matrix by a scalar.
50 '
60 '       Enter the scalar on line 120,
70 '       the number of rows on line 130,
80 '       the number of columns on line 140,
90 '       and the matrix in the DATA.
100 '
110 DIM M(10,10)
120 S = 7
130 R = 3
140 C = 3
150 FOR I = 1 TO R
160    FOR J = 1 TO C
170      READ M(I,J)
180      S(I,J) = M(I,J) * S
190      LPRINT M(I,J);
200    NEXT J
210    LPRINT
220 NEXT I
230 LPRINT
240 FOR I = 1 TO R
250    FOR J = 1 TO C
260      LPRINT S(I,J);
270    NEXT J
280    LPRINT
290 NEXT I
300 DATA 7, -3, 7
310 DATA -2, 9, 5
320 DATA 9, 1, 0
330 END
```

Exercise 7, Page 303

1. $\begin{pmatrix} 0.000 & -0.200 \\ 0.125 & 0.100 \end{pmatrix}$

3. $\begin{pmatrix} 0.118 & 0.059 \\ -0.020 & 0.157 \end{pmatrix}$

5. $\begin{pmatrix} 0.07377 & 0.16803 & 0.05738 \\ 0.14754 & -0.16393 & 0.11475 \\ 0.09016 & 0.09426 & -0.04098 \end{pmatrix}$

7. $\begin{pmatrix} 0.500 & 0.000 & -0.500 & 0.000 \\ 10.273 & -0.545 & -13.091 & 1.364 \\ -4.727 & 0.455 & 5.909 & -0.636 \\ -4.591 & 0.182 & 5.864 & -0.455 \end{pmatrix}$

Chapter Test, Page 305

1. $(2 \quad 12 \quad 10 \quad -2)$

3. 86

5. 20

7. 30

9. $x = -4, y = -3, z = 2, w = 1$

11. $x = 2, y = -1, z = 1$

13. $x = 3, y = 5$

15. $(54 \quad 27 \quad -18 \quad 72)$

17. $\begin{pmatrix} 0.063 & 0.250 \\ -0.188 & 0.250 \end{pmatrix}$

CHAPTER TEN: RADICALS AND RADICAL EQUATIONS

Exercise 1, Page 312

1. $\sqrt[4]{a}$ **3.** $\sqrt[4]{z^3}$ **5.** $\sqrt{m - n}$ **7.** $\sqrt[3]{\dfrac{y}{x}}$

9. $b^{1/2}$ **11.** y **13.** $(a + b)^{1/n}$ **15.** xy

17. $3\sqrt{2}$ **19.** $3\sqrt{7}$ **21.** $-2\sqrt[3]{7}$ **23.** $a\sqrt{a}$

25. $6x\sqrt{y}$ **27.** $2x^2\sqrt[3]{2y}$ **29.** $6y^2\sqrt[5]{xy}$ **31.** $a\sqrt{a - b}$

33. $3\sqrt{m^3 + 2n}$ **35.** $\dfrac{\sqrt{21}}{7}$ **37.** $\dfrac{\sqrt[3]{2}}{2}$ **39.** $\dfrac{\sqrt[3]{6}}{3}$

41. $\dfrac{\sqrt{2x}}{2x}$ **43.** $\dfrac{a\sqrt{15ab}}{5b}$ **45.** $\dfrac{x\sqrt{3}}{2}$ **47.** $s = 4t\sqrt{16t^2 - 120t + 325}$

Exercise 2, Page 316

1. $\sqrt{6}$ **3.** $\sqrt{6}$ **5.** $25\sqrt{2}$ **7.** $-7\sqrt[3]{2}$

9. $-5\sqrt[3]{5}$ **11.** $-\sqrt[4]{3}$ **13.** $10x\sqrt{2y}$ **15.** $\dfrac{11\sqrt{6}}{10}$

17. $(a + b)\sqrt{x}$ **19.** $(x - 2a)\sqrt{y}$ **21.** $-4a\sqrt{3x}$ **23.** 45

25. $16\sqrt{5}$ **27.** $24\sqrt[3]{5}$ **29.** $6\sqrt[6]{72}$ **31.** $10\sqrt[6]{72}$

33. $6x\sqrt{3ay}$ **35.** $2\sqrt[5]{x^3y^3}$ **37.** $\sqrt[4]{a^2b}$ **39.** $2xy\sqrt[6]{x^4yz^3}$

41. $9y$ **43.** $9x^3\sqrt[3]{4x}$ **45.** $9 - 30\sqrt{a} + 25a$ **47.** $250x\sqrt{2x}$

49. $250ax$ **51.** $\dfrac{3}{4}$ **53.** $\dfrac{\sqrt[6]{18}}{4}$ **55.** $\dfrac{3\sqrt[3]{2a}}{a}$

57. $\dfrac{4\sqrt[3]{18}}{3} + \sqrt[3]{4} + 3$ **59.** $\dfrac{3}{2}$ **61.** $\dfrac{2\sqrt[6]{a^5b^2c^3}}{ac}$ **63.** $\dfrac{1}{2c}\sqrt[12]{a^2bc^{10}}$

65. $\dfrac{8 + 5\sqrt{2}}{2}$ **67.** $\dfrac{4\sqrt{ax}}{a}$

Exercise 3, Page 320

1. 36 **3.** 4 **5.** 8 **7.** 8

9. 7 **11.** 4.79 **13.** 3 **15.** $C = \dfrac{1}{\omega^2 L - \omega\sqrt{Z^2 - R^2}}$

Chapter Test, Page 320

1. $2\sqrt{13}$ **3.** $3\sqrt[3]{6}$ **5.** $\sqrt[3]{2}$

7. $9x\sqrt[4]{x}$ **9.** $(a - b)x\sqrt[3]{(a - b)^2x}$ **11.** $\dfrac{\sqrt{a^2 - 4}}{a + 2}$

13. $x^2\sqrt{1 - xy}$ **15.** $-\dfrac{1}{46}(6 + 15\sqrt{2} - 4\sqrt{3} - 10\sqrt{6})$ **17.** $\dfrac{2x^2 + 2x\sqrt{x^2 - y^2} - y^2}{y^2}$

19. $26\sqrt[3]{x^2}$ **21.** $\sqrt[6]{a^3b^2}$ **23.** $9 + 12\sqrt{x} + 4x$

25. $7\sqrt{2}$ **27.** $\frac{25\sqrt{2}}{12}$ **29.** $\frac{\sqrt{2b}}{2b}$ **31.** $72\sqrt{2}$

33. 10 **35.** 14 **37.** 15.2

CHAPTER ELEVEN: QUADRATIC EQUATIONS

Exercise 1, Page 325

1. $a = 1, b = -9, c = -5$ **3.** $a = 3, b = 4, c = -21$ **5.** $a = 6, b = -5, c = -6$
7. checks

Exercise 2, Page 328

1. 0, 2/5 **3.** 0, 9/2 **5.** ± 3 **7.** ± 1
9. 0, 0.355 **11.** 3, -5 **13.** -4, 5 **15.** -1, 2
17. -1, -2 **19.** -2, 9 **21.** -4, -8 **23.** -4, 16
25. 5, 8 **27.** 3/2, -4 **29.** 1/5, -3 **31.** -1, 2
33. 1/6, $-5/4$ **35.** 6, -3 **37.** 1/2, $-2/3$ **39.** 9, -4
41. 1, $-1/6$ **43.** 4, $-2/3$ **45.** 5, $-17/3$ **47.** ± 6
49. 1, $-13/6$ **51.** 6, 15/4 **53.** 4 **55.** 2

Exercise 3, Page 331

1. $\frac{-7 \pm \sqrt{61}}{2}$ **3.** $4 \pm \sqrt{14}$ **5.** $\frac{3 \pm \sqrt{89}}{8}$ **7.** $\frac{-3 \pm \sqrt{21}}{3}$

9. $\frac{-7 \pm \sqrt{129}}{8}$

Exercise 4, Page 335

1. 8.83, 3.17 **3.** 3.89, -4.87 **5.** 1.96, -5.96 **7.** 0.401, -0.485
9. 0.170, -0.599 **11.** 2.87, 0.465 **13.** 0.907, -2.57 **15.** 5.87, -1.87
17. 2.31, 0.625 **19.** 5.08, 10.9 **21.** 0.132, -15.1 **23.** 12.0, -14.0
25. real and unequal **27.** not real **29.** not real

31.
```
10 '    QUAD-1
20 '
30 '    This program calculates the two roots of a quadratic using the
40 '    quadratic equation.
50 '
60 '    Run the program and enter A, B, and C in the input statements.
70 '
80  INPUT "What is the coefficient of the x squared term(a)"; A
90  INPUT "What is the coefficient of the x term(b)"; B
100 INPUT "What is the constant term(c)"; C
110 D = B^2 - 4 * A * C
120 IF D > 0 THEN GOTO 150
130 PRINT "The roots of this equation are imaginary."
140 GOTO 190
150 X1 = (-B + D^.5) / (2* A)
160 X2 = (-B - D^.5) / (2* A)
170 PRINT "The roots of this equation are "; X1 ;"and "; X2
180 END
```

Exercise 5, Page 337

1. 2/3 or 3/2 **3.** 4 and 11 **5.** 5 and 15

7. 4m × 6m **9.** 15 × 30 cm **11.** 162 × 200 m

13. 32 × 48 ft **15.** 6.47 in. **17.** 146 mi/h

19. 80 km/h **21.** 9 h/box and 12 h/box **23.** 7 m/day

25. $x = l$ or 0 **27.** 0.630 s and 8.385 s **29.** 125 Ω and 665 Ω

31. 0.381 A and 0.769 A **33.** −0.5 A and 0.1 A **35.** −1.1 V

Exercise 6, Page 341

1.

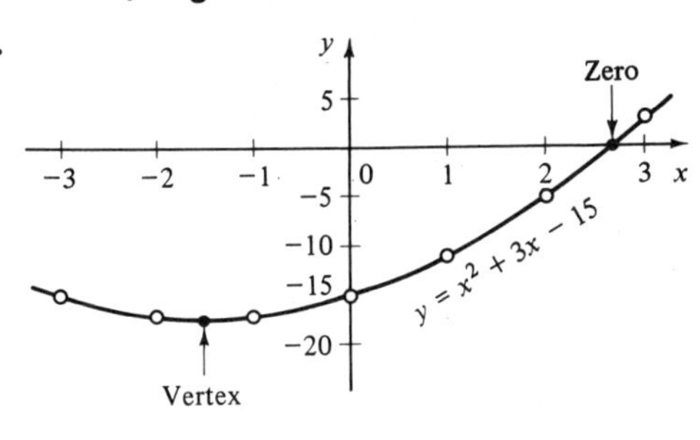

3.

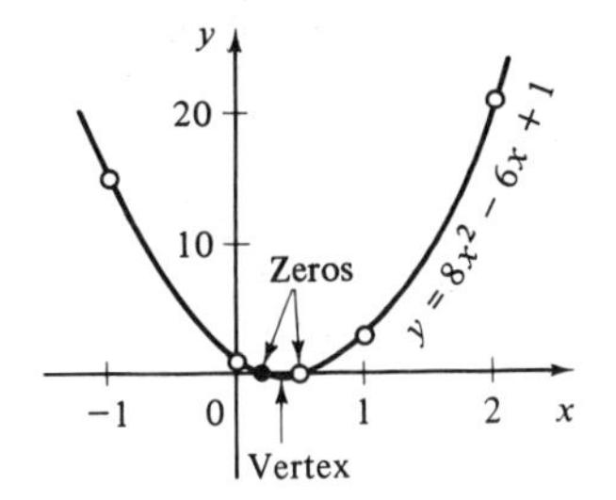

7.

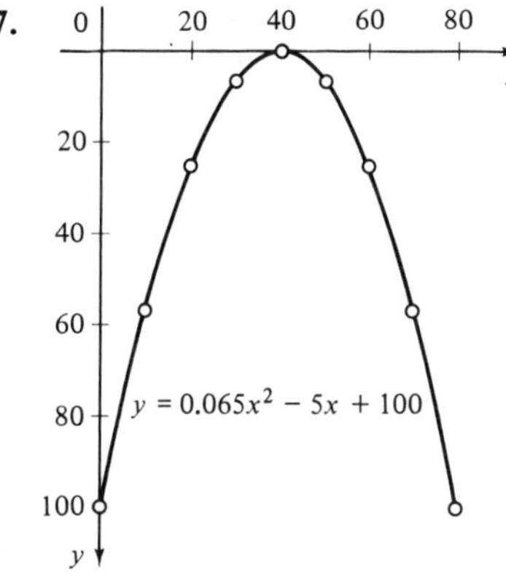

9. 20 cm

11. 6.33 ft

13.

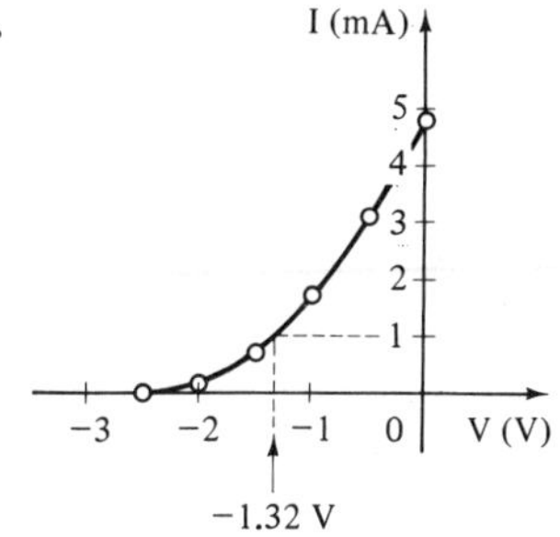

Exercise 7, Page 346

1. $\sqrt[3]{2}, \sqrt[3]{4}$
3. 1
5. 1
7. $-1, -1/27$
9. $1, -64$
11. $\dfrac{\sqrt[3]{2}}{8}$
13. 625
15. 496

Exercise 8, Page 349

1. $1, 3, -5$
3. $1, 2, -1$
5. $3/2, -4, -1$
7. $4/3, -8/3, -1$
9.

```
10 '   SIMPL-IT
20 '
30 ' This program finds roots by simple iteration.
40 '
50 ' Enter the equation, solved for x, on line 70.
60 '
70 DEF FNA(X) = .05882*X^3 + .05882*X^2 + .8823
80 X = 3
90 LET X1 = FNA(X)
100 LPRINT X
110 IF ABS(X - X1) < .00001 THEN 140
120 X = X1
130 GOTO 90
140 END
```

```
3
2.99982
2.999471
   ⋮
1.000385
1.000053
```

Exercise 9, Page 352

1. (6, 2)
3. (3, 2) and $(-5, 6)$
5. (2, 3) and $(-46, 15)$
7. $(2.61, -1.2)$ and $(-2.61, -1.2)$
9. (4.86, 0.845) and $(4.86, -0.845)$
11. $(7, -4)$ and $(4, -7)$

Chapter Test, Page 352

1. $6, -1$
3. $\pm 3, \pm 2$
5. 0, 5
7. $0, -2$
9. $5, -2/3$
11. $\frac{1}{2}, 1$
13. ± 3
15. $1.79, -2.79$
17. 4, 9
19. $0.692, -0.803$
21. $\pm\sqrt{10}$
23. $-2, \frac{1}{2}$
25. 0, 9
27. 1, 512
29. ± 5
31. $1, -3$
33. 200 bags
35. 15 ft × 30 ft
37. $4, -3, 2$
39. $(7.41, -0.41)$ and $(-0.41, 7.41)$

CHAPTER TWELVE: THE RIGHT TRIANGLE

Exercise 1, Page 358

1. 0.485 rad
3. 0.615 rad
5. 3.5 rad
7. 0.7595 rev
9. 0.191 rev
11. 0.215 rev
13. 162°
15. 171°
17. 29.45°
19. 244°57′45″
21. 161°54′36″
23. 185°58′19″

25.
```
10 '  DEG-MIN
20 '
30 '   This program accepts a decimal angle
40 '   and converts to deg, min, & sec.
50 '
60  INPUT "Enter the angle in decimal degrees"; A
70  LET B = A - INT(A)
80  LET B = B * 60
90  LET C = B - INT(B)
100 LET C = C * 60
110 PRINT INT(A); "DEG, "; INT(B); "MIN, "; C; "SEC"
120 END
```

Exercise 2, Page 365

9. $\sin\theta = 0.528$ $\cos\theta = 0.849$ $\tan\theta = 0.622$
$\cot\theta = 1.61$ $\sec\theta = 1.18$ $\csc\theta = 1.89$

11. $\sin\theta = 0.447$ $\cos\theta = 0.894$ $\tan\theta = 0.500$

13. $\sin\theta = 0.832$ $\cos\theta = 0.555$ $\tan\theta = 1.50$

15. $\sin\theta = 0.491$ $\cos\theta = 0.871$ $\tan\theta = 0.564$

17. 0.979 **19.** 3.04 **21.** 0.9319 **23.** 1.024

25. $\sin\theta = 0.955$ $\cos\theta = 0.296$ $\tan\theta = 3.23$

27. $\sin\theta = 0.546$ $\cos\theta = 0.838$ $\tan\theta = 0.652$

29. $\sin\theta = 0.677$ $\cos\theta = 0.736$ $\tan\theta = 0.920$

31. 30° **33.** 31.9° **35.** 28.9°

37. $\sin A = 3/5$ $\cos A = 4/5$ $\tan A = 3/4$

39. $\sin A = 5/13$ $\cos A = 12/13$ $\tan A = 5/12$

41. 75.0° **43.** 40.5° **45.** 13.6°

47.
```
10 ' COS-X
20 '
30 '   This program uses series
40 '   to calculate cosines.
50 '
60 '   Enter the angle in degrees.
70 '
80  INPUT "What is the value of x"; X
90  R = X * (3.1416 / 180)
100 D = 1 : S = 1
110 FOR N = 2 TO 8 STEP 2
120  FOR F = 1 TO N
130   D = D * F
140  NEXT F
150  C = (R^N / D) * (-1)^S + C
160  S = S + 1
170  D = 1
180 NEXT N
190 A = C + 1
200 PRINT A
210 END
```

Exercise 3, Page 371

1. $B = 47.1°$, $b = 167$, $c = 228$ **3.** $b = 1.08$, $A = 58.1°$, $c = 2.05$

5. $B = 0.441$ rad, $b = 134$, $c = 314$ **7.** $c = 470$, $A = 54.3°$, $B = 35.7°$

9. $a = 2.80$, $A = 35.2°$, $B = 54.8°$ **11.** $b = 25.6$, $A = 46.9°$, $B = 43.1°$

13. cos 52° **15.** cot 71° **17.** tan 26.8°

19. cot 54°46′ **21.** cos 0.696 rad **23.** 1690 m

25. 2760 ft **27.** $\theta = 30.2°$, $AB = 20.5$ ft **29.** 38°

31. 63° and 117° **33.** 125 cm **35.** 9.398 cm

37. 7.91 cm **39.** $p = 0.866$ cm, $r = 0.433$ cm

41.
```
10 '    RIGHT-TR
20 '
30 ' This program solves a right triangle when two
40 ' sides, or one side and one angle are known.
50 '
60  INPUT "Which method do you wish to use(SS, AS)"; A$
70  IF A$ = "AS" THEN GOTO 170
80  INPUT "What are the lengths of the sides a and b"; A, B
90  PRINT " Angle A", " Angle B", " Side a"," Side b"," Side c"
100 C = (A^2 + B^2)^(1/2)
110 R1 = ATN(A / B)
120 R2 = ATN(B / A)
130 A1 = R1 * (180 / 3.1416)
140 A2 = R2 * (180 / 3.1416)
150 PRINT A1, A2, A, B, C
160 STOP
170 INPUT "What is the length of side a"; A
180 INPUT "What is angle A, in degrees"; A1
190 R1 = A1 * (3.1416 / 180)
200 B = A / TAN(R1)
210 R2 = ATN(B / A)
220 A2 = R2 * (180 / 3.1416)
230 C = (A^2 + B^2)^(1/2)
240 PRINT " Angle A", " Angle B", " Side a", " Side b", "Side c"
250 PRINT A1, A2, A, B, C
260 END
```

Exercise 4, Page 381

1. $V_x = 3.28$, $V_y = 3.68$ **3.** $V_x = 0.9917$, $V_y = 1.602$ **5.** $R = 616$, $\theta = 51.7°$
7. $R = 8811$, $\theta = 56.70°$ **9.** $R = 2.42$, $\theta = 31.1°$ **11.** $A = 8.36$, $\theta = -35.3°$
13. $A = 4.57$, $\theta = 240.2°$ **15.** $R = 6.76$, $\theta = 70.6°$ **17.** $R = 1070$, $\theta = -106°$
19. $R = 8.70$, Angle $= 23.5°$ **21.** $R = 1.836$, Angle $= 6°06'$ **23.** $76\bar{0}0$ lb
25. 25.5 ft **27.** 3760 N **29.** 5.70 m/min, 1.76 min
31. 115 mi/h **33.** $Z = 3774\ \Omega$, $R = 2690\ \Omega$ **35.** $X = 458\ \Omega$, $R = 861\ \Omega$
37. $R = 4427\ \Omega$, $\phi = 37.2°$

Chapter Test, Page 383

1. 38°12′, 0.667 rad, 0.106 rev. **3.** 157.3°, 157°17′, 0.4369 rev
5. $\sin\theta = 0.919, \cos\theta = 0.394, \tan\theta = 2.33, \cot\theta = 0.429, \sec\theta = 2.54, \csc\theta = 1.09, \theta = 66.8°$
7. $\sin\theta = 0.600, \cos\theta = 0.800, \tan\theta = 0.75, \cot\theta = 1.33, \sec\theta = 1.25, \csc\theta = 1.67, \theta = 36.9°$
9. 0.956, 0.294, 3.25, 0.308, 3.40, 1.05. **11.** 0.867, 0.498, 1.74, 0.574, 2.01, 1.15
13. 34.5° **15.** 70.7° **17.** 65.88°
19. $B = 61.5°$, $a = 2.02$, $c = 4.23$ **21.** $V_x = 356$, $V_y = 810$
23. $R = 473$, $\theta = 35.5°$ **25.** 7.29 ft

CHAPTER THIRTEEN: OBLIQUE TRIANGLES

Exercise 1, Page 394

1. II **3.** IV **5.** I
7. II **9.** II or III **11.** IV

13. III

15. pos

17. pos

19. pos

21. neg

	sin	cos	tan	cot	sec	csc
23.	+	−	−			
25.	−	+	−			
27.	+	−	−			
29.	0.947	−0.316	−3	−0.333	−3.16	1.05
31.	−0.471	−0.882	0.533	1.88	−1.13	−2.13
33.	0	1	0	−	1	−
35.	−3/5	−4/5	3/4			
37.	$-2/\sqrt{5}$	$-1/\sqrt{5}$	2			
39.	2/3	$\sqrt{5}/3$	$2/\sqrt{5}$			
41.	0.9816	−0.1908	−5.145			
43.	−0.4848	0.8746	−0.5543			
45.	−0.8898	0.4563	−1.950			
47.	0.8090	−0.5878	−1.376			
49.	0.9108	−0.4128	−2.206			
51.	−0.2264	−0.9740	0.2324			
53.	0.7917	−0.6109	−1.296			
55.	0.9738	−0.2274	−4.283			

57. 0.8192

59. 4.511

61. 1.711

63. 1.141

65. 135° and 315°

67. 33.2° and 326.8°

69. 81.1° and 261.1°

71. 90° and 270°

73. 227.4°, 312.6°

75. 34.2°, 325.8°

77. 102.6°, 282.6°

79. 280

81. 15.2°

Exercise 2, Page 401

1. $A = 32°48'$, $C = 100°57'$, $c = 413$

3. $A = 1.88$ rad, $B = 0.51$ rad, $a = 29$

5. $B = 1.24$ rad, $b = 8.20$, $c = 6.34$

7. $A = 103.6°$, $a = 21.7$, $c = 15.7$

9. $A = 23°$, $a = 43$, $b = 90$

11. $A = 106°25'$, $a = 90.74$, $c = 29.55$

13. $C = 5\pi/6$, $b = 34.16$, $c = 82.12$

15.
```
10 ' SINE-LAW
20 '
30 ' This program solves a triangle
40 ' by the law of sines.
50 '
60  INPUT "Which method (AAS, SSA, ASA)"; A$
70  IF A$ = "SSA" THEN 230
80  IF A$ = "ASA" THEN 390
90  INPUT "The two angles, in degrees"; A1, A2
100 INPUT "What is the length of side a"; A
110 R1 = A1 * (3.1416 / 180)
120 R2 = A2 * (3.1416 / 180)
130 B = (A * SIN(R2)) / SIN(R1)
140 R3 = 3.1416 - R1 - R2
150 C = (A * SIN(R3)) / SIN(R1)
```

```
160 A3 = R3 * (180 / 3.1416)
170 PRINT " Angle A", " Angle B", " Angle C"
180 PRINT A1, A2, A3
190 PRINT
200 PRINT " Side a", " Side b", " Side c"
210 PRINT A, B, C
220 STOP
230 INPUT "Lengths of the sides a and b"; A, B
240 INPUT "What is angle A, in degrees"; A1
250 R1 = A1 * (3.1416 / 180)
260 S = (B * SIN(R1)) / A
270 R2 = ATN(S / (1 - S^2)^.5)
280 A2 = R2 * (180 / 3.1416)
290 R3 = 3.1416 - R1 - R2
300 C = (A * SIN(R3)) / SIN(R1)
310 A3 = R3 * (180 / 3.1416)
320 PRINT " Angle A", " Angle B", " Angle C"
330 PRINT A1, A2, A3
340 PRINT
350 PRINT
360 PRINT " Side a", " Side b", " Side c"
370 PRINT A, B, C
380 STOP
390 INPUT "Angles A and B, in degrees"; A1, A2
400 INPUT "What is the length of side c"; C
410 R1 = A1 * (3.1416 / 180)
420 R2 = A2 * (3.1416 / 180)
430 R3 = 3.1416 - R1 - R2
440 A3 = R3 * (180 / 3.1416)
450 B = (C * SIN(R2)) / SIN(R3)
460 A = (C * SIN(R1)) / SIN(R3)
470 PRINT " Angle A", " Angle B", " Angle C"
480 PRINT A1, A2, A3
490 PRINT
500 PRINT " Side a", " Side b", " Side c"
510 PRINT A, B, C
520 END
```

Exercise 3, Page 406

1. $A = 56.9°$, $B = 95.8°$, $c = 70.1$

3. $B = 48.7°$, $C = 63.0°$, $a = 22.6$

5. $A = 44.2°$, $B = 29.8°$, $c = 21.7$

7. $B = 25.5°$, $C = 19.5°$, $a = 452$

9. $A = 45.6°$, $B = 80.4°$, $C = 54.0°$

11. $B = 30°41'$, $C = 34°01'$, $a = 82.9$

13. $A = 1.20$ rad, $B = 0.77$ rad, $c = 8.95$

15. $B = 0.18$ rad, $C = 0.21$ rad, $a = 3.69$

17. $A = 66.21°$, $B = 72.02°$, $c = 1052$

19. $A = 69.7°$, $B = 51.7°$, $C = 58.6°$

21.

```
10 ' COS-LAW
20 '
30 ' This program uses the law of cosines
40 ' to solve an oblique triangle.
50 '
60  INPUT "Which method do you wish to use (SAS, SSS)"; A$
70  IF A$ = "SSS" THEN GOTO 240
80  INPUT "What is the length of the side opposite the acute angle(a)"; A
90  INPUT "What is the length of the other side(b)"; B
```

```
100 INPUT "What is angle C, in degrees"; A3
110 R3 = A3 * (3.1416 / 180)
120 C = (A^2 + B^2 - 2 * A * B * COS(R3))^.5
130 S = (A * SIN(R3)) / C
140 R1 = ATN(S / (1 - S^2)^.5): 'We must use a trig-identity here.
150 A1 = R1 * (180 / 3.1416)
160 R2 = 3.1416 - R1 - R3
170 A2 = R2 * (180 / 3.1416)
180 PRINT " Angle A", " Angle B", " Angle C"
190 PRINT A1, A2, A3
200 PRINT
210 PRINT " Side a", " Side b", " Side c"
220 PRINT A, B, C
230 STOP
240 INPUT "What is one side opposite an acute angle(a)"; A
250 INPUT "What is the other side opposite an acute angle(b)"; B
260 INPUT "What is the final side (c)"; C
270 S = (B^2 + C^2 - A^2) / (2 * B * C)
280 R1 = ATN((1 - S^2)^.5 / S) : ' This is another trig-identity.
290 A1 = R1 * (180 / 3.1416)
300 S = (B * SIN(R1)) / A
310 R2 = ATN(S / (1 - S^2)^.5)
320 A2 = R2 * (180 / 3.1416)
330 R3 = 3.1416 - R1 - R2
340 A3 = R3 * (180 / 3.1416)
350 PRINT " Angle A", " Angle B", " Angle C"
360 PRINT A1, A2, A3
370 PRINT
380 PRINT " Side a", " Side b", " Side c"
390 PRINT A, B, C
400 END
```

Exercise 4, Page 409

1. 30.8 m, 85.6 m
3. 5.56 m
5. 32.3°, 60.3°, 87.4°
7. N48°48′W
9. S59°24′E
11. 598 km
13. 77.3 m and 131 m
15. 107 ft
17. 337 mm
19. 33.7 cm
21. 73.3 in.
23. 53.8 mm, 78.2 mm
25. 419
27. $A = 45°$, $B = 60°$, $C = 75°$, $a = 994$, $b = 1220$, $c = 1360$
29. 21.9 ft

Exercise 5, Page 416

1. $R = 521$ at 10.0°
3. $R = 87.1$ at 31°54′
5. $R = 6690$ at 0.720 rad
7. $R = 37.3$ N at 20.5°
9. 121 N, N59°58′W
11. $R = 1715$ N at 29.8°
13. 20.7° and 44.3°
15. Airplane S84.8°W, wind S37.4°E
17. 413 km/h, N41°12′E
19. 632 km/h, 3.09°
21. 26.9 at 32.4°
23. 39.8 at 26.2°
25.

```
10 '  VECTORS
20 '
30 '  This program accepts the direc-
40 '  tion and magnitude of any number
50 '  of vectors and computes the
60 '  magnitude and direction of the
70 '  resultant.
80 '
```

```
90  '  Enter the vectors in the data
100 '  statements, first magnitude,
110 '  then direction in degrees
120 '
130 READ M, T1
140 IF M = 0 THEN 210
150 LET T = T1 * (3.1416 / 180)
160 LET X = M * COS(T)
170 LET Y = M * SIN(T)
180 LET X1 = X1 + X
190 LET Y1 = Y1 + Y
200 GOTO 130
210 LET R = SQR(X1^2 + Y1^2)
220 LET T2 = ATN(Y1 / X1)
230 LET T3 = T2 * (180 / 3.1416)
240 PRINT R; " at "; T3; "degrees."
250 DATA 273, 34
260 DATA 179,143
270 DATA 203,225
280 DATA 138,314
290 DATA 0, 0
300 END
```

Chapter Test, Page 419

1. $A = 24.1°$, $B = 20.9°$, $c = 77.6$
3. $A = 61.6°$, $C = 80.0°$, $b = 1.30$
5. $B = 20.2°$, $C = 27.8°$, $a = 82.4$
7. IV **9.** II **11.** negative
13. negative **15.** negative
17. -0.800, -0.600, 1.33, 0.750, -1.67, -1.25
19. 1200 at 36.3° **21.** 22.1 at 121° **23.** 1.11 km
25. 0.0872, -0.9962, -0.0875, -11.43, -1.004, 11.47
27. 0.7948, -0.6069, -1.310, -0.7636, -1.648, 1.258
29. 0.4698 **31.** 1.469 **33.** S3.5°E
35. 130.8°, 310.8° **37.** 47.5°, 132.5° **39.** 80.0°, 280.0°
41. 695 lb at 17.0°

CHAPTER FOURTEEN: ADDITIONAL TOPICS IN TRIGONOMETRY

Exercise 1, Page 425

1. $\frac{\pi}{3}$ **3.** $\frac{11\pi}{30}$ **5.** $\frac{7\pi}{10}$ **7.** $\frac{13\pi}{30}$
9. $\frac{20\pi}{9}$ **11.** $\frac{9\pi}{20}$ **13.** 22.5° **15.** 147°
17. 20.0° **19.** 158° **21.** 24.0° **23.** 15.0°
25. 0.8660 **27.** 0.7071 **29.** -0.3090 **31.** 3.864
33. -2.747 **35.** -0.8090 **37.** 1.366 **39.** 0.5000
41. 0.1585 **43.** 6.07 **45.** $23\bar{0}$ **47.** 43.5 in.
49. 2.21 rad **51.** 1.24 rad **53.** 0.824 **55.** 125 ft
57. 3790 mi **59.** 589 m **61.** 24.5 cm **63.** 2.0×10^8 mi
65. 0.251 m **67.** 607 cm^2 **69.** 69.4°, 69.4°, 41.2° **71.** 3640 km
73. 5.17 in.
75.

```
10 ' SM-ANGLE
20 '
30 ' This program computes sines and
40 ' cosines of small angles in radians.
50 '
60  LPRINT "Radians", "Sine", "Tangent"
70  FOR A = 0 TO 10 STEP .5
80  R = A * (3.1416 / 180)
```

Radians	Sine	Tangent
0.00000	0.00000	0.00000
0.00873	0.00873	0.00873
0.01745	0.01745	0.01746
0.02618	0.02618	0.02619
⋮	⋮	⋮
0.16581	0.16505	0.16734
0.17453	0.17365	0.17633

```
90  S = SIN(R)
100 T = TAN(R)
110 A$="#.#####     #.#####        #.#####"
120 LPRINT USING A$; R; S; T
130 NEXT A
140 END
```

Exercise 2, Page 430

1. 194 rad/s, 11,100 deg/s
3. 12.9 rev/min, 1.35 rad/s
5. 8.02 rev/min, 0.840 rad/s
7. 4350 deg/s
9. 0.00749 s
11. 353 rev/min
13. 1.62 m/s
15. 1037 mi/h
17. 57.6 rad/s
19. 155 rev/min
21. 336 mi/h

Exercise 3, Page 440

1. 1 s, 7

3.

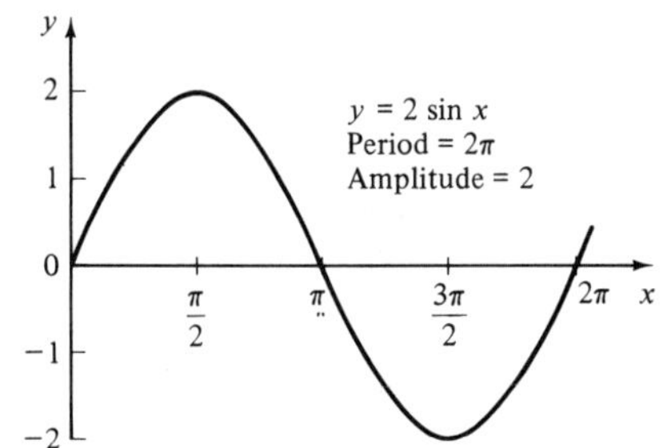

5.

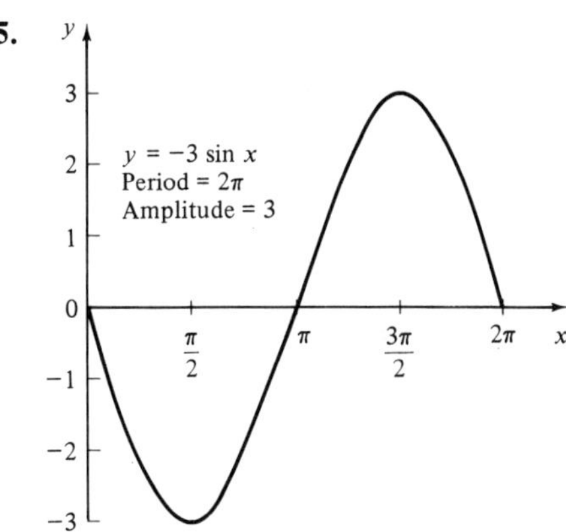

7.

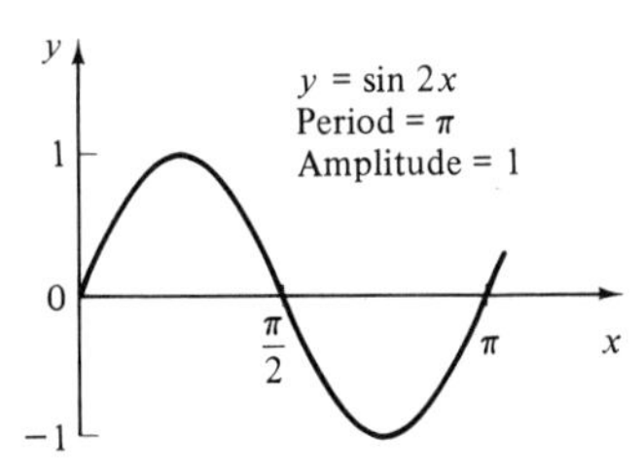

9.

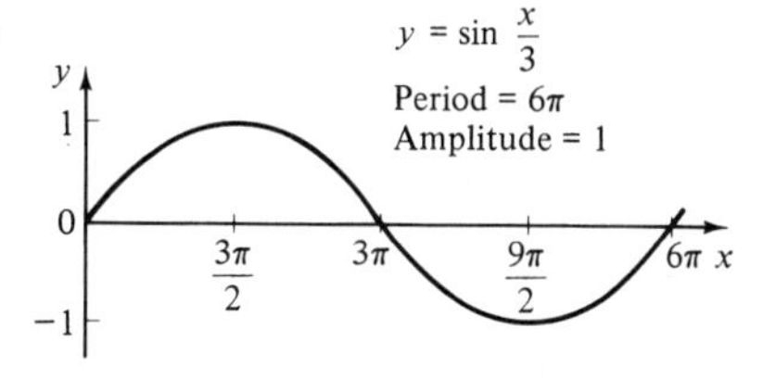

11.

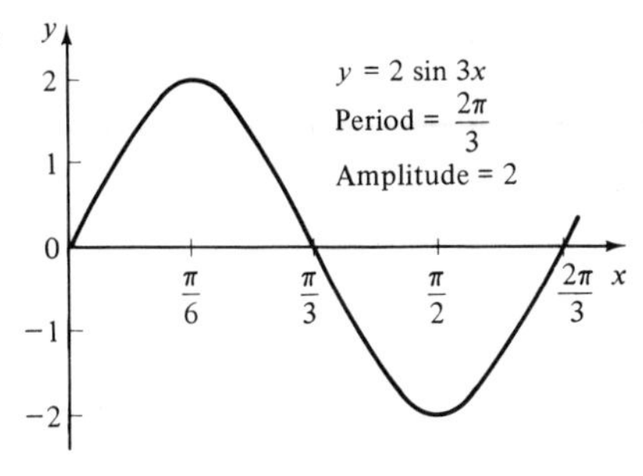

13.

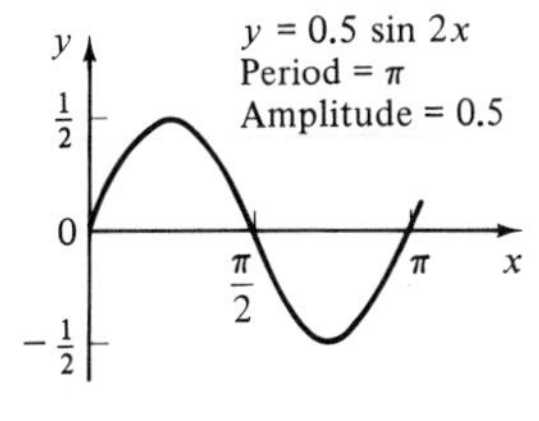

15.

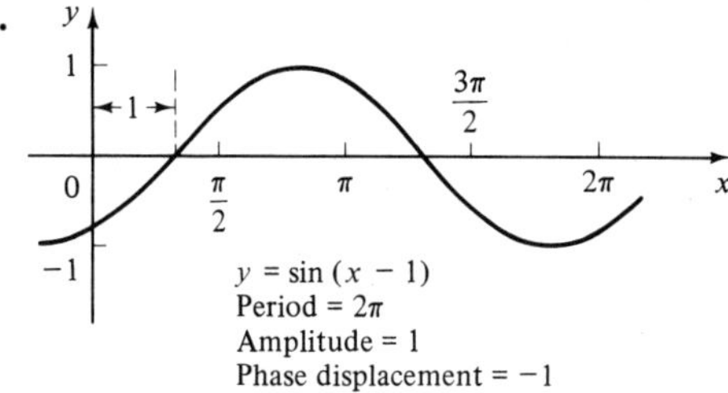

17.

19.

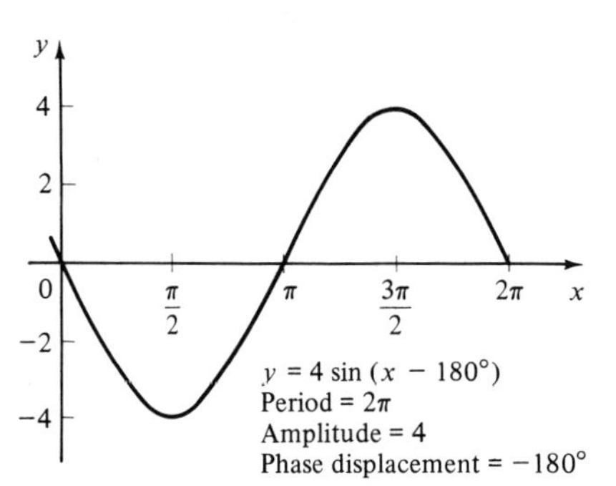

21.

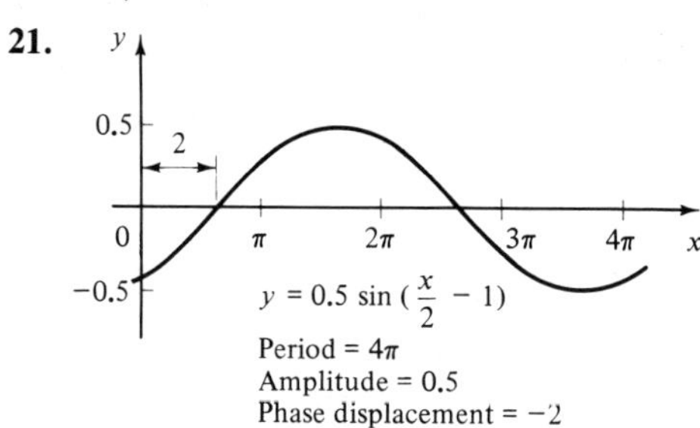

23. **(a)** $y = \sin 2x$

(b) $y = \sin\left(x - \frac{\pi}{2}\right)$

(c) $y = 2 \sin (x - \pi)$

25. $y = -2 \sin\left(\frac{x}{3} + \frac{\pi}{12}\right)$

Exercise 4, Page 443

1.

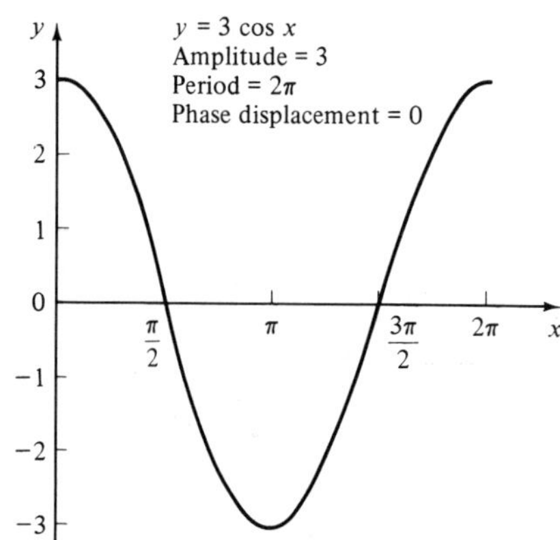

3.

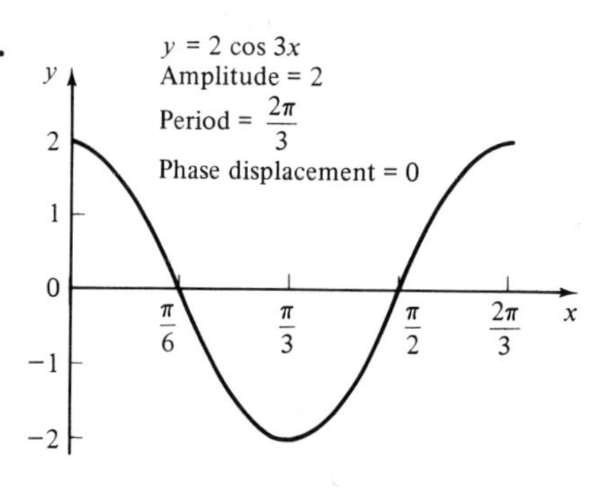

5.

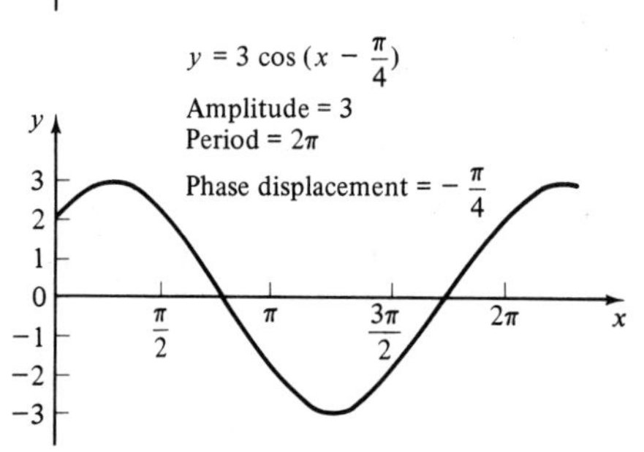

Exercise 5, Page 447

1. 0.0147 s, 427 rad/s
3. 0.0002 s, 31,400 rad/s
5. 8 Hz, 50.3 rad/s
7. 3.33 s
9. 0.0138 s, 72.4 Hz
11. 0.0126 s, 79.6 Hz
13. 400 ms, 10, 1.1 rad
15. $y = 5 \sin (750t + 15°)$

17.

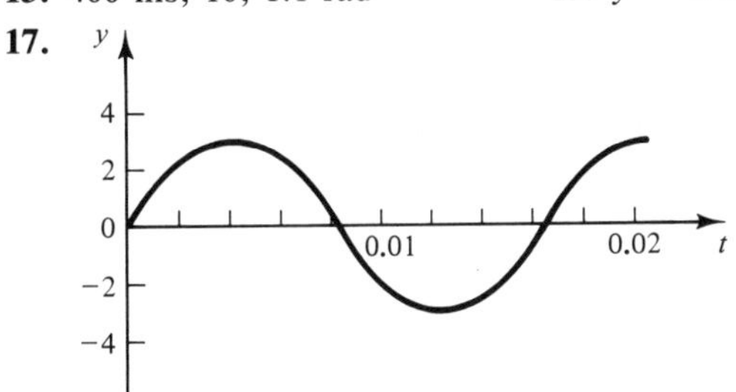

19.

21. $v = 155 \sin (377t + 22°)$

Exercise 6, Page 451

Problems **1, 3, 5, 7, 9,** and **11** plotted on graph.

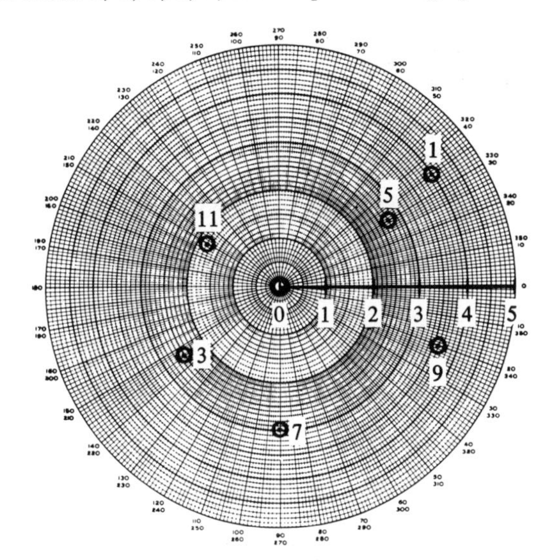

13. (6.71, 63.4°)
15. (5, 143°)
17. (7.61, 231°)
19. (597, 238°)
21. (3.41, 3.66)
23. (298, −331)
25. (2.83, −2.83)
27. (12.3, −8.60)
29. (−7.93, 5.76)

Exercise 7, Page 455

1. $\tan x - \sec x = \dfrac{\sin x}{\cos x} - \dfrac{1}{\cos x} = \dfrac{\sin x - 1}{\cos x}$

3. $\tan \theta \csc \theta = \dfrac{\sin \theta}{\cos \theta} \cdot \dfrac{1}{\sin \theta} = \dfrac{1}{\cos \theta}$

5. $\dfrac{\tan \theta}{\csc \theta} + \dfrac{\sin \theta}{\tan \theta} = \dfrac{\sin \theta / \cos \theta}{1/\sin \theta} + \dfrac{\sin \theta}{\sin \theta / \cos \theta} = \dfrac{\sin^2 \theta}{\cos \theta} + \cos \theta = \dfrac{\sin^2 \theta + \cos^2 \theta}{\cos \theta} = \dfrac{1}{\cos \theta}$

7. $1 - \sec^2 x = 1 - (1 + \tan^2 x) = -\tan^2 x$

9. $\dfrac{\cos\theta}{\cot\theta} = \dfrac{\cos\theta}{\dfrac{\cos\theta}{\sin\theta}} = \sin\theta$

11. $\tan\theta\csc\theta = \dfrac{\sin\theta}{\cos\theta}\cdot\dfrac{1}{\sin\theta} = \dfrac{1}{\cos\theta} = \sec\theta$

13. $\sec x \sin x = \dfrac{1}{\cos x}\cdot\sin x = \tan x$

15. $\csc\theta\tan\theta - \tan\theta\sin\theta = \tan\theta(\csc\theta - \sin\theta) = \dfrac{\sin\theta}{\cos\theta}\left(\dfrac{1}{\sin\theta} - \dfrac{\sin\theta}{1}\right) = \dfrac{1}{\cos\theta} - \dfrac{\sin\theta\sin\theta}{\cos\theta} = \dfrac{1-\sin^2\theta}{\cos\theta} = \dfrac{\cos^2\theta}{\cos\theta} = \cos\theta$

17. $\cot\theta\tan^2\theta\cos\theta = \dfrac{\cos\theta}{\sin\theta}\left(\dfrac{\sin\theta}{\cos\theta}\right)^2\cos\theta = \sin\theta$

19. $\dfrac{\sin\theta}{\cos\theta\tan\theta} = \dfrac{\tan\theta}{\tan\theta} = 1$

21. $\dfrac{1}{\sec^2 x} + \dfrac{1}{\csc^2 x} = \cos^2 x + \sin^2 x = 1$

23. $\csc x - \cot x\cos x = \dfrac{1}{\sin x} - \dfrac{\cos x\cos x}{\sin x} = \dfrac{1-\cos^2 x}{\sin x} = \dfrac{\sin^2 x}{\sin x} = \sin x$

25. $\dfrac{\sec x - \csc x}{1 + \cot x} = \dfrac{\dfrac{1}{\cos x} - \dfrac{1}{\sin x}}{1 + \dfrac{\cos x}{\sin x}} = \dfrac{(\sin x + \cos x)/\cos x\sin x}{(\sin x + \cos x)/\sin x} = \dfrac{1}{\cos x} = \sec x$

27. $\sec^2 x(1 - \cos^2 x) = \sec^2 x(\sin^2 x) = \sin^2 x/\cos^2 x = \tan^2 x$

29. $\cos\theta\sec\theta - \dfrac{\sec\theta}{\cos\theta} = \cos\theta\cdot\dfrac{1}{\cos\theta} - \dfrac{1}{\cos\theta\cos\theta} = 1 - \sec^2\theta = -\tan^2\theta$

31. $\tan x\cos x = \sin x$
$\dfrac{\sin x}{\cos x}\cdot\cos x = \sin x$
$\sin x = \sin x$

33. $\dfrac{\sin x}{\csc x} + \dfrac{\cos x}{\sec x} = 1$
$\dfrac{\sin x}{\dfrac{1}{\sin x}} + \dfrac{\cos x}{\dfrac{1}{\cos x}} = 1$
$\sin^2 x + \cos^2 x = 1$
$1 = 1$

35. $(\cos^2\theta + \sin^2\theta)^2 = 1$
$1^2 = 1$
$1 = 1$

37. $\dfrac{\csc\theta}{\sec\theta} = \cot\theta$
$\dfrac{\dfrac{1}{\sin\theta}}{\dfrac{1}{\cos\theta}} = \cot\theta$
$\dfrac{\cos\theta}{\sin\theta} = \cot\theta$
$\cot\theta = \cot\theta$

39. $\cos x + 1 = \dfrac{\sin^2 x}{1 - \cos x}$
$\cos x + 1 = \dfrac{1 - \cos^2 x}{1 - \cos x}$
$\cos x + 1 = 1 + \cos x$

41. $\cot^2 x - \cos^2 x = \cos^2 x\cot^2 x$
$\left(\dfrac{\cos x}{\sin x}\right)^2 - \cos^2 x = \cos^2 x\cot^2 x$
$\dfrac{\cos^2 x - \sin^2 x\cos^2 x}{\sin^2 x} = \cos^2 x\cot^2 x$
$\dfrac{\cos^2 x}{\sin^2 x}(1 - \sin^2 x) = \cos^2 x\cot^2 x$
$\cot^2 x(\cos^2 x) = \cos^2 x(\cot^2 x)$

43. $1 = (\csc x - \cot x)(\csc x + \cot x)$
$1 = \csc^2 x - \cot^2 x$

45. $\dfrac{\tan x + 1}{1 - \tan x} = \dfrac{\sin x + \cos x}{\cos x - \sin x}$

$\dfrac{(\sin x/\cos x) + 1}{1 - (\sin x/\cos x)} = \dfrac{\sin x + \cos x}{\cos x - \sin x}$

$\dfrac{\dfrac{\sin x + \cos x}{\cos x}}{\dfrac{\cos x - \sin x}{\cos x}} = \dfrac{\sin x + \cos x}{\cos x - \sin x}$

$\dfrac{\sin x + \cos x}{\cos x - \sin x} = \dfrac{\sin x + \cos x}{\cos x - \sin x}$

$1 = 1 + \cot^2 x - \cot^2 x$
$1 = 1$

47. $\dfrac{\sin\theta + 1}{1 - \sin\theta} = (\tan\theta + \sec\theta)^2$

$\dfrac{\sin\theta + 1}{1 - \sin\theta} = \left(\dfrac{\sin\theta}{\cos\theta} + \dfrac{1}{\cos\theta}\right)^2$

$\dfrac{\sin\theta + 1}{1 - \sin\theta} = \left(\dfrac{\sin\theta + 1}{\cos\theta}\right)^2$

$\dfrac{\sin\theta + 1}{1 - \sin\theta} = \dfrac{(\sin\theta + 1)(\sin\theta + 1)}{\cos^2\theta}$

$\dfrac{\sin\theta + 1}{1 - \sin\theta} = \dfrac{(\sin\theta + 1)(\sin\theta + 1)}{(1 - \sin\theta)(1 + \sin\theta)}$

$\dfrac{\sin\theta + 1}{1 - \sin\theta} = \dfrac{\sin\theta + 1}{1 - \sin\theta}$

49. $(\sec\theta - \tan\theta)(\tan\theta + \sec\theta) = 1$
$\sec^2\theta - \tan^2\theta = 1$
$(1 + \tan^2\theta) - \tan^2\theta = 1$
$1 = 1$

Exercise 8, Page 459

1. 60°, 120°, 240°, 300°
3. 15°, 135°
5. 30°, 150°, 210°, 330°
7. 45°, 135°, 225°, 315°
9. 0, 45°, 180°, 225°
11. 60°, 120°, 240°, 300°
13. 30°, 90°, 150°
15. 120°, 240°
17. 0°, 60°, 120°, 180°, 240°, 300°
19. 135°, 315°
21. 60°, 300°

Chapter Test, Page 460

1.

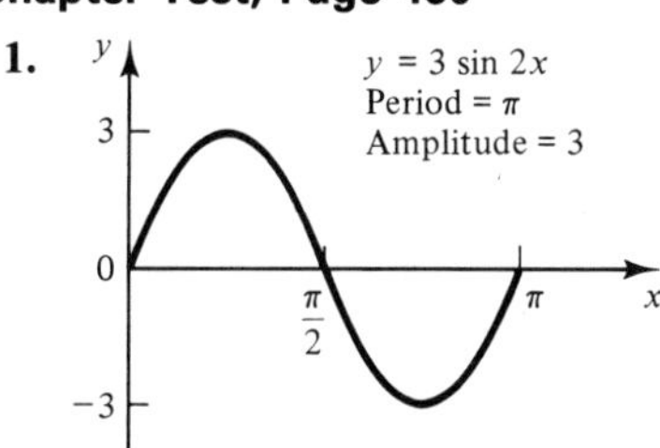

3.

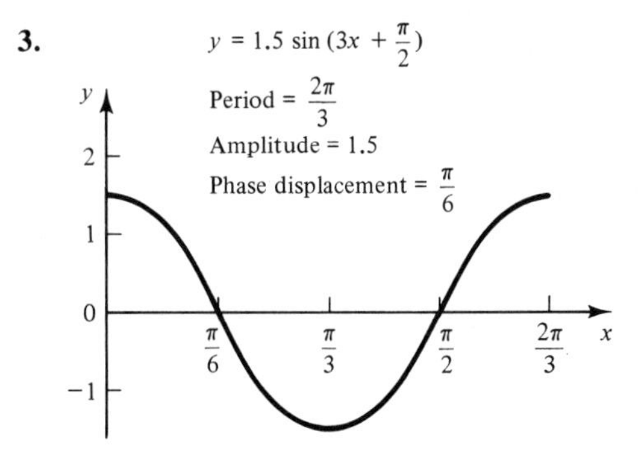

5.

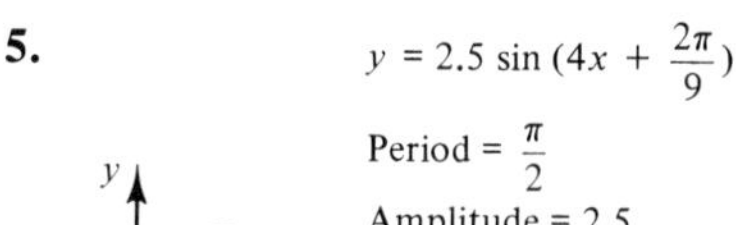

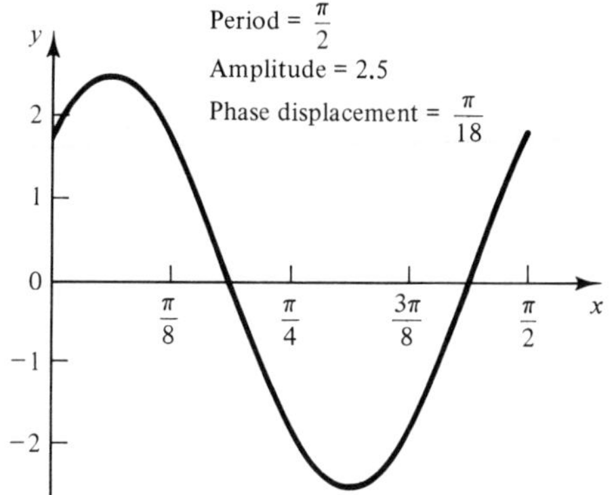

7. $\sec^2\theta + \tan^2\theta = \sec^4\theta - \tan^4\theta$
$\sec^2\theta + \tan^2\theta = (\sec^2\theta + \tan^2\theta)(\sec^2\theta - \tan^2\theta)$
$\sec^2\theta + \tan^2\theta = (\sec^2\theta + \tan^2\theta)(1)$
$\sec^2\theta + \tan^2\theta = \sec^2\theta + \tan^2\theta$

9. $\tan^2\theta \sin^2\theta = \tan^2\theta - \sin^2\theta$
$\tan^2\theta \sin^2\theta = (\tan\theta + \sin\theta)(\tan\theta - \sin\theta)$
$\tan^2\theta \sin^2\theta = (\tan\theta + \tan\theta\cos\theta)(\tan\theta - \tan\theta\cos\theta)$
$\tan^2\theta \sin^2\theta = \tan\theta(1 + \cos\theta)\tan\theta(1 - \cos\theta)$
$\tan^2\theta \sin^2\theta = \tan^2\theta(1 - \cos^2\theta)$
$\tan^2\theta \sin^2\theta = \tan^2\theta(\sin^2\theta)$

11. $\tan\theta + \cot\theta = \cot\theta \sec^2\theta$
$$\frac{\sin\theta}{\cos\theta} + \frac{\cos\theta}{\sin\theta} = \frac{\cos\theta}{\sin\theta}\frac{1}{\cos^2\theta}$$
$$\frac{\sin^2\theta + \cos^2\theta}{\sin\theta\cos\theta} = \frac{1}{\sin\theta\cos\theta}$$
$$\frac{1}{\sin\theta\cos\theta} = \frac{1}{\sin\theta\cos\theta}$$

13. $$\frac{\sin^3\theta + \cos^3\theta}{\sin\theta + \cos\theta} = 1 - \sin\theta\cos\theta$$
$$\frac{(\sin\theta + \cos\theta)(\sin^2\theta - \sin\theta\cos\theta + \cos^2\theta)}{\sin\theta + \cos\theta} = 1 - \sin\theta\cos\theta$$
$\sin^2\theta - \sin\theta\cos\theta + \cos^2\theta = 1 - \sin\theta\cos\theta$
$\sin^2\theta + \cos^2\theta - \sin\theta\cos\theta = 1 - \sin\theta\cos\theta$
$1 - \sin\theta\cos\theta = 1 - \sin\theta\cos\theta$

15. $y = 5\sin\left(\frac{2x}{3} + \frac{\pi}{9}\right)$
17. 405°
19. 72°
21. 0.800 rad
23. 770 m/min

25. $\sec^2\theta - \sin^2\theta\sec^2\theta = \sec^2\theta(1 - \sin^2\theta) = \frac{1}{\cos^2\theta}(\cos^2\theta) = 1$

27. $$\frac{\sin^3\theta + \cos^3\theta}{\sin\theta + \cos\theta} = \frac{\sin\theta(1 - \cos^2\theta) + \cos\theta(1 - \sin^2\theta)}{\sin\theta + \cos\theta} =$$
$$\frac{(\sin\theta + \cos\theta)(\sin^2\theta - \sin\theta\cos\theta + \cos^2\theta)}{\sin\theta + \cos\theta} = 1 - \sin\theta\cos\theta$$

29. $(\sec\theta - 1)(\sec\theta + 1) = \sec^2\theta - 1 = \tan^2\theta$

31. $$\frac{\cos\theta}{1 - \tan\theta} + \frac{\sin\theta}{1 - \cos\theta} = \frac{\cos\theta}{1 - \frac{\sin\theta}{\cos\theta}} + \frac{\sin\theta}{1 - \frac{\cos\theta}{\sin\theta}} = \frac{\cos^2\theta}{\cos\theta - \sin\theta} + \frac{\sin^2\theta}{\sin\theta - \cos\theta} =$$
$$\frac{\cos^2\theta - \sin^2\theta}{\cos\theta - \sin\theta} = \frac{(\cos\theta - \sin\theta)(\cos\theta + \sin\theta)}{\cos\theta - \sin\theta} = \sin\theta + \cos\theta$$

33. $\tan^2\theta(1 + \cot^2\theta) = \tan^2\theta\csc^2\theta = \frac{\sin^2\theta}{\cos^2\theta} \cdot \frac{1}{\sin^2\theta} = \sec^2\theta$

35.

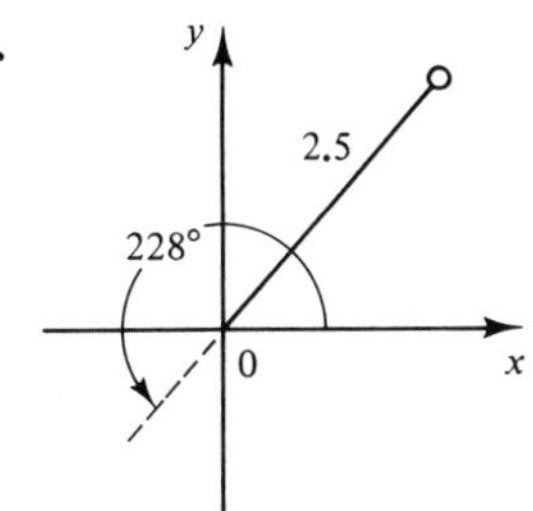

37. 30°, 90°, 150° **39.** 45°, 90°, 135°, 225°, 270°, 315° **41.** 63.4°, 71.6°, 234.4°, 251.6°
43. 30°, 90°, 150°, 210°, 270°, 330° **45.** 0°, 90°, 270° **47.** $7.62\angle 23.2°$
49. $57\angle 245°$ **51.** (−61, 22)

CHAPTER FIFTEEN: EXPONENTIAL AND LOGARITHMIC FUNCTIONS

Exercise 1, Page 470

1.

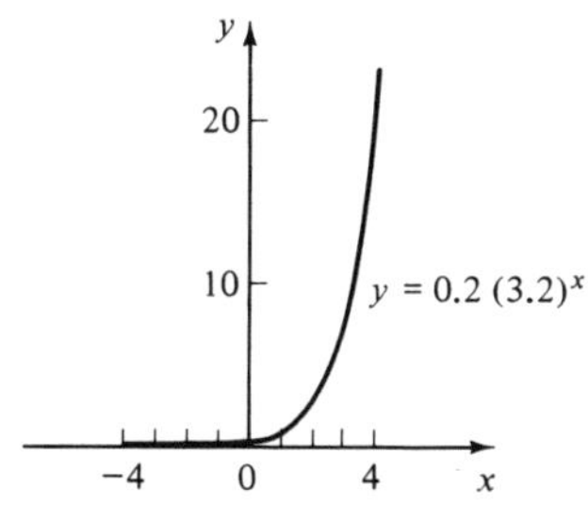

3.

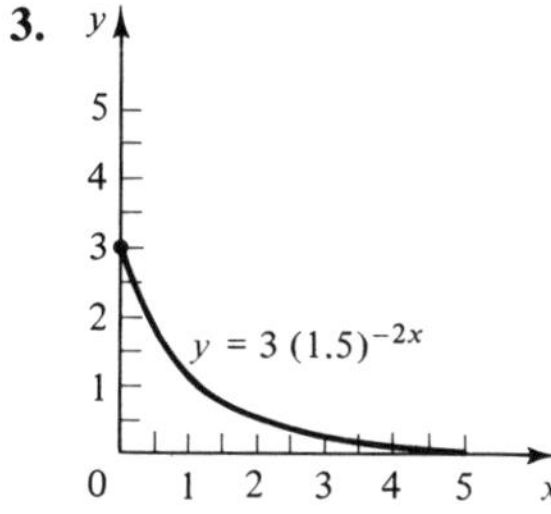

5.

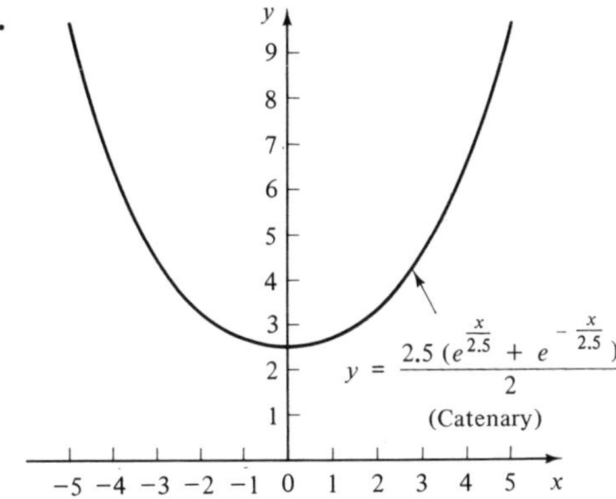

7. \$4851.94

9. (a) \$6.73
(b) \$7.33
(c) \$7.39
(d) \$7.39

11. \$1822.67 **13.** 284 units **15.** 5.16 billion
17. 80 million **19.** 506°C **21.** 0.101 A
23. 1048°F **25.** 9.604 in. Hg **27.** 72 mA
29. 30%

31.
```
10 ' E-SERIES
20 '
30 ' This program approximates
40 ' e with an infinite series.
50 '
60  D = 1 : E = 2
70  FOR N = 2 TO 6
80    FOR C = 1 TO N
90      D = D * C
100   NEXT C
110   E = E + 1 / D
120   D = 1
130 NEXT N
140 PRINT "E is equal to"; E
150 END
```

Exercise 2, Page 477

1. $\log_3 81 = 4$
3. $\log_4 4096 = 6$
5. $\log_x 995 = 5$
7. $\log_y 73.8 = 4$
9. $\log_7 83 = x$
11. $\log_{10} 4000 = x$
13. $\log_{10} b = x$
15. $\log_x b = a$
17. $2^4 = 16$
19. $4^5 = 1024$
21. $x^{285} = 54$
23. $e^{1.378} = x$
25. $a^c = b$
27. $10^y = x$
29. $e^x = 74.8$
31.

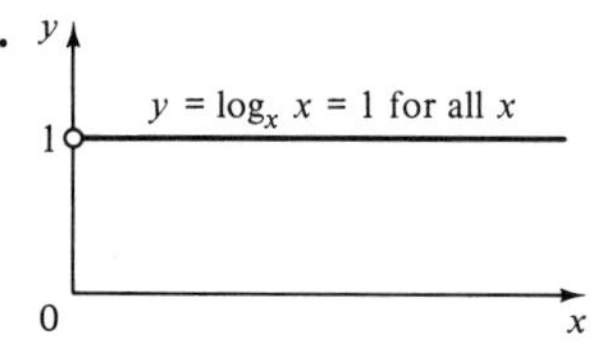

33.

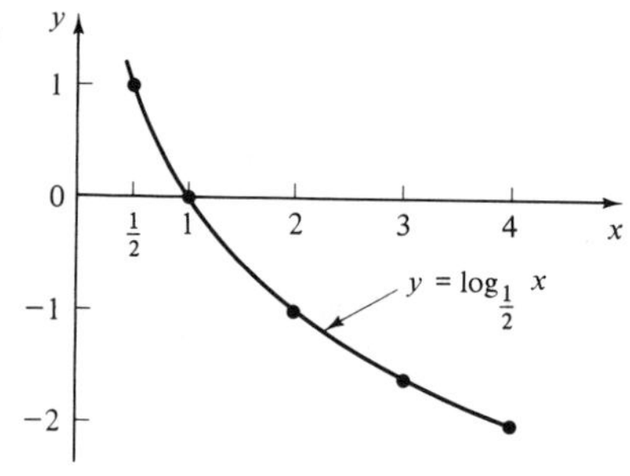

35.

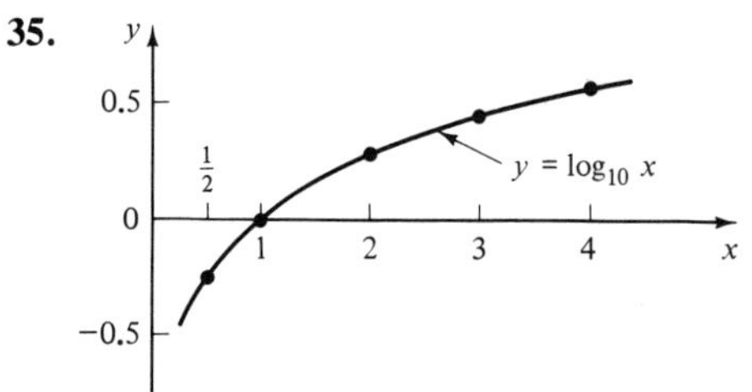

37. 2
39. 1/3
41. 2/3
43. 2/3
45. 2
47. 3
49. 25
51. 6
53. 3/2
55. 1/3
57. −3/2
59. 125

Exercise 3, Page 482

1. $\log 2 - \log 3$
3. $\log a + \log b$
5. $\log x + \log y + \log z$
7. $\log 3 + \log x - \log 4$
9. $-\log 2 - \log x$
11. $\log a + \log b + \log c + \log d$
13. $\log 12$
15. $\log 3/2$
17. $\log 27$
19. $\log \dfrac{a^3c^4}{b^2}$
21. $\log\left(\dfrac{xy^2z^3}{ab^2c^3}\right)$
23. $\log[\sqrt{x+2}(x - 2)]$
25. $2^x = xy^2$
27. $a^2 + 2b^2 = 0$
29. $p - q = 100$

Exercise 4, Page 485

1. 0.7731
3. 1.6839
5. 2.9222
7. 1.4378
9. −2.1325
11. 4.5437
13. 38.4
15. 187,500
17. 1.39×10^{-3}
19. 5.93×10^{-3}
21. 9528
23. 1.07×10^{-4}
25. 549.0
27. 322
29. 2.735
31. 12.58
33. 3.634

Exercise 5, Page 488

1. 3.8774
3. 7.7685
5. −0.1684
7. −4.7843
9. −9.9825
11. 10.5558
13. 17.22
15. 2.41
17. 0.694
19. 2.55
21. 7.572×10^6
23. 1.063×10^{-8}
25. 3.63
27. 1.639
29. 2.28
31. 195
33. −8.796
35. 5.46
37.

```
10 ' LOG
20 '
30 ' This program uses an infinite
40 ' series to calculate ln x.
50 '
60  INPUT "WHAT IS X"; X
70  A = (X - 1) / (X + 1)
80  FOR N = 1 TO 11 STEP 2
90     L = L + 2 * A^N / N
100 NEXT N
110 PRINT "ln"; X;" is equal to"; L
120 END
```

Exercise 6, Page 492

1. 2.81
3. 16.1
5. 2.46
7. 1.17
9. 5.10
11. 1.49
13. 0.239
15. −3.44
17. 1.39
19. 0
21. 7.79
23. 22.4 s
25. 1.5 s
27. 28,600 ft
29. 4.62 rad
31. 2.41 yrs
33. 7.95 yrs
35. 9.9 yrs
37. 3.5%
39.

```
10  '  OIL
20  '
30  '  This program calculates the
40  '  depletion of the world's
50  '  oil reserves.
60  '
70  LPRINT "Year", "Ann'l Cons.", "Oil Left"
80  LET Y = 0: C = 17: R = 1700
90  LPRINT Y, C, R
100 FOR Y = 1 TO 100
110    LET C = C*1.05
120    LET R = R - C
130    IF R < 0 THEN 160
140    LPRINT Y, C, R
150 NEXT Y
160 LPRINT "Oil reserves are gone"
170 END
```

Year	Ann'l Cons.	Oil Left
0	17	1700
1	17.85	1682.15
2	18.7425	1663.408
3	19.67962	1643.728
⋮	⋮	⋮
33	85.05404	270.8633
34	89.30673	181.5565
35	93.77206	87.78446

Oil reserves are gone

Exercise 7, Page 496

1. 47.5
3. −11.0, 9.05
5. 10/3
7. 22
9. 0.928
11. 0.916
13. 12
15. 101
17. 50,000
19. 7 years
21. 317°F
23. 2773 ft
25. 13 dB
27. 2.5×10^{-3} W
29. 164 Btu/h
31. 2.24×10^{-5}
33. 3.16×10^{-4}. It is the square root of the ion concentration at a pH of 7.

Exercise 8, Page 506

1.

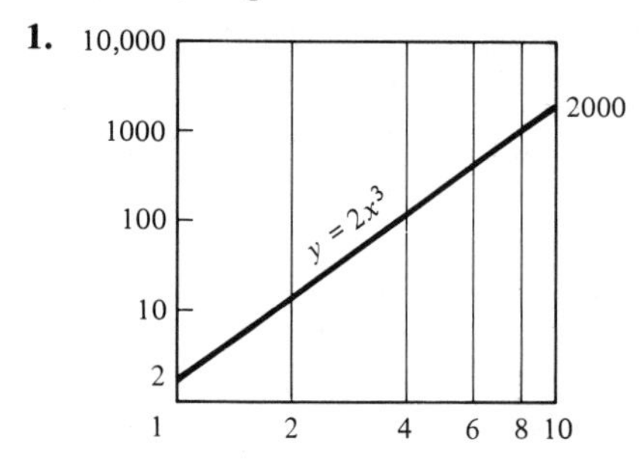

3.

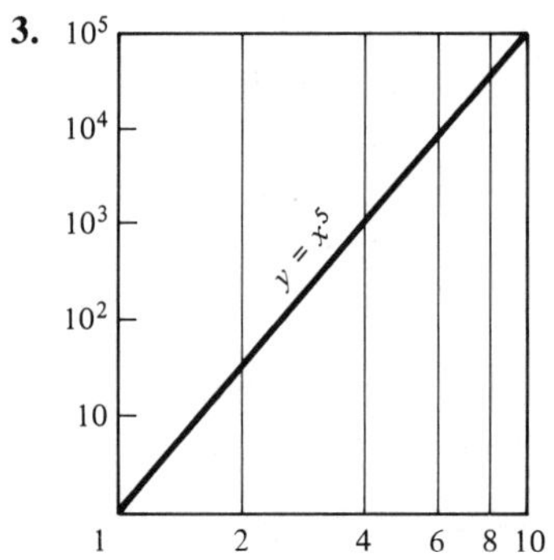

5.

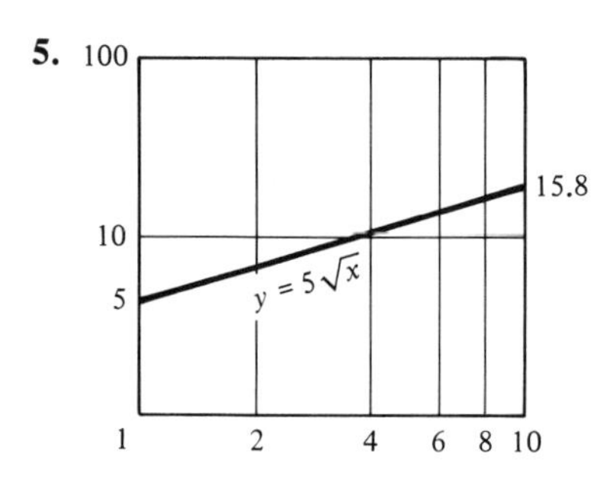

7.

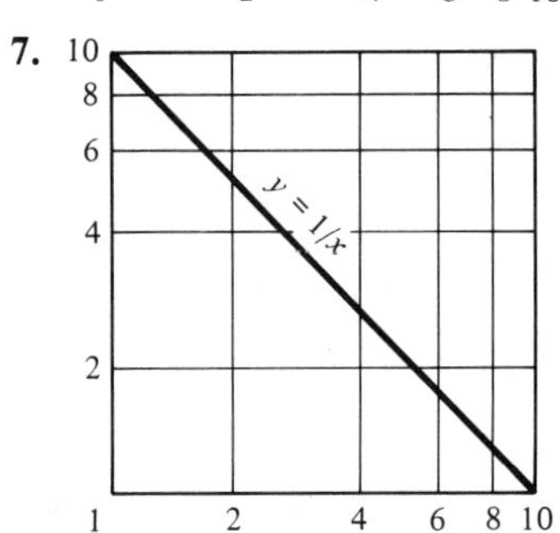

9.

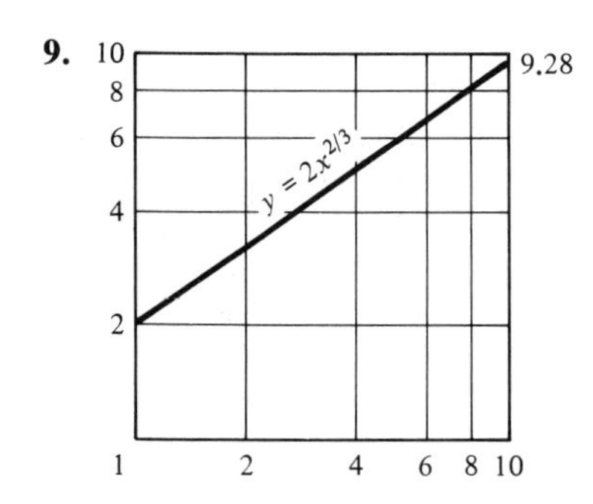

11. $y = 2000x^{-2.97}$

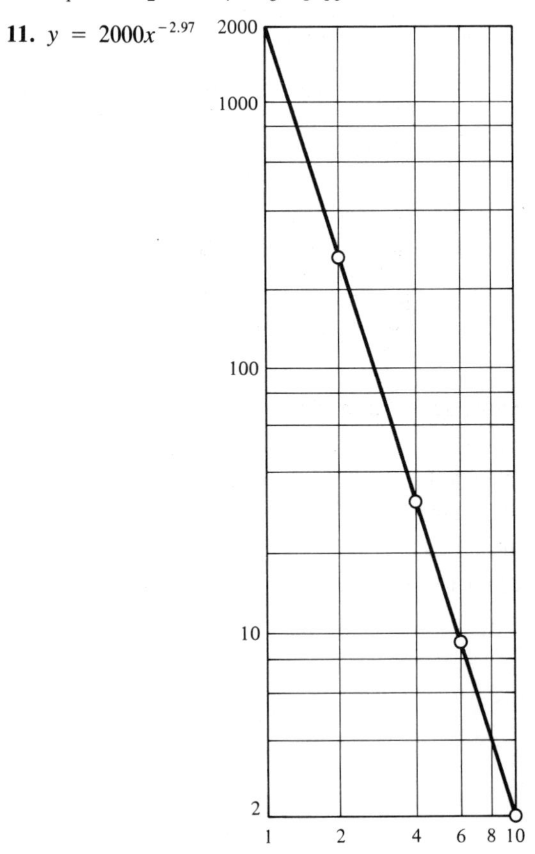

Chapter Test, Page 507

1. $\log_x 352 = 5.2$ **3.** $\log_{24} x = 1.4$ **5.** $x^{124} = 5.2$ **7.** $\frac{3}{4}$
9. 1/128 **11.** $\log 3 + \log x - \log z$ **13.** log 10 **15.** $\log \frac{\sqrt{p}}{\sqrt[4]{q}}$
17. 2.5611 **19.** 701.5 **21.** 0.3365 **23.** 2.943
25. 4.4394 **27.** −4.7410 **29.** 4.362 **31.** 2.101
33. 13.44 **35.** 3.17 **37.** 10^{-3} **39.** $2070.63
41. 828 rev/min **43.** 23 years

CHAPTER SIXTEEN: COMPLEX NUMBERS

Exercise 1, Page 514

1. $j3$ **3.** $j/2$ **5.** $4 + j2$ **7.** $-5 + j3/2$
9. $-j$ **11.** -1 **13.** -1 **15.** $-j3a$
17. $-7 - j3$ **19.** $(p + q) + j(p + q)$ **21.** $-112 + j19$ **23.** $-2\sqrt{3}$
25. $3 + j$ **27.** $13 - j13$ **29.** -25 **31.** $12 + j30$
33. $-5 + j7$ **35.** $6 + j5$ **37.** $5 + j8$ **39.** $3/j$
41. $\frac{32 + j8}{17}$ **43.** $-\frac{27}{13} + \frac{j8}{13}$ **45.** $-1.5 - j1.5$ **47.** $-0.75 \pm j1.39$
49. $-0.125 \pm j1.41$

Exercise 2, Page 516

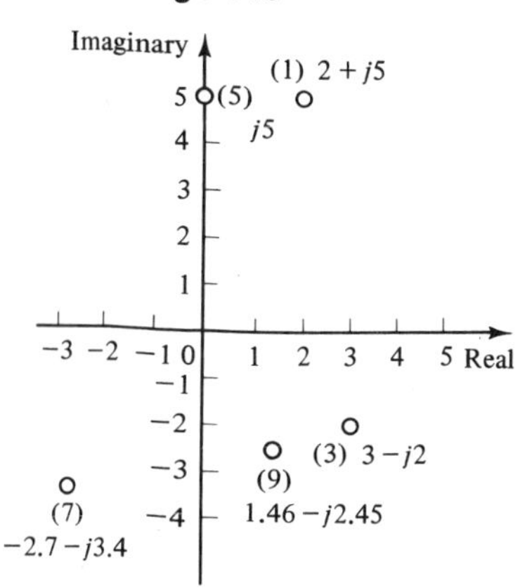

Exercise 3, Page 519

1. $6.4\angle 38.7°$, $6.4(\cos 38.7° + j \sin 38.7°)$
3. $5\angle 323°$, $5(\cos 323° + j \sin 323°)$
5. $5.39\angle 202°$, $5.39(\cos 202° + j \sin 202°)$
7. $10.3\angle 209°$, $10.3(\cos 209° + j \sin 209°)$
9. $8.06\angle 240°$, $8.06(\cos 240° + j \sin 240°)$
11. $1.63 + j2.52$, $3\angle 57°$ **13.** $-7.79 + j4.50$, $9\angle 150°$ **15.** $3.70 + j4.01$, $5.46\angle 47.3°$
17. $4.64 + j7.71$, $9(\cos 59° + j \sin 59°)$

13.

10,000
1000
100
10
1
0 1 2 3 4 5
3125

15.

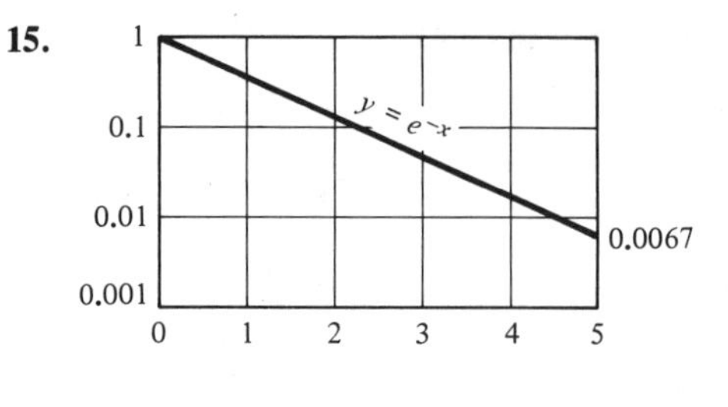

17.

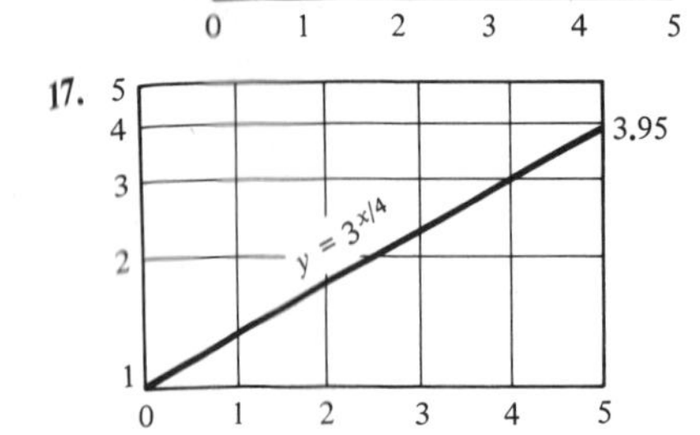

19.

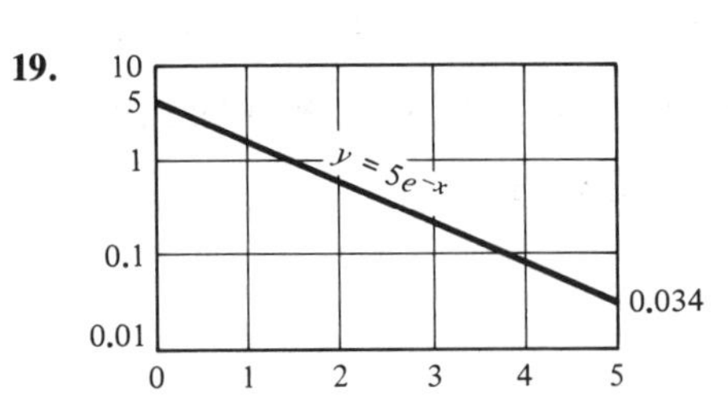

21. $y = 2.5^x$

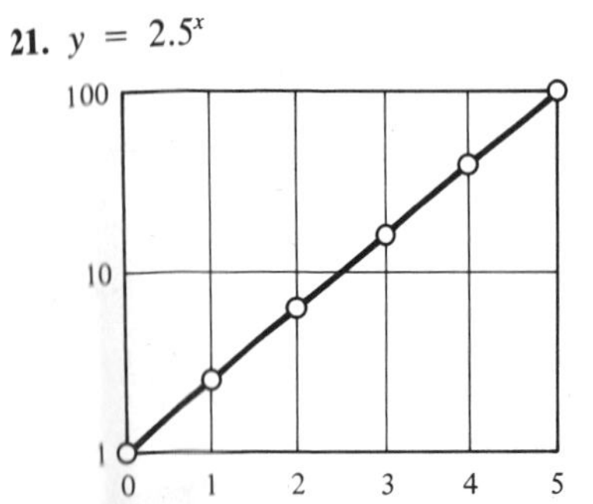

23. $y = 16.5x^{0.598}$

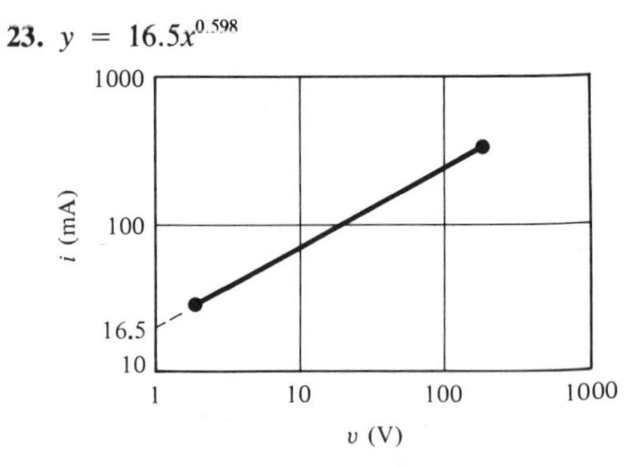

25. $y = 460x^{-1.07}$

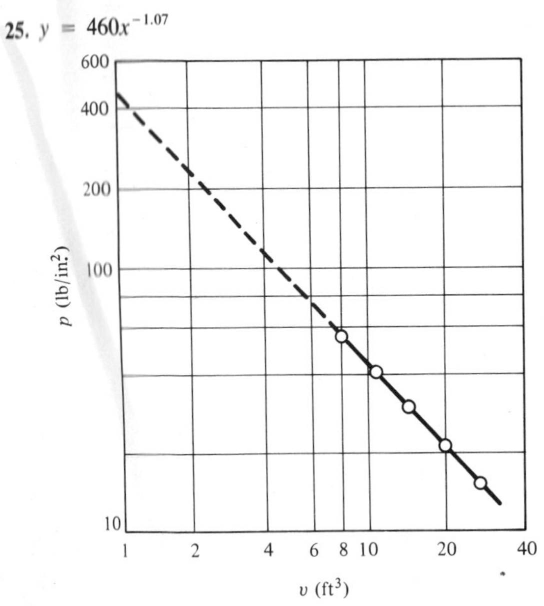

19. $4.21 - j5.59$, $7(\cos 307^\circ + j \sin 307^\circ)$
21. $15(\cos 40^\circ + j \sin 40^\circ)$
23. $16.1(\cos 54.9^\circ + j \sin 54.9^\circ)$
25. $56\angle 60^\circ$
27. $2(\cos 25^\circ + j \sin 25^\circ)$
29. $4.70(\cos 50.2^\circ + j \sin 50.2^\circ)$
31. $10\angle 60^\circ$
33. $8(\cos 45^\circ + j \sin 45^\circ)$
35. $49\angle 40^\circ$

Exercise 4, Page 522

1. $4.59 + j5.28$, $7\angle 49^\circ$
3. $54.8 + j185$, $193\angle 73.5^\circ$
5. $-31.2 + j23.3$, $39\angle 143^\circ$
7. $8 - j2$
9. $38.2 + j60.3$
11. $8.67 + j6.30$
13. $29 - j29$
15. $12(\cos 32^\circ - j \sin 32^\circ)$
17. $3.08\angle 32^\circ$

Exercise 5, Page 526

1. $250\angle 25^\circ$
3. $57\angle 270^\circ$
5. $144\angle 0^\circ$
7. $150 \sin \omega t$
9. $300 \sin (\omega t - 90^\circ)$
11. $7.5 \cos \omega t$
13. $155 + j0 = 155\angle 0^\circ$
15. $0 - j18 = 18\angle 270^\circ$
17. $72 - j42 = 83.4\angle -30.3^\circ$
19. $552 - j572 = 795\angle -46^\circ$
21. **(a)** $v = 603 \sin (\omega t + 85.3^\circ)$
(b) $v = 293 \sin (\omega t - 75.5^\circ)$

Chapter Test, Page 528

1. $3 + j2 = 3.61\angle 33.7^\circ = 3.61(\cos 33.7^\circ + j \sin 33.7^\circ)$
3. $-2 + j7 = 7.28\angle 106^\circ = 7.28(\cos 106^\circ + j \sin 106^\circ)$
5. j
7. $9 + j2$
9. $60.8 + j45.6$
11. $15.3\angle 177^\circ$
13. $33/17 - j21/17$
15. $2(\cos 45^\circ + j \sin 45^\circ)$

Problem **17** on graph.

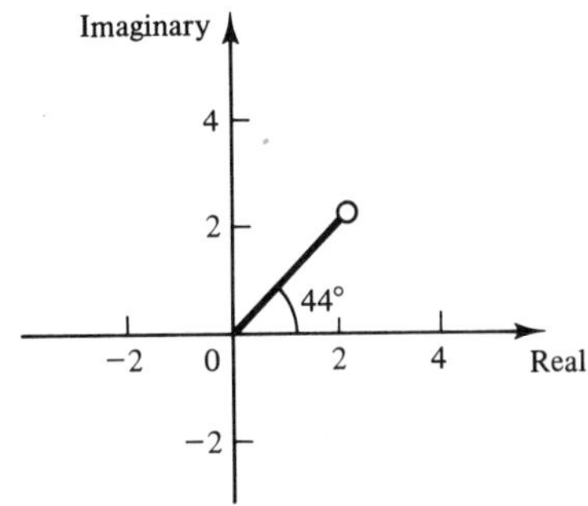

19. $7 - j24$
21. $125(\cos 30^\circ + j \sin 30^\circ)$
23. $V = 283 \sin (\omega t - 22^\circ)$

CHAPTER SEVENTEEN: LINEAR PROGRAMMING

Exercise 1, Page 535

1. $x > 5$

0 1 2 3 4 5 6 7 x

3. $x \geq -2$

−3 −2 −1 0 1 2 3 4 x

5. 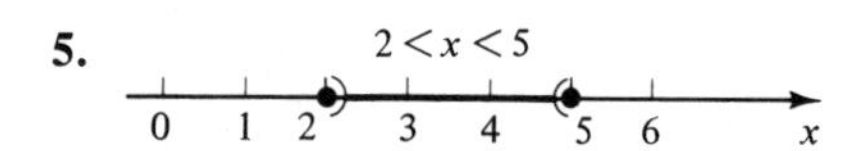

7. 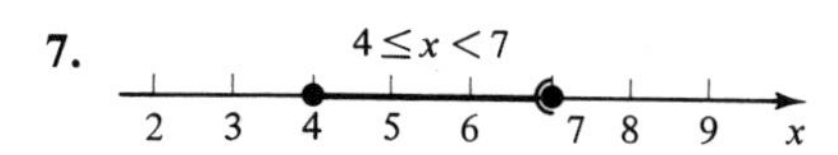

9. 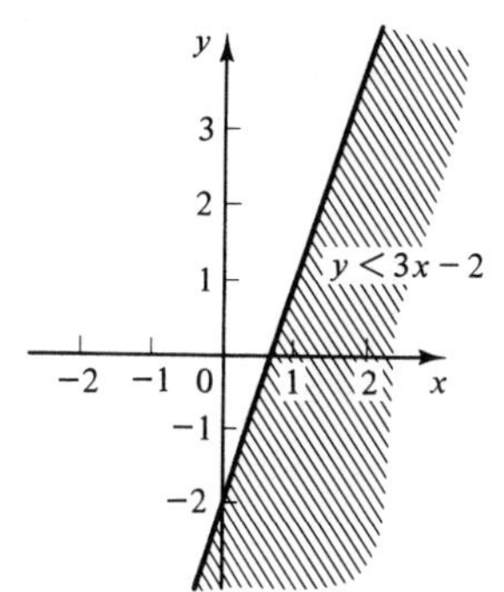

11. 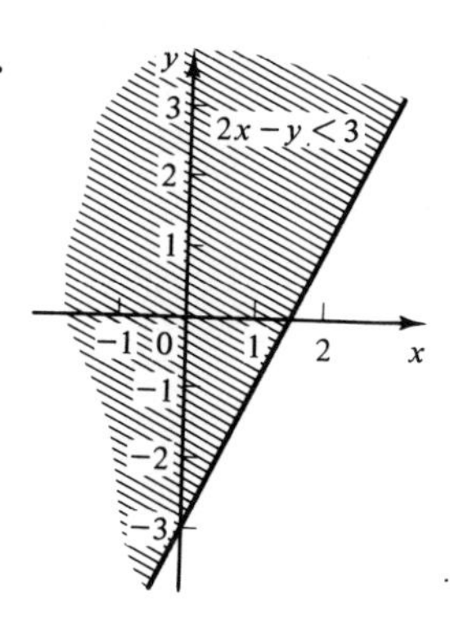

13. 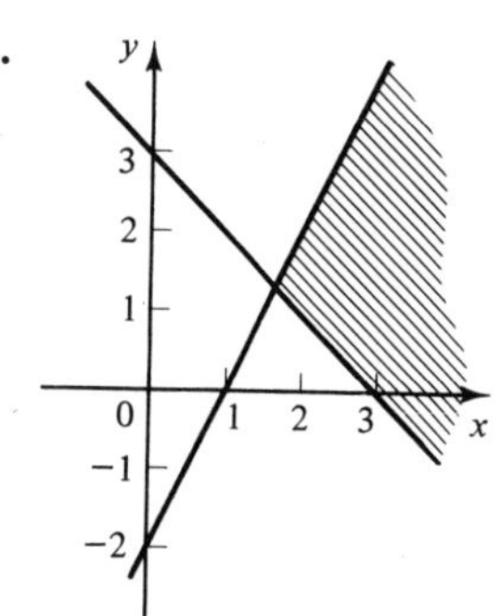

15. 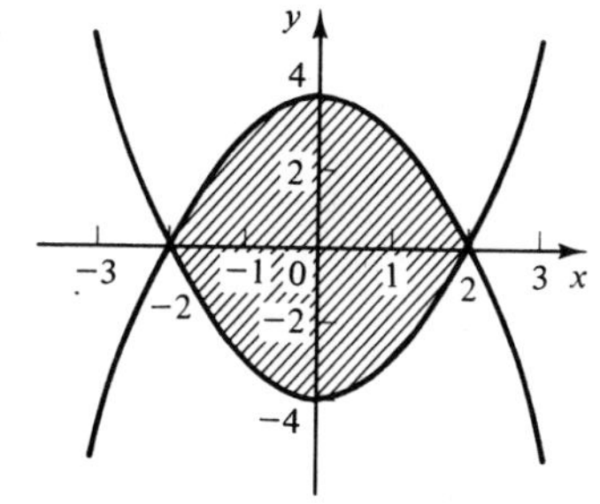

17.

$2.5x + 3.1y \leq 8$

Exercise 2, Page 539

1. $x > 9$
3. $-3 \leqq x \leqq 3$
5. $3/2 < x < 1$
7. $x \geqq 40/7$
9. $-3 > x > 3$
11. $-8/5 > x > 8/5$
13. $10/3 \leqq x \leqq -2$
15. $x \leqq 7.8\ h$

Exercise 3, Page 545

1. $x = 1.5$, $y = 3.5$
3. $x = 4.59$, $y = 0.724$
5. 69 stereos, 68 TVs
7. 7.69 drives, 3.84 monitors

Chapter Test, Page 547

1.

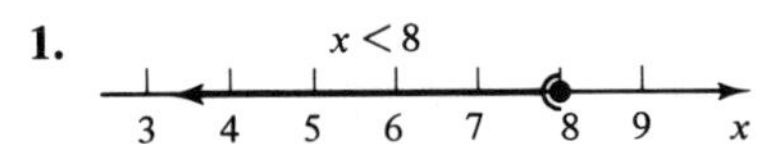

3.

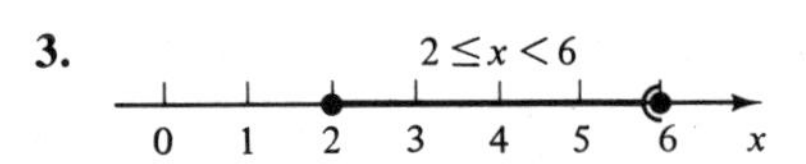

5.

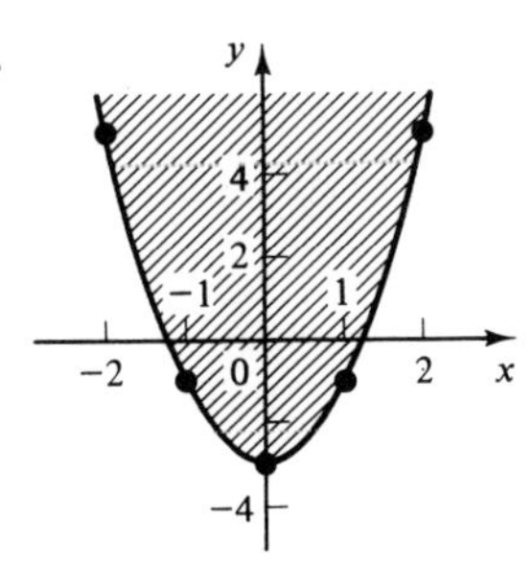

7. $x > 8$

9. $-3 > x > 11$

11. $x = 2.5,\ y = 0$

CHAPTER 18: BOOLEAN ALGEBRA

Exercise 1, Page 555

1. 0

3. 0

5. 1

7.

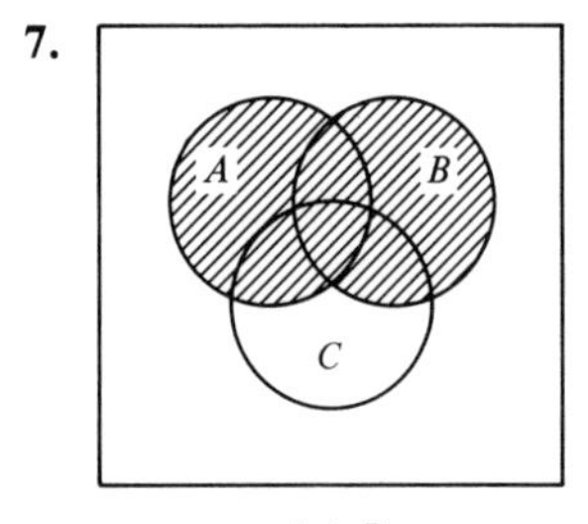

$A + B$

9.

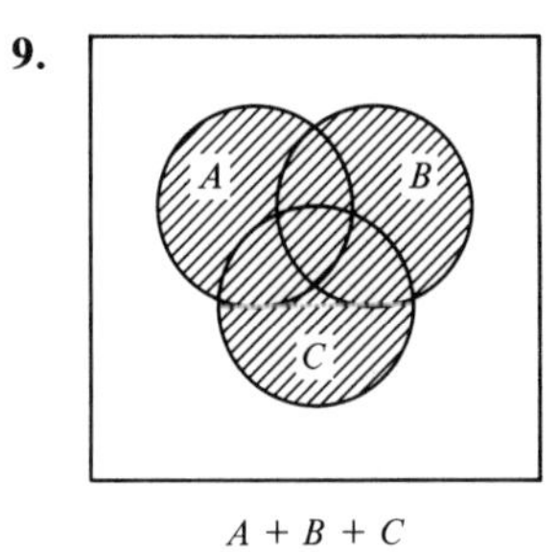

$A + B + C$

11.

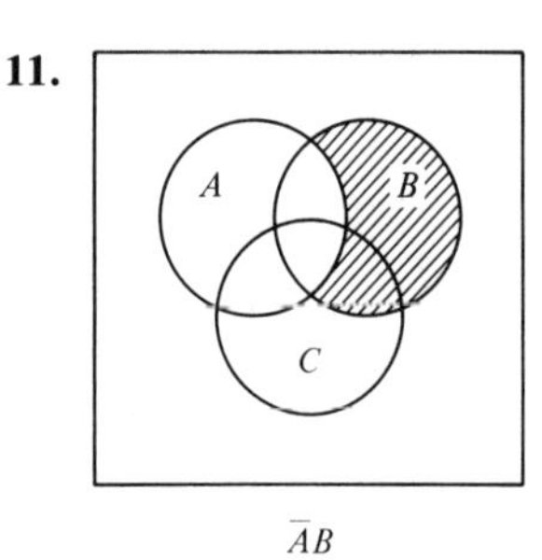

$\overline{A}B$

13.

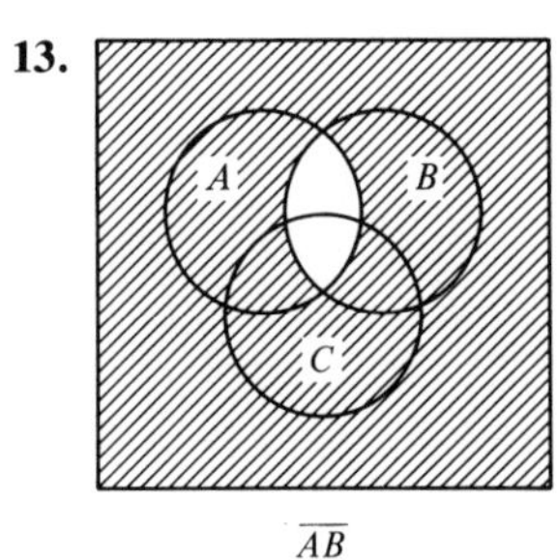

$\overline{AB}$

15. $X + Y + Z$

17. $X\overline{Y}(\overline{X} + Y)$

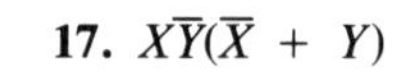

19.

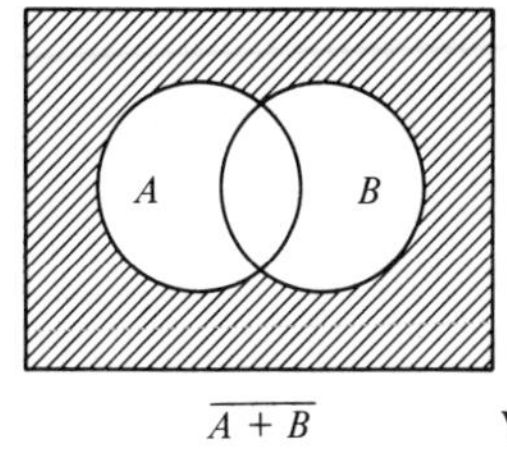

=

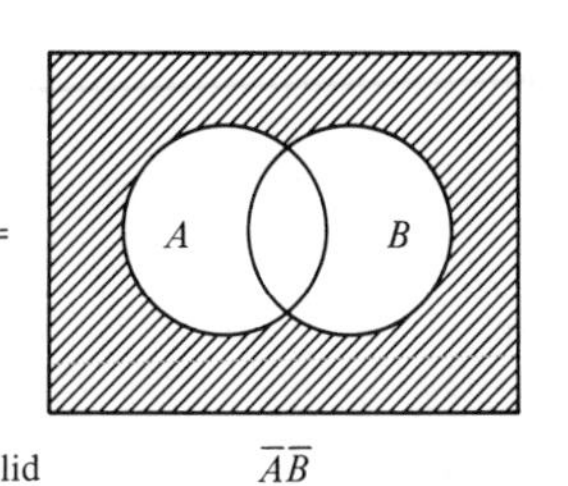

$\overline{A + B}$ Valid $\overline{A}\,\overline{B}$

21.

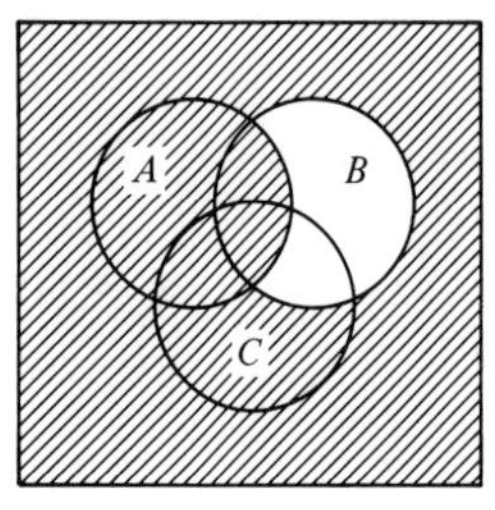

$A + \overline{B}$

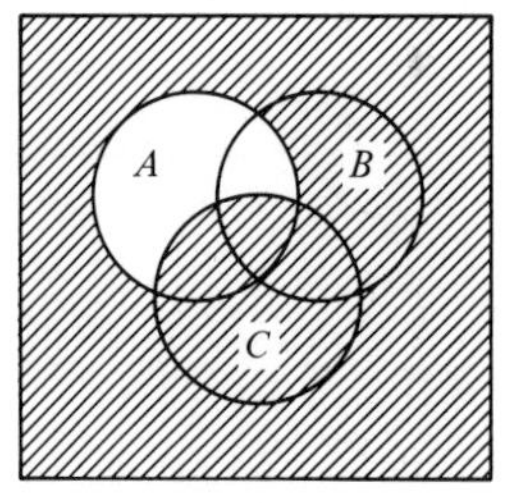

$\overline{A} + C$

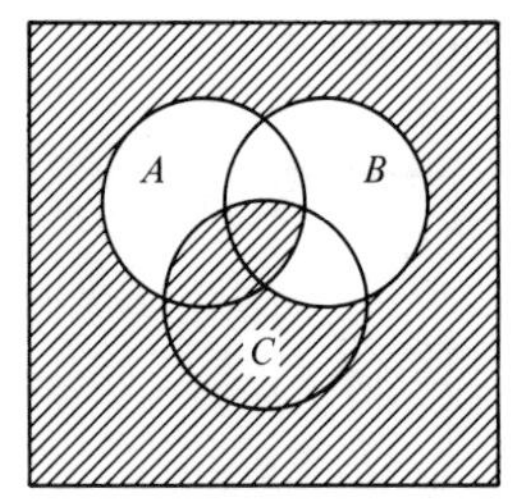

$\overline{A}\,\overline{B} + AC$

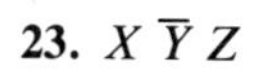

23. $X\ \overline{Y}\ Z$

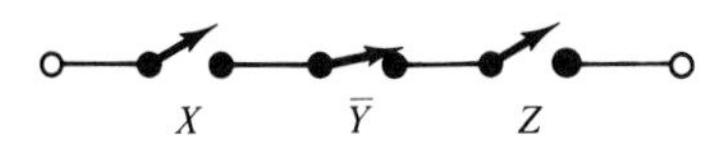

Exercise 2, Page 564

1.

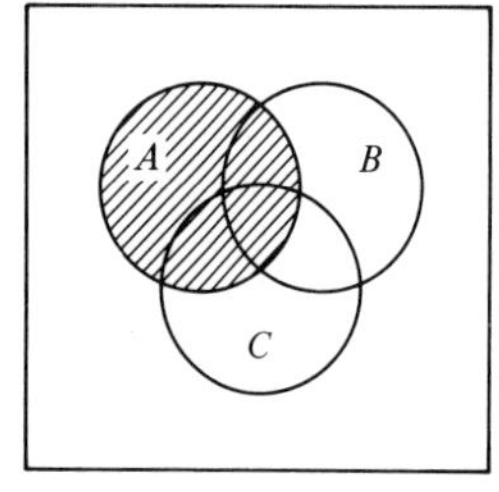

A

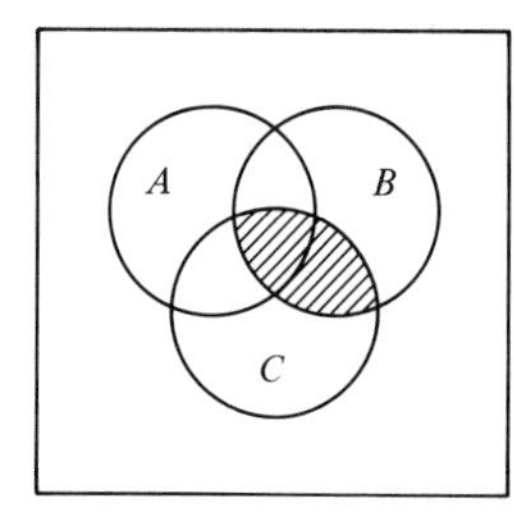

BC

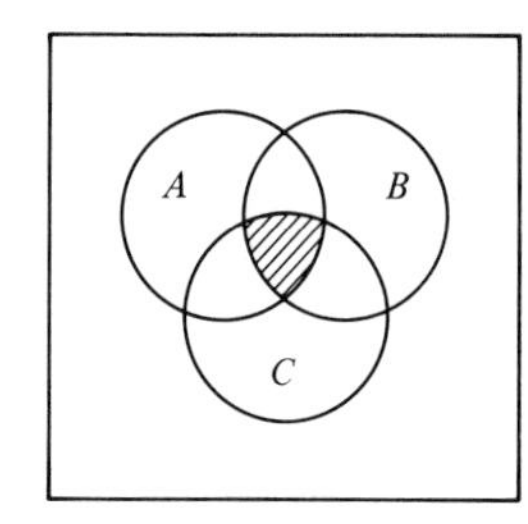

$A(BC)$

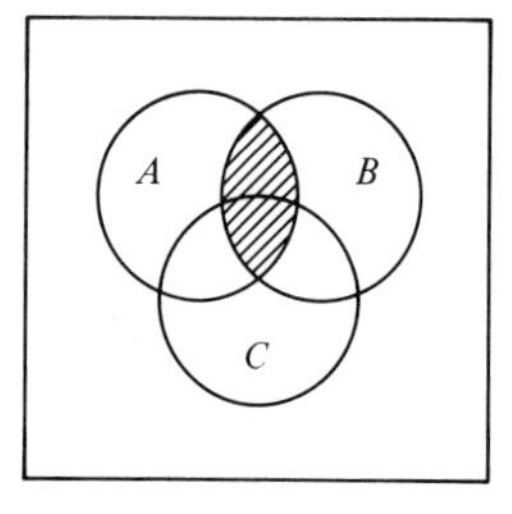

AB

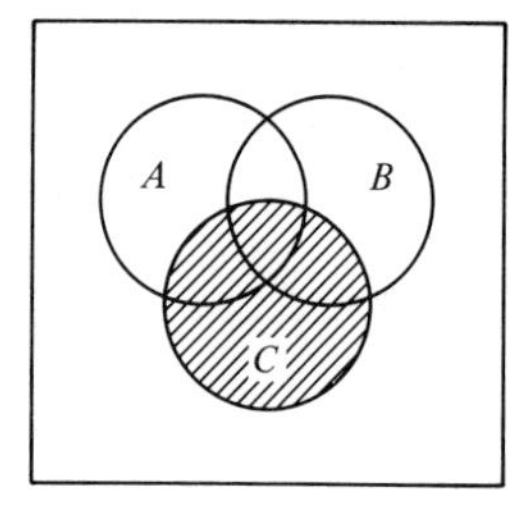

C

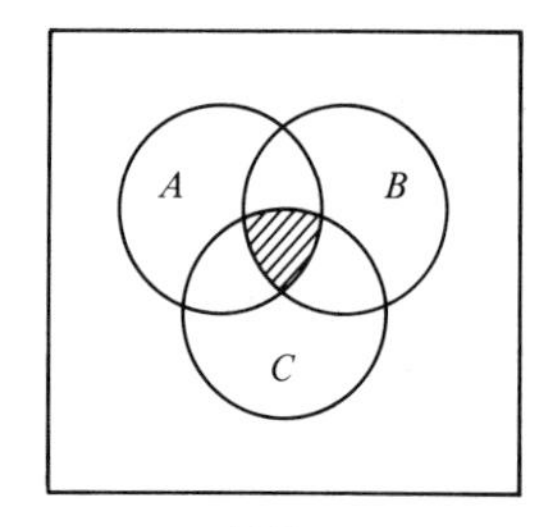

$(AB)C$

3. $A + AB = A$

law 234(a) $A(1 + B) = A$

law 229(b) $A(1) = A$

law 230(a) $A = A$

Exercise 3, Page 567

1. $\overline{B}$
3. B
5. XZ
7. X
9. Y
11. X
13. $X + Y + Z$
15. X
17. WYZ
19. $WX\overline{Y}Z$
21. $XY + \overline{X}\,\overline{Y} + Z$

Exercise 4, Page 571

1. $(X + Y)(\overline{Y} + Z)$
3. $(X + Y)\overline{Y} + ZY$

X

5. X

7. XZ

X Z

9.

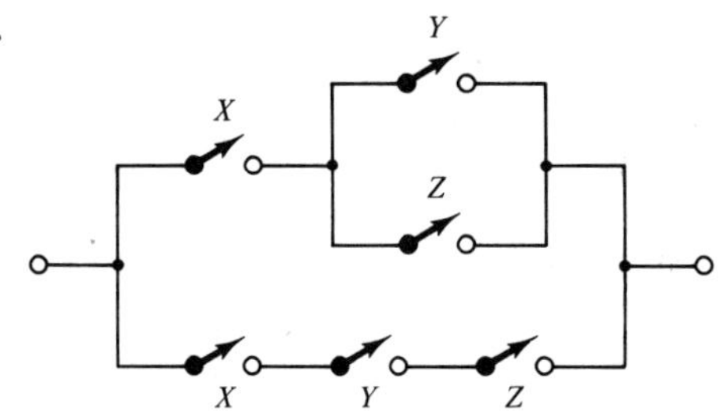

11.

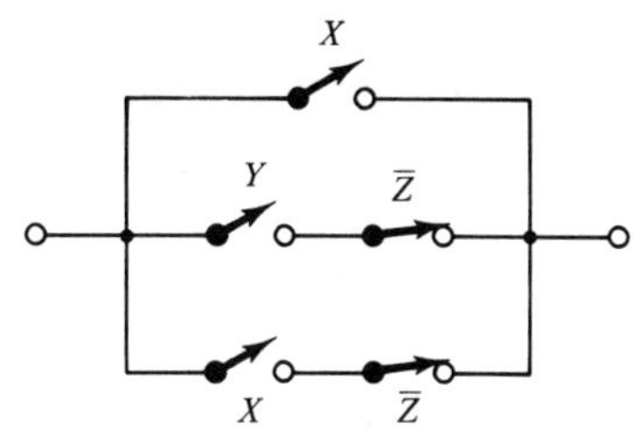

13.

X	Y	Z	XY	XZ	$XY + XZ$
0	0	0	0	0	0
0	0	1	0	0	0
0	1	0	0	0	0
0	1	1	0	0	0
1	0	0	0	0	0
1	0	1	0	1	1
1	1	0	1	0	1
1	1	1	1	1	1

15.

X	Y	Z	$X + Y$	$\overline{Y}$	$\overline{Y} + Z$	$(X + Y)(\overline{Y} + Z)$
0	0	0	0	1	1	0
0	0	1	0	1	1	0
0	1	0	1	0	0	0
0	1	1	1	0	1	1
1	0	0	1	1	1	1
1	0	1	1	1	1	1
1	1	0	1	0	0	0
1	1	1	1	0	1	1

17.

X	Y	Z	$X + Y$	$\overline{Y}$	$(X + Y)\overline{Y}$	ZY	$(X + Y)\overline{Y} + ZY$
0	0	0	0	1	0	0	0
0	0	1	0	1	0	0	0
0	1	0	1	0	0	0	0
0	1	1	1	0	0	1	1
1	0	0	1	1	1	0	1
1	0	1	1	1	1	0	1
1	1	0	1	0	0	0	0
1	1	1	1	0	0	1	1

Exercise 5, Page 579

1.

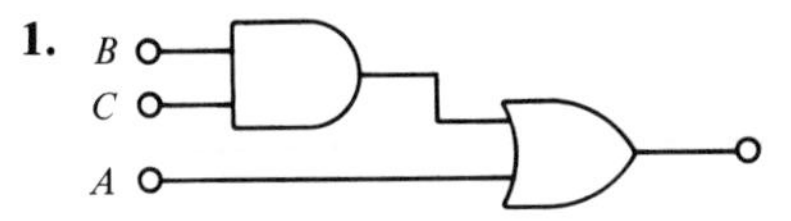

3.

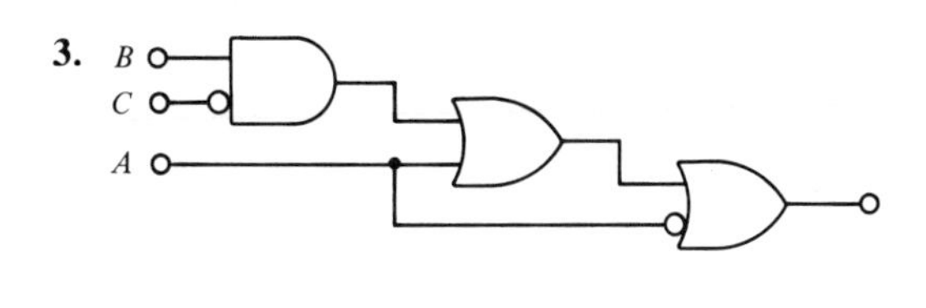

5.

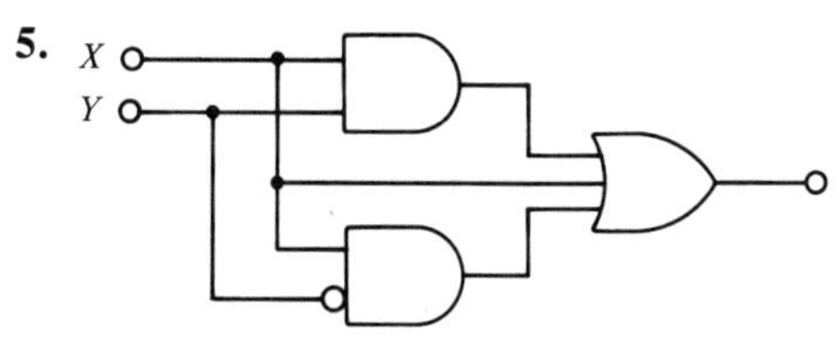

7. 1

9. $\overline{Y}Z$

11. $XY + Z$

13. $(\overline{\overline{W}X + Y})Z$

Chapter Test, Page 580

1. 0

3.

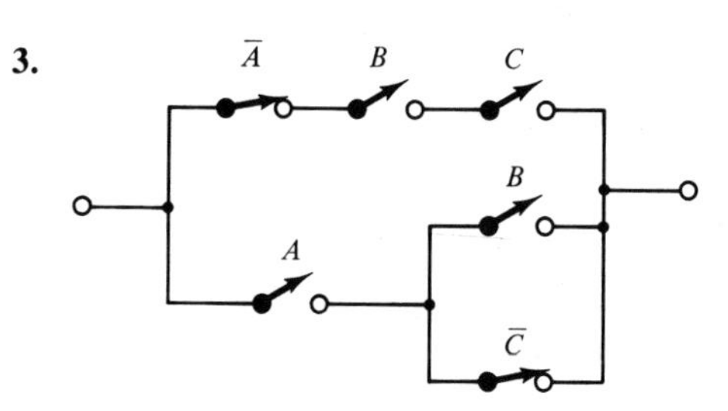

5. $X(\overline{X}Z + Y)Y$

7. 1

9. 0 (open circuit)

11. $\overline{W} + \overline{Z} + X\overline{Y}$

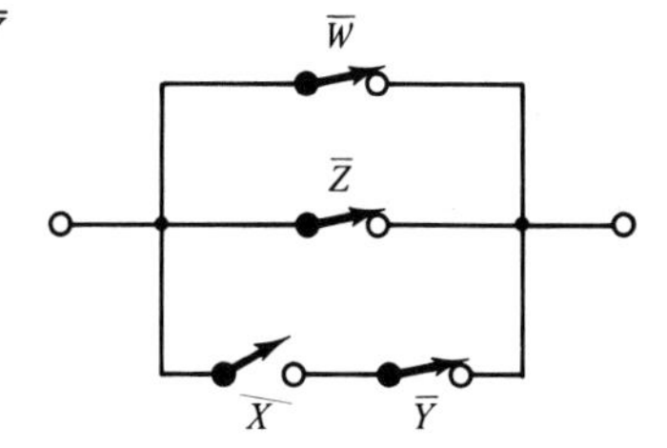

13. $\overline{X}$

15. 1

Index

INDEX TO APPLICATIONS

ENERGY

FINANCIAL

FLUIDS

GEODESY

LIGHT

MACHINE SHOP

MATERIALS

MECHANISMS

NAVIGATION

STATICS

STRENGTHS OF MATERIALS

STRUCTURES

SURVEYING

THERMAL

GENERAL INDEX